Pitman Research Notes in Mathematics Series

Submission of proposals for consideration

Suggestions for publication, in the form of outlines and representative samples, are invited by the Editorial Board for assessment. Intending authors should approach one of the main editors or another member of the Editorial Board, citing the relevant AMS subject classifications. Alternatively, outlines may be sent directly to the publisher's offices. Refereeing is by members of the board and other mathematical authorities in the topic concerned, throughout the world.

Preparation of accepted manuscripts

On acceptance of a proposal, the publisher will supply full instructions for the preparation of manuscripts in a form suitable for direct photo-lithographic reproduction. Specially printed grid sheets can be provided and a contribution is offered by the publisher towards the cost of typing. Word processor output, subject to the publisher's approval, is also acceptable.

Illustrations should be prepared by the authors, ready for direct reproduction without further improvement. The use of hand-drawn symbols should be avoided wherever possible, in order to maintain maximum clarity of the text.

The publisher will be pleased to give any guidance necessary during the preparation of a typescript, and will be happy to answer any queries.

Important note

In order to avoid later retyping, intending authors are strongly urged not to begin final preparation of a typescript before receiving the publisher's guidelines. In this way it is hoped to preserve the uniform appearance of the series.

Addison Wesley Longman Ltd
Edinburgh Gate
Harlow, Essex, CM20 2JE
UK
(Telephone (0) 1279 623623)

Titles in this series. A full list is available from the publisher on request.

Marek Niezgódka and Pawel Strzelecki

ICM, Warsaw University, Poland

(Editors)

Free boundary problems, theory and applications

Proceedings of the Zakopane Congress '95

 LONGMAN

Addison Wesley Longman Limited
Edinburgh Gate, Harlow
Essex CM20 2JE, England
and Associated Companies throughout the world.

Published in the United States of America
by Addison Wesley Longman Inc.

First published 1996

AMS Subject Classifications   (Main) 35
                              (Subsidiary) 49

ISSN 0269-3674

ISBN 0 582 30593 4

British Library Cataloguing in Publication Data

A catalogue record for this book is
available from the British Library

Library of Congress Cataloging-in-Publication data

A catalog record for this book is available

Printed and bound in Great Britain
by Biddles Ltd, Guildford and King's Lynn

Table of contents

Part 5. FBP's in Science 357

Preface

This volume contains (some of) the papers and lectures presented at the Congress *Free Boundary Problems, Theory and Applications* which was held in Zakopane (Poland, Tatra Mountains), June 11–18, 1995, following an interdisciplinary tradition of former meetings in Montecatini, Maubuisson, Irsee, Montreal, and Toledo[1]. The papers are devoted to various aspects of nonlinear Partial Differential Equations, including both theory and applications. Among the questions addressed by the authors, one can find classical existence and uniqueness questions, asymptotic behaviour of solutions, regularity of solutions and interfaces, presence and description of singularities, numerical analysis of examples, and applications to diverse problems of phase change, phase transition and phase separation.

The contents of the volume is arranged into five parts:

1. Stefan-like problems.

2. Free boundary problems in phase transitions, phase separation and related topics.

3. Mathematical miscellania related to free boundary problems.

4. Free boundary problems in Environment and Technology.

5. Free boundary problems in Science.

Important applications of mathematical techniques presented in these proceedings are shown not solely by mathematicians, but also by other scientists — coming from biophysics, chemistry, astronomy, materials science. One might be tempted to claim that the mathematics in these applications has been pretty standard for about last twenty years or even more. However, we think this is not true; a careful reader will no doubt come to the conclusion that contributions of non-mathematicians to this volume contain (among other things) a vast number of mathematical challenges.

Concluding, we believe that this volume tries to show some (applicable or applied) mathematics as it should be presented — viewed from many different points, not only by people who produce new results in mathematics but also by people who work on *real* problems to which mathematics is being applied.

The Editors

---

[1]For more details (presented in a lively style) on the tradition and scope of these meetings the reader is referred to the paper by J.R. Ockendon and M. Primicerio *The Conference Series on FBP: Theory and Applications*, in *FBP News* no. 1. (April 1993), available also on the net at http://www.lmc.fc.ul.pt/fbp/library/conf-ser/ock_prim.html .

Part 1.

Stefan-like problems

T AIKI AND H IMAI

Behaviour of blow-up solutions to one-phase Stefan problems with Dirichlet boundary conditions

Abstract We consider the one-phase Stefan problem for the equation $u_t = u_{xx} + u^{1+\alpha}$ ($\alpha > 0$) with homogeneous Dirichlet boundary condition in one-dimensional space. Clearly, our problem has a blow-up solution, because a solution of the initial-boundary value problem for above equation in bounded domain blows up at finite time for large initial data. In this paper, we show behavior of free boundaries of blow-up solutions at finite time. Precisely speaking, under the some assumptions for initial data if the solution u blows up at (T^*, x), $0 < T^* < \infty$ and $0 < x < \infty$, then the free boundary ℓ converges. In our proof we shall use the techniques investigated by Friedman and McLeod.

1   Introduction

In this paper we consider the following one-phase Stefan problem $SP := SP(u_0, \ell_0)$ with Dirichlet boundary condition in one-dimensional space: Find a curve (a free boundary) $x = \ell(t) > 0$ on $[0, T]$, $0 < T < +\infty$ and a function $u = u(t, x)$ on $Q_\ell(T) := \{(t, x); 0 < t < T, 0 < x < \ell(t)\}$ satisfying that

(1.1) $$u_t = u_{xx} + u^{1+\alpha} \quad \text{in } Q_\ell(T),$$

(1.2) $$u(0, x) = u_0(x) \quad \text{for } 0 \le x \le \ell_0,$$

(1.3) $$u(t, 0) = 0 \quad \text{for } 0 < t < T,$$

(1.4) $$u(t, \ell(t)) = 0 \quad \text{for } 0 < t < T,$$

(1.5) $$\ell'(t) = -u_x(t, \ell(t)) \quad \text{for } 0 < t < T,$$

(1.6) $$\ell(0) = \ell_0,$$

where α and ℓ_0 are given positive constants and u_0 is a given initial function.

For the initial data $\ell_0 > 0$ and $u_0 \in C^2((0, \ell_0)) \cap C^1([0, \ell_0])$, we suppose that the following conditions (H1) $\sim$ (H4) hold;

(H1) $u_0(x) > 0$ for $x \in (0, \ell_0)$,

(H2) $u_0(0) = u_0(\ell_0) = 0$,

(H3) $u_{0,x} > 0$ on $[0, x_0)$ and $u_{0,x} < 0$ on $(x_0, \ell_0]$ for some $x_0 \in (0, \ell_0)$,

(H4) $u_{0,xx}(x) + u_0^{1+\alpha}(x) \ge 0$ for $x \in (0, \ell_0)$.

In [1] Fasano-Primicerio established the local existence in time and the uniqueness for solutions to the above SP in the classical formulation (which means that u_t and u_{xx} are continuous functions). Besides, for the solution of SP in the distribution sense (which means that u_t and u_{xx} belong to L^2-class) the existence, the comparison and the behavior were studied by Aiki-Kenmochi [2, 3, 4]. In particular, we showed the following global existence result of a solution to SP in the distribution sense. Let $[0, T^*)$ be the maximal interval of existence of the solution to SP, we see (cf. [3]) that the following cases (a) or (b) must occur:

(a) $T^* = +\infty$;
(b) $T^* < +\infty$ and $|u|_{L^\infty(Q_\ell(t))} \to +\infty$ as $t \uparrow T^*$.

The purpose of the present paper is to establish the precise behavior of a solution to SP at finite blow-up time T^* in the classical formulation. Our main result is stated as follows:

If $T^* < +\infty$ then one and only one of the following cases (A) and (B) always happens;
(A) $\ell(t) \to \ell_\infty \in (0, \infty)$ as $t \uparrow T^*$ and there exists only one blow-up point $x^* \in (0, \ell_\infty)$;
(B) $\ell(t) \to +\infty$ as $t \uparrow T^*$ and the set $\{|u(t, x)|\}$ is bounded on $\{(t, x); 0 \le t < T^*, 0 \le x \le \min\{a, \ell(t)\}\}$ for each $a > 0$.

In Aiki [5] we investigated the one-phase Stefan problem with Neumann boundary condition, $\partial u / \partial x(t, 0) = 0$. It was mentioned that the set of the blow-up points consists of a single point $x = 0$ and the free boundary converges to some point as t tends to blow-up time.

Imai and Kawarada (cf. [6]) already studied a similar one-dimensional one-phase Stefan problem with Dirichlet boundary condition to our problem SP, numerically. Taking into account of the blow-up solutions to fixed boundary problems [7] they thought out similar system to SP and carried out numerical simulations. Their purpose was to obtain the possibility of blow-up phenomena of the free boundary, however they drew a conclusion from numerical results that in the system the free boundary does not blow up although the unknown function u does. Saying it differently, the above case (B) does not occur. So, we tried to prove that this conjecture is true, theoretically, but we can not show impossibility for free boundary to blow up at finite time because of difficulty in understanding the place of blow-up point.

In order to analyze the behavior of solutions we shall use the strong maximal principle so that smoothness of solutions is required in our problem.

Blow-up solutions to a semilinear parabolic equation have been studied by many authors (cf. [8, 9, 10] and etc.). Particularly, the following initial boundary value problem was discussed in Friedman-McLeod [9]: $Q(T) := (0, T) \times (0, \ell_0), 0 < T < +\infty$,

$$v_t = v_{xx} + v^{1+\alpha} \quad \text{in } Q(T),$$

$$v(t,0) = v(t,\ell_0) = 0 \quad \text{for } 0 < t < T,$$

$$v(0,x) = v_0(x) \quad \text{for } 0 \le x \le \ell_0.$$

If v_0 satisfies (H1) $\sim$ (H4) then the set of blow-up points consists of a single point. The proof of our main result is done with help of ideas of [9, 10].

2   The main result and notations

Throughout this paper, we assume that $\alpha > 0$ and $\ell_0 > 0$, and put $Q_\ell(T; \cdot, a) = \{(t,x) \in Q_\ell(T); x < a\}$ and $Q_\ell(T; a, \cdot) = \{(t,x) \in Q_\ell(T); a < x\}$, for each positive number a. We begin with the precise definition of a solution to $SP(u_0, \ell_0)$. Let $C^{1,0}(\overline{Q_\ell(T)})$ be the set of functions which are continuous on $\overline{Q_\ell(T)}$ with their $x-$derivatives.

Definition 2.1.   A couple $\{u, \ell\}$ of functions $u = u(t,x)$ and $x = \ell(t)$ is said to be a solution of $SP := SP(u_0, \ell_0)$ on a compact interval $[0, T]$, $0 < T < +\infty$, if the following conditions (S1) and (S2) are satisfied:

(S1) $\ell \in C^1([0, T])$, and $u \in C^{1,0}(\overline{Q_\ell(T)})$, u_{xx} and u_t are continuous in $Q_\ell(T)$;

(S2)   (1.1) $\sim$ (1.6) hold in the classical sense.

Also, we call a couple $\{u, \ell\}$ is a solution of SP on an interval $[0, T')$, $0 < T' \le \infty$, if it is a solution of SP on $[0, T]$ in the above sense for any $0 < T < T'$.

First, we recall the theorem concerned with local existence of solutions to the above SP.

Theorem 2.1.   (cf. [1; Theorem 1]) We assume that $u_0 \in C^1([0, \ell_0])$, $u_0 \ge 0$ on $[0, \ell_0]$, $u_0(0) = u_0(\ell_0) = 0$. Then there exists a positive number T_0 depending only on $|u_0|_{C^1([0,\ell_0])}$, ℓ_0 and α such that problem $SP(u_0, \ell_0)$ has a unique solution $\{u, \ell\}$ on $[0, T_0]$.

For the problem $SP(u_0, \ell_0)$, we say that $[0, T)$, $0 < T \le +\infty$, is the maximal interval of existence of the solution, if the problem has a solution on time-interval $[0, T']$, for every T' with $0 < T' < T$ and the solution can not be extended in time beyond T.

The main results stated as follows.

Theorem 2.2.   Assume that (H1) $\sim$ (H4) hold. Let $[0, T^*)$ be the maximal interval of existence of the solution $\{u, \ell\}$ to SP. If T^* is finite, then either the following cases (A) or (B) always happens:

(A) $\ell(t) \to \ell_\infty$ as $t \uparrow T^*$ where ℓ_∞ is some positive number, there exists one and only one point $x^* \in (0, \ell_\infty)$ such that $u(t, x^*) \to +\infty$ as $t \uparrow T^*$ and for $x \in (0, \ell_\infty)$ with $x \ne x^*$ there is a positive constant $M_1(x)$ such that $|u(t,x)| \le M_1(x)$ for t with $(t,x) \in Q_\ell(T^*)$;

5

(B) $|u(t)|_{L^\infty(0,\ell(t))} \to \infty$, $\ell(t) \to +\infty$ as $t \uparrow T^*$ and for any $x > 0$ there is a positive number $M_2(x)$ satisfying that $|u(t,\xi)| \le M_2(x)$ for $(t,\xi) \in Q_\ell(T^*; \cdot, x)$.

Let us introduce the notations of spaces and norms to be used in this paper.

For any $\nu \in (0,1)$ and $0 \le a < b < \infty$, $z \in H^\nu([a,b])$ means that z is Hölder continuous, exponent ν, on $[a,b]$, that is, the following norm is finite;

$$|z|_{H^\nu([a,b])} = \sup_{t,t' \in [a,b]} \frac{|z(t) - z(t')|}{|t - t'|^\nu}.$$

For $0 \le t < T < \infty$, let us consider the parabolic domain $D(t,T) = (t,T) \times (0,1)$. Take $P_1 = (t_1, x_1)$, $P_2 = (t_2, x_2) \in D(t,T)$ and define the distance

$$\overline{P_1 P_2} = (|x_1 - x_2|^2 + |t_1 - t_2|)^{1/2}.$$

Let u be a continuous function on $\overline{D(t,T)}$. We say that $u \in C^\nu(\overline{D(t,T)})$ for a given $\nu \in (0,1)$, if

$$|u|_{C^\nu(\overline{D(t,T)})} = |u|_{C(\overline{D(t,T)})} + \sup_{P_1,P_2 \in \overline{D(t,T)}} \frac{|u(P_1) - u(P_2)|}{\overline{P_1 P_2}^\nu}$$

is finite.

In the present paper we define a blow-up point in the following way:

Definition 2.2. Let $\{u, \ell\}$ be a solution of SP, $[0, T^*)$ be a maximal interval of existence of solution to SP and $T^* < \infty$. A point $x \in (0, \infty)$ is said to be a blow-up point if there is a sequence $\{(t_n, \xi_n)\} \subset Q_\ell(T^*)$ such that

$$t_n \uparrow T^*, x_n \to x \text{ and } u(t_n, x_n) \to \infty \quad \text{as } n \to \infty.$$

The proof of Theorem 2.2 is done in the following way.

First, we shall show continuation of solutions of our problem SP. Precisely speaking, for $0 < T_0 < \infty$ if a solution $\{u, \ell\}$ to SP on $[0, T_0)$ satisfies that

$$|u(t,x)| \le K \quad \text{for } (t,x) \in \overline{Q_\ell(T_0)} \cap \{t < T_0\}$$

where K is a positive constant, then the solution $\{u, \ell\}$ is extended in time beyond T_0.

Next, we assume that (B) in the assertion of Theorem 2.2 does not hold, that is, (A1) or (A2) (stated below) is valid:

(A1) There is a number $x_1 \in (0, \infty)$ satisfying that for some sequence $\{(t_n, \xi_n)\} \subset Q_\ell(T^*; \cdot, x_1)$ $u(t_n, \xi_n) \to \infty$ as $n \to \infty$;

(A2) there exists a positive constant L_0 such that $\ell(t) \le L_0$ for any $t \in [0, T^*)$. Therefore, it is sufficient to show that (A1) is a sufficient condition for (A) in the statement of Theorem 2.2 and (A2) is, too.

6

Under the condition (A1) by the similar argument to [9] we infer that the set of blow-up points consists of only one point $x_1 \in (0, \infty)$. Clearly, $u(t, x_1) \to +\infty$ as $t \uparrow T^*$ and for some $x_2 > x_1$ there is a positive constant M such that

(2.1) $$|u(t, x)| \leq M \quad \text{for any } (t, x) \in Q_\ell(T^*; x_2, \cdot).$$

By using the estimate (2.1) and a comparison theorem for solutions to one-phase Stefan problems we conclude that (A) holds. Similarly, we can prove that (A2) implies (A).

3   Known results

One of the purposes of this section is to show comparison and global existence results for a solution to one-phase Stefan problem we introduce a solution (in the distribution sense) of the following problem $CSP = CSP(a; g, k; u_0, \ell_0)$. The problem CSP is to find a curve $x = \ell(t) > a$ on $[0, T]$, $0 < T < +\infty$, $a \geq 0$ and a function $u = u(t, x)$ on $(0, T) \times (a, \infty)$ satisfying,

(3.1) $$u_t = u_{xx} + g \quad \text{in } Q_\ell(T; a, \cdot),$$

(3.2) $$u(t, a) = k(t) \quad \text{for } t \in [0, T],$$

(3.3) $$u(t, x) = 0 \quad \text{for } 0 < t < T \text{ and } x \geq \ell(t),$$

(3.4) $$\ell'(t) = -u_x(t, \ell(t)-) \quad \text{for a.e. } t \in [0, T],$$

(3.5) $$u(0, x) = u_0(x) \quad \text{for } x \in [a, \ell_0],$$

(3.6) $$\ell(0) = \ell_0$$

where g is a given function on $(0, T) \times (a, \infty)$; k is a given function on $[0, T]$; ℓ_0 is a number with $\ell_0 > a$; u_0 is a given function on $[a, \infty)$.

Definition 3.1.   We say that a couple $\{u, \ell\}$ is a solution of CSP on $[0, T]$, $0 < T < \infty$, if the following properties (i) and (ii) are fulfilled:

(i) $u \in W^{1,2}(0, T; L^2(a, \infty)) \cap L^\infty(0, T; W^{1,2}(a, \infty))$, and $\ell \in W^{1,2}(0, T)$ with $a \leq \ell < +\infty$ on $[0, T]$,

(ii) (3.1) holds in the sense of $\mathcal{D}'(Q_\ell(T; a, \cdot))$, and (3.2) $\sim$ (3.6) are satisfied.

Also, we call a couple $\{u, \ell\}$ is a solution of CSP on an interval $[0, T')$, $0 < T' \leq \infty$, if it is a solution of CSP on $[0, T]$ for any $0 < T < T'$.

The following propositions are concerned with the comparison and global existence of a solution to CSP.

Proposition 3.1.   (cf. [2; Theorem 5.1]) Let a be any non-negative number. Assume that for $i = 1, 2$, $\ell_{0,i} > a$, $u_{0,i} \in W^{1,2}([a, \infty))$, $u_{0,i} \geq 0$ on $[a, \infty)$, $u_{0,i} = 0$ on $[\ell_{0,i}, \infty)$ and for $0 < T < \infty$, $g_i \in L^\infty((0, T) \times (a, \infty))$, g_i is non-negative, $k_i \in W^{1,2}(0, T)$, k_i is non-negative and $u_{0,i}(a) = k_i(0)$. For $i = 1, 2$, let $\{u_i, \ell_i\}$ be the solution of $CSP(a; g_i, k_i; u_{0,i}, \ell_{0,i})$ on $[0, T]$. If $\ell_{0,1} \le \ell_{0,2}$, $u_{0,1} \le u_{0,2}$ on $[0, \infty)$, $g_1 \le g_2$ a.e. on $(0, T) \times (a, \infty)$ and $k_1 \le k_2$ on $[0, T]$ then

$$\ell_1 \le \ell_2 \quad on \ [0, T], \quad u_1(t, x) \le u_2(t, x) \quad for \ 0 \le t \le T \ and \ x \ge a.$$

Proposition 3.2. *(cf. [5; Proposition 2.2]) Let a be any non-negative number and ℓ_0 be any positive number with $\ell_0 > a$. We suppose that $g \in L^\infty((0, \infty) \times (a, \infty))$, $g \ge 0$ a.e. on $(0, \infty) \times (a, \infty)$, $k \in L^2_{loc}([0, \infty))$, $k \ge 0$ on $(0, \infty)$ and $u_0 \in W^{1,2}([a, \infty))$ with $u_0 \ge 0$ on $[a, \infty)$ and $u_0 = 0$ on $[\ell_0, \infty)$. Then there exists one and only one solution $\{u, \ell\}$ of CSP on $[0, \infty)$.*

In the rest of this section we show a proposition about continuation of a solution to SP.

Proposition 3.3. *Assume that all the conditions of Theorem 2.2 hold. Let T_0 be any positive number and $\{u, \ell\}$ be the solution of $SP(u_0, \ell_0)$ on an interval $[0, T_0)$. Suppose that for a positive constant K*

$$(3.7) \qquad |u(t, x)| \le K \quad for \ (t, x) \in \overline{Q_\ell(T_0)} \cap \{t < T_0\}.$$

Then for any $\nu \in (0, 1)$ there are constants N_0 and N_1 depending only on K, T_0, u_0, ℓ_0, ν and α such that

$$(3.8) \qquad \ell(t) \le N_0 \quad for \ 0 \le t < T_0,$$

$$(3.9) \qquad |u_x(t)|_{H^\nu([0, \ell(t)])} \le N_1 t_0^{-1/2} \quad for \ 0 < t_0 < t < T_0.$$

Moreover, the solution $\{u, \ell\}$ is extended in time beyond time T_0.

The proof of this proposition is standard and quite similar to that of [5; Prop. 3.1]. So, we omit its proof.

4   The set of blow-up points

The purpose of this section is to obtain that the set of blow-up points consists of only one point. We shall prove this result with help of ideas of [9; section 5].

Throughout this section let $\{u, \ell\}$ be a solution of SP and $[0, T^*)$ be the maximal interval of existence of the solution of SP and we assume that (H1) $\sim$ (H4) hold and $T^* < \infty$.

Immediately, by using the strong maximum principle we see that

$$(4.1) \qquad u_x(t, 0) < 0 \ and \ u_x(t, \ell(t)) > 0 \ for \ 0 < t < T^*.$$

It follows from (4.1) that $\ell' > 0$ on $(0, T^*)$. Hence there exists $\ell_\infty \in (0, \infty]$ such that $\ell(t) \to \ell_\infty$ as $t \uparrow T^*$.

Also, we recall the following lemma in [5] concerned with a sign property of u_t.

Lemma 4.1. (cf. [5; Lemma 4.1]) If $\{u, \ell\}$ is a solution of SP on $[0, T]$, $0 < T < \infty$, then $u_t \geq 0$ on $Q_\ell(T)$.

Lemma 4.2. (cf. [9; Lemma 5.2]) There exist curves $x = s^\pm(t)(0 < t < T^*)$ such that for any $0 < t < T^*$

$$u_x(t, x) > 0 \quad \text{if and only if} \quad 0 < x < s^-(t),$$
$$u_x(t, x) < 0 \quad \text{if and only if} \quad s^+(t) < x < \ell(t).$$

This is a direct consequence of [9; Lemma 5.2], so we omit the proof. For simplicity, we put $s^+ = \limsup_{t \uparrow T^*} s^+(t)$ and $s^- = \liminf_{t \uparrow T^*} s^-(t)$.

We shall show two lemmas concerned with the places of s^- and s^+.

Lemma 4.3.

$$s^- > 0.$$

Proof. By assumption (H3) we can choose a positive number β such that $u_{0,x} > 0$ on $[0, 2\beta]$. Here, we put

$$w(t, x) = u(t, x) - u(t, 2\beta - x) \quad \text{for } 0 < t < T^* \text{ and } 0 \leq x \leq \beta.$$

Clearly, for any $(t, x) \in (0, T^*) \times (0, \beta)$ we have

$$
\begin{aligned}
w_t(t, x) &= u_t(t, x) - u_t(t, 2\beta - x) \\
&= u_{xx}(t, x) + u^{1+\alpha}(t, x) - u_{xx}(t, 2\beta - x) - u^{1+\alpha}(t, 2\beta - x) \\
(4.2) \qquad &= w_{xx}(t, x) + \psi(t, x)w(t, x)
\end{aligned}
$$

where ψ is a suitable continuous function on $(t, x) \in [0, T^*) \times [0, \beta]$.

Also, $w(t, 0) < 0$ for $0 < t < T^*$, $w(0, x) < 0$ for $0 \leq x \leq \beta$ and $w(t, \beta) = 0$ for $0 < t < T^*$. By applying the maximum principle to (4.2) we infer that $w_x(t, \beta) > 0$ for $0 < t < T^*$, that is, $u_x(\cdot, \beta) > 0$ on $(0, T^*)$. Hence, this implies that $s^- \geq \beta > 0$. $\square$

Lemma 4.4. If $\ell_\infty < \infty$, then we have $s^+ < \ell_\infty$.

Proof. There is a number T_0 such that

$$\ell_\infty - \ell(t) < \frac{\ell_\infty}{3} \quad \text{for } t \in [T_0, T^*)$$

since $\ell(t) \uparrow \ell_\infty$ as $t \uparrow T^*$. Moreover, (4.1) implies that for some positive number $\eta_0 \in (0, \ell_\infty/3)$

$$u_x(T_0, x) < 0 \quad \text{for } \ell(T_0) - 2\eta_0 \leq x \leq \ell(T_0).$$

Here, we put $\beta = \ell(T_0) - \eta_0$,

$$w(t, x) = u(t, y) - u(t, 2\beta - x) \quad (t, x) \in \{(t, x); T_0 \leq t < T^*, \beta \leq x \leq \ell(t)\}.$$

9

It is clear that

$$w_t = w_{xx} + \psi w \quad \text{in } \{(t,x); T_0 < t < T^*, \beta < x < \ell(t)\}$$

where ψ is a suitable continuous function on $\{(t,x); T_0 < t < T^*, \beta < x < \ell(t)\}$.

Therefore, by the similar argument to the proof of Lemma 4.3 we conclude that this lemma is true. $\qquad\square$

Lemma 4.5. *(cf. [10, 5]) If a point $x \in (0, \ell_\infty)$ is a blow-up point, then $s^- \le x \le s^+$.*

Proof. We assume that x_* is a blow-up point with $x_* < s^-$. By Definition 2.2 there is a sequence $\{(t_n, x_n)\} \subset Q_\ell(T^*)$ such that $(t_n, x_n) \to (T^*, x_*)$ and $u(t_n, x_n) \to \infty$ as $n \to \infty$. Also, it follows from the definition of s^- and Lemma 4.2 that for any $r \in (x_*, s^-)$ there is a positive number δ such that

$$(4.3) \qquad u_x > 0 \text{ in the rectangle } \Lambda := \{(t,x); 0 < x < r, T^* - \delta < t < T^*\}.$$

For any $x_b \in (x_*, r)$ we have $x_n < x_b$ for sufficiently large n so that

$$(4.4) \qquad\qquad\qquad u(t, x_b) \uparrow \infty \quad \text{as } t \uparrow T^*,$$

since $u_t \ge 0$ in $Q_\ell(T^*)$ and $u_x > 0$ in Λ.

We fix $y \in (x_b, r)$ and put $\Omega(\tau) = (\tau, T^*) \times (x_b, y)$, $r = \pi/(x_b - y)$, $d(t,x) = \sin(r(x - y))$, $E(\xi) = \xi^{(1+\alpha)\sigma}$, $F(\xi) = \xi^{(1+\alpha)}$ and

$$J(t,x) = -u_x(t,x) + \eta d(t,x) E(u(t,x))$$

where $\sigma = 3/4$ and η is an arbitrary positive constant.

Clearly, we see that $J(t, y) < 0$ and $J(t, x_b) < 0$ for any $t \in (T^* - \delta, T^*)$. Moreover, we observe that

$$J_t - J_{xx}$$

$$= \quad -u_{xt} + \eta d u_t E'(u) + u_{xxx} - \eta d_{xx} E(u)$$

$$\quad -2\eta d_x u_x E'(u) - \eta d u_{xx} E'(u) - \eta d (u_x)^2 E''(u)$$

$$= \quad -u_x(F'(u) + 2\eta d_x E'(u)) + \eta d E'(u) F(u) - \eta d_{xx} E(u) - \eta d (u_x)^2 E''(u)$$

By putting $b(t,x) = F'(u(t,x)) + 2\eta d_x(t,x) E'(u(t,x))$, we infer that

$$J_t - J_{xx} - bJ$$

$$= \quad \eta d \{ E'(u)F(u) - E(u)F'(u) + 2\eta d_x E(u) E'(u) - \frac{d_{xx}}{d} E(u) \} - \eta (u_x)^2 E''(u)$$

$$(4.5) \le \quad \eta d \{ (E'(u)F(u) - E(u)F'(u)) + 2\eta d_x E(u) E'(u) + r^2 d E(u) \}.$$

10

As a consequence of (4.4), we obtain that there is a number $\tau_1 \in (T^* - \delta, T^*)$ such that

$$u(t,x) \geq \left(\frac{2r^2}{(1+\alpha)(1+\sigma)} \right)^{1/\alpha} \qquad \text{for } (t,x) \in \Omega(\tau_1),$$

since $u_x < 0$ and $u_t \geq 0$ on $\Omega(\tau_1)$. Hence, we have

$$(4.6) \qquad E(u)F'(u) - E'(u)F(u) \geq 2r^2 E(u) \qquad \text{on } \Omega(\tau_1).$$

Also, by taking a number $\tau_2 \in (\tau_1, T^*)$ satisfying that

$$u(t,x) \geq \left(\frac{\sigma}{1-\sigma} \right)^{\frac{1}{(1+\alpha)(1-\sigma)}} \qquad \text{for } (t,x) \in \Omega(\tau_2),$$

we have

$$E(u)F'(u) - E'(u)F(u) \geq E'(u)E(u) \qquad \text{on } \Omega(\tau_2).$$

Further, choosing $\eta_1 > 0$ such that $1 > 4\eta|r|$ we infer that for $0 < \eta < \eta_1$

$$(4.7) \qquad E(u)F'(u) - E'(u)F(u) \geq E'(u)E(u) \qquad \text{on } \Omega(\tau_2).$$

It follows from $(4.5) \sim (4.7)$ that

$$(4.8) \qquad J_t - J_{xx} - bJ \leq 0 \qquad \text{on } \Omega(\tau_2).$$

According to (4.3) we see that $\max\{u_x(\tau_2, x); x_b \leq x \leq y\}(:= \lambda) > 0$. Hence, for

$$0 < \eta < \min\left\{ \eta_1, \frac{\lambda}{2(\max\{E(u(\tau_2,x)); x_b \leq x \leq y\} + 1)} \right\}(:= \eta_2)$$

$$(4.9) \qquad J(\tau_2, x) \leq \lambda + \eta \sin(\delta(x-y))E(u(\tau_2,x)) < 0 \qquad \text{for } x_b < x < y.$$

By applying strong maximum principle to J we conclude that for $0 < \eta < \eta_2$

$$(4.10) \qquad J(t,x) < 0 \qquad \text{for } (t,x) \in \Omega(\tau_2).$$

Here, we put $\varphi(z) = \int_z^\infty E^{-1}(s)ds$. Obviously,

$$(4.11) \qquad \varphi(z) \to 0 \text{ if and only if } z \to +\infty.$$

Noting that $u(t,y) > 0$ for $t \in (\tau_2, T^*)$ we infer from (4.10) that

$$\varphi(u(t,x)) > \varphi(u(t,x)) - \varphi(u(t,y))$$

$$= -\int_x^y \frac{d}{d\xi}\varphi(u(t,\xi))d\xi$$

$$= \int_x^y \frac{u_x(t,\xi)}{E(u(t,\xi))}d\xi$$

$$\geq \eta \int_y^x \sin(\delta(\xi - y))d\xi > 0 \qquad \text{for } (t,x) \in \Omega(\tau_2).$$

This inequality contradicts (4.4) and (4.11) so that $x_* \geq s^-$.

Similarly, we can show that $x_* \leq s^+$. We accomplish the proof of Lemma 4.5. $\square$

Lemma 4.6. *We suppose that s^- is finite. Then the each point $x \in [s^-, s^+)$ is a blow-up point. Moreover, if $s^+ < \infty$ then s^+ is a blow-up point, too.*

Proof. First, we assume that $x_b \in (s^-, s^+)$. By definitions of s^- and s^+ we can choose a sequence $\{t_n\}$ with $t_n \uparrow \infty$ such that

$$s^-(t_n) \leq x_b \leq s^+(t_n) \quad \text{for each } n.$$

Here, we note that for any $y \in [s^-(t), s^+(t)]$ $u(y,t) = \max_{0 \leq x \leq \ell(t)} u(t,x)(:= U(t))$ and $U(t) \to +\infty$ as $t \uparrow T^*$. Therefore, we observe that

$$u(t_n, x_b) \to +\infty \quad \text{as } t \uparrow T^*,$$

that is, x_b is a blow-up point.

Next, we put $x_b = s^-$ and suppose that the point x_b is not a blow-up point. Hence, for some positive number M the following inequality holds:

$$0 < u(t, x_b) \leq M \quad \text{for } 0 \leq t < T^*.$$

If there exists a sequence $\{t_n\}$ such that $t_n \uparrow T^*$ as $n \to \infty$ and $s^-(t_n) \leq s^-$ for each n. This implies that $u_x(t_n, x_b) = 0$ so that $u(t_n, x_b) = U(t_n) \to \infty$ as $n \to \infty$. This is a contradiction. Hence, we conclude that for some $T_0 \in (0, T^*)$ $s^-(t) > s^-$ for $t \geq T_0$. It is clear that $u(t, s^-(t)) = U(t) \to +\infty$ as $t \uparrow T^*$. Accordingly, there exists a number $T_1 \in (T_0, T^*)$ such that

$$u(t, s^-(t)) \geq M + 1 \quad \text{for } t \geq T_1.$$

We put $x_a = s^-(T_1)$, $\beta = (x_a + x_b)/2$ and

$$w(t,x) = u(t,x) - u(t, 2\beta - x) \quad \text{for } T_1 \leq t < T^* \text{ and } x_b \leq x \leq \beta.$$

By elementary calculation we get the following relations:

(4.12) $$w_t = w_{xx} + \psi w \quad \text{in } (T_1, T^*) \times (x_b, \beta),$$

(4.13) $$w(t, x_b) < 0 \quad \text{for } T_1 \leq t < T^*,$$

(4.14) $$w(t, \beta) = 0 \quad \text{for } T_1 \leq t < T^*,$$

(4.15) $$w(T_1, x) < 0 \quad \text{for } x_b < x < \beta$$

where ψ is a continuous function on $[T_1, T^*) \times [x_b, \beta]$.

By applying strong maximum principle to (4.13) $\sim$ (4.15) we obtain that $w_x(t, \beta) > 0$ for $T_1 \leq T < T^*$. Consequently, we have $u_x(\cdot, \beta) > 0$ on $[T_1, T^*)$. Hence, $\beta \leq s^-$. This is a contradiction. Therefore, x_b is a blow-up point.

12

In case $s^+ < \infty$ we can prove that s^+ is a blow-up point by using the above argument. Thus this lemma has been proved. $\qquad\square$

Lemma 4.7. *If* $s^- < \infty$, *then* $s^- = s^+$.

Proof. We assume that $s^+ > s^-$ and choose numbers $r \in (s^-, s^+)$, $\varepsilon \in (0, \frac{r-s^-}{2})$, and put $x_* = s^- - \varepsilon$. By virtue of Lemma 4.6 we see that x_* is not a blow-up point. Hence, there is a positive constant C_0 such that

$$(4.16) \qquad\qquad u(t, x_*) \leq C_0 \quad \text{for } 0 \leq t < T^*.$$

Also, by Lemma 4.6 r is a blow-up point. Consequently, there exists a number $T_0 \in (0, T^*)$ satisfying that

$$(4.17) \qquad\qquad s^-(t) > x_* \text{ and } u(t, r) \geq C_0 \quad \text{for } t \in [T_0, T^*).$$

Moreover, by definition of s^+ we can take a number $t_0 \in (T_0, T^*)$ such that $s^+(t_0) > r$. Here, we put $\alpha = (x_* + r)/2$. Then, it is obvious that

$$(4.18) \qquad u(t_0, x) \leq u(t_0, 2\alpha - x) \quad \text{for } x_* < x < \alpha,$$

$$(4.19) \qquad u(t_0, \cdot) - u(t_0, 2\alpha - \cdot) \text{ is not a constant function on } (x_*, \alpha)$$

We put

$$\hat{w}(t, x) = u(t, 2\alpha - x) - u(t, x) \quad \text{for } (t, x) \in [t_0, T^*) \times [x_*, \alpha].$$

It is clear from (4.18) and (4.19) that $\hat{w}(t_0, x) \geq 0$ for $x_* \leq x \leq \alpha$, $\hat{w}(t_0, \cdot)$ is not a constant function on (x_*, α), $\hat{w}(t, x_*) > 0$ for $t \in [t_0, T^*)$, $\hat{w}(t, \alpha) = 0$ for $t \in [t_0, T^*)$ and

$$\hat{w}_t - \hat{w}_{xx} - \hat{\psi}\hat{w} = 0 \quad \text{in } (t_0, T^*) \times (x_*, \alpha),$$

where $\hat{\psi}$ is a continuous function on $[t_0, T^*) \times [x_*, \alpha]$.

By applying the strong maximum principle to these facts, we infer that $\hat{w}(t, x) > 0$ in $(t_0, T^*) \times (x_*, \alpha)$, that is, $u(t, 2\alpha - x) > u(t, x)$ in $(t_0, T^*) \times (x_*, \alpha)$. Therefore, for any $t \in [t_0, T^*)$, the maximum value of $u(t, x)$ for $x_* \leq x \leq r$ can not be attained for $x \in [x_*, \alpha)$. It follows that $s^-(t) \geq \alpha$ for $t \in [t_0, T^*)$. Accordingly, $s^- \geq \alpha$. We note that

$$s^- \leq \alpha = \frac{x_* + r}{2} < s^-.$$

This is a contradiction. Hence, we conclude that $s^- = s^+$. $\qquad\square$

5 The proof of Theorem 2.2

Throughout this section we assume that the conditions (H1) $\sim$ (H4) hold and $T^* < \infty$ and use the same notations as in the statement of Theorem 2.2 and Section 1.

First, it follows from Proposition 3.3 that

$$|u(t)|_{L^\infty(0, \ell(t))} \to \infty \quad \text{as } t \uparrow \infty.$$

If (B) does not hold, there are two possibilities (A1) and (A2).

(A1) There are a positive number x' and a sequence $\{(t_n, \xi_n)\} \subset Q_\ell(T^*; \cdot, x')$ satisfying that $t_n \uparrow T^*$ as $n \to \infty$ and $|u(t_n, \xi_n)| \geq n$ for each n.

(A2) There exists a positive constant L_0 such that $\ell(t) \leq L_0$ for any $t \in [0, T^*)$.

Case (A1): We obtain by taking subsequence if necessary that $\xi_n \to x_1$ for some $x_1 \in (0, \infty)$, that is, x_1 is a blow-up point. It is clear that for some $T_0 \in (0, T^*)$ $x_1 < \ell$ on $[T_0, T^*)$ because of the positivity of ℓ'. According to Lemmas 4.5 and 4.7 we infer that $s^- \leq x_1 < \infty$ and $s^- = x_1 = s^+$. We choose a number $x_2 \in (x_1, \ell_\infty)$ where $\ell_\infty = \lim_{t \uparrow T^*} \ell(t) \leq +\infty$. We conclude that there is a positive constant M such that

(5.1) $$u(t, x) \leq M \quad \text{for } (t, x) \in Q_\ell(T^*; x_2, \cdot).$$

In fact, by using Lemma 4.7, again, x_2 is not a blow-up point. Hence, there exist positive constants M and $T_1 \in [T_0, T^*)$ such that

$$u(t, x_2) \leq M \quad \text{for } (t, x_2) \in Q_\ell(T^*) \text{ and } s^+(t) < x_2 \quad \text{for } t \geq T_1.$$

Obviously, the above facts are reduced to (5.1).

Here, we can take a number $a \in (x_2, \ell_\infty)$ such that $u(\cdot, a) \in W_{loc}^{1,2}([\tau_0, T^*))$ where τ_0 is positive. Clearly, $\{u, \ell\}$ is a solution of $CSP(a; u^{1+\alpha}, u(\cdot, a); u(\tau_0, \cdot), \ell(\tau_0))$ on $[\tau_0, T^*)$. Besides, by Proposition 3.2 we can obtain a solution $\{\overline{u}, \overline{\ell}\}$ of $CSP(a; M^{1+\alpha}, M; \overline{u}_0, \ell(\tau_0) + 1)$ on $[\tau_0, \infty)$ where $\overline{u}_0 \in C^\infty([a, \ell(\tau_0) + 1])$ satisfying that $\overline{u}_0 = M$ on $[a, \ell(\tau_0)]$, $\overline{u}_0 > 0$ on $[\ell(\tau_0), \ell(\tau_0) + 1)$ and $\overline{u}_0(\ell(\tau_0) + 1) = 0$.

Therefore, there is a positive number $\overline{L}$ such that $\overline{\ell}(t) \leq \overline{L}$ for $t \in [\tau_0, T^*]$. Hence, Proposition 3.1 implies that the statement (A) is true.

Case (A2): Immediately, we have $\ell_\infty < \infty$ and $s^- < \infty$. It follows from Lemmas 4.3, 4.4, 4.6 and 4.7 that $s^- = s^+ \in (0, \ell_\infty)$, that is, (A) holds.

Thus we accomplish the proof of Theorem 2.2.

References

[1] A. Fasano and M. Primicerio, Free boundary problems for nonlinear parabolic equations with nonlinear free boundary conditions, *J. Math. Anal. Appl.* **72**, 247–273(1979).

[2] T. Aiki, The existence of solutions to two-phase Stefan problems for nonlinear parabolic equations, *Control Cyb.* **19**, 41–62(1990).

[3] T. Aiki and N. Kenmochi, Behavior of solutions to two-phase Stefan problems for nonlinear parabolic equations, *Bull. Fac. Education, Chiba Univ.* **39**, 15–62(1991).

[4] N. Kenmochi, A new proof of the uniqueness of solutions to two-phase Stefan problems for nonlinear parabolic equations, in *Free Boundary Problems, Inter. Ser. Num. Math.*, Vol. 95, pp. 101–126(1990).

[5] T. Aiki, Behavior of free boundaries blow-up solutions to one-phase Stefan problems, to appear in *Nonlinear Anal. TMA*.

[6] H. Imai and H. Kawarada, Numerical analysis of a free boundary problem involving blow-up phenomena, *Proc. Joint Symposium on Applied Mathematics*, 277 –280(1987). (in Japanese).

[7] H. Fujita, On the blowing up of solutions of the Cauchy problem for $u_t = \Delta u + u^{1+\alpha}$, *J. Fac. Sci. Univ. Tokyo Sect. I* **13**, 109–124(1966).

[8] M. Tsutsumi, Existence and nonexistence of global solutions for nonlinear parabolic equations, *Publ. Res. Inst. Math. Sci.* **8**, 211–229(1972).

[9] A. Friedman and B. McLeod, Blow-up of positive solutions of semilinear heat equations, *Indiana Univ. Math. J.* **34**, 425–447(1985).

[10] H. Fujita and Y.-G. Chen, On the set of blow-up points and asymptotic behaviours of blow-up solutions to a semilinear parabolic equation, in *Analyse Mathématique et Applications*, Gauthier-Villars, Paris, pp. 181–201(1988).

Toyohiko AIKI
Department of Mathematics, Faculty of Education, Gifu University
Gifu 501-11, Japan

Hitoshi IMAI
Faculty of Engineering, Tokushima University
Tokushima 770, Japan

P COLLI AND G SAVARÈ

Time discretization of Stefan problems with singular heat flux

Abstract. In this paper we study the approximation of singular Stefan problems via finite differences in time and establish convergence results. Taking advantage of an intermediate regularization of the heat flux, we recover the optimal error estimates holding in the standard setting.

1. Introduction and results.

Consider the following initial–boundary value problem

$$(\vartheta + \chi)_t - \Delta\alpha(\vartheta) = g \quad \text{in } Q := \Omega \times (0, T), \tag{1.1}$$

$$\chi \in H(\vartheta - \vartheta_c) \quad \text{in } Q, \tag{1.2}$$

$$-\frac{\partial\alpha(\vartheta)}{\partial\nu} = \alpha(\vartheta) - \alpha(\vartheta_b) \quad \text{in } \Sigma := \Gamma \times (0, T), \tag{1.3}$$

$$(\vartheta + \chi)(\cdot, 0) = \eta_0 \quad \text{in } \Omega, \tag{1.4}$$

$T > 0$ being fixed and Ω denoting a bounded smooth domain of $\mathbb{R}^N$ with boundary Γ and outward unitary normal ν. The pair of equations (1.1–2) yields the well–known weak formulation of the Stefan problem (see, e.g., [13] and references therein) in terms of the absolute temperature $\vartheta : Q \to \mathbb{R}$ and the phase variable $\chi : Q \to [0, 1]$ (representing, e.g., the local fraction of water in a water–ice system), with $g : Q \to \mathbb{R}$ standing for the (known) heat supply, H specifying the Heaviside graph, and ϑ_c being the critical (constant) temperature of phase change. While (1.1) results from the balance of internal energy with the heat flux given by $-\nabla\alpha(\vartheta)$, the inclusion (1.2) says that $\chi = 1$ where $\vartheta > \vartheta_c$ (liquid region), $\chi = 0$ where $\vartheta < \vartheta_c$ (solid region), and $\chi \in [0, 1]$ where $\vartheta = \vartheta_c$ (mushy region). The data $\vartheta_b : \Sigma \to]0, +\infty[$ in (1.3) and $\eta_0 : \Omega \to]0, +\infty[$ in (1.4) play as outside temperature on the boundary and initial enthalpy, respectively. In particular, (1.3) asserts the heat flux is proportional to a suitable variation of the absolute temperature between the exterior and the interior of the body. Similar conditions can be found in [6] and also in [5], to which we especially refer for a more detailed presentation of (1.1–4). In this connection, we should note that here we have taken all the physical constants (except obviously ϑ_c) equal to 1 for the sake of simplicity.

Let us focus now on the function $\alpha(\vartheta)$. The classical choice of α linear or, more generally, globally bi–Lipschitz continuous is motivated by the Fourier law. This paper is instead characterized by constitutive relations for the heat flux becoming singular as ϑ approaches the zero value. Namely, still keeping α strictly increasing and smooth, we require that

$$\alpha :]0, +\infty[\rightarrow \mathbb{R}, \quad \alpha(\vartheta) \searrow -\infty \text{ as } \vartheta \searrow 0.$$

In addition, α can be presented as a maximal monotone graph from $\mathbb{R}$ to $\mathbb{R}$ with domain $D(\alpha) =]0, +\infty[$ and with range not bounded from below. Remarking that α may be surjective or not according to its behaviour for large temperatures, we notice in advance that the former case turns out to be simpler from the viewpoint of the mathematical treatment since well-established theories (like, for instance, a suitable modification of the Brézis argument reported in [2, Ex. 5, p. 67]) can be applied.

In the framework of α singular there is the good property of an evident maximum principle in the energy balance equation (1.1), independent of the sign of g, whereas, even for the classical heat equation, to prove that the (absolute) temperature is non negative one has to ask for $g \geq 0$. Moreover, such an approach includes the position $\alpha(\vartheta) = -1/\vartheta$ commonly employed in a series of recent studies on the thermodynamically consistent phase–field model by Penrose and Fife (see, e.g., [12,6,8]). In a very simple situation, the Penrose–Fife model gives rise to a system coupling (1.1) with the following modification of (1.2), that is,

$$\delta \chi_t - \varepsilon \Delta \chi + H^{-1}(\chi) \ni \frac{1}{\vartheta_c} - \frac{1}{\vartheta} \tag{1.5}$$

for small positive parameters δ and ε. By supplying (1.5) with the homogeneous Neumann boundary condition and an initial condition for χ and collecting also (1.1) and (1.3)–(1.4), it has been shown [5] that the solution of the resulting initial and boundary value problem weakly converges in appropriate spaces to the solution of (1.1–4) as δ and ε tend to 0 (without any order relation between the two parameters). Error estimates have been subsequently proved in [7] with the further investigation of a time discrete version. However, we aim to point out that only rather special initial values η_0 are compatible with the limit procedure detailed in [5,7]. In addition, the approximation of (1.2) by (1.5), even if it provides smooth components χ, does not preserve the pointwise monotonicity of χ (and consequently of $\vartheta + \chi$) with respect to ϑ.

This paper is concerned with another kind of approximation of (1.1–4), obtained substituting the time derivative in (1.1) with a finite difference and dealing with the correponding implicit Euler scheme. The procedure allows more generality on the data and, as you will see, leads to worthwhile convergence results. From a qualitative point of view, our solutions can be compared with the variational solution proposed in [9, Thm. 3.3, p. 201] for the classical Stefan problem (regarding that, some hystorical important references can be found in the related "Commentaires" at p. 306 of [9]). Moreover, our analysis covers both possibilities of α surjective or not analysis and, via another regularization in the latter case, it can fully exploit the abstract estimates of [10,11] (which extend previous theorems of [1] deduced in a slightly different setting).

Before discussing the discrete problem, let us rewrite (1.1–4) in an equivalent form involving just one maximal monotone graph. Set $u = \alpha(\vartheta) - \alpha(\vartheta_c)$ and $e = \vartheta - \vartheta_c + \chi$ so that, owing to (1.2) and to the strict monotonicity of α, one has $\chi \in H(u)$ and $e \in \gamma(u)$, being

$$\gamma(v) = \alpha^{-1}(v + \alpha(\vartheta_c)) - \vartheta_c + H(v) \quad \text{for } v \in]-\infty, -\alpha(\vartheta_c) + \sup \alpha[,$$

i.e., in the effective domain. Moreover, observe that $\gamma(0) \ni 0$ and the same holds for the inverse graph $\beta = \gamma^{-1}$ which, restricted to $]-\vartheta_c, +\infty[$, is a function with one horizontal segment. Thus, setting $h = \alpha(\vartheta_b) - \alpha(\vartheta_c)$ and $e_0 = \eta_0 - \vartheta_c$, in terms of the new variables the system (1.1–4) becomes

$$e_t - \Delta u = g \quad \text{and} \quad e \in \gamma(u) \quad \text{in } Q, \tag{1.6}$$

$$\frac{\partial u}{\partial \nu} + u = h \quad \text{in } \Sigma, \tag{1.7}$$

$$e(\cdot, 0) = e_0 \quad \text{in } \Omega, \tag{1.8}$$

where (see [4, Ex. 2.8.1, p. 43] for properties of monotone graphs)

$\beta, \gamma : \mathbb{R} \to 2^{\mathbb{R}}$ are maximal monotone operators related by $\gamma = \beta^{-1}$, (1.9)

$D(\beta) =]-\vartheta_c, +\infty[$, β is a continuous function in $D(\beta)$, and $\beta(0) = 0$. (1.10)

In order to specify the hypotheses on the data, it will be useful to introduce the "primitive" functions j and ϕ of β and γ, respectively. Then

$j, \phi : \mathbb{R} \mapsto]-\infty, +\infty]$ are convex lower semicontinuous functions (1.11)

fulfilling

$$\beta = \partial j, \quad \gamma = \partial \phi, \quad \text{and} \quad j(0) = \phi(0) = 0. \tag{1.12}$$

At this point, we can assume that

$$g \in L^2(Q), \quad h \in L^2(\Sigma) \quad \text{with} \quad \phi(h) \in L^1(\Sigma), \tag{1.13}$$

$$e_0 \in L^2(\Omega) \quad \text{with} \quad j(e_0) \in L^1(\Omega). \tag{1.14}$$

Coming back to (1.1–4), observe that (1.13) and (1.14) are ensured provided, for instance, ϑ_b and η_0 lie between two positive constants almost everywhere in Σ and Ω, respectively (cf. the framework of [5]).

We give a variational formulation of (1.6–8) in which e, u are viewed as time dependent functions with values in the triple of densely embedded Hilbert spaces $H^1(\Omega)$, $L^2(\Omega)$, $(H^1(\Omega))'$. To fix notation let $\langle \cdot, \cdot \rangle$ indicate either the duality pairing between $(H^1(\Omega))'$ and $H^1(\Omega)$ or the scalar product in $L^2(\Omega)$ and take

$$((v, w)) := \int_\Omega \nabla v \cdot \nabla w + \int_\Gamma vw, \quad v, w \in H^1(\Omega), \tag{1.15}$$

as scalar product in $H^1(\Omega)$, with the associated Riesz isomorphism

$$L : H^1(\Omega) \to (H^1(\Omega))', \quad \langle Lv, w \rangle := ((v, w)) \quad \text{for } v, w \in H^1(\Omega). \tag{1.16}$$

Let also $F \in L^2(0, T; (H^1(\Omega))')$ represent the family of functionals

$$\langle F(t), v \rangle := \int_\Omega g(\cdot, t)v + \int_\Gamma h(\cdot, t)v \quad \forall\, v \in H^1(\Omega), \text{ for a.e. } t \in\,]0, T[, \tag{1.17}$$

naturally associated with the data g and h. Then the notion of weak solution to (1.6–8) is readily settled.

Problem (P). It consists in seeking a pair (e, u) such that $e \in H^1(0, T; (H^1(\Omega))') \cap L^\infty(0, T; L^2(\Omega))$, $u \in L^2(0, T; H^1(\Omega))$, and

$$e'(t) + Lu(t) = F(t) \quad \text{for a.e. } t \in\,]0, T[, \tag{1.18}$$
$$e \in \gamma(u) \quad \text{a.e. in } Q, \tag{1.19}$$
$$e(0) = e_0, \tag{1.20}$$

(1.18) and (1.20) having an obvious meaning in $(H^1(\Omega))'$.

Referring to the next section for the control of the well–posedness of (P), we shall see that the uniqueness for (P) relies on the monotonicity of γ and on the linearity of L. We are interested to the following time discretization of (P), in which $\tau = T/n$ ($n \in \mathbb{N}$) denotes the time step and

$$F_\tau^i := \frac{1}{\tau} \int_{(i-1)\tau}^{i\tau} F(t)dt \quad \text{is an element of } (H^1(\Omega))'$$

connected with (cf. (1.17)) the analogously defined g_τ^i and h_τ^i for $i = 1, \ldots, n$.

19

Problem (P_τ). Find the vectors $(e_\tau^1, \ldots, e_\tau^n) \in (L^2(\Omega))^n$ and $(u_\tau^1, \ldots, u_\tau^n) \in (H^1(\Omega))^n$ satisfying

$$\frac{e_\tau^i - e_\tau^{i-1}}{\tau} + Lu_\tau^i = F_\tau^i, \tag{1.21}$$

$$e_\tau^i \in \gamma(u_\tau^i) \qquad \text{a.e. in } \Omega \tag{1.22}$$

if $i = 1, \ldots, n$, where $e_\tau^0 := e_0$.

In the discrete problem (P_τ) it is even easier to show uniqueness than in (P). The existence of the discrete and continuous solutions as well as the convergence properties of the approximation scheme will be recovered with the involvement of an additional regularization process. This is carried out in order to prevent the (possible) lack of coerciveness for the convex function j or, equivalently, the lack of surjectivity for β (cf., e.g., [4, Rem. 2.3, p. 43]). Actually, we regularize (P) and (P_τ) replacing β by

$$\beta_\varepsilon := \beta + \varepsilon I, \qquad \varepsilon > 0, \qquad I \text{ identity in } \mathbb{R}, \tag{1.23}$$

and define consequently

$$\gamma_\varepsilon := \beta_\varepsilon^{-1}, \qquad j_\varepsilon(r) := j(r) + \frac{\varepsilon}{2} r^2 \text{ and } \phi_\varepsilon(r) := \int_0^r \gamma_\varepsilon(s) ds \quad \forall\, r \in \mathbb{R}. \tag{1.24}$$

Note that γ_ε is Lipschitz continuous with Lipschitz constant $1/\varepsilon$ in all of $\mathbb{R}$, whence the solution of the next discrete problem is certainly smoother in its e component.

Problem ($P_{\varepsilon\tau}$). Determine the two vectors $(e_{\varepsilon\tau}^1, \ldots, e_{\varepsilon\tau}^n)$ and $(u_{\varepsilon\tau}^1, \ldots, u_{\varepsilon\tau}^n)$ belonging to $(H^1(\Omega))^n$ and fulfilling

$$\frac{e_{\varepsilon\tau}^i - e_{\varepsilon\tau}^{i-1}}{\tau} + Lu_{\varepsilon\tau}^i = F_\tau^i, \tag{1.25}$$

$$e_{\varepsilon\tau}^i = \gamma_\varepsilon(u_{\varepsilon\tau}^i) \qquad \text{a.e. in } \Omega \tag{1.26}$$

for $i = 1, \ldots, n$, where $e_{\varepsilon\tau}^0 := e_0$.

In this approximating setting it is possible to use the abstract theory of [10,11] which will be addressed to ($P_{\varepsilon\tau}$) and its continuous version.

Problem (P_ε). Find $e_\varepsilon \in H^1(0, T; (H^1(\Omega))') \cap L^2(0, T; H^1(\Omega))$ and $u_\varepsilon \in L^2(0, T; H^1(\Omega))$ such that

$$e_\varepsilon'(t) + Lu_\varepsilon(t) = F(t) \qquad \text{for a.e. } t \in {]0, T[}, \tag{1.27}$$

$$e_\varepsilon = \gamma_\varepsilon(u_\varepsilon) \qquad \text{a.e. in } Q, \tag{1.28}$$

$$e_\varepsilon(0) = e_0, \tag{1.29}$$

We can finally state our main results involving the solutions of the four listed problems. Henceforth we maintain the notations (e, u) and $(e_\varepsilon, u_\varepsilon)$ for the solutions of (P) and (P$_\varepsilon$), while, with obvious position, e_τ, u_τ, $e_{\varepsilon\tau}$, $u_{\varepsilon\tau}$ stand for the piecewise constant functions attaining the respective values e_τ^i, u_τ^i, $e_{\varepsilon\tau}^i$, $e_{\varepsilon\tau}^i$ in the time interval $](i-1)\tau, i\tau]$, for $i = 1, \ldots, n$.

Theorem 1. *Under the assumptions (1.9–14), as ε and τ tend to 0, the one independent from the other, we have that*

$$e_{\varepsilon\tau} \to e \text{ weakly star in } L^\infty(0, T; L^2(\Omega)) \text{ and strongly in } L^\infty(0, T; H^{-s}(\Omega))$$
$$(1.30)$$

for $s > 0$ and

$$u_{\varepsilon\tau} \to u \text{ strongly in } L^2(0, T; H^1(\Omega)). \tag{1.31}$$

Of course the same convergences hold with e_τ, u_τ in place of e, u as only ε goes to 0 ($\tau > 0$ fixed) and with e_ε, u_ε in place of e, u as only τ goes to 0 ($\varepsilon > 0$ fixed). In particular this theorem implies the existence of the solutions to the problems (P) and (P$_\tau$). The statement below provides suitable error estimates.

Theorem 2. *Assume that (1.9–14) hold. Then there is some constant C, depending only on Ω and T, such that*

$$\|e_{\varepsilon\tau} - e\|_{L^\infty(0,T;(H^1(\Omega))')}^2$$
$$\leq C(\tau + \varepsilon) \left\{ \|g\|_{L^2(Q)}^2 + \|h\|_{L^2(\Sigma)}^2 + \|\phi(h)\|_{L^1(\Sigma)} + \|e_0\|_{L^2(\Omega)}^2 + \|j(e_0)\|_{L^1(\Omega)} \right\}$$
$$(1.32)$$

for any choice of the parameters ε, τ. Moreover, under the additional assumptions

$$g \in W^{1,1}(0, T; L^2(\Omega)), \quad h \in W^{1,1}(0, T; H^{-1/2}(\Gamma)), \quad u_0 = \beta(e_0) \in H^1(\Omega), \quad (1.33)$$

we have that

$$\|e_\tau - e\|_{L^\infty(0,T;(H^1(\Omega))')}^2 + \tau \|u_\tau - u\|_{L^2(0,T;H^1(\Omega))}^2$$
$$\leq C\tau^2 \left\{ \|g\|_{W^{1,1}(0,T;L^2(\Omega))}^2 + \|h\|_{W^{1,1}(0,T;H^{-1/2}(\Gamma))}^2 + \|e_0\|_{L^2(\Omega)}^2 + \|u_0\|_{H^1(\Omega)}^2 \right\}$$
$$(1.34)$$

where the time regularity $W^{1,1}$ of g and h in (1.33–34) can be reduced to BV (bounded variation) or substituted with H^s for $s > 1/2$.

2. Proofs.

This section contains the basic details of the arguments. For the sake of convenience we split the matter into different steps.

Step 1. *Uniqueness of the solution of* (P).

Let (e_i, u_i), $i = 1, 2$, be two pairs solving (P) with respect to the same data. We integrate the difference of the two equations (1.18) from 0 to $s \in \,]0, T[$ and we test it (via duality pairing) by $u_1 - u_2$. As $\langle e_1 - e_2, u_1 - u_2 \rangle \geq 0$ by the monotonicity of γ, another integration allows to get the equality of the two terms $\int_0^s u_i(t) dt$ for any $s \in [0, T]$. Then it suffices to make use of (1.18) and (1.20) to conclude that (e_i, u_i) coincide. ∎

Step 2. *A priori estimates.*

¿From now on we denote by C all the constants which are independent of the data and of the parameters ε, τ. Besides, we use the notation $\|\{e_0, g, h\}\|$ for

$$\|\{e_0, g, h\}\|^2 := \|e_0\|^2_{L^2(\Omega)} + \|g\|^2_{L^2(Q)} + \|h\|^2_{L^2(\Sigma)} = \|\{e_0, g, h\}\|^2_{L^2(\Omega) \times L^2(Q) \times L^2(\Sigma)},$$

and set

$$J(v) := \begin{cases} \int_\Omega j(v) & \text{if } v \in L^1(\Omega) \text{ and } j(v) \in L^1(\Omega) \\ +\infty & \text{otherwise} \end{cases}$$

as well as

$$\Phi^\Gamma(w) := \begin{cases} \int_\Gamma \phi(w) & \text{if } w \in L^1(\Gamma) \text{ and } \phi(w) \in L^1(\Gamma) \\ +\infty & \text{otherwise} \end{cases},$$

with analogous definitions for J_ε, Φ^Γ_ε. It is well known that (see, e.g., [2,4]) J, J_ε and Φ^Γ, Φ^Γ_ε are convex and lower semicontinuous in $L^2(\Omega)$ and $L^2(\Gamma)$, respectively. Moreover, their subdifferential mappings in these Hilbert spaces are nothing but $\partial j = \beta$, $\partial j_\varepsilon = \beta_\varepsilon$ applied almost everywhere in Ω and $\partial \phi = \gamma$, $\partial \phi_\varepsilon = \gamma_\varepsilon$ applied almost everywhere in Γ.

It will be clear from the fourth step that Problem ($P_{\varepsilon\tau}$) admits a unique solution. Now we establish some basic a priori estimates on the discrete solution.

Claim. *For every* $\varepsilon \in \,]0, 1]$ *and every time step* τ *we have that*

$$J(e_{\varepsilon\tau}(T)) + \frac{\varepsilon}{2}\|e_{\varepsilon\tau}(T)\|^2_{L^2(\Omega)} + \|u_{\varepsilon\tau}\|^2_{L^2(0,T;H^1(\Omega))}$$

$$\leq J(e_0) + \frac{\varepsilon}{2}\|e_0\|^2_{L^2(\Omega)} + \int_0^T \langle F_\tau(t), u_{\varepsilon\tau}(t)\rangle dt \tag{2.1}$$

and

$$\sup_{t\in[0,T]} \|e_{\varepsilon\tau}(t)\|^2_{L^2(\Omega)} + \varepsilon\|e_{\varepsilon\tau}\|^2_{L^2(0,T;H^1(\Omega))} + \int_0^T \Phi^\Gamma_\varepsilon(u_{\varepsilon\tau})$$

$$\leq C\Big\{ \|e_0\|^2_{L^2(\Omega)} + \|g\|^2_{L^2(Q)} + \int_0^T \Phi^\Gamma(h) \Big\}$$

(2.2)

Proof of (2.1). Take the duality pairing of the equation (1.25) multipied by τ and of $u^i_{\varepsilon\tau}$, then sum with respect to i from 1 to n. Remarking that

$$\langle e^i_{\varepsilon\tau} - e^{i-1}_{\varepsilon\tau}, u^i_{\varepsilon\tau} \rangle = \langle e^i_{\varepsilon\tau} - e^{i-1}_{\varepsilon\tau}, \beta_\varepsilon(e^i_{\varepsilon\tau}) \rangle \geq J_\varepsilon(e^i_{\varepsilon\tau}) - J_\varepsilon(e^{i-1}_{\varepsilon\tau})$$

thanks to (1.26) and (1.24), we easily obtain

$$J_\varepsilon(e^n_{\varepsilon\tau}) + \tau \sum_{i=1}^n \|u^i_{\varepsilon\tau}\|^2_{H^1(\Omega)} \leq J_\varepsilon(e_0) + \tau \sum_{i=1}^n \langle F^i_{\varepsilon\tau}, u^i_{\varepsilon\tau} \rangle,$$

that is (2.1). As a by–product, with the help of (1.17) and (1.13–14) we deduce the uniform boundedness of $\{u_{\varepsilon\tau}\}$ in $L^2(0,T;H^1(\Omega))$, namely,

$$\|u_{\varepsilon\tau}\|^2_{L^2(0,T;H^1(\Omega))} \leq 2\,J(e_0) + \|\{e_0, g, h\}\|^2. \quad \blacksquare$$

(2.3)

Proof of (2.2). Test (1.25) by $\tau e^i_{\varepsilon\tau}$ which belongs to $H^1(\Omega)$ since γ_ε is Lipschitz continuous. Recalling (1.15–16) and (1.23), we infer that

$$\frac{1}{2}\|e^i_{\varepsilon\tau}\|^2_{L^2(\Omega)} - \frac{1}{2}\|e^{i-1}_{\varepsilon\tau}\|^2_{L^2(\Omega)} + \tau \int_\Omega \left(\beta'(e^i_{\varepsilon\tau}) + \varepsilon \right) |\nabla e^i_{\varepsilon\tau}|^2$$

$$+ \tau \int_\Gamma \left(u^i_{\varepsilon\tau} - h^i_\tau \right) \gamma_\varepsilon(u^i_{\varepsilon\tau}) \leq \tau \int_\Omega g^i_\tau e^i_{\varepsilon\tau}$$

(2.4)

and we control the integral on the boundary Γ by exploiting (1.24) and (1.11–12). In fact, it is straightforward to check that $\phi_\varepsilon \leq \phi$ and to derive

$$\tau \int_\Gamma \left(u^i_{\varepsilon\tau} - h^i_\tau \right) \gamma_\varepsilon(u^i_{\varepsilon\tau}) \geq \tau \Phi^\Gamma_\varepsilon(u^i_{\varepsilon\tau}) - \tau \Phi^\Gamma(h^i_\tau).$$

Therefore, summing up in (2.4) and using the Hölder and elementary Young inequalities to estimate the term $\int_0^T \|g_\tau\|_{L^2(\Omega)} \|e_\tau\|_{L^2(\Omega)}$, it is not difficult to achieve (2.2) for some constant C depending on T. $\quad \blacksquare$

Step 3. *The discrete problem and weak convergence results.*

We denote by $\hat{e}_{\varepsilon\tau}$ the piecewise linear interpolant of the values $\{e^0_{\varepsilon\tau}, e^1_{\varepsilon\tau}, \dots, e^n_{\varepsilon\tau}\}$ at the nodes $\{0, \tau, \dots, n\tau\}$ so that

$$\hat{e}'_{\varepsilon\tau}(t) \equiv \frac{e^i_{\varepsilon\tau} - e^{i-1}_{\varepsilon\tau}}{\tau} \quad \text{if } t \in \,](i-1)\tau, i\tau[, \ i = 1, \dots, n.$$

Then, from (1.25–26) it turns out that the following discrete version of the problem (P_ε)

$$\hat{e}'_{\varepsilon\tau}(t) + Lu_{\varepsilon\tau}(t) = F_\tau(t) \quad \text{for a.e. } t \in]0, T[,$$

$$e_{\varepsilon\tau} \in \gamma_\varepsilon(u_{\varepsilon\tau}) \quad \text{a.e. in } Q,$$

$$e_{\varepsilon\tau}(\cdot, 0) = e_0.$$

is satisfied. In addition, owing to (2.2) and (2.3) there is a constant C_1, independent of ε and τ, such that

$$\|\hat{e}_{\varepsilon\tau}\|^2_{H^1(0,T;(H^1(\Omega))')\cap C^0([0,T];L^2(\Omega))} + \varepsilon\|e_{\varepsilon\tau}\|^2_{L^2(0,T;H^1(\Omega))} + \|u_{\varepsilon\tau}\|^2_{L^2(0,T;H^1(\Omega))} \leq C_1. \tag{2.5}$$

By applying standard monotonicity and compactness arguments (see, e.g., [4,9]) and reminding that (P) has at most one solution, (2.5) enables us to pass to the limit in the above equations and conditions as ε and τ go to 0, the weak star convergences holding for the whole sequences. The limits e of either $\{\hat{e}_{\varepsilon\tau}\}$ or $\{e_{\varepsilon\tau}\}$ (note that $(\hat{e}_{\varepsilon\tau} - e_{\varepsilon\tau}) \to 0$ strongly in $L^\infty(0,T;(H^1(\Omega))')$) and u of $\{u_{\varepsilon\tau}\}$ solve Problem (P). At this point, to complete the proof of Theorem 1 we just have to show (1.30) and (1.31). ∎

Step 4. *The formulation of $(P_{\varepsilon\tau})$ in $(H^1(\Omega))'$ and the abstract theory.*

On account of (1.16), observe that the linear isomorphism L^{-1} maps $(H^1(\Omega))'$ to $H^1(\Omega)$ and induces a scalar product in $(H^1(\Omega))'$ through

$$((v, w))_* := \langle v, L^{-1}w \rangle, \quad v, w \in (H^1(\Omega))'. \tag{2.6}$$

Hence the duality pairing of (1.25) and $L^{-1}w$ yields

$$\tau^{-1}((e^i_{\varepsilon\tau} - e^{i-1}_{\varepsilon\tau}, w))_* + \langle u^i_{\varepsilon\tau}, w \rangle = ((F^i_\tau, w))_* \quad \forall w \in (H^1(\Omega))'.$$

Incidentally, let us point out that the continuous version (1.27) could be rewritten in a similar way. As (1.26) alternatively reads $u^i_{\varepsilon\tau} = \beta_\varepsilon(e^i_{\varepsilon\tau})$ a.e. in Ω, choosing w of the form $e^i_{\varepsilon\tau} - \eta$ (with $\eta \in L^2(\Omega)$) and using the subdifferential property (in $\mathbb{R}$) of β_ε lead to

$$((\frac{e^i_{\varepsilon\tau} - e^{i-1}_{\varepsilon\tau}}{\tau} - F^i_\tau, e^i_{\varepsilon\tau} - \eta))_* + J_\varepsilon(e^i_{\varepsilon\tau}) \leq J_\varepsilon(\eta) \quad \forall \eta \in L^2(\Omega). \tag{2.7}$$

Since $J_\varepsilon(v) \geq \varepsilon\|v\|^2_{L^2(\Omega)}$ for any $v \in L^2(\Omega)$, the convex function J_ε is coercive in $L^2(\Omega)$ and consequently it is lower semicontinuous also in $(H^1(\Omega))'$. The corresponding subdifferential $\partial^* J_\varepsilon$ is then the maximal monotone operator (from $(H^1(\Omega))'$ to $(H^1(\Omega))'$) supplied by [2, Prop. 2.10, p. 67] and specified by

$$w \in \partial^* J_\varepsilon(v) \quad \text{if and only if} \quad v \in L^2(\Omega), \ \beta_\varepsilon(v) \in H^1(\Omega), \ w = L(\beta_\varepsilon(v)). \tag{2.8}$$

Therefore we can conclude that (1.25–26), the inequality (2.7) and the equation

$$\tau^{-1}(e_{\varepsilon\tau}^i - e_{\varepsilon\tau}^{i-1}) + \partial^* J_\varepsilon(e_{\varepsilon\tau}^i) \ni F_\tau^i, \tag{2.9}$$

are all equivalent and all admit the same unique solution $e_{\varepsilon\tau}^i$.

In addition, as τ tends to 0 we have that (see, for instance, [4]) $e_{\varepsilon\tau}$ strongly converges in $L^\infty(0,T;(H^1(\Omega))')$ to the solution e_ε of the problem (P_ε), that fulfils

$$e'_\varepsilon(t) + \partial^* J_\varepsilon e_\varepsilon(t) \ni F(t) \quad \text{for a.e. } t \in [0,T], \qquad e_\varepsilon(0) = e_0. \tag{2.10}$$

Finally, the abstract results of [10,11] allow us to deduce the uniform estimate

$$\|e_{\varepsilon\tau} - e_\varepsilon\|^2_{L^\infty(0,T;(H^1(\Omega))')} \le C\tau\left\{\|\{e_0, g, h\}\|^2 + J(e_0)\right\}, \tag{2.11}$$

which remains valid for the difference $e_\tau - e$, weak star limit of the sequence $e_{\varepsilon\tau} - e_\varepsilon$ in the space $L^\infty(0,T;(H^1(\Omega))')$ when $\varepsilon \searrow 0$ (cf. Step 3).

Step 5. *The non–coercive case.*

Compared to J_ε, in general J is no longer lower semicontinuous in $(H^1(\Omega))'$ and the previous construction cannot be directly applied. However, one can suitably modify J in order to obtain a new function $\tilde{J}$, lower semicontinuous in $(H^1(\Omega))'$ and coinciding with J on $L^2(\Omega)$. To this aim, let us consider the convex function $\Phi : L^2(\Omega) \to]-\infty, +\infty]$ defined as Φ^Γ, but with Ω instead of Γ. Recalling (1.12) and the fact that β is the inverse graph of γ, from standard items on convex conjugate functions we infer that

$$J(v) = \sup_{p \in L^2(\Omega)} \left(\langle v, p \rangle - \Phi(p)\right) \quad \forall\, v \in L^2(\Omega). \tag{2.12}$$

As the restriction of Φ to $H^1(\Omega)$ is still lower semicontinuous, (2.12) suggests us to put

$$\tilde{J}(v) := \sup_{p \in H^1(\Omega)} \left(\langle v, p \rangle - \Phi(p)\right) \quad \text{for } v \in (H^1(\Omega))'. \tag{2.13}$$

Thanks to, e.g., [2, Prop. 2.3, p. 52], $\tilde{J}$ is proper, convex, and lower semicontinuous on the dual of $H^1(\Omega)$. Further, it is clear that

$$\text{if } v \in L^2(\Omega) \text{ then } \tilde{J}(v) \le J(v). \tag{2.14}$$

What it is actually interesting for us is to get the equality in (2.14).

Claim. For any $v \in L^2(\Omega)$ we have that $J(v) = \tilde{J}(v)$.

Proof. It suffices to verify that $J(v) \geq \tilde{J}(v)$. We remark that

$$\tilde{J}(v) \geq \sup_{p \in H_0^1(\Omega)} \left(\langle v, p \rangle - \Phi(p) \right).$$

On the other hand, a theorem shown by Brezis [3, Sec. 2, Thm. 1] asserts that

$$\sup_{p \in H_0^1(\Omega)} \left(\langle v, p \rangle - \Phi(p) \right) = \int_{\Omega} \phi^*(v(x)) dx \quad \forall\, v \in L^2(\Omega),$$

where $\phi^* : \mathbb{R} \to]-\infty, +\infty]$ is the conjugate of ϕ. Having already noticed that $\phi^* = j$, the statement is proved. ∎

The property $\tilde{J} = J$ on $L^2(\Omega)$ yields the following corollary.

Claim. *If $v \in L^2(\Omega)$, $z \in H^1(\Omega)$, and $z \in \beta(v)$ a.e. in Ω, then $Lz \in \partial^* \tilde{J}(v)$.*

Proof. By (1.9), (1.12), and (2.12) one realizes that $\langle v, z \rangle = \Phi(z) + J(v)$. Thus, also from (2.13) it follows that

$$((\eta - v, Lz))_* = \langle \eta - v, z \rangle = \langle \eta, z \rangle - \Phi(z) - J(v) \leq \tilde{J}(\eta) - J(v)$$

for all $\eta \in (H^1(\Omega))'$, which entails $Lw \in \partial^* \tilde{J}(v)$. ∎

Thanks to the above conclusion, the a priori estimates of Step 2 ensure that (P_τ) and (P) can be set the forms (2.9) and (2.10), respectively (of course, without the index ε and with J replaced by $\tilde{J}$).

Consequently, we can recover the error estimate (1.34) by adapting the abstract framework of [11, Thm. 4] to our situation.

Step 6. *Strong convergences.*

In view of the formulations of the problems (P_ε) and (P) and of the relationships $\gamma_\varepsilon^{-1} = \beta_\varepsilon = \beta + \varepsilon I$ and $\gamma^{-1} = \beta$, it results that $u_\varepsilon - u = \beta(e_\varepsilon) - \beta(e) + \varepsilon e_\varepsilon$ a.e. in Q. Therefore, subtracting (1.18) from (1.27), testing by $L^{-1}(e_\varepsilon - e)(t)$, and integrating in time, with the help of (2.6), (1.20), and (1.29) one checks that

$$\|(e_\varepsilon - e)(t)\|_{(H^1(\Omega))'}^2 + 2 \int_0^T \langle \beta(e_\varepsilon) - \beta(e), e_\varepsilon - e \rangle = -2\varepsilon \int_0^T \langle e_\varepsilon - e, e_\varepsilon \rangle$$

for any $t \in [0, T]$. The pointwise monotonicity of β enables us to obtain

$$\|e_\varepsilon - e\|_{L^\infty(0,T;(H^1(\Omega))')}^2 \leq 2\varepsilon \int_0^T \|e_\varepsilon(t)\|_{H^1(\Omega)} \|(e_\varepsilon - e)(t)\|_{(H^1(\Omega))'} dt,$$

whence, by (2.2),

$$\|e_\varepsilon - e\|^2_{L^\infty(0,T;(H^1(\Omega))')} \leq C\varepsilon\Big\{\|e_0\|^2_{L^2(\Omega)} + \|g\|^2_{L^2(Q)} + \int_0^T \Phi^\Gamma(h)dt\Big\}.$$

Using this inequality, it is now straightforward to derive (1.32) from (2.11) and subsequently get (1.30) by interpolation.

It remains to deduce (1.31). Multiplying (1.18) by $u = \beta(e) \in L^{-1}\partial^*\tilde{J}(e)$ and integrating from 0 to T lead to

$$\tilde{J}(e(T)) + \int_0^T \|u(t)\|^2_{H^1(\Omega)}dt = \tilde{J}(e_0) + \int_0^T \langle F(t), u(t)\rangle dt. \qquad (2.15)$$

On account of the previous step, since e is weakly continuous from $[0, T]$ to $L^2(\Omega)$, we can substitute $\tilde{J}$ with J in (2.15). Moreover, in view of (2.2) and (2.5) we have that

$$e_{\varepsilon\tau}(T) \to e(T) \quad \text{weakly in} \quad L^2(\Omega)$$

and, by the lower semicontinuity of J, that

$$\liminf_{\varepsilon,\tau\searrow 0} J(e_{\varepsilon\tau}(T)) \geq J(e(T)).$$

Then, comparing (2.15) with (2.1) we can conclude that

$$\limsup_{\varepsilon,\tau\searrow 0} \|u_{\varepsilon\tau}\|_{L^2(0,T;H^1(\Omega))} \leq \|u\|_{L^2(0,T;H^1(\Omega))},$$

which implies the strong convergence (1.31) in virtue of the weak convergence previously established in Step 3. ∎

REFERENCES

[1] C. Baiocchi, *Discretization of evolution variational inequalities*, in "Partial differential equations and the calculus of variations," F. Colombini, A. Marino, L. Modica, and S. Spagnolo (eds.), Birkäuser, Boston, 1989, pp. 59–92.

[2] V. Barbu, "Nonlinear semigroups and differential equations in Banach spaces," Academiei, Bucureşti / Noordhoff, Leyden, 1976.

[3] H. Brézis, *Integrales convexes dans les espaces de Sobolev*, Israel J. Math., **13** (1973), 9-23.

[4] H. Brézis, "Opérateurs maximaux monotones et semi–groupes de contractions dans les espaces de Hilbert," North–Holland Math. Studies, **5**, North–Holland, Amsterdam, 1973.

[5] P. Colli and J. Sprekels, *Stefan problems and the Penrose–Fife phase field model*, Report N. **945**, Istituto di Analisi Numerica del C.N.R., Pavia 1994, pp. 1–23.

[6] N. Kenmochi and M. Niezgódka, *Systems of nonlinear parabolic equations for phase change problems*, Adv. Math. Sci. Appl., **3** (1993/94), 89–117.

[7] O. Klein, *A semidiscrete scheme for a Penrose–Fife system and some Stefan problems in* $\mathbb{R}^3$, Adv. Math. Sci. Appl., to appear.

[8] Ph. Laurençot, *Solutions to a Penrose–Fife model of phase–field type*, J. Math. Anal. Appl., **185** (1994), 262–274.

[9] J. L. Lions, "Quelques méthodes de résolution des problèmes aux limites non linéaires," Dunod Gauthier–Villars, Paris, 1969.

[10] J. Rulla, *Error analyses for implicit approximations to solutions to Cauchy problems*, SIAM J. Numer. Anal., to appear.

[11] G. Savarè, *Weak solutions and maximal regularity for abstract evolution inequalities*, Adv. Math. Sci. Appl., to appear.

[12] J. Sprekels and S. Zheng, *Global smooth solutions to a thermodynamically consistent model of phase–field type in higher space dimensions*, J. Math. Anal. Appl., **176** (1993), 200–223.

[13] C. Verdi, *Numerical aspects of parabolic free boundary and hysteresis problems*, in "Phase transitions and hysteresis," A. Visintin (ed.), Lecture Notes in Math., **1584**, Springer–Verlag, Berlin, 1994, pp. 213–284.

Pierluigi Colli
Dipartimento di Matematica
Università di Torino
Via Carlo Alberto 10
10123 Torino, Italy

Giuseppe Savarè
Istituto di Analisi Numerica del C.N.R.
Via Abbiategrasso 209
27100 Pavia, Italy

I RUBINSTEIN AND B ZALTZMAN

Morphological instability of the similarity solutions to the Stefan problem with undercooling and surface tension

In our paper we address the morphological stability of a similarity solution to the Stefan problem with surface tension and initial supercooling.

The study of shape instabilities in solidification with supercooling is important, in particular, because these instabilities initiate the appearance of dendrites.

Following the excellent review by Langer *Instabilities and pattern formation in crystal growth*, let us recall two well-known examples of morphological instabilities. The first concerns the quasi-steady-state growth of a spherical germ into a supercooled melt (Mullins and Sekerka, 1963). The respective model problem reads:

$$u_{rr} + \frac{2}{r}u_r = 0 \qquad \forall r \neq R,$$

$$u|_{r=R} = -\frac{2\gamma}{R}, \qquad \gamma \text{ --- surface tension,}$$

$$D[u_r]|_{r=R} = \frac{dR}{dt},$$

$$u(\infty) = -\delta.$$

Here, $[\ ]$ corresponds to the jump of the interior function across the boundary from solid to liquid.

There is a simple solution to this radial problem according to which a germ with an initial radius bigger than some critical value R_0 goes on growing monotonically. The linear stability analysis of this solution yields that there exists another critical value of the germ radius $R_{cr} > R_0$, uniquely determined by γ and δ, above which the solution is unstable.

The next example we would like to recall concerns the travelling wave solution to the one-dimensional Stefan problem which, after transform to a coordinate system moving with the wave speed v reads

$$u_{zz} + vu_z = 0 \qquad \forall z \neq z_0,$$

$$u|_{z=z_0} = 0,$$

$$[u_z]_{z_0+0}^{z_0-0} = v,$$

$$u(\infty) = u_\infty,$$

$$u(-\infty) = 0.$$

A one-parameter family of solutions (with arbitrary v) to this problem exists for only one critical value of undercooling: $u_\infty = -1$. For different undercoolings no planar

29

traveliing-wave solution exist. The linear stability of this solution yields a critical wave length of perturbation, above which the solution is unstable. This critical wave length depends on the wave speed, unspecified by undercooling. This sets ground for the "selection" problem, resulting here from particularity of the basic solution concerned. In this sense, the aforementioned sphere growing into undercooled melt, represents a fairly exceptional example of an analytic solution to be subject of stability analysis and completely determined by the data of the problem (undercooling).

Another explicit solution of the Stefan problem, specified completely by the undercooling is the famous planar similarity solution of the two-phase Stefan problem. An additional importance of this solution lies in the fact that it represents the long-time asymptotics of the solutions to the one-dimensional Cauchy-Stefan problems with constant temperatures at infinities, whenever supercooling is below the critical value. In this solution speed of free boundary varies and tends to zero as time goes on. Stability of the planar similarity solution with respect to multidimensional perturbations of finite wave-length has been proved by L. Rubinstein for the normal (non-supercooled) Stefan problem without surface tension. J. Chadam and P. Ortoleva addressed stability of the respective solutions for the Stefan problem with surface tension and initial supercooling. They found that similarity solution is unstable without surface tension and asymptotically stable when surface tension, however small, is present. More precisely, they proved that, with a nonvanishing surface tension, every perturbation mode with a finite non-zero wave number asymptotically decays in time.

We wish, however, to reexamine here the conclusion of overall asymptotic stability of the planar similarity solution with supercooling in the presence of surface tension. Our ultimate claim is that a proper handling of the long-wave components of the perturbation results in instability for any given surface tension.

The formulation of the problem is as follows. We consider a two-dimensional Stefan problem with surface tension:

$$u_t = \Delta u, \quad S(\underline{x}, t) \neq 0, \quad t > t_0,$$

$$u = -\gamma K, \quad [\nabla u \nabla S] = S_t, \quad \text{for } S(\underline{x}, t) = 0, \quad t > t_0,$$

$$u(\underline{x}, T) = u_0(\underline{x}), \quad S(\underline{x}, T) = S_0(\underline{x}),$$

$$u_0(\underline{x}) \to u_\infty \in (-1, 0) \quad \text{as } x \to \infty, \quad u_0(\underline{x}) \to 0 \quad \text{as } x \to -\infty.$$

We are about to analyse the perturbation of the following similarity solution to the aforementioned problem:

$$S(\underline{x}, t) = \{x = R_S(t) = 2\alpha\sqrt{t}\},$$

$$u_S = \begin{cases} u_\infty + 2\alpha \exp(\alpha^2)\left(1 - \operatorname{erf}\left(\frac{x}{2\sqrt{t}}\right)\right), & x > 2\alpha\sqrt{t}, \\ 0, & x < 2\alpha\sqrt{t}. \end{cases}$$

Here α is the root of the transcendental equation

$$2\alpha \exp(\alpha^2) \int_\alpha^\infty \exp(-y^2)\, dy = -u_\infty.$$

30

Let us consider a non-planar perturbation of the similarity solution of the form:

$$u_\epsilon(x, y, t) = u_S(s, t) + \epsilon u(x, y, t) + O(\epsilon^2),$$

$$S(\underline{x}, t) = \{(x, y, t): \ x = R_\epsilon(y, t)\}, \qquad R_\epsilon(y, t) = R_S(t) + \epsilon R(y, t) + O(\epsilon^2)$$

with $R_S(t) = 2\alpha\sqrt{t}$. The linearized initial-boundary value problem for perturbation reads:

$$U_t = \Delta u, \qquad x > R_S(t) = 2\alpha\sqrt{t}, \ t > T,$$

$$u = -\frac{\partial u_S}{\partial x} + \frac{\gamma}{2} R_{yy}, \quad u_x = -R_t - \frac{\partial^2 u_S}{\partial x^2} R, \qquad \text{on } x = 2\alpha\sqrt{t}, t > T,$$

$$u(x, y, T) = \phi(x, y), \qquad \text{for } x > 2\alpha\sqrt{T}, \ \phi(x, y) \to 0 \text{ as } x \to \infty.$$

Following the Lev's Rubinstein idea we make the change of the dependent variable

$$v = u + \frac{\partial u_S}{\partial x} R - \frac{\gamma}{2} R_{yy}$$

followed by the Fourier transform in y:

$$w^l = \int_{-\infty}^{\infty} v(x, y, t) \exp(-ily) \, dy \, \exp(l^2 t),$$

$$f^l(t) = \int_{-\infty}^{\infty} R(y, t) \exp(-ily) \, dy.$$

This yields the following one-dimensional time-dependent initial-boundary value problem for w^l and f^l:

$$w_t^l = w_{xx}^l + \left(u_{Sx} + \frac{\gamma l^2}{2}\right)(\dot{f}^l + l^2 f^l) \exp(l^2 t), \qquad x > 2\alpha\sqrt{t}, \ t > T,$$

$$w^l = 0, \quad w_x^l = -\dot{f}^l \exp(l^2 t), \qquad \text{for } x = 2\alpha\sqrt{t}, \ t > T,$$

$$w^l(x, T) = \phi^l(x), \quad f^l(T) = F(l).$$

It may be shown, using the respective Green's function of the heat equation, that the solution satisfies the following integro-differential equation

$$\dot{f}^l(t) = -I(t) \exp(-l^2 t) - \int_T^t (\dot{f}^l(t) + l^2 f^l(t)) \exp(-l^2(t - \tau))(P(t, \tau) + \Gamma(t, \tau)) \, d\tau.$$

Here $I(t) = U_{1x}(2\alpha\sqrt{t}, t)$ and U_1 is a solution of the following problem

$$U_{1t} = U_{1xx}, \qquad x > 2\alpha\sqrt{t}, \ t > T,$$

$$U_1(2\alpha\sqrt{t}, t) = 0, \qquad t > T,$$

$$U_1(x, T) = \phi^l(x), \qquad x > 2\alpha\sqrt{T}.$$

Next, $P(t,\tau) = U_{2x}(2\alpha\sqrt{t},\tau)$ and U_2 is a solution of the problem

$$U_{2t} = U_{2xx}, \qquad x > 2\alpha\sqrt{t}, \quad t > \tau,$$

$$U_1(2\alpha\sqrt{t},t) = 0, \qquad t > \tau,$$

$$U_1(x,\tau) = u_{Sx}(x,\tau), \qquad x > 2\alpha\sqrt{\tau}.$$

Finally, $\Gamma(t,\tau)$ is defined by replacing $u_{Sx}(x,\tau)$ by $\frac{\gamma l^2}{2}$ in the definition of $P(t,\tau)$.

Using the asymptotic properties of the kernels $P(t,\tau)$, $\Gamma(t,\tau)$ and the function $I(t)$ we obtain the following asymptotic equation for the function f, valid for $l^2 t \gg 1$:

$$(f^l \exp(b^2 t))' \left(1 + b^2 - \frac{\alpha}{2\sqrt{l^2 - b^2}}\right) = f^l \exp(b^2 t)\frac{\alpha\sqrt{l^2 - b^2}}{2\sqrt{t}} + I(t)(b^2 - l^2)t \quad \text{for } lt \gg 1.$$

Here $8b^2 = -\gamma^2 l^4 + l^3\sqrt{16 + \gamma^2 l^2}$. For large t and small l the approximate solution of this equation reads

$$f^l \approx \text{const}(l)\exp(\alpha l\sqrt{t})\exp(-\gamma l^3 t)(l^2 t)^{\alpha^2}.$$

Using this equality Chadam and Ortoleva concluded that for every $\gamma, l > 0$ the similarity solution is stable, since $f \to 0$ when $t \to \infty$. One may, however, distinguish two time scales as defined by this equality: for t such that $l^3 t \ll 1$, f^l grows exponentially in $\sqrt{t}$, whereas for t such that $l^4 t \gg 1$ it decays exponentially. Thus, for l sufficiently small the stabilization time is very large.

This is the motivation for the following hypothesis: the neighbourhood of zero of the perturbation Fourier spectrum may have effect upon the solution stability. To analyse this effect, we have to study not a fixed mode of perturbation, but rather investigate all modes of perturbation, especially the long-wave part of the Fourier spectrum, uniformly in time.

For this purpose, one should look for a suitable scaling of the time and space variables. Let us define the scaled variables as

$$\tau = t l^2, \quad y = x l, \quad W^l = \frac{w^l}{l}.$$

The function W^l satisfies the following initial-boundary value problem:

$$W^l_\tau = W^l_{yy} + u_{Sy}(\dot{f}^l + f^l)\exp(\tau) + \frac{\gamma l}{2}(\dot{f}^l + f^l)\exp(\tau), \qquad y > 2\alpha\sqrt{\tau}, \ \tau > l^2 T,$$

$$W^l = 0, \quad W^l_y = -\dot{f}^l(\tau)\exp(\tau) \qquad \text{for } y = 2\alpha\sqrt{\tau}, \ \tau > l^2 T,$$

$$W^l(y, l^2 T) = \frac{\phi^l(y)}{l}, \quad f^l(l^2 T) = F(l).$$

The number l enters crucially this scaled formulation through:

1. The initial time dependence $l^2 T$.

2. The inhomogeneous term $\frac{\gamma l}{2}(\dot{f}^l + f^l)\exp(\tau)$ in the equation.

The possibility to aproximate the solution by the limiting one for $l \to 0$ is proved for $\tau \ll 1/l$.

Theorem 1. *If* $\phi^l(y) = O(l^2)$ *then for* $\forall \tau \in [0, \frac{1}{l^{1-\epsilon}}]$, $\epsilon > 0$,

$$W^l = F(l)W_0(1 + O(l^\epsilon)), \quad f^l = F(l)f_0(\tau)(1 + O(l^\epsilon)).$$

Here, W_0 and f_0 are the solutions of the following limiting initial-boundary-value problem independent of l:

$$W_{0\tau} = W_{0yy} + u_{Sy}(\dot{f}_0 + f_0)\exp(\tau) \qquad y > 2\alpha\sqrt{\tau}, \quad \tau > 0,$$

$$W_0 = 0, \quad W_{0y} = -\dot{f}(\tau)\exp(\tau) \qquad for\ y = 2\alpha\sqrt{\tau}, \quad \tau > 0,$$

$$W_0(y, 0) = 0, \quad f_0(0) = 1.$$

Making use of equality

$$f^l(\tau) \approx \mathrm{const}(l)\exp(\alpha\sqrt{\tau})\exp\left(-\frac{\gamma l \tau}{2}\right)(\tau)^{\alpha^2},$$

asymptotically valid for $\tau \gg 1$, and returning to the original variables we prove the following main theorem.

Theorem 2. *Let* $F(l) = \int_{-\infty}^{\infty} R(y, T)\exp(-ily)\,dy$ *and* $F(0) = 0$, $F(\infty) = 0$, $F^{(n)}(0) \neq 0$ *for some* n. *Then if* $\phi^l(x) = O(l^2)$, *the perturbation of the free boundary grows unboundedly in time.*

The idea of the proof is as follows. The overlap of the two aforementioned time intervals yields the Fourier transform of the free boundary $f^l(t) = \int_{-\infty}^{\infty} R(y, t)e^{-ily}\,dy$ in the form

$$f^l(t) \approx \mathrm{const}\, F(l)\exp(\alpha l\sqrt{t})\exp\left(-\frac{\gamma l^3 t}{2}\right)(l^2 t)^{\alpha^2}$$

with the constant factor independent of l. Making use of the inverse Fourier transform one proves then the statement of Theorem 2.

The following example has been considered: let the initial front displacement be $R(y, T) = y \exp(-y^2/2)$. The graphs of $R(y, t)$ for $t = T = 10$, $T + 100$, $T + 1000$, $T + 5000$ are presented in the Fig. 1.

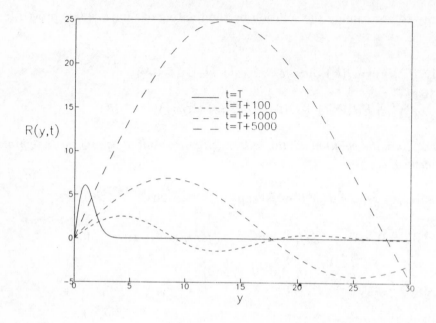

Figure 1. Graphs of $R(y, t)$.

Conclusions: The planar solidification is morphologically unstable also in the presence of finite surface tension. The necessary condition for stability is vanishing at zero wave number of the Fourier transform of the initial perturbation of the free boundary together with all its derivatives. The respective sufficient condition is the absence of the long-wave modes in the initial free boundary perturbation.

D A TARZIA

A steady-state two-phase Stefan–Signorini problem with mixed boundary data

Abstract. We consider a steady-state heat conduction problem in a multidimensional bounded domain Ω which has a regular boundary Γ composed by the union of two parts Γ_1 and Γ_2. We assume, without loss of generality, that the melting temperature is zero degree centigrade. We consider a source term g in the domain Ω. On the boundary Γ_2 we have a positive heat flux q and on the boundary Γ_1 we have a Signorini type condition with a positive external temperature b.

We obtain sufficient conditions on data q, g, b to obtain a change of phase (steady-state, two-phase, Stefan-Signorini problem) in Ω, that is a temperature of non-constant sign in Ω. We use the elliptic variational inequalities theory. We also find that the solution of the corresponding elliptic variational inequality is differentiable with respect to the Neumann datum q on Γ_2. Several properties already obtained for variational equalities can also be generalized for variational inequalities.

Moreover, by using the finite element method, we also obtain sufficient conditions on data to obtain a steady-state, two-phase, discretized Stefan-Signorini problem in the corresponding discretized domain, that is a discrete temperature of non-constant sign in Ω.

1. Introduction.

We consider a bounded domain $\Omega \subset \mathbf{R}^n$ with regular boundary $\Gamma = \Gamma_1 \cup \Gamma_2$ with $|\Gamma_2| = \text{meas}(\Gamma_2) > 0$ and $|\Gamma_1| > 0$. We suppose that $\Gamma_1 = \Gamma_{1_t} \cup \Gamma_{1_s}$ with $|\Gamma_{1_i}| > 0$ for $i = t, s$.

We consider a steady-state heat conduction problem in Ω. We assume, without loss of generality, that the melting temperature is zero degree centigrade. We consider a source term g in the domain Ω. On the boundary Γ_2 we have a positive heat flux q and on the boundary Γ_{1_t} we impose a positive temperature b. On the boundary Γ_{1_s} we have a Signorini type condition with a positive external temperature b. If θ is the temperature of the material we can consider the new unkown function in Ω defined by[Du, Ta1]

$$(1) \qquad u = k_2\theta^+ - k_1\theta^-$$

where $k_i > 0$ is the thermal conductivity of the phase i ($i = 1$: solid phase, $i = 2$: liquid phase). Let $B = k_2b > 0$ where $b > 0$ is the temperature imposed on Γ_{1_t}.

We consider the following steady-state Stefan-Signorini free boundary problem

$$(2) \qquad -\triangle u = g \quad \text{in} \quad \Omega$$

(3)
$$-\frac{\partial u}{\partial n}/\Gamma_2 = q \qquad \text{on} \qquad \Gamma_2$$

(4)
$$u/\Gamma_{1_t} = B \qquad \text{on} \qquad \Gamma_{1t}$$

(5)
$$u \geq B, \quad \frac{\partial u}{\partial n} \geq 0, \quad (u - B)\frac{\partial u}{\partial n} = 0 \qquad \text{on} \qquad \Gamma_{1_s}.$$

The goal of this paper is to find sufficient conditions on $q = \text{Const.} > 0$ on Γ_2 to obtain a temperature u of non-constant sign in Ω, that is a steady-state, two-phase, Stefan-Signorini problem. When $\Gamma_1 = \Gamma_{1_t}$ (i.e. $\Gamma_{1_s} = \emptyset$) the corresponding free boundary problem without Signorini boundary conditions was studied in [GaTa]. We follow a method similar to the one developed in [BoShTa, GaTa, GoTa, Sa, Ta1, Ta2, Ta3].

We shall present some theoretical and numerical (by finite element approximation) results through variational inequalities and the corresponding related estimates in terms of the finite element approximation parameter h.

2. Continuous analysis.

The variational formulation of the problem (2)-(5) is given by

(6)
$$\begin{cases} a(u, v - u) \geq L(v - u), \quad \forall v \in K_B \\ u \in K_B \end{cases}$$

where

(7)
$$\begin{cases} V = H^1(\Omega), \\ W_0 = \{v \in V/\ v/\Gamma_{1_t} = v/\Gamma_{1_s} = 0\} \subset V_0 = \{v \in V/\ v/\Gamma_{1_t} = 0\} \\ K_B = \{v \in V/\ v/\Gamma_{1_t} = B, \ v/\Gamma_{1_s} \geq B\} = B + K_0 \\ K_0 = \{v \in V/\ v/\Gamma_{1_t} = 0, \ v/\Gamma_{1_s} \geq 0\} \supset W_0 \end{cases}$$

and

(8)
$$\begin{cases} a(u, v) = \int_\Omega \nabla u . \nabla v \, dx \\ L(v) = L_{qg}(v) = \int_\Omega gv \, dx - \int_{\Gamma_2} qv \, d\gamma. \end{cases}$$

For $g \in L^2(\Omega)$, we have a unique solution $u = u_{qgB}$ (it will be denoted by u_q) of the variational inequality (6) [KiSt, Ta1].

We obtain for u_q the following properties:

Lemma 1. We have

(i)(9)
$$\alpha \left\| u_{q_2} - u_{q_1} \right\|_V^2 \leq a\left(u_{q_2} - u_{q_1}, u_{q_2} - u_{q_1}\right) \leq (q_1 - q_2)\int_{\Gamma_2}\left(u_{q_2} - u_{q_1}\right) d\gamma$$

where $\alpha > 0$ is the coercive constant of the bilinear form a.

36

(ii)(10)
$$
\begin{cases}
\left\| u_{q_2} - u_{q_1} \right\|_V \le \dfrac{|\Gamma_2|^{\frac{1}{2}} \|\gamma_0\|}{\alpha} |q_2 - q_1| \\[3mm]
\left\| u_{q_2} - u_{q_1} \right\|_{L^2(\Gamma_2)} \le \dfrac{|\Gamma_2|^{\frac{1}{2}} \|\gamma_0\|}{\alpha} |q_2 - q_1|
\end{cases}
$$

where γ_0 is the trace operator.

(iii) The function $\mathbf{R}^+ \to \mathbf{R}$,

(11)
$$ q \to \int_{\Gamma_2} u_q \, d\gamma $$

is a continuous and strictly decreasing function. Moreover, we have

$$ q_1 \le q_2 \implies u_{q_2} \le u_{q_1} \quad \text{in} \quad \overline{\Omega} \quad \text{and} \quad \Gamma_2 \int_{\Gamma_2} u_{q_2} \, d\gamma \le \int_{\Gamma_2} u_{q_1} \, d\gamma . $$

(iv) There exists $u_q' \in V_0$ such that:

(12)
$$
\begin{cases}
(i) \ \dfrac{u_{q+\delta} - u_q}{\delta} \rightharpoonup u_q' \text{ in } V - \text{weak} , \text{ when } \delta \to 0 \\[3mm]
(ii) \ \dfrac{u_{q+\delta} - u_q}{\delta} \rightharpoonup u_q' \text{ in } L^2(\Gamma_2) - \text{weak} , \text{ when } \delta \to 0
\end{cases}
$$

and

(13)
$$ a(u_q, u_q') = L(u_q') \left(= \int_\Omega g u_q' \, dx - q \int_{\Gamma_2} u_q' \, d\gamma \right) . $$

The element $u,$ unique solution of (6), is also characterized by the following minimization problem:

(14)
$$
\begin{cases}
J(u) \le J(v) , \quad \forall v \in K_B \\[2mm]
u \in K_B
\end{cases}
$$

where

(15)
$$ J(v) = J_{qg}(v) = \tfrac{1}{2} a(v.v) - L_{qg}(v). $$

We can define the real function $f : \mathbf{R}^+ \to \mathbf{R}$ in the following way [GaTa, Ta2]

(16)
$$ f(q) = J(u_q) = \tfrac{1}{2} a(u_q, u_q) - \int_\Omega g u_q \, dx + q \int_{\Gamma_2} u_q \, d\gamma $$

where u_q is the unique solution of the variational inequality (6) for each heat flux q> 0.

Theorem 2. The function f is differentiable. Moreover, f' is a continuous and strictly decreasing function, and it is given by the following expression

(17)
$$ f'(q) = \int_{\Gamma_2} u_q \, d\gamma. $$

Proof.- We use (6), (13) and the definition of f'.

Corollary 3. We have the following properties:

(18) $$\frac{d}{dq}\left[\int_\Omega g u_q \, d\gamma\right] = \int_\Omega g u'_q \, dx$$

(19) $$\frac{d}{dq}[a(u_q, u_q)] = 2a(u_q, u'_q)$$

(20) $$f''(q) = \int_{\Gamma_2} u'_q \, d\gamma \ .$$

Theorem 4. The element u'_q does not depend on q, that is $u'_q = \eta \in K_0$ where η is the unique solution of the following elliptic variational inequality:

(21) $$\begin{cases} a(\eta, v - \eta) \geq -\int_{\Gamma_2} (v - \eta) \, d\gamma, \quad \forall v \in K_0 \\ \eta \in K_0 \ . \end{cases}$$

Moreover, $\eta/_{\Gamma_2} \leq 0$ with

(22) $$-\int_{\Gamma_2} \eta \, d\gamma \geq a(\eta, \eta) \geq \alpha \, \|\eta\|_V^2 > 0.$$

Corollary 5. We have the following properties :
(i) The element u_q can be written by

(23) $$u_q = u_{qgB} = B + U_g + q\eta$$

where U_g is the unique solution of the following elliptic variational equality

(24) $$\begin{cases} a(U_g, v) = \int_\Omega g v \, dx, \quad \forall v \in W_0 \\ U_g \in W_0 \ . \end{cases}$$

(ii) We have

(25) $$f'(q) = (B \, |\Gamma_2| + C_g) - Dq, \qquad f''(q) = \int_{\Gamma_2} \eta \, d\gamma$$

where

(26) $$C_g = \int_{\Gamma_2} U_g \, d\gamma, \qquad D = -\int_{\Gamma_2} \eta \, d\gamma > 0.$$

We can define the real function $R = R(B, g)$ in the following way

(27) $$R(B, g) = \frac{B \, |\Gamma_2| + C_g}{D}.$$

Theorem 6. For $B > 0$ and $g \in L^2(\Omega)$, we have:

(28) $$q > R(B, g) \Rightarrow u \text{ is of non-constant sign in } \Omega,$$

i.e. there exists a steady-state two-phase Stefan-Signorini problem.

Proof. The result (28) is obtained by considering the following equivalence

(29) $$q > R(B, g) \Longleftrightarrow f'(q) = \int_{\Gamma_2} u_q \, d\gamma < 0.$$

38

3. Numerical analysis.

We suppose that $\Omega \subset \mathbf{R}^n$ is a convexe polygonal bounded domain. We consider τ_h, a regular triangulation of the polygonal domain Ω with Lagrange triangles of type 1, constituted by affine-equivalent finite element of class C^0, where $h > 0$ is a parameter which goes to zero. We can take h equal to the longest side of the triangles $T \in \tau_h$ [BrSc, Ci, GlLiTr]. We follow a method similar to the one developed in [Ta3] to obtain the discrete equivalent of the continuous result (28).

The variational formulation of the continuous problem (6) is given by

$$(30) \qquad \begin{cases} a(u_h, v_h - u_h) \geq L(v_h - u_h), & \forall v_h \in K_{B_h} \\ u_h \in K_{B_h} \end{cases}$$

where

$$(31) \qquad \begin{cases} K_{B_h} = B + K_{0_h} \subset K_B, \ P_1 = \text{set of the polynomials of degree} \leq 1 \\ K_{0_h} = \left\{ v_h \in C^0(\overline{\Omega}) / \ v_h/_T \in P_1(T), \forall T \in \tau_h, v_h/_{\Gamma_{1t}} = 0, v_h/_{\Gamma_{1s}} \geq 0 \right\} \\ V_{0_h} = \left\{ v_h \in C^0(\overline{\Omega}) / \ v_h/_T \in P_1(T), \forall T \in \tau_h, v_h/_{\Gamma_{1t}} = 0 \right\} \\ W_{0_h} = \left\{ v_h \in C^0(\overline{\Omega}) / \ v_h/_T \in P_1(T), \forall T \in \tau_h, v_h/_{\Gamma_{1t}}, v_h/_{\Gamma_{1s}} = 0 \right\} \end{cases}$$

with

$$(32) \qquad \begin{cases} W_{0_h} \subset K_{0_h} \subset V_{0_h} \\ W_{0_h} \subset W_0, \ K_{0_h} \subset K_0, \ V_{0_h} \subset V_0. \end{cases}$$

For $g \in L^2(\Omega)$, the unique solution of the variational inequality (30) will be denoted by $u_h = u_{h_q}$. The element u_h is also characterized by the minimization problem:

$$(33) \qquad \begin{cases} J(u_h) \leq J(v_h), & \forall v_h \in K_{B_h} \\ u_h \in K_{B_h}. \end{cases}$$

For each $h > 0$, we define the real function $f_h : \mathbf{R}^+ \to \mathbf{R}$ in the following way

$$(34) \qquad f_h(q) = J(u_{h_q}) = \tfrac{1}{2} a(u_{h_q}, u_{h_q}) - L_q(u_{h_q}).$$

We obtain the following properties for the discrete solution u_{h_q} of the elliptic variational inequality (30).

Theorem 7. We have the following properties:
(i) There exists an element $u'_{h_q} \in V_0$ such that

$$(35) \qquad \frac{u_{h_q+\delta} - u_{h_q}}{\delta} \rightharpoonup u'_{h_q} \ \text{in} \ V-\text{weak} \ , \text{when} \ \delta \to 0$$

$$(36) \qquad \frac{u_{h_q+\delta} - u_{h_q}}{\delta} \rightharpoonup u'_{h_q} \ \text{in} \ L^2(\Gamma_2)-\text{weak} \ , \text{when} \ \delta \to 0$$

(37)
$$a(u_{h_q}, u'_{h_q}) = \int_{\Omega} g u'_{h_q} \, dx - q \int_{\Gamma_2} u'_{h_q} \, d\gamma.$$

(ii) The function $\mathbf{R}^+ \to \mathbf{R}$, $q \to \int_{\Gamma_2} u_{h_q} d\gamma$ is a continuous and strictly decreasing function.

(iii) The function f_h is differentiable. Moreover, we have the following expressions:

(38)
$$f'_h(q) = \int_{\Gamma_2} u_{h_q} \, d\gamma \quad , \quad f''_h(q) = \int_{\Gamma_2} u'_{h_q} \, d\gamma.$$

(iv) The element u_{h_q} can be written as

(39)
$$u_{h_q} = B + U_{h_g} + q\eta_h \quad , \quad \eta_h = u'_{h_q} \in K_{0_h}$$

where U_{h_g} and η_h are respectively the unique solutions of the variational equality (40) and inequality (41), that is:

(40)
$$\begin{cases} a(U_{h_g}, v_h) = \int_{\Omega} g v_h \, dx \, , \quad \forall v_h \in W_{0_h} \\ \\ U_{h_g} \in W_{0_h} \end{cases}$$

(41)
$$\begin{cases} a(\eta_h, v_h - \eta_h) \geq - \int_{\Gamma_2} (v_h - \eta_h) \, d\gamma, \quad \forall v_h \in K_{0_h} \\ \\ \eta_h \in K_{0_h} \, . \end{cases}$$

(v) We have that $\eta_{h/\Gamma_2} < 0$ and

(42)
$$- \int_{\Gamma_2} h \eta_h \, d\gamma \geq a(\eta_h, \eta_h) \geq \alpha \, \|\eta_h\|_V^2 > 0.$$

(vi) Also, we have

(43)
$$f'_h(q) = \left(B \, |\Gamma_2| + C_{h_g} \right) - D_h q, \qquad f''_h(q) = \int_{\Gamma_2} \eta_h \, d\gamma < 0 \quad ,$$

where

(44)
$$C_{h_g} = \int_{\Gamma_2} U_{h_g} d\gamma \quad , \quad D_h = - \int_{\Gamma_2} \eta_h \, d\gamma > 0.$$

(vii) If, for each $h > 0$, we define the real function

(45)
$$R_h(B, g) = \frac{B \, |\Gamma_2| + C_{h_g}}{D_h}$$

then we obtain that

(46)
$$q > R_h(B, g) \Rightarrow u_h \text{ is of non-constant sign in } \Omega,$$

i.e. there exists a discrete steady-state two-phase Stefan-Signorini problem.

Proof. We use a method similar to the one developed in [Ta3].

4. Error bounds.

Let Π_h be the corresponding linear interpolation operator for the finite element approximation. There is a constant $C_0 > 0$ (independent of h) such that [BrSc, Ci]

(47)
$$\|v - \Pi_h v\|_V \leq C_0 h^{r-1} \|v\|_{r,\Omega} \quad , \qquad \forall v \in H^r(\Omega) \quad , \qquad r > 1.$$

40

If we suppose the regularity property:

$$(48) \qquad u_1 \in H^r(\Omega) \quad , \quad \eta \in H^r(\Omega)$$

we obtain the following error estimates.

Theorem 8. We have

$$(49) \qquad \|u_1 - u_{1_h}\|_V \le O\left(h^{r-1}\right)$$

$$(50) \qquad 0 < C_1 - C_{1_h} = O\left(h^{2r-2}\right) \quad , \quad 0 < q_{0_h}(B) - q_0(B) = O\left(h^{2r-2}\right)$$

$$(51) \qquad \left|C_{h_g} - C_g\right| = O\left(h^{r-1}\right)$$

$$(52) \qquad \|\eta - \eta_h\|_V \le O\left(h^{\frac{r-1}{2}}\right) \quad , \quad |R_h(B,g) - R(B,g)| = O\left(h^{\frac{r-1}{2}}\right).$$

Acknowledgments. This paper has been sponsored by the Project "Aplicaciones de Problemas de Frontera Libre" from CONICET, Rosario-Argentina.

References

[BrSc] S.C.Brenner - L.R.Scott, "The mathematical theory of finite element methods", Springer Verlag, New York (1994).

[BoShTa] J.E.Bouillet - M.Shillor - D.A.Tarzia, "Critical outflow for a steady-state Stefan problem", Applicable Analysis, 32 (1989), 31-51.

[Ci] P.G.Ciarlet, "The finite element method for elliptic problem", North Holland, Amsterdam (1978).

[Du] G. Duvaut, "Problèmes à frontière libre en théorie des milieux continus", Rapport de Recherche # 185, INRIA (ex LABORIA-IRIA), Rocquencourt (1976).

[GaTa] G.G.Garguichevich - D.A.Tarzia, "The steady-state two-phase Stefan problem with an internal energy and some related problems", Atti Sem. Mat. Fis. Univ. Modena, 39 (1991), 615-634.

[GlLiTr] R.Glowinski - J.L.Lions - R.Tremolières, "Analyse numérique des inéquations variationelles", Dunod, Paris (1976).

[GoTa] R.L.V.Gonzalez - D.A.Tarzia, "Optimization of the heat flux in domain with temperature constrains", J. Optimization Th. Appl., 65 (1990), 245-256.

[KiSt] D. Kinderlehrer - G. Stampacchia, "An introduction to variational inequalities and their applications", Academic Press, New York (1980).

[Sa]   M.C.Sanziel, "Análisis numérico de un problema estacionario de Stefan a dos fases con energía interna", To appear.

[Ta1]   D.A.Tarzia, "The two-phase Stefan problem and some related conduction problems", Reuniões em Matemática Aplicada e Computacão Científica, Vol.5, SBMAC, Rio de Janeiro (1987).

[Ta2]   D.A.Tarzia, "An inequality for the constan heat flux to obtain a stady-state two-phase Stefan problem", Engineering Analysis, 5 (1988), 177-181.

[Ta3]   D.A.Tarzia, "Numerical analysis for the heat flux in mixed elliptic problem to obtain a discrete steady-state two-phase Stefan problem", SIAM J. Numerical Analysis, 33 (1996), To appear.

Domingo Alberto Tarzia
Depto. de Matemática,
FCE, Universidad Austral,
Paraguay 1950,
(2000) Rosario, Argentina.
E-mail: TARZIA@UAUFCE.EDU.AR

D A TARZIA AND C V TURNER

The one-phase supercooled Stefan problem

Abstract: We consider the supercooled one-phase Stefan problem with convective boundary condition at the fixed face. We analyse the relation between the heat transfer coefficient and the possibility of continuing the solution for arbitrarily large time intervals.

1. Introduction.

The classical one-dimensional Stefan problem has been studied since 1831(see [11]); it models conductive heat transfer on either side of a phase boundary in pure material on the assumptions (i) that the temperature at the phase boundary is constant, say zero, (ii) that there is a release of latent heat at the boundary on solidification, and an uptake on melting, and (iii) that the material on the solid and liquid sides of the phase boundary has negative and positive temperature, respectively.

With these assumptions the problem has a weak formulation and a global solution is known to exist ([4]). If the data are such that just one phase boundary exists the problem has also been shown to be well-posed in the classical sense [5,7].

But if the initial and/or boundary data violate the sign requirement (iii), i.e., if the liquid is supercooled or the solid is superheated, a solution still may exist, at least formally, but the result is generally only local in time and finite time blow-up can easily occur ([8]).

In this paper we consider this kind of problem in the following setting:

Problem I:

Find $\theta(y, \tau)$ the temperature and $r(\tau)$ the free-boundary such that:

$r(\tau)$ is Lipschitz continuous for $\tau \geq 0$;

$\dot{r}(\tau)$ is continuous for $\tau > 0$;

$\theta(y, \tau)$ is continuous for $\tau > 0$ and $0 \leq y \leq r(\tau)$;

$\theta_\tau(y, \tau)$, $\theta_{yy}(y, \tau)$ are continuous for $\tau > 0$ and $0 < y < r(\tau)$;

$\theta_y(y, \tau)$ is continuous for $\tau > 0$, $0 \leq y \leq r(\tau)$;

$r(\tau)$ and $\theta(y, \tau)$ obey the conditions:

$$\theta_\tau = \alpha\theta_{yy} \quad 0 < y < r(\tau), \quad 0 < \tau < \tau_0$$
$$\theta(r(\tau), \tau) = 0 \quad 0 < \tau < \tau_0$$
$$k\theta_y(r(\tau), \tau) = -\rho\lambda\dot{r}(\tau), \quad 0 < \tau < \tau_0$$
$$k\theta_y(0, \tau) = h(\theta(0, \tau) - g(\tau)) \quad 0 < \tau < \tau_0$$
$$\theta(y, 0) = \theta_0(y) \quad 0 \le y \le b$$
$$r(0) = b$$

The parameters are
$\alpha = \frac{k}{\rho c}$ material thermal diffusivity (m^2/s)
$k =$ material thermal conductivity (KJs^0C/m)
$\rho =$ material density (Kg/m^3)
$\lambda =$ latent heat of melting (KJ/Kg)
$h =$ fluid to material surface heat transfer coefficient (KJs^0C/m^2)
$g(\tau) =$ ambient fluid temperature (^0C)
$c =$ specific heat (KJ^0C/Kg),

The melting front at time τ is $r(\tau)$ while $\theta(y, \tau)$ is the temperature at position y and time τ.

It is known that a solution to Problem I exists [1], for suitable τ_0 'sufficiently small'. This problem is often referred to as a mathematical scheme for the freezing of a supercooled liquid (although this simple scheme for such a non-equilibrium phenomenon is far from being satisfactory)[10].

The freezing of a supercooled liquid is due to convective heat transfer from a fluid with ambient temperature $g(\tau)$ flowing across the face $x = 0$.

This problem has been studied in [2],[3],[6] and [12].

The adimensional problem is obtained by the following transforms

$$x = \frac{y}{b} \qquad t = \frac{k\tau}{\rho c b^2}$$

$$z(x, t) = \frac{c}{\lambda}\theta(y, \tau) \qquad s(t) = \frac{r(\tau)}{b}$$

Then the variables (T, s, z) satisfy the problem:

Problem II:
(1.1) $z_{xx} = z_t$, in D_T;
(1.2) $s(0) = 1$;
(1.3) $z(s(t), t) = 0$, $0 < t < T$;
(1.4) $z_x(s(t), t) = -\dot{s}(t)$, $0 < t < T$;
(1.5) $z(x, 0) = \varphi(x)$, $0 < x < 1$;

(1.6) $z_x(0,t) = \beta z(0,t) - G(t)], 0 < t < T.$

where $\beta = \frac{h}{kb}$ is an adimensional parameter, and

$$D_T = \{(x,t)|0 < x < s(t), 0 < t < T\}$$

$$G(t) = \frac{c}{\lambda} g\left(\frac{b^2 \rho c t}{k}\right).$$

2. The one-phase supercooled Stefan problem

In this section we consider the following hypotheses

$$\varphi(x) \le 0, 0 < x < 1 \quad \text{and} \quad G(t) \le 0, \ t > 0$$

and the compatibility condition

$$\varphi'(0) = \beta[\varphi(0) - G(0)].$$

The first simple properties of the solution of (1.1)-(1.6) are summarized in the following proposition :

Proposition 2.1. *If (T, s, z) is a solution of Problem II, then*
i) $z \le 0$ in D_T.
ii) $\dot{s}(t) < 0, t > 0$.
iii) $\dot{G}(t) \le 0, \varphi(x) \ge G(0) = \max_{t>0} G(t)$, then $z \ge G(t)$ in D_T.
iv) $\varphi' \ge 0$, then $z_x \ge 0$ in D_T.
v) $\dot{G} \ge 0, \varphi'' > 0$ then $z_t > 0$ in D_t.
vi)

$$(2.1) \qquad s(t)\left[1 + \frac{\beta}{2}s(t)\right] = 1 + \frac{\beta}{2} + \int_0^1 (1 + \beta x)\varphi(x)\, dx + \int_0^t \beta G(\tau)\, d\tau$$

$$- \int_0^{s(t)} (1 + \beta x)z(x,t)\, dx$$

Proof.
The proof is obtained by using the maximum principle and Green's identity.

Remark 1: In the following sections we denote

$$(2.2) \qquad Q(t) = 1 + \frac{\beta}{2} + \int_0^1 (1 + \beta x)\varphi(x)\, dx + \int_0^t \beta G(\tau)\, d\tau$$

If $\varphi(1) = 0$, $\varphi(x)$ is Hölder continuous for $x = 1$ and $G(t)$ is a piecewise continuous on every interval $(0,t)$, $t > 0$, this problem possess one solution for suitable T

45

"sufficiently small" (see [1], [5], [6] where uniqueness and continuous dependence are also discussed).

Moreover, if a solution exists, then three cases can occur (see [6], Theorem 8 and [2]).

(A) The problem has a solution with arbitrarily large T.

(B) There exists a constant $T_B > 0$ such that $\lim_{t \to T_B} s(t) = 0$.

(C) There exists a constant $T_C > 0$ such that $\inf_{t \in (0,T_C)} s(t) > 0$ and $\lim_{t \to T_C} \dot{s}(t) = -\infty$.

We shall investigate the occurrence of these cases in conection with the behavior of the initial data φ, the adimensional temperature G of the external fluid and the adimensional coefficient β, (see [12]).

Our next aim will be to look for some conditions on φ, G and β giving an a priori caracterization of cases (A), (B) and (C).

Proposition 2.2. *If $\dot{G} \leq 0$, $\varphi(x) \geq G(0)$ and the solution (T, s, z) of Problem II is case (B), then $Q(T_B) = 0$.*

Proof. Setting $t \longrightarrow T_B$ in (2.3) and using the boundedness of z obtained in Proposition 2.1 we conclude the result. $\square$

Proposition 2.3. *If (T, s, z) is a solution of problem P II, and the initial and boundary data satisfy the following hypotheses:*

i) $\varphi(x) \geq M(x - 1)$, $0 \leq x \leq 1$, $0 < M < 1$;
ii) $G(t) \geq -M$

and it exists a time T_B such that $Q(T_B) = 0$ then the solution (T_B, s, z) is case (B).

Proof. First we prove that $z(x, t) \geq M(x - 1)$. This is easily followed from the maximum principle applied to $w = z - M(x - 1)$.

We replace this inequality in (2.1) for $t = T_B$, then $s(T_B)$ satisfies the following inequality

$$s(T_B) \left[(1 - M) + s(T_B) \left[\frac{\beta(1 - M) + M}{2} \right] + \beta s^2(T_B) \frac{M}{3} \right] \leq 0.$$

The quadratic form in brackets has coefficients $1 - M > 0$ and $\frac{\beta(1-M)+M}{2} > 0$, then $s(T_B) = 0$. $\square$

Proposition 2.4. *Suppose that, $t_0 < T$ and $\lim_{t \to t_0} s(t) > 0$. φ satisfies the hypotheses iv) of Proposition 2.1. Moreover $Q(t) > 0$ for all $t \leq t_0$. Then if we define a function*

$$\eta(t) = \begin{cases} \max\{x \in [0, s(t)] | z(x, t) \leq -1\} \\ 0 \quad \text{if } z(x, t) > -1, \ x \in [0, s(t)] \end{cases}$$

then it follows

$$\lim_{t \to t_0} \eta(t) < \lim_{t \to t_0} s(t)$$

Proof. The proof is similar that of Proposition 2.3 in [2].

Proposition 2.5. *Let (T, s, z) be a solution of Problem II such that $S_T = \inf_{t \in (0,T)} s(t) > 0$. If there exist two constants $d \in (0, S_T)$, $z_0 \in (0,1)$ such that $Md \geq z_0$, and*

$$z(s(t) - d, t) \geq -z_0, \quad 0 \leq t \leq T,$$

then

$$\dot{s}(t) \geq \frac{\ln(1 - z_0)}{d}.$$

Proof. It is the same that of the Lemma 2.4 in [2] (See also [6]). □

Proposition 2.6. *Let be (T, s, z) a solution of Problem II and φ satisfies the hypotheses of Proposition 2.1 iv), then if the solution is case (C), then $Q(T_C) \leq 0$.*

Proof. Suppose $Q(T_C) > 0$, then from the Proposition 2.5 the isotherm $z = -1$ is separated from the free-boundary. Using the Proposition 2.5 $\dot{s}$ has a lower bound, which contradicts the case (C). □

Corollary 2.7. *If (T, s, z) is a solution of Problem II and φ, G satisfy the following hypotheses:*
  i) $\varphi(x) \geq M(x - 1)$, $0 \leq x \leq 1$;
  ii) $G(t) \geq -M$, $0 < M < 1$.
  iii) $\dot{\varphi}(x) \geq 0$, $0 \leq x \leq 1$.
  And the solution is case (C), then $Q(T_C) < 0$.

Proof. It follows from Propositions 2.3 and 2.6. □

Proposition 2.8. *Let (T, s, z) be a solution of Problem II, φ and G satisfy the following hypotheses:*
  i) $\varphi(x) \geq M(x - 1)$, $M > 0$, $0 \leq x \leq 1$;
  ii) $G \in L^1(0, \infty)$.
  If the solution is case (A), then $Q(t) \geq 0$, $t > 0$. Moreover, if $G(t) \geq -M$, $(M > 0)$, $\forall t > 0$, then case (A) implies that $Q(t) > 0$, $\forall t > 0$.

Proof. This proof can be seen in [12].

3. Asymptotic behavior of the solution

Proposition 3.1. *Let (T, s, z) be a solution of Problem II of case (A) under the hypotheses of Proposition 2.9 and (iii) of Proposition 2.1. Moreover, we assume that the limit of $G(t)$ when $t \to \infty$ exists. If we denote $Q_\infty = \lim_{t \to \infty} Q(t)$ and $s_\infty = \lim_{t \to \infty} s(t)$, then s_∞ is given by*

(3.1)
$$s_\infty = \frac{-1 + \sqrt{1 + 2\beta Q_\infty}}{\beta}$$

47

Proof. The existence of the limit of $G(t)$ when $t \to \infty$ and $G \in L^1(0, \infty)$ assure that $\lim_{t \to \infty} G(t) = 0$.

We denote z_∞ the limit of z when t tends to infinity. The existence of $\lim_{t \to \infty} z(x, t)$ is due to Proposition 2.1 and [6,Chapter 6]. The function z_∞ satisfies: $z_\infty'' = 0$ in $(0, s_\infty)$, $z_\infty(s_\infty) = 0$, $z_\infty'(0) = \beta z_\infty(0)$, then $z_\infty(x) = 0$, $0 < x < s_\infty$.

Taking limit when $t \longrightarrow \infty$ in (2.3), then

$$s_\infty \left[1 + \beta \frac{s_\infty}{2} \right] - Q_\infty = 0$$

That means that $s_\infty \in (0, 1)$ is the root of the above equation, that is 3.1.

Moreover, we have $s_\infty < 1$ since

$$s_\infty < 1 \Longleftrightarrow 1 + 2\beta Q_\infty < (1 + \beta)^2 \Longleftrightarrow 2Q_\infty - 2 - \beta < 0.$$

By taking limit when $t \longrightarrow \infty$ in (2.3) the last inequality holds always due to the following expresion

$$2Q_\infty - 2 - \beta = 2 \int_0^1 (1 + \beta x)\varphi(x) \, dx - 2\beta \|G\|_1 < 0$$

where $\|G\|_1 = - \int_0^\infty G(\tau)d\tau$ □

4 The oxygen-comsumption problem

As in [8] we are interested in the dependence on the heat transfer coefficient h or its adimensional coefficient β.If, in Problem II we perform the classical transformation

$$u(x, t) = \int_x^{s(t)} \left\{ \int_\gamma^{s(t)} [1 + z(\alpha, t)] \, d\alpha \right\} d\gamma$$

then we obtain the following oxygen-comsumption problem.

Problem III:

$u_{xx} - u_t = 1, \quad$ in D_t;
$s(0) = 1$;
$u(s(t), t) = u_x(s(t), t) = 0, \quad t > 0$;
$u(x, 0) = H(x), \quad 0 \le x \le 1$;
$u_x(0, t) - H'(0) = \beta[u(0, t) - H(0) + \|G\|_{1,t}], \quad t > 0$,
where

$$H(x) = \int_x^1 \int_\gamma^1 (1 + \varphi(\alpha)) \, d\alpha d\gamma$$

From now on, in this section, we consider the following hypotheses for φ

$$-1 < \varphi(x) \le 0, \quad 0 \le x \le 1.$$

then

$$H(x) > 0, 0 \le x \le 1; \; H'(x) < 0, 0 \le x \le 1; \; H''(x) > 0, 0 \le x \le 1.$$

We now address the question of how the solution to Problem III depends upon $G(t)$.

Proposition 4.1. *The solution (T, s, u) of Problem III depends monotonically on G. In particular if (T_i, s_i, u_i), $i = 1, 2$ are the solutions for G_1 and G_2 respectively, and if $G_1(t) < G_2(t)$, then $s_1(t) \le s_2(t)$ and $u_1(x, t) \le u_2(x, t)$ whatever they are both defined.*

Proof. This is seen by considering the difference

$$v(x, t) = u_2(x, t) - u_1(x, t)$$

at the points where they are both defined.

Let $t^* = \sup\{t > 0 | u_2(0, t) > u_1(0, t)\}$ and $t^{**} = \sup\{t > 0 | s_2(t) > s_1(t)\}$. Let us suppose that both t^* and t^{**} are finite. By definition v satisfies the following problem

$$v_{xx} = v_t, x \in (0, s_1(t)), t \in (0, t^{**});$$

$$v(x, 0) = 0;$$

$$v(s_1(t), t) = u_2(s_1(t), t) > 0;$$

$$v_x(0, t) = \beta \left[v(0, t) + (||G_2||_{1,t} - ||G_1||_{1,t}) \right].$$

Claim 1 : $t^* \ne t^{**}$.

In order to prove that t^* and t^{**} are different, let us suppose that they are equal, then
a) $s_1(t^*) = s_2(t^*)$
b) $\dot{s}_1(t^*) > \dot{s}_2(t^*)$
c) $v(s_1(t^*), t^*) = u_2(s_1(t^*), t^*) = u_2(s_2(t^*), t^*) = 0$
Morever $u_2(0, t) > u_1(0, t)$ for $t < t^*$, then

$$v(0, t) > 0, \quad t < t^*$$

and

$$v(s_1(t), t) = u_2(s_1(t), t) > 0.$$

49

Since v has the minimum value zero at $(s_1(t^*), t^*)$, the minimum principle to v in $D_{t^*}^1$, we get $v_x(s_1(t^*), t^*) < 0$ which is a contradiction by (a) to

$$v_x(s_1(t^*), t^*) = u_{2x}(s_1(t^*), t^*) = u_{2x}(s_2(t^*), t^*) = 0$$

Then $t^* \neq t^{**}$.

Claim 2 : $t^* < t^{**}$ is impossible:

On $[0, t^*]$, $s_1(t) < s_2(t)$, whence $v(s_1(t), t) > 0$. By definition $v(0, t) > 0$ for $t < t^*$ and $v(0, t^*) = 0$. That implies $v(0, t^*)$ is a minimum value up to time t^* whence $v_x(0, t^*) > 0$, which contradicts

$$v_x(0, t^*) = \beta \left[v(0, t^*) + (\|G_2\|_{1,t^*} - \|G_1\|_{1,t^*}) \right] = \beta \left[\|G_2\|_{1,t^*} - \|G_1\|_{1,t^*} \right] < 0$$

Claim 3 : $t^{**} < t^*$ is impossible since:

Let be $t^{**} < t^*$, and since $v(0, t) > 0$, $v(s_1(t), t) = u_2(s_1(t), t) > 0$, for $t < t^{**}$, the point $(s_1(t^{**}), t^{**})$ is a minimum point for v because $v(s_1(t^{**}), t^{**}) = u_2(s_1(t^{**}), t^{**}) = u_1(s_1(t^{**}), t^{**}) = 0$.

By the corner minimum principle

$$v_x(s_1(t^{**}), t^{**}) < 0$$

which contradicts

$$v_x(s_1(t^{**}), t^{**}) = u_{2x}(s_2(t^{**}), t^{**}) = 0.$$

Thus the proposition is proved. $\square$

References.

1. J. R. CANNON, C. D. HILL, *Remarks on a Stefan problem*, J. Math. Mech. **17** (1967), 433–441.

2. E. COMPARINI, R. RICCI, D.A. TARZIA, *Remarks on a one dimentional Stefan problem related to the diffusion-comsumption model*, Z. Angew. Math. Mech. **64** (1984), 543–550.

3. E.COMPARINI, D.TARZIA, *A Stefan problem for the heat equation subject to an integral condition*, Rend Sem Matem Univ Padova **73** (1985), 119-136.

4. M.ELLIOT, J.R.OCKENDON, *Weak and variational methods for moving boundary problems,,* Res. Notes Math. Pitman, Boston-London **59** (1982).

5. A. FASANO, M. PRIMICERIO, J. Math. Anal Appl. I ;II: **58** (1977), 202-231 **57** (1977), 694–723 General free-boundary problems for the heat equation.

6. A. FASANO, M. PRIMICERIO, *New results on some classical parabolic free-boundary problems*, Quarterly of Applied Mathematics **38** (1981), 439–460.

7. A. FASANO,A. S. KAMIN, M. PRIMICERIO, *Regularity of weaks solutions of one-dimensional two-phase Stefan Problems,,* Ann. Mat. Pura Appl. **115,** (1977), 341-348.

8. A. FASANO, S.D. HOWINSON J. OCKENDON M. PRIMICERIO, *Some remarks on the regularization of supercooled one-phase Stefan problems in one dimension*, Quaterly Appl. Math **48** (1990), 153-168.

9. A. FRIEDMAN, *Partial Differential Equations of Parabolic Type*, Prentice Hall, Englewood Cliffs, N J, 1964.

10. A. D. SOLOMON, V. ALEXIADES, D. G. WILSON, *The Stefan problem with a convective boundary condition*, Quarterly of Applied Mathematics **40** (1982), 203–217.

11. D.A. TARZIA, *A bibliography on Moving-Free boundary problems for the heat Diffusion equation, Istituto Matematico Ulisse Dini, Firenze*, 1988.

12. D.A. TARZIA ,C.TURNER, *The one-phase supercooled Stefan problem with convective boundary condition*, to appear in Quat. Appl. Math.

Domingo Alberto Tarzia
Facultad de Ciencias Empresariales
Universidad Austral
Paraguay 1950
2000- Rosario
Argentina
e-mail: tarzia@uaufce.edu.ar

Cristina Vilma Turner
Facultad de Mat. Fís. y Astr.
Universidad Nacional Córdoba
Ciudad Universitaria
5000-Córdoba
Argentina
e-mail: turner@mate.uncor.edu

T TIIHONEN
Stefan problem with non-local radiation condition

Abstract: We consider the Stefan problem for a non-convex body or a collection of disjoint conductive bodies with Stefan-Boltzmann radiation condition on the surface. The main virtue of the resulting problem is non-locality of the boundary condition due to self-illuminating radiation on the surface. Moreover, the problem is non-linear, non-monotone and in the general case also non-coercive. We show that the non-local boundary value problem has a maximum principle. Hence, we can prove the existence of a weak solution assuming the existence of upper and lower solutions.

1. Introduction

Radiative heat exchange plays an important role in many situations. It has to be taken into account in general always, when the temperature on a visible surface of the system is high enough, or when other heat transfer mechanisms are not present (like in vacuum, for example). Apart from some simple cases such as a convex radiating body with known irradiation from infinity, we have to take into account the radiative heat exchange between different parts of the surface of our system. This leads to a non-local boundary condition on the radiating part of the boundary. While the theory of radiative heat transfer is well established in the engineering literature, see [2], for example, the mathematical analysis of the resulting equations seems to be almost non-existent.

In this paper we shall consider heat radiation in presence of phase transitions. The most simple non-trivial case is studied. Namely, the case where the phase transition takes place in an opaque material whose shape remains fixed. The surfaces are assumed to be diffuse and gray emitters and reflectors. This means that only the intensity of the radiation, not its wavelength (color) or direction has to be resolved. The material surrounding the opaque bodies is assumed non-participating, i.e. non-conductive and non-absorbing.

For simplicity we assume that the radiative properties of the materials do not depend on temperature. Reader is, however, encouraged to think of challenges arising from situations where the phase transition also changes the optical properties of the material, say melting of snow for example.

In Chapter 2 we introduce the model for heat transfer in presence of surface radiation. Then some main properties of the integral operators arising from the model are recalled. In the fourth chapter we discuss the solvability of the model which is non-local, non-linear, non-monotone and in general case even non-coercive. In a general geometry which contains enclosures we can prove solvability only provided suitable sub- and supersolutions exist.

NOTATIONS. Throughout the text we shall use the standard Lebesgue and Sobolev

52

spaces with the following notations: By $L^p(\Omega)$ we denote the space of measurable functions v for which $\|v\|_{L^p(\Omega)} = |\int_\Omega |v|^p|^{1/p}$ is finite. By $W^{m,p}(\Omega)$ we denote the space of $L^p(\Omega)$ functions whose generalized derivatives up to order m are in $L^p(\Omega)$. The space $W^{m,p}(\Omega)$ is equipped with the norm $\|v\|_{m,p,\Omega} = \sum_{k=0}^{m} |v|_{k,p,\Omega}$. Here $|v|_{k,p,\Omega} = \|D^k v\|_{L^p(\Omega)}$ is a seminorm of $W^{k,p}(\Omega)$. We denote $W^{k,2}$ by H^k and finally $H^1(\Omega;\Gamma)$ stands for the space $\{v \in H^1(\Omega) \mid v = 0 \text{ on } \Gamma\}$. In the general case Ω can be disconnected. Then the above spaces are to be understood as product of corresponding spaces defined in the components of Ω.

2. Model problem with surface radiation

We denote by Ω a union of conductive, opaque objects and note that in general Ω is not a connected set. We assume that the boundary of Ω, $\partial\Omega$ can be represented as $\partial\Omega = \Gamma \cup \Sigma_c \cup \Sigma$ where Γ denotes the part of the boundary on which we assume non-radiative boundary conditions (Dirichlet, Neumann, Robin) to be given. In the sequel we shall mainly concentrate to questions related to the radiating boundaries. Hence, to keep notations simpler we omit the terms related to Γ. By Σ_c we denote the set $\partial\Omega \cap \partial\Omega_c \setminus \Gamma$ where Ω_c is the convex hull of Ω. The set Σ forms the rest of $\partial\Omega$. By definition, any point on Σ is an interior point of Ω_c. Consequently, it will 'see' some other points of Σ, that is, for any point $s \in \Sigma$ there exists a set $\Sigma_s \subset \Sigma$ defined by $\Sigma_s = \{z \in \Sigma \mid \overline{sz} \cap \Omega = \emptyset\}$. Obviously, $z \in \Sigma_s$ implies that $s \in \Sigma_z$. Thus Σ can be naturally decomposed into disjoint subsets Σ_i which have the property $\Sigma_i = \cup_{s \in \Sigma_i} \Sigma_s$.

Under the above assumptions the heat equation for Ω combined with heat radiation on $\Sigma_c \cup \Sigma$ results in a coupled system for the absolute temperature T and radiosity λ (intensity of total radiation leaving the surface). Namely, we have that in the body

$$(2.1) \qquad E_t - \nabla \cdot (vE) - k\Delta T \ni \hat{f} \qquad \text{in } \Omega \times]0, \tau[,$$

where k is the coefficient of heat conductivity and $\hat{f}$ the intensity of internal heat source. By v we denote the convection velocity and by E the volumetric internal energy. We suppose that $E = E(T)$ is a maximally monotone graph with linear growth when T approaches infinity.

On Σ_c the body radiates heat to infinity, which results to heat balance

$$(2.2) \qquad k\frac{\partial T}{\partial n} + \epsilon\sigma T^4 = \epsilon\lambda^\infty \qquad \text{on } \Sigma_c \times]0, \tau[.$$

Here λ^∞ denotes the intensity of radiation coming from outside of the system. By ϵ we denote the emissivity of the surface. On Σ_i we have

$$(2.3) \qquad k\frac{\partial T}{\partial n} + (I - K_i)\lambda_i = \lambda_i^\infty \qquad \text{on } \Sigma_i \times]0, \tau[$$

which means that part of radiation coming to Σ_i originates from Σ_i itself. Here λ_i^∞ is the intensity of the radiation coming outside of the system to Σ_i. λ_i is the radiosity of the surface Σ_i. It depends on the surface temperature through the relation

$$(2.4) \qquad \epsilon\sigma T^4 + (I - (1 - \epsilon)K_i)\lambda_i = (1 - \epsilon)\lambda_i^\infty \qquad \text{on } \Sigma_i \times]0, \tau[.$$

The self illumination which is due to non-convexity of the surface Σ_i is described by the integral operator K_i,

$$(2.5) \qquad (K_i\lambda)(z) = \int_{\Sigma_i} \frac{n_z \cdot (s-z)\, n_s \cdot (z-s)}{\pi\|s-z\|^4} \Xi(z,s)\lambda(s)ds.$$

where $\Xi(z,s) = 1$ if s and z can see each other ($\overline{zs} \cap \Omega = \emptyset$). Otherwise $\Xi(z,s) = 0$.

The initial condition is set for E by assuming that $E = E_0$ at time 0.

Before introducing the variational form of our problem we shall make some notational simplifications. First of all, we formulate our problem in the case where $\Gamma = \emptyset$ to shorten the notations. Inclusion of traditional boundary conditions can be made in a straightforward way. Further, as the Stefan-Boltzmann law is physically meaningful only for positive temperatures we shall monotonize it by replacing σT^4 with

$$(2.6) \qquad h(T) = \sigma|T|^3 T.$$

Final simplification is obtained by solving (formally) λ_i from (2.4) as a function of $h(T)$. When the value of λ_i is inserted in (2.3) we obtain

$$(2.7) \qquad k\frac{\partial T}{\partial n} + G_i(h(T)) = \hat{\lambda}_i^\infty.$$

Here $G_i = (I - K_i)(I - (1-\epsilon)K_i)^{-1}\epsilon$. This is justified in the next chapter.

Writing (2.1)-(2.4) in variational form we get (formally) the problem

$$(2.8) \qquad \begin{aligned} \langle E(T)_t, \phi\rangle_{V'\times V} &+ \int_\Omega \nabla\cdot(E(T)v)\phi + \int_\Omega k\nabla T\nabla\phi + \int_{\Sigma_c} \epsilon h(T)\phi + \\ &\sum_i \int_{\Sigma_i} G_i(h(T))\phi \ni \langle f,\phi\rangle_{V'\times V} \quad \forall\phi \in V, \end{aligned}$$

for the absolute temperature T. Here f depends on $\hat{f}$ and on λ^∞. More precise formulation will be given in Chapter 4. A proper choice of spaces is not trivial because to make the non-linear terms well defined (in three dimensional case) we need L^5-integrability for the boundary temperatures which is not guaranteed automatically for $H^1(\Omega)$ functions in three dimensional case. In two dimensions the standard H^1 set up can be used.

3. Properties of radiation integral operators

In this chapter we shall recall some properties of the operators K_i and the corresponding kernel. Unless stated otherwise we consider Ω to be a three dimensional body in what follows. In order to simplify the notations in the sequel we shall denote the integral kernel in (2.5) as $w(z,s) = w^*(z,s)\Xi(z,s)$ where

$$(3.1) \qquad w^*(z,s) = \frac{n_z \cdot (s-z)\, n_s \cdot (z-s)}{\pi\|s-z\|^4}.$$

First, if the surface Σ_i is smooth enough we have the following result from [3].

Lemma 1. *Let Σ be a piecewise $C^{1,\delta}$-surface which is also a Lipschitz surface. Denote by S the set of non-smooth points on Σ. Then the integral $\int_{\Sigma} w(z,s)ds$ exists for $z \in \Sigma \setminus S$ and its value is less than or equal to one.*

In the sequel we assume that Σ satisfies the above conditions even if we do not state them explicitly.

Lemma 2. *The operator K_i is non-negative, i.e. it has a non-negative kernel. K_i maps $L^p(\Sigma_i)$ into itself compactly and $\|K_i\| \leq 1$. If there exist a constant $k < 1$ such that $\int_{\Sigma_i} w(z,s)ds \leq k \quad \forall z \in \Sigma_i$, then $\|K_i\| \leq k$.*

PROOF: [3]

Let us now consider the equation (2.4) We denote by E the operator induced by multiplication with ϵ. Then it holds that λ_i can be solved from (2.4), which justifies the use of operators G_i defined in the introduction.

Lemma 3. *The operator $I - (I - E)K_i$ from $L^p(\Sigma_i)$ into itself is invertible with non-negative inverse whenever $0 < \epsilon_0 \leq \epsilon \leq 1$.*

PROOF: [3]

We notice that the operators G_i can be written in several forms like

$$(3.2) \qquad G_i = (I - K_i)(I - (I - E)K_i)^{-1}E$$
$$(3.3) \qquad = (I - EK_i(I - (I - E)K_i)^{-1})E$$
$$(3.4) \qquad = I - (I - E + EK_i(E^{-1}(I - K_i) + K_i)^{-1}).$$

For the operator G_i we have

Lemma 4. *The symmetric part of operator G_i, $G_i + G_i^*$ from $L^2(\Sigma_i)$ into itself is positively semidefinite. Moreover, if $\rho(K_i) < 1$, then $G_i + G_i^*$ is positively definite.*

PROOF: [3]

Finally, we shall need the result

Lemma 5. *The operator G_i can be written as $G_i = I - H_i$ where $H_i \geq 0$ and $\rho(H_i) \leq 1$. Moreover, if ϵ is constant, then $\|H_i\| \leq 1$ as a mapping from $L^p(\Sigma_i)$ into itself.*

PROOF: [3]

The above Lemma 5 is one of the main tools in the sequel as it suggests the use of maximum principle. Namely, the structure of G_i is analogous to that of an M-matrix in the finite dimensional case.

4. Existence result

Let us introduce some notations. We write

$$(4.1) \qquad \tilde{a}(T, \phi) = a_0(T, \phi) + a_c(T, \phi),$$

where

$$(4.2) \qquad a_0(T, \phi) = \int_\Omega k \nabla T \nabla \phi,$$

$$(4.3) \qquad a_c(T, \phi) = \int_{\Sigma_c} \epsilon h(T) \phi.$$

Similarly, we write

$$(4.4) \qquad b(T, \phi) = \sum_i \int_{\Sigma_i} h(T) \phi$$

$$(4.5) \qquad c(T, \phi) = \sum_i \int_{\Sigma_i} (h(T)) H_i^* \phi.$$

We shall denote the space-time domain by $Q = \Omega \times]0, \tau[$ where τ is the length of the time interval. The solution T will eventually be searched for in the space $X = L^2([0, \tau] : H^1(\Omega)) \cap H^1(]0, \tau] : (H^1(\Omega))')$.

Before defining the solution we shall make some assumptions on the data.

1. There exists initial temperature $T_0 \in L^\infty(\Omega)$ so that $E_0 \in E(T_0)$.

2. The right hand side $f \in L^2([0, \tau] : (H^1(\Omega))')$.

3. $v \in H^1(\Omega)$ with $\nabla \cdot v = 0$ and $v \cdot n = 0$ on $\partial\Omega$.

4. There exist super and subsolutions (ψ, e_ψ) and (ϕ, e_ϕ) in $(L^\infty(Q) \cap X) \times L^2(Q)$ such that $e_\phi(0, x) \leq E_0(x) \leq e_\psi(0, x)$ for a.e. $x \in \Omega$ and

$$(4.6)$$
$$\int_Q -e_\phi(w_t + v \cdot \nabla w) + \int_0^\tau \tilde{a}(\phi, w) + b(\phi, w) - c(\phi, w) \leq \int_0^\tau \langle f, w \rangle + \int_\Omega e_\phi(0) w(0)$$

$$(4.7)$$
$$\int_Q -e_\psi(w_t + v \cdot \nabla w) + \int_0^\tau \tilde{a}(\psi, w) + b(\psi, w) - c(\psi, w) \geq \int_0^\tau \langle f, w \rangle + \int_\Omega e_\psi(0) w(0)$$

with $e_\phi \in E(\phi)$, $e_\psi \in E(\psi)$ and for any $w \in H^1(Q)$, $w \geq 0$ and $w(\tau) = 0$.

We can now truncate the non-linearity. If we denote by $[T] := min(\psi, max(\phi, T))$ the truncated temperature, we can introduce the truncated problem

$$(4.8)$$
$$\int_Q -e_T(w_t + v \cdot \nabla w) + \int_0^\tau \tilde{a}(T, w) + b([T], w) - c([T], w) = \int_0^\tau \langle f, w \rangle + \int_\Omega E_0 w(0)$$

for all $w \in H^1(Q)$ with $w(\tau) = 0$ and for some $e_T \in E(T)$, $e_T \in L^2(Q)$.

Theorem 1. *Assume that the above mentioned assumptions are verified. Then the problem (4.8) has a solution $(T, e_T) \in X \times L^2(Q)$. Moreover, $\phi \leq T \leq \psi$ so that (T, e_T) solves also the original problem.*

PROOF: (Sketch) We define a sequence $\{(T_n, e_n)\}$ in the following way. We set $T_1 = \psi$, $e_1 = e_\psi$ and when T_{n-1} is known we define T_n as the solution of the problem

(4.9)
$$\int_Q -e_n(w_t + v \cdot \nabla w) + \int_0^\tau \tilde{a}(T_n, w) + b([T_n], w)$$
$$= \int_0^\tau c(T_{n-1}, w) + \int_0^\tau \langle f, w \rangle + \int_\Omega E_0 w(0)$$

where $e_n \in E(T_n)$. We claim that

1. The problem (4.9) has a solution.

2. If (T_{n-1}, e_{n-1}) is a supersolution, then $T_n \leq T_{n-1}$, $e_n \leq e_{n-1}$ and (T_n, e_n) is also a supersolution.

3. $T_n \geq \phi$ and $e_n \geq e_\phi$ for all n.

From above we can then conclude that the sequence (T_n, e_n) is monotone and bounded. Hence it has a limit which is a solution of problem (4.8). So to prove the theorem we only have to show that the above claims are true.

The problem (4.9) is a Stefan-problem with monotone non-linear boundary condition. Its solvability has been studied in many references, for example in [1], where also the weak maximum principle has been proved.

Next we must show that $T_n \leq T_{n-1}$. As (T_{n-1}, e_{n-1}) is a supersolution we have that

(4.10)
$$-\int_Q (e_n - e_{n-1})(w_t + v \cdot \nabla w) + \int_0^\tau (\tilde{a}(T_n - T_{n-1}, w) + b([T_n], w) - b(T_{n-1}, w)) \leq 0$$

for all $w \geq 0$, $w(\tau) = 0$. Now from the weak maximum principle for the Stefan problem without the non-local boundary condition we obtain that $T_n \leq T_{n-1}$ and $e_n \leq e_{n-1}$. Moreover, as d has a non-negative kernel we get that $d(T_n, w) \leq d(T_{n-1}, w)$ for $w \geq 0$. Hence (T_n, e_n) is also a supersolution. By similar argument we can also show that $T_n \geq \phi$ for all n. Hence the proof is complete.

As a simple example of application of the above result we can state

Corollary 1. *Assume that we have no internal heat source $\hat{f}$ and that the exterior radiation sources λ^∞ and λ_i^∞ correspond to radiation from a black surface with bounded temperature. Moreover, if Σ_i is an enclosure $\lambda_i^\infty = 0$. Then the problem (4.8) has a solution.*

PROOF: It is sufficient to choose constant sub- and supersolutions ϕ and ψ such that ψ is bigger than the temperatures corresponding to λ^∞ and λ_i^∞: and that $e_\psi(0) \geq E_0$. For ϕ we have analogous requirements.

Let us check that ψ is a supersolution. First of all,

$$(4.11) \qquad \tilde{a}(\psi, w) = \int_{\Sigma_c} \epsilon h(\psi) w \geq \int_{\Sigma_c} \epsilon \lambda^\infty w$$

by construction. If Σ_i is an enclosure, then constant functions are in the kernel of G_i. It remains to consider Σ_i:s that are not enclosures. If now λ_i^∞ corresponds to uniform temperature ψ at infinity we see, by considering an enclosure constructed by 'closing' Σ_i that

$$(4.12) \qquad h(\psi) - H_i^* h(\psi) - \lambda_i^\infty = 0 \qquad \text{a. e. on } \Sigma_i.$$

As actually λ_i^∞ can be smaller we get that

$$(4.13) \qquad b(\psi, w) - c(\psi, w) \geq \langle f, w \rangle \qquad \forall w \geq 0.$$

References

[1] J.-F. Rodrigues. Variational methods in the Stefan problem. In A. Visintin, editor, *Phase Transitions and Hysteresis*. Springer-Verlag, Berlin, 1994.

[2] E.M. Sparrow and R.D. Cess. *Radiation Heat Transfer*. Hemisphere, 1978.

[3] T. Tiihonen. Stefan-Boltzmann radiation on non-convex surfaces. Math. Meth. in Appl. Sci (to appear).

T. Tiihonen
University of Jyväskylä
Laboratory of Scientific Computing
P.O. Box 35
FIN–40351 Jyväskylä, Finland
E-Mail: tiihonen@math.jyu.fi

Part 2.

Free boundary problems in phase transitions, phase separation and related topics

G CAGINALP AND W XIE

Mathematical analysis of phase memory alloys

Abstract: We present a mathematical theory of the phase field thermal binary alloy model derived in Caginalp and Jones [1995]. Formal asymptotics of the system of parabolic differential equations leads to new interface relations as part of macroscopic model which arises in the limit of vanishing interface thickness. Under suitable conditions we prove that the phase field system has a unique solution which converges to the limiting macroscopic solution. The concentration and phase are monotonic across the interface for a simplified system. Trasition layers in concentration are induced due to the change in phase and the change in material diffusion across the interface. The material diffusion asympotically vanishes in the solid phase, thereby creating mathematically challenging problems, particularly in the limit as interface thickness vanishes. We prove existence and convergence theorems for some model problems that contain the key aspects of the concentration problem.

1. Introduction

The mathematical modeling of free boundary problems that arise from solidification problems has been of interest for many years, both from the phase field (or diffused interface) and sharp interface approaches. Pattern formation has been a main focus of the study of these problems. From a practical metallurgical perspective, however, a most important aspect of these patterns is the trail of impurities that are left behind in most industrial applications. For example, in forming a sheet of aluminum by solidification the dendrites that appear at regularly spaced intervals can result in tones of impurities that implies a brittle, mechanically weak sheet of metal. In other words, the interaction between the curvature and velocity of the interface with the temperature and concentration at the interface is a crucial aspect of this problem. Moreover, since the impurities that move freely in the liquid are essentially frozen into the solid, obtaining an accurate value of the concentration near the interface becomes of paramount importance. If a sharp interface approach is used then one must generalize the temperature, curvature, velocity relation (i.e. Gibbs-Thompson type). Of course, even if this is done properly the sharp interface models have the limitation that they cease to be valid when the interface self-intersects. (Note that a mathematical regularization does not guarantee a physically valid conditiion). Since it is impossible to determine the distribution of impurities without pursuing the problem through complete soidification (thereby involving self intersections) the sharp interface approach is inadequate without fundamental advances.

Consequently, the development of a phase field model for alloy solidification through a unified thermodynamic perspective is of value in terms of obtaining a set of smooth parabolic equations and also in obtaining a reliable set of interface conditions upon taking the limit as interface thickness vanishes. The interface conditions are particularly useful for stability studies of the interface.

The standard phase field equations for a pure material are written in terms of temperature, T, and phase, φ, as

$$C_p T_t + \frac{l}{2} \phi_t = K \Delta T \tag{1.1}$$

$$\alpha \epsilon^2 \phi_t = \epsilon^2 \Delta \phi + \frac{1}{2}(\phi - \phi^3) + \epsilon \frac{[s]_E}{\sigma}(T - T_M) \tag{1.2}$$

in a region $\Omega \subseteq R^d$. The main feature of these equations is the use of a phase or order parameter, ϕ, ($\phi = \pm 1$ corresponds to the two phases and $\phi = 0$ to the interface) in addition to the (absolute) temperature T. Here $C_p :=$ specific heat, $l :=$ latent heat, $K :=$ thermal conductivity, $\sigma :=$ surface tension, $[s]_E :=$ entropy density difference between phases, $T_M :=$ (absolute) melting temperature and ϵ is the width of the interface. The phase field equations identify each of the physical parameters explicitly so that quantitative comparison is possible in analytic or numerical study.

A broad spectrum of sharp interface problems can be obtained rigorously from these equations in distinguished limits as ϵ (and various other parameters in each case) vanish [C2, CC]. Alternatively it has also been demonstrated that ϵ can be varied by many orders of magnitude without a significant change in the computed motion of the interface (see [CS] and references therein).

This feature is a necessary condition for the feasibility of numerical calculcsations, since the physical interface width is of order 10^{-8} cm and the domain size is at least 10^{-1} cm. Since self-intersections are also naturally handled by the phase field approach, these methods are quite suitable for realistic computations.

In particular, the phase field methods are ideal for studying interface problems involving alloys. Toward this end systems of equations have been derived and studied in [WBM 1,2] for the isothermal problem and in [CJ], [CX]. In this paper we develop the mathematical theory by proving existence, uniqueness and other properties. We prove in simple geometric settings that the solutions converge to those of the appropriate sharp interface models. Furthermore, the transition due to the change in material diffusion introduces an addtional (dynamic) tranition layer that intersects with the (steady state) transition induced by the change in phase. By considering a material diffusion constant that vanishes in the solid phase, one can address the fundamental issues involved in the "freezing in" of the impurities into the solid at the interface during solidification.

The layout of this paper is as follows. In Section 2 we review the traditional models for alloys and present the phase field approach to interface in binary mixtures. In Section 3 we perform an asymptotic analysis in order to derive the limiting sharp interface problems (in the limit as interface thickness vanishes) which involve new interface relations and reduce to the traditional models in the limit of small concentration. In Section 4 we discuss the general leading order inner expansion in which interface thickness depends on concentration. In Section 5 we examine rigorously the question of the dynamic and static transition layers which are formed in the concentration variable. The interaction between these layers is the basis for the trapping of solute in the solid. Finally, in Sections 6 we present a rigorous proof of some of the assertions of the asymptotics in one-dimensional space at constant temperature. Concluding remarks are in Section 7.

2. A Thermal-Alloy Phase Field Model.

We consider a model based on the phase field ideas of Section 1 which incorporates the physics of binary alloys in addition to thermal properties. Our aim here is to study the simplest set of equations which with (a) exhibit the proper behavior as the interface thickness, ϵ, approaches zero, and (b) identify all of the material parameters.

We begin by considering a free energy, $\mathcal{F}$, which describes the intermediate phase and concentration,

$$
\begin{aligned}
(2.6) \quad \mathcal{F}(\phi, c, T) = \int d^N x \Big\{ & \frac{\xi_A^2}{2} (\nabla \phi)^2 c + \frac{\xi_B^2}{2} (\nabla \phi)^2 (1 - c) + \frac{1}{8 a_A} (\phi^2 - 1)^2 c \\
& + \frac{1}{8 a_B} (1 - c)(\phi^2 - 1)^2 - \frac{[s]_A}{2} (T - T_A) \phi c \\
& - \frac{[s]_B}{2} (T - T_B) \phi (1 - c) + V c (1 - c) \\
& + RT \{ c \ln c + (1 - c) \ln(1 - c) \} - C_V T \ln T \Big\}.
\end{aligned}
$$

Here, the subscripts A and B denote the two materials, respectively. Briefly, the free energy is constructed by considering the analogous terms for each of the pure materials. For each term we assume a linear crossover between the two terms. For the sake of simplicity, we have omitted the temperature dependence in the $\xi_{A,B}$ and $a_{A,B}$ terms [CJ] which would provide additional nonlinear temperature terms.

The free energy can be interpreted geometrically in a four dimensional space $(\phi, c, T, \mathcal{F})$. For the pure phases, $\phi = \pm 1$, the free energy is experimentally well established for all concentrations. For the pure materials, $c = 0, 1$, one has (for the entire range of ϕ) the usual phase field equations for a pure material. Thus, mathematical modelling of the free energy involves a derivation of $\mathcal{F}$ in the intermediate regions $|\phi| < 1$, representing the four dimensional space within the more established three dimensional planes.

The first two terms involve the gradient of ϕ terms, which arise from the microscopic interaction strengths. The coefficients ξ_A and ξ_B are characteristics of the two materials, respectively, as are the 'well- depths' of the double-well potentials, denoted a_A and a_B. The third and fourth terms are thus linear combinations of the analogous terms for the homogeneous materials. The next two terms involve the entropy differences between the two phases for the pure materials, denoted $[s]_A$ and $[s]_B$. The temperature T_A and T_B are the equilibrium melting temperatures of the two materials. The quadratic term in c arises from the differences in the bonding energies between the A atoms and the B atoms [Co], so that V is an energy density. The terms containing the logarithm of c constitute the entropy of mixing [Co]. The last term involves the specific heat per unit volume, C_V. By setting c at 0 or 1, one has the free energy of the pure A or B material respectively.

We write three equations describing (ϕ, c, T). The concentration c satisfies a conserved variational formulation

$$(2.7) \qquad \tau_1 c_t = \nabla \cdot K_2(\phi, c) c(1 - c) \nabla \frac{\delta F}{\delta c},$$

while the usual phase field (nonconserved) applies to ϕ,

$$(2.8) \qquad \tau_2 \phi_t = -\frac{\delta F}{\delta \phi}.$$

Here, τ_1 and τ_2 are relaxation constants while $K_2(\phi, c) c(1 - c)$ is the mobility term. In general, K_2 may depend on phase and concentration, but usually varies most significantly as phase changes. The term $c(1 - c)$ used by [WBM] is a standard approximation reflecting the fact that mobility vanishes in the two phases and attains a peak at equal concentrations of the two materials. The energy conservation equation arises [CJ] from thermodynamic balance equations as

$$(2.9) \qquad C_V T_t + \frac{1}{2}\left(l(c)\phi_t - (Q + [l]_{B,A}\phi)c_t \right) = \nabla \cdot K_1 \nabla T,$$

where

$$(2.10) \qquad l(c) \equiv \left(T_A[s]_A c + T_B[s]_B(1 - c) \right),$$

$$(2.11) \qquad [l]_{B,A} \equiv \left(T_B[s]_B - T_A[s]_A \right),$$

$$(2.12) \qquad Q \equiv 2(T_B - T_A)\left(C_V + \frac{s^l + s^s}{2} \right),$$

and C_V and K_1 are the specific heat and the thermal conductivity, respectively.

Note that $l(c)$ is the analog of the usual latent heat since a basic thermodynamic relation implies latent heat of the pure A material is $T[s]_A$. In the limit of a pure material, equation (2.9) clearly reduces to the usual heat equation.

Similarly, the concentration equation (2.7) may be written as

$$(2.13) \qquad \tau_1 c_t = \nabla \cdot K_2(\phi, c) c(1 - c) \nabla \sum_i \frac{\partial f_i}{\partial c},$$

where

(2.14)
$$\frac{\partial f_1}{\partial c} \equiv (\xi_A^2 - \xi_B^2)(\nabla\phi)^2,$$

(2.15)
$$\frac{\partial f_2}{\partial c} \equiv \frac{1}{8}(\phi^2 - 1)\left(\frac{1}{a_A} - \frac{1}{a_B}\right),$$

(2.16)
$$\frac{\partial f_3}{\partial c} \equiv V - 2Vc,$$

(2.17)
$$\frac{\partial f_4}{\partial c} \equiv \frac{1}{2}\left\{-[s]_A(T - T_A) + [s]_B(T - T_B)\right\}\phi,$$

(2.18)
$$\frac{\partial f_5}{\partial c} \equiv RT \ln \frac{c}{1-c}.$$

The phase equation (2.8) can be written as

(2.19)
$$\frac{2}{3}\frac{\epsilon}{\sigma}\tau_2\phi_t = \epsilon^2\Delta\phi + \frac{1}{2}(\phi - \phi^3)$$
$$+ \frac{2}{3}\frac{\epsilon}{\sigma}\left\{[s]_A(T - T_A)c + [s]_B(T - T_B)(1 - c)\right\},$$

where

(2.20)
$$\sigma(c) \equiv \frac{2}{3}\left\{\xi_A^2 c + \xi_B^2(1 - c)\right\}^{1/2}\left\{\frac{c}{a_A} + \frac{1-c}{a_B}\right\}^{1/2},$$

(2.21)
$$\epsilon(c) \equiv \left\{\xi_A^2 c + \xi_B^2(1 - c)\right\}^{1/2}\left\{\frac{c}{a_A} + \frac{1-c}{a_B}\right\}^{-1/2}.$$

This equation differs from the pure version in that the entropy term is a linear combination of the entropy terms for each of the pure materials and ϵ and σ both depend on c.

Hence, equations (2.19), (2.9), and (2.13) are the system of equations to be studied subject to approriate initial and boundary conditions. Note that if the total mass of each material is constant, then Neumann conditions for concentration are appropriate, and the equation (2.13) is in full conservation form. The boundary conditions for temperature are imposed in accordance with the thermal environment, e.g., Dirichlet conditions if the alloy is in a container whose walls remain at constant temperature, or Neumann conditions if it is insulated.

3. Asymptotic Analysis

We perform an asymptotic analysis for the thermal-alloy phase field model discussed in Section 2. In particular we study the system.

(3.1)
$$\alpha\epsilon^2(c)\phi_t = \epsilon^2(c)\Delta\phi + \frac{1}{2}(\phi - \phi^3)$$

$$+ \frac{\epsilon(c)}{3\sigma(c)}\left\{[s]_A(T - T_A)c + [s]_B(T - T_B)(1 - c)\right\}$$

(3.2)
$$C_V T_t + \frac{1}{2}\left(l(c)\phi_t - (Q + [l]_{B,A}\phi)c_t\right) = \nabla \cdot K_1\nabla T$$

(3.3)
$$\tau_1 c_t = \nabla \cdot K_2(\phi)c(1 - c)\nabla\left[(M + M_1 T)\phi + RT\ln\frac{c}{1 - c}\right]$$

where

(3.4)
$$K_2(\phi) = K_2^- + \frac{1}{2}(K_2^+ - K_2^-)(1 + \phi).$$

Here $\sigma(c)$, $\epsilon(c)$, $l(c)$, $[l]_{B,A}$ and Q are given by (2.15), (2.16), (2.5), (2.6) and (2.7), respectively, and the other physical parameters are also defined in Section 2.

Note that we have

(3.5)
$$\epsilon^2(c) = \left\{\xi_A^2 c + \xi_B^2(1 - c)\right\}\left(\frac{c}{a_A} + \frac{1 - c}{a_B}\right)^{-1}$$
$$= \xi_B^2 a_B \frac{1 - c + (\xi_A^2/\xi_B^2)c}{1 - c + (a_B/a_A)c} = \xi_A^2 a_A \frac{c + (\xi_B^2/\xi_A^2)(1 - c)}{c + (a_A/a_B)(1 - c)}$$

and

(3.6)
$$a(c) = \left(\frac{c}{a_A} + \frac{1 - c}{a_B}\right)^{-1} = \frac{a_B}{1 - c + (a_B/a_A)c} = \frac{a_A}{c + (a_A/a_B)(1 - c)}$$

so the system (3.1)-(3.3) can be rewritten as follows.

(3.7)
$$\alpha\epsilon^2 E(c)\phi_t = \epsilon^2 E(c)\Delta\phi + \frac{1}{2}(\phi - \phi^3)$$
$$+ \frac{\epsilon}{3\sigma(c)}\left\{[s]_A(T - T_A)c + [s]_B(T - T_B)(1 - c)\right\}$$

(3.8)
$$C_V T_t + \frac{1}{2}\left(l(c)\phi_t - (Q + [l]_{B,A}\phi)c_t\right) = \nabla \cdot K_1\nabla T$$

(3.9)
$$\tau_1 c_t = \nabla \cdot K_2(\phi)c(1 - c)\nabla\left[(M + M_1 T)\phi + RT\ln\frac{c}{1 - c}\right]$$

where $E(c)$ and $\sigma(c)$ are two bounded functions with positive lower bound, ϵ in (3.7) is a (small) positive constant and the other constants are the same as in (3.1)-(3.3).

The following two asymptotic results can be verified by the method employed in [CX].

Asymptotic Analysis for the Simplest System. We first study the system (3.7)-(3.9) when $E(c) = 1$, $\sigma(c)$ and $l(c)$ are constants independent of c and $[s]_A = [s]_B = [s]_E$. We set τ_1 to unity by incorporating it into K_2. For mathematical convenience we assume that thermal gradients are not very large (except for

66

small neighborhoods) and neglect the $M_1 T\phi$ term in (3.9) and incorporate RT into M, thereby replacing RT by unity. One then has the following system

$$
\text{(3.10)} \quad
\begin{cases}
\alpha\epsilon^2\phi_t = \epsilon^2\Delta\phi + \dfrac{1}{2}(\phi - \phi^3) + \dfrac{\epsilon}{3\sigma}[s]_E\Big\{T - T_A c - T_B(1 - c)\Big\} \\[2ex]
C_V T_t + \dfrac{l}{2}\phi_t = \nabla \cdot K_1 \nabla T \\[2ex]
c_t = \nabla \cdot K_2(\phi)c(1 - c)\nabla\Big[M\phi + \ln \dfrac{c}{1 - c}\Big]
\end{cases}
$$

and the asymptotic result is as follows.

Proposition 3.1. *In limit as $\epsilon \to 0$, the first order approximation (ϕ^0, T^0, c^0) of the solution (ϕ, T, c) to (3.10) is governed by the following sharp interface model*

$$
\text{(3.11)} \quad
\begin{cases}
C_V T_t^0 = K_1 \Delta T^0 \quad \text{in} \quad \Omega \setminus \Gamma(t) \\[1.5ex]
c_t^0 = K_2^{\pm}\Delta c^0 \quad \text{in} \quad \Omega \setminus \Gamma(t) \\[1.5ex]
[T^0]_-^+ = 0 \quad \text{on} \quad \Gamma(t) \\[1.5ex]
[K_1 T_r^0]_-^+ = -l v^0 \quad \text{on} \quad \Gamma(t) \\[1.5ex]
[\ln \dfrac{c^0}{1 - c^0}]_-^+ = -2M \quad \text{on} \quad \Gamma(t) \\[1.5ex]
[K_2 c_r^0]_-^+ = -[c^0]_-^+ v^0 \quad \text{on} \quad \Gamma(t) \\[1.5ex]
(\alpha v^0 + \kappa^0) = \dfrac{[s]_E}{-\sigma}\Big[T^0 - T_B - \dfrac{T_A - T_B}{2M}\ln\dfrac{1 - c_+^0}{1 - c_-^0}\Big] \quad \text{on} \quad \Gamma(t).
\end{cases}
$$

Asymptotic Analysis for the System (3.7)-(3.9). Setting τ_1 to unity in (3.9) one has the asymptotic result below.

Proposition 3.2. *In limit as $\epsilon \to 0$, the first order approximation (ϕ^0, T^0, c^0) of the solution (ϕ, T, c) to (3.7)-(3.9) is governed by the following sharp interface model*

$$
\text{(3.12)} \quad
\begin{cases}
C_V T_t^0 - \dfrac{1}{2}\Big(Q \pm [l]_{B,A}\Big)c_t^0 = K_1 \Delta T^0 \quad \text{in} \quad \Omega \setminus \Gamma(t) \\[2ex]
c_t^0 = \nabla \cdot K_2^{\pm} c^0(1 - c^0)\nabla\Big[RT^0 \ln \dfrac{c^0}{1 - c^0} \pm M_1 T^0\Big] \quad \text{in} \quad \Omega \setminus \Gamma(t)
\end{cases}
$$

and on the interface $\Gamma(t)$

$$
\text{(3.13)} \quad
\begin{cases}
[T^0]_-^+ = 0 \\[1.5ex]
[K_1 T_r^0]_-^+ = \dfrac{1}{2}\Big(Q[c^0]_-^+ + [l]_{B,A}(c_+^0 + c_-^0) - 2T_B[s]_B\Big)v^0 \\[1.5ex]
RT^0[\ln \dfrac{c^0}{1 - c^0}]_-^+ = -2(M + M_1 T^0) \\[1.5ex]
[K_2 RT^0 c_r^0]_-^+ + [RK_2 c^0(1 - c^0)\Big(\ln \dfrac{c^0}{1 - c^0}\Big)T_r^0]_-^+ \\[1.5ex]
\quad + K_2^+ M_1 c_+^0(1 - c_+^0)(T_r^0)_+ + K_2^- M_1 c_-^0(1 - c_-^0)(T_r^0)_- = -[c^0]_-^+ v^0
\end{cases}
$$

subject to a Gibbs-Thomson type relation on $\Gamma(t)$

(3.14)
$$\int_{R^1} (H_1 + H_2)\psi(z)dz = 0$$

where H_1, H_2 are given by

(3.15)
$$
\begin{cases}
H_1 \equiv \left(\alpha v^0 + \Delta r^0\right)\hat{\phi}_z^0 + \dfrac{1}{3E(\hat{c}^0)\sigma(\hat{c}^0)}\left(a + b\hat{c}^0\right) \\[4mm]
H_2 \equiv \dfrac{E'(\hat{c}^0)\hat{\phi}_{zz}^0}{\bar{R}E(\hat{c}^0)}\hat{c}^0(1 - \hat{c}^0)\left[-v^0\displaystyle\int_0^z \dfrac{dy}{K_2(\hat{\phi}^0)(1 - \hat{c}^0)} \right. \\[4mm]
\qquad\qquad \left. + c_4(s,t)\displaystyle\int_0^z \dfrac{dy}{K_2(\hat{\phi}^0)\hat{c}^0(1 - \hat{c}^0)} + c_6(s,t)\right].
\end{cases}
$$

and $\psi = \hat{\phi}_z^0$, where $\hat{\phi}^0$ is a solution of the system

(3.16)
$$
\begin{cases}
E(\hat{c}^0)\hat{\phi}_{zz}^0 + \dfrac{1}{2}\left(\hat{\phi}^0 - (\hat{\phi}^0)^3\right) = 0 \\[4mm]
\left\{K_2(\hat{\phi}^0)\hat{c}^0(1 - \hat{c}^0)\left(\bar{A}\hat{\phi}^0 + \bar{R}\ln\dfrac{\hat{c}^0}{1 - \hat{c}^0}\right)_z\right\}_z = 0 \\[4mm]
\hat{\phi}^0(-\infty) = -1, \quad \hat{\phi}^0(0) = 0, \quad \hat{\phi}^0(\infty) = 1 \\[2mm]
\hat{c}^0(-\infty) = c_-^0, \quad \hat{c}^0(\infty) = c_+^0,
\end{cases}
$$

$\bar{A} = M + M_1\hat{T}^0$, $\bar{R} = R\hat{T}^0$, $c_4(s,t)$ and $c_6(s,t)$ are constants independent of z, and $a + bc \equiv [s]_A(\hat{T}^0 - T_A)c + [s]_B(\hat{T}^0 - T_B)(1 - c)$. More discussion for the system (3.16) is presented in next section.

In particular, if $E(c) = 1$, then the above system (3.16) can be solved explicitly and the integral in (3.14) can be evaluated. This leads to the interface condition on $\Gamma(t)$

(3.17)
$$\alpha v^0 + \kappa^0 = \frac{RT^0}{2(M + M_1T^0)}\left[[s]_A(T^0 - T_A)\int_{c_-^0}^{c_+^0} \frac{du}{\sigma(u)(1 - u)}\right.$$

$$\left. + [s]_B(T^0 - T_B)\int_{c_-^0}^{c_+^0} \frac{du}{\sigma(u)u}\right].$$

4. A System From Inner Expansion

We discuss, in this section, a system from the inner expansion of Section 3, where the solvability condition for a second order equation is used in order to obtain the Gibbs-Thomson relation across the interface (see (3.14)). Specifically, we shall consider the following system.

$$(4.1) \qquad E(c)\phi_{zz} + \frac{1}{2}(\phi - \phi^3) = 0,$$

$$(4.2) \qquad \left\{ c(1-c)\left(A\phi + \ln \frac{c}{1-c}\right)_z \right\}_z = 0,$$

$$(4.3) \qquad E(c)\psi_{zz} + \frac{1}{2}\left[1 - 3\phi^2 - 2AE'(c)\phi_{zz}c(1-c)\right]\psi = 0,$$

with boundary conditions

$$(4.4) \qquad \phi(-\infty) = -1, \quad \phi(0) = 0, \quad \phi(\infty) = 1,$$

$$(4.5) \qquad c(-\infty) = c_-, \quad c(\infty) = c_+,$$

$$(4.6) \qquad \psi(-\infty) = 0, \quad \psi(\infty) = 0,$$

where A, c_- and c_+, the latter two positive, are constants (the constant A is related to the temperature in inner expansion), and $E(c) \geq \theta > 0$ is a smooth function.

Let T be the operator

$$(4.7) \qquad T(c, \phi)u \equiv u_{zz} + \frac{1}{2E(c)}\left[1 - 3\phi^2 - 2AE'(c)\phi_{zz}c(1-c)\right]u.$$

We ivestigate the spectrum of the operator T, and in particular, show that 0 is a simple eigenvalue of T and $\dim N(T) = 1$.

4.1. Existence of solution for (4.1), (4.2), (4.4) and (4.5). As we study the existence of bounded solutions for (4.1), (4.2), (4.4) and (4.5), the functions ϕ and c satisfy

$$(4.8) \qquad \phi_{zz} + \frac{1}{2E(c)}(\phi - \phi^3) = 0,$$

$$(4.9) \qquad c_z + Ac(1-c)\phi_z = 0$$

with boundary conditions (4.4) and (4.5). By solving (4.9) and using the boundary condition (4.5), we have

$$(4.10) \qquad c = \frac{e^{B-A\phi}}{1 + e^{B-A\phi}},$$

where

$$(4.11) \qquad B = \frac{1}{2}\left(\ln \frac{c_-}{1-c_-} + \ln \frac{c_+}{1-c_+}\right).$$

Substituting c into (4.8), we have the following ϕ equation

$$(4.12) \qquad \phi_{zz} + \frac{1}{2\hat{E}(\phi)}(\phi - \phi^3) = 0,$$

where $\hat{E}(\phi) = E(c(\phi))$.

Existence of the solution for the system (4.10) and (4.12) with conditions (4.4) can be proven by a standard argument and this will not be stated here for brevity.

4.2. Monotonicities of the solution $\phi(z)$ and $c(z)$. We now show that the solution ϕ is monotonically increasing and c is also monotonic function (increasing if $A < 0$ and decreasing if $A > 0$).

To show that the solution ϕ is monotonically increasing, we define

$$\Phi^R(z) = \phi(z + R) - \phi(z),$$

where R is a real number. The function $\Phi^R(z)$ satisfies an equation

$$(4.13) \qquad \Phi^R_{zz} + \left\{ \int_0^1 F'\Big(\tau\phi(z+R) + (1-\tau)\phi(z)\Big)d\tau \right\}\Phi^R = 0$$

where

$$(4.14) \qquad F(\phi) = \frac{1}{2\hat{E}(\phi)}(\phi - \phi^3).$$

Let $\delta > 0$ be a small constant such that

$$(4.15) \qquad F'(\phi) < 0 \quad \text{as} \quad 0 \le 1 \pm \phi \le \delta.$$

Since $\lim_{z\to-\infty} \phi(z) = -1$ and $\lim_{z\to\infty} \phi(z) = 1$, there exists a constant $N = N(\delta)$ such that

$$(4.16) \qquad \begin{cases} \phi(z) \le -1 + \delta & \text{if} \quad z \le -N, \\ \phi(z) \ge 1 - \delta & \text{if} \quad z \ge N, \\ \dfrac{\delta}{2} \le \phi(z) \le \delta & \text{in a neighborhood of} \quad z = -N. \end{cases}$$

Now we choose $R_0 > 0$ sufficiently large such that $R_0 > 2N$ and

$$(4.17) \qquad \begin{cases} \phi(z) < \dfrac{\delta}{2} & \text{as} \quad z \le -N - R_0, \\ \phi(z + R_0) \ge \phi(z) & \text{as} \quad -N \le z \le N, \\ \displaystyle\sup_{z \le -N-R_0} \phi(z) \le \min_{-N \le z \le -N+R_0} \phi(z). \end{cases}$$

Then we have the inequalities $\Phi^{R_0}(-N - R_0) \ge 0$ and $\Phi^{R_0}(z) \ge 0$ as $-N \le z \le N$, so the maximum principle yields $\Phi^{R_0}(z) \ge 0$ as $z \le -N - R_0$.

Let

$$(4.18) \qquad \Psi^R(z) \equiv \phi(z - R - N + R_0) - \phi(z),$$

for $-N - R_0 < z < R < -N$. Then $\Psi(z)$ satisfies

$$(4.19) \qquad \Psi_{zz}^R + \left(\int_0^1 F'(\xi(\tau)) d\tau \right) \Psi^R = 0,$$

where $\xi(\tau) = \tau \phi(z - R - N + R_0) + (1 - \tau) \phi(z)$.

Noting that, from (4.16) and (4.17), $\Psi^R(-N - R_0) \geq 0$, and the boundedness of $F'(\xi(\tau))$ in the L^∞−norm, we can use the method of the sliding domain and the maximum principle to obtain that $\Psi^R(z) \geq 0$ for every R, with $-N - R_0 < R < -N$. In particular, we have $\Phi^{R_0}(z) \geq 0$ as $\quad -N - R_0 \leq z \leq -N$. Thus we obtain

$$(4.20) \qquad \Phi^{R_0}(z) \geq 0 \quad \text{as} \quad z \leq N.$$

With the inequality (4.20), we can easily prove, by the maximum principle, that

$$(4.21) \qquad \Phi^{R_0}(z) \geq 0 \quad \text{as} \quad -\infty < z < \infty,$$

since the negative minimum for $\Phi^{R_0}(z)$ cannot be attained in the interval $z \geq N$.

Now starting with this (possibly large number) R_0 for which (4.21) holds, we decrease it to a value r which is defined by

$$r \equiv \inf\{ R \mid \Phi^R(z) \geq 0 \}.$$

Clearly, we have $-\infty < r \leq R_0$. We now show that $r = 0$.

Indeed, if we assume $r \neq 0$ $(-\infty < r \leq R_0)$, we will show that $\Phi^r(z) \equiv 0$.

First, suppose $\Phi^r(z) \not\equiv 0$. Then we have, by the maximum principle, that $\Phi^r(z) > 0$. On the other hand, by the definition of r, the exists a monotonic increasing sequence $\{r_n\}$ with $r_n \nearrow r$ such that for each r_n, there exists z_n so that $\Phi^{r_n}(z_n) < 0$. Note that since one has $\Phi^{r_n}(z) \to 0$ as $z \to \pm\infty$, we can assume, without loss of generality, that

$$\Phi^{r_n}(z_n) = \inf_{-\infty < z < \infty} \Phi^{r_n}(z).$$

Obviously, we have $|z_n + r| < N$. Thus we may suppose that (at least for a subsequence), $z_n \to z_0$ and z_0 is a finite point. By taking the limit as $n \to \infty$, we obtain

$$\lim_{n \to \infty} \Phi^{r_n}(z_n) = \Phi^r(z_0) \leq 0.$$

This clearly contradicts the fact $\Phi^r(z) > 0$ as $-\infty < z < \infty$.

Therefore $\Phi^r(z) \equiv 0$, so the function $\Phi(z)$ is a periodic function with period $|r|$. But this is impossible since we have $\lim_{z \to \pm\infty} \phi(z) = \pm 1$. This final contradiction allows us to assert that $r = 0$.

Hence we have that for any $r \geq 0$ $\phi(z + r) \geq \phi(z)$ in $z \in (-\infty, \infty)$, which implies that $\phi(z)$ is monotonically increasing with $\phi'(z) > 0$.

71

Remark: Monotonicity of $\phi(z)$ in this case can be shown in an easy way by verifying that solutions to equations with given conditions are between -1 and 1 and that solutions to equations are symmetric with respect to maxima and minima, if the extrema exists. The boundary conditions (4.4) rules out this possibility and monotonicity follows. However, our method above can be used below (in section 4.3) to show that the operator T defined in (4.7) is of $\dim N(T) = 1$.

From (4.10), we have

$$(4.22) \qquad c'(z) = \frac{-Ae^{\bar{A}-A\phi(z)}}{\left(1 + e^{\bar{A}-A\phi(z)}\right)^2} \phi'(z).$$

Consequently the function $c(z)$ is monotonically decreasing if the constant A is a positive and it is monotonically increasing if A is a negative constant.

4.3. Solvability conditions for the equation $T(c, \phi)\psi = f$. We now discuss the solvability condition for the equation

$$(4.23) \qquad\qquad T(c, \phi)\psi = f$$

where the operator $T(c, \phi)$ is defined in (4.7).

When c and ϕ are given, the operator $T(c, \phi)$ is self-adjoint. We can also verify that ϕ_z is a solution of $T(c, \phi)\psi = 0$. With the monotonicity of ϕ, we are going to show that $N(T) = Span\{\phi_z\}$. To do that, we need only to show that if ψ is a solution of $T(c, \phi)\psi = 0$, then $\psi = \mu\phi_z$ for some constant $\mu \in R^1$.

In fact, let

$$\Phi^r \equiv \psi + r\phi_z$$

and $M > 0$ is a large number such that

$$g(c, \phi) \leq 0 \quad \text{as} \quad |z| \geq M,$$

where

$$g(c, \phi) = \frac{1}{2E(c)}\left[1 - 3\phi^2 + \frac{AE'(c)}{E(c)}(\phi - \phi^3)c(1 - c)\right].$$

Using the assumption $\phi_z > 0$ and a similar argument employed in §4.2, we can show that there exists a number $r > 0$ such that $\Phi^r(z) \geq 0$ for all $z \in R^1$. By reducing r, we can show, by the maximum principle and the same method employed in section 4.2, that

$$\Phi^{r_0}(z) \equiv 0,$$

where $r_0 = \inf\{r \mid \Phi^r(z) \geq 0\}$. Letting $\mu = r_0$, we obtain $\psi = \mu\phi_z$.

Thus we proved that 0 is a simple eigenvalue of the operator $T(c, \phi)$, ϕ_z is its eigenfuction and $\dim N(T) = 1$, i.e. $T(c, \phi)$ is a Fredholm operator with index zero.

Hence the equation (4.23) is solvable if and only if f is orthogonal to ϕ_z (see also (3.14)).

5. Interaction Between Dynamic and Static Transition Layers

We now compare two equations, one having only a dynamic transition layer, and the other having both dynamic and static layers. The two equations, which contain the crucial parts of (3.3), are

$$(5.1) \qquad u_t = \frac{\partial}{\partial x}\left\{D(\phi^\epsilon)u_x\right\},$$

$$(5.2) \qquad c_t = \frac{\partial}{\partial x}\left\{D(\phi^\epsilon)\left(c_x + Ac(1-c)\phi_x^\epsilon\right)\right\},$$

with appropriate boundary and initial conditions, where

$$(5.3) \qquad \phi^\epsilon(x) \equiv \tanh\frac{x}{\epsilon}, \qquad \epsilon > 0.$$

The first of these equations is just the heat equation with a variable diffusion coefficient, D, exhibiting a transition layer near $x = 0$. In particular, we can imagine a thin rod consisting of two materials, with very different heat conduction properties, fused together at $x = 0$. The second is just the simplest concentration equation with a change of phase at $x = 0$. This means that thermal conditions are balanced so that the interface remains fixed.

We will consider the cases when

$$(5.4) \qquad D(\phi^\epsilon) = \frac{1}{2}(3 + \phi^\epsilon), \quad \text{and}$$

$$(5.5) \qquad D(\phi^\epsilon) = \frac{1}{2}(1 + \phi^\epsilon)$$

respectively. In particular, the form of D expressed by (5.5) involves a degeneracy in that D approaches zero as x approaches $-\infty$. In each of these equations, a transition layer arises in u and c, respectively, even if the initial conditions do not involve a steep gradient (i.e., of order $1/\epsilon$). In equation (5.2) a transition layer in c occurs even if $D = 1$, according to the formal asymptotics of Section 5.2. One of our goals is to obtain a rigorous understanding of the similarities and differences in the transition layers which arise from various physical effects, particularly the following three situations:

(A) Equation (5.2) with $A = 0$ (namely equation (5.1)) subject to (5.4) or (5.5).
(B) Equation (5.2) with $D = 1$.
(C) Equation (5.2) subject to (5.4) or (5.5).

The transition layer which arises in (A) is a consequence of the change in the diffusion coefficient. The physical origin of this is in the differences of heat diffusion in the two materials (in (5.1)) or of mass diffusion in the two phases. This is clearly a dynamical effects and the transition layer forms, then disappears in finite time. On the other hand the transition layer in (B) occurs even in equilibrium and has no analog for thermal diffusion. It arises from the particular nature of the phase diagram in which the liquidus and solidus are separated by a finite distance at a fixed temperature ($|c_l - c_s|$) which is in fact the size of the jump in (5.2) as $t \to \infty$. The transition layer in (C) combines both of these effects and permits the study of the interaction between the development and disappearance of the diffusion induced layer with the evolution of the phase diagram layer.

(1a). We first consider the simpler (nondegenerate case) when $D(\phi^\epsilon)$ is given by (5.4). The precise problem which we consider here is

(5.6)
$$\begin{cases} u_t = \dfrac{\partial}{\partial x}\left\{D(\phi^\epsilon)u_x\right\}, & x \in (-1,1),\ t > 0, \\ u_x(-1,t) = u_x(1,t) = 0, & t > 0, \\ u(x,0) = u_0(x), & -1 < x < 1, \end{cases}$$

where $u_0(x)$ is a given function ($u_0(x)$ is continuous on the interval $(-1,1)$, say).

For any $\epsilon > 0$, the problem (5.6) possesses a unique solution $u^\epsilon(x,t)$, which is smooth in $Q_T \equiv (-1,1) \times (0,T)$, and $u \in C^{2,1}(\overline{Q_T})$ if we assume that $u_0(x) \in C^2[-1,1]$ and the consistency condition $u_0'(\pm1) = 0$. As $\epsilon \to 0^+$, we can show that the solution $u^\epsilon(x,t) \to u(x,t)$, which satisfies

(5.7)
$$\begin{cases} u_t = u_{xx} & \text{if } -1 < x < 0, t > 0, \\ u_t = 2u_{xx} & \text{if } 0 < x < 1, t > 0, \\ u(x,0) = u_0(x) & \text{in } -1 < x < 1, \\ u_x(-1,t) = u_x(1,t) = 0 & \text{as } t > 0, \end{cases}$$

and on the interface $x = 0$, one has

(5.8)
$$\begin{cases} u^- = u^+, \\ u_x^- = 2u_x^+. \end{cases}$$

To summarize, we state the result without the proof as

Theorem 5.1. *Let $u_0(x) \in C^2[-1,1]$ satisfy consistency conditions $u_{0x}(\pm1) = 0$ and let the function $D(\phi^\epsilon)$ be given by (5.4). Then the problem (5.6) possesses a unique solution $u^\epsilon(x,t)$ for any $\epsilon > 0$. Moreover, the solution $u^\epsilon(x,t)$ converges to a function $u(x,t)$ as $\epsilon \to 0$, and the limit function $u(x,t)$ is a unique solution to the problem (5.7) with interface conditions (5.8).*

Hence, we have proven that the diffused interface problem (5.1), (5.4) has solutions which are governed to leading order by the sharp interface problem (5.7) and (5.8).

74

In physical terms, the problem in which two materials (with different diffusion coefficients) are joined together with a sharp boundary has a temperature distribution which is close to the problem in which the two materials are fused together over a distance ϵ.

(1b). Again, we assume that $D(\phi^\epsilon)$ is given by (5.4) and consider the problem

(5.9)
$$\begin{cases} c_t = \dfrac{\partial}{\partial x}\Big\{D(\phi^\epsilon)\Big(c_x + Ac(1-c)\phi_x^\epsilon\Big)\Big\} & \text{in} \quad -1 < x < 1, t > 0, \\ c_x(-1,t) = c_x(1,t) = 0, \\ c(x,0) = c_0(x), \end{cases}$$

where $c_0(x)$ is a given function with $0 < c_0(x) < 1$ on $-1 \le x \le 1$, and A is a constant.

Problem (5.9) describes the simplest possible concentration problem with a stationary interface. When $A = 0$, we are again back to problem (5.6). When A is nonzero, the factor ϕ_x^ϵ induces (for small value of the parameter ϵ) a transition layer in concentration c, so that the constant A characterizes the amptitude of the jump in concentration.

For any $\epsilon > 0$, the problem (5.9) is a standard parabolic boundary value problem so that there exists a unique solution $c^\epsilon(x,t)$ that is smooth in Q_T. As in the problem for (5.6), we are interested in the behavior of $c^\epsilon(x,t)$ as $\epsilon \to 0^+$.

First, by use of a comparison principle and the condition $0 < c_0(x) < 1$, we can prove

(5.10)
$$0 < c^\epsilon(x,t) < 1.$$

Noting that $c^\epsilon(x,t)$ satisfies

(5.11)
$$c_t^\epsilon = \frac{\partial}{\partial x}\Big\{D(\phi^\epsilon)c^\epsilon(1-c^\epsilon)\frac{\partial}{\partial x}\Big(\ln\frac{c^\epsilon}{1-c^\epsilon} + A\phi^\epsilon\Big)\Big\}.$$

we let

(5.12)
$$f(\phi, c) = A\phi + \ln\frac{c}{1-c}, \quad \text{and}$$

(5.13)
$$F(\phi, c) = \int_0^c f(\phi, y)dy = A\phi c + c\ln c + (1-c)\ln(1-c).$$

Then we have

$$\int_0^T\int_{-1}^1 f(\phi^\epsilon, c^\epsilon)\frac{\partial c^\epsilon}{\partial t}dxdt = \int_0^T\int_{-1}^1 \frac{\partial}{\partial t}F(\phi^\epsilon, c^\epsilon)dxdt = \int_{-1}^1 F(\phi^\epsilon, c^\epsilon)\Big|_{t=0}^{t=T}dx.$$

Thus the quantity $\int_0^T \int_{-1}^1 f(\phi^\epsilon, c^\epsilon) c_t^\epsilon dx dt$ is bounded and the bound is independent of ϵ. Since one has

$$\int_0^T \int_{-1}^1 \frac{\partial}{\partial x} \left\{ D(\phi^\epsilon) c^\epsilon (1 - c^\epsilon) \frac{\partial}{\partial x} f(\phi^\epsilon, c^\epsilon) \right\} f(\phi^\epsilon, c^\epsilon) dx dt$$

$$= \int_0^T D(\phi^\epsilon) c^\epsilon (1 - c^\epsilon) f(\phi^\epsilon, c^\epsilon) \frac{\partial}{\partial x} f(\phi^\epsilon, c^\epsilon) \Big|_{x=-1}^{x=1} dt$$

$$- \int_0^T \int_{-1}^1 D(\phi^\epsilon) c^\epsilon (1 - c^\epsilon) \left(\frac{\partial}{\partial x} f(\phi^\epsilon, c^\epsilon) \right)^2 dx dt,$$

and the first term of the right side of the above equality is also bounded by a constant independent of ϵ. Then we obtain the estimate

$$(5.14) \qquad \int_0^T \int_{-1}^1 D(\phi^\epsilon) c^\epsilon (1 - c^\epsilon) \left(\frac{\partial}{\partial x} f(\phi^\epsilon, c^\epsilon) \right)^2 dx dt \leq C,$$

where the constant C is independent of ϵ.

Multiplying a test function $\zeta \in C^\infty(\overline{Q_T})$ with $\zeta(x, T) = 0$ on both sides of equation (5.11), and integrating by parts, we obtain

$$\int_0^T \int_{-1}^1 \left[-c^\epsilon \zeta_t + D(\phi^\epsilon) c^\epsilon (1 - c^\epsilon) \left(\frac{\partial}{\partial x} f(\phi^\epsilon, c^\epsilon) \right) \zeta_x \right] dx dt$$

$$+ \int_0^T D(\phi^\epsilon) c^\epsilon (1 - c^\epsilon) \left(\frac{\partial}{\partial x} f(\phi^\epsilon, c^\epsilon) \right) \zeta \Big|_{x=-1}^{x=1} dt = \int_{-1}^1 c_0(x) \zeta(x, 0) dx,$$

and by the estimate (5.14), we have that the quantity

$$\int_0^T \int_{-\epsilon|\ln \epsilon|}^{\epsilon|\ln \epsilon|} D(\phi^\epsilon) c^\epsilon (1 - c^\epsilon) \left(\frac{\partial}{\partial x} f(\phi^\epsilon, c^\epsilon) \right) \zeta_x dx dt$$

approaches to zero as $\epsilon \to 0^+$.

By taking the limit $\epsilon \to 0^+$, we see that there exists a limit function $c(x, t)$ such that, at least for a subsequence,

$$c^\epsilon(x, t) \to c(x, t) \quad \text{in} \quad L^\infty \quad \text{weak} *,$$

and in the region $|x| \geq 2\epsilon|\ln \epsilon|$, one has

$$c^\epsilon(x, t) \to c(x, t) \quad \text{weakly in} \quad H^1.$$

Thus, we have
(5.15)

$$\int_0^T \int_{-1}^1 -c \zeta_t dx dt + \int_0^T \int_{-1}^{0^-} c_x \zeta_x dx dt + \int_0^T \int_{0^+}^1 2 c_x \zeta_x dx dt = \int_0^1 c_0(x) \zeta(x, 0) dx,$$

for any $\zeta \in C^\infty(\overline{Q_T})$ with $\zeta(\cdot, T) = 0$.

Again, by a test function argument, we see that the limit function $c(x, t)$ satisfies

(5.16)
$$\begin{cases} c_t = c_{xx}, & -1 < x < 0, t > 0, \\ c_t = 2c_{xx}, & 0 < x < 1, t > 0, \\ c(x, 0) = c_0(x), & -1 < x < 1, \\ c_x(-1, t) = c_x(1, t) = 0, & t > 0. \end{cases}$$

To specify the interface condition on $x = 0$, we notice that, (5.15) implies

(5.17)
$$\int_0^T \int_{-1}^{0-} \left[-(\zeta c)_t + (\zeta c_x)_x \right] dx dt + \int_0^T \int_{0+}^1 \left[-(\zeta c)_t + (\zeta c_x)_x \right] dx dt$$
$$= \int_0^1 c_0(x) \zeta(x, 0) dx.$$

Taking $\zeta \in C^\infty(\overline{Q_T})$ with $\zeta(x, 0) = \zeta(x, T) = \zeta(\pm 1, t) = 0$ in (5.17), we obtain

$$\int_0^T \zeta(0, t) \left(c_x^- - 2c_x^+ \right) dt = 0.$$

This implies that

(5.18)
$$c_x^- = 2c_x^+,$$

on the interface $x = 0$. Another interface condition (which is the same for the dynamics as for equlibrium) follows from the estimate (5.14) and $c(x, t) > 0$, i.e.,

$$\ln \frac{c^-}{1 - c^-} - A = \ln \frac{c^+}{1 - c^+} + A,$$

or equivalently

(5.19)
$$\frac{c^-(1 - c^+)}{c^+(1 - c^-)} = e^{2A}.$$

To summarize, we have

Theorem 5.2. *Let $c_0(x) \in C^1[-1, 1]$ satisfy $0 < c_0(x) < 1$ and consistency conditions $c_{0x}(\pm 1) = 0$ and the function $D(\phi^\epsilon)$ is given by (5.4). Then the problem (5.9) possesses a unique solution $c^\epsilon(x, t)$ for any $\epsilon > 0$. Moreover, the solution $c^\epsilon(x, t)$ converges to a function $c(x, t)$ as $\epsilon \to 0$, and the limit function $c(x, t)$ is a unique solution to the problem (5.16) with interface conditions (5.18) and (5.19).*

Thus, we have proven that solutions of the diffused interface problem (5.9) converge to the sharp interface problem (5.16), (5.18) and (5.19). Physically, this is the alloy problem in which the interface remains fixed due to thermal conditions.

Our result shows once again the proximity between the phase field and the sharp interface models. However, for complex geometries, the phase field approach, (5.9), will remain valid when the sharp interface model does not apply.

Remark 5.1. We can take $\phi^\epsilon(x,t) = \tanh\{(x - vt)/\epsilon\}$ instead of $\phi^\epsilon(x)$ as given by (5.3). Then the same arguments can be carried out as above. In this case the interface condition (5.18) on $x = vt$ for concentration equation is replaced by

$$(5.20) \qquad c_x^- - 2c_x^+ = -v(c^- - c^+),$$

whereas the interface conditions on the interface $x = vt$ for u equation remains unchanged.

(2a). We now turn to consider the case when $D(\phi^\epsilon)$ is given by (5.5) and investigate the behavior of the solution $u^\epsilon(x,t)$ of the problem (5.6) as small parameter $\epsilon \to 0^+$. Note that in this case we have

$$D(\phi^\epsilon) \sim \epsilon^2 + O(\epsilon^4) \quad \text{as} \quad x < -\epsilon|\ln \epsilon|.$$

Unlike the situation discussed in (1a), we have degeneracy in one phase here, so the type of equation will change in the limit as $\epsilon \to 0$.

By the same method we employed above, we can show that the limit function $u(x,t)$ satisfies

$$(5.21) \qquad \begin{cases} u_t = 0 & \text{if} \quad -1 < x < 0, \\ u_t = u_{xx} & \text{if} \quad 0 < x < 1, \\ u(x,0) = u_0(x) & \text{in} \quad -1 < x < 1, \\ u_x(0^+,t) = u_x(1,t) = 0 & \text{as} \quad t > 0. \end{cases}$$

It may be worth noting here that unlike case (1a), the solution $u(x,t)$ of (5.21) may not be necessarily continuous across the interface $x = 0$. This is due to the degeneracy of the original equation.

Remark 5.2. If we take $\phi^\epsilon(x,t) = \tanh\{(x - vt)/\epsilon\}$, then in the case $v < 0$, the problem (5.21) is satisfied by the limit function $u(x,t)$ in the domains $\{x < vt\}$ and $\{x > vt\}$ respectively, and $u_x^+ = 0$ on the interface $x = vt$. However, if $v > 0$, we have non-uniqueness in the region $0 < x < vt$, $t > 0$ for the limit function $u(x,t)$.

(2b). The problem (5.9) can be treated in a fashion similar to problem (5.6) when $D(\phi^\epsilon)$ is given by (5.5). For brevity we will not duplicate this arguement here. Instead, we remark here that the limit function $c(x,t)$ satisfies the same equations as $u(x,t)$ does, regardless of the constant A vanishing or not in the original equation. Again, this fact is a consequence of the degeneracy.

6. A Steady State Problem with Constant Temperature

Having obtained formal asymptotic results in Section 3 and rigorous results for the transition layer of a single equation in Section 5, we now discuss a coupled system (ϕ, c) in a one-dimensional steady state at constant temperature. We thereby resolve the basic rigorous issues involving planar equilibrium and prove the proximity of ϕ to a tanh function as well as the interface conditions which define equilibrium.

The time-independent, one-dimensional version of (3.10) subject to constant temperature has the form:

$$(6.1) \qquad \epsilon^2 \phi'' + 2\phi(1 - \phi^2) + 2\epsilon(a + bc) = 0 \quad \text{in} \quad (-1, 1)$$

$$(6.2) \qquad c'' + A\Big(c(1 - c)\phi'\Big)' = 0 \quad \text{in} \quad (-1, 1).$$

By integrating equation (6.2), we find that c satisfies

$$(6.3) \qquad c' + Ac(1 - c)\phi' = B,$$

where B is a constant. If we limit our attention to solutions ϕ and c which satisfy $c'(-1) = 0$ and $\phi'(-1) = 0$ or $c'(-1) = 0$ and $c'(-1) + Ac(-1)(1 - c(-1))\phi'(-1) = 0$, then the constant B vanishes in (6.3) and so the equation for c can be reduced to

$$c' + Ac(1 - c)\phi' = 0 \quad \text{in} \quad (-1, 1).$$

In general, if we consider the boundary conditions $c(-1) = \sigma_-$ and $c'(-1) = 0$, then the constant B is proportional to $\phi'(-1)$ and modification of arguments is needed in this case to obtain a similar result.

The precise problem we study in this section can then be stated as follows:

$$(6.4) \qquad \begin{cases} \epsilon^2 \phi'' + 2\phi(1 - \phi^2) + 2\epsilon(a + bc) = 0 \quad \text{in} \quad (-1, 1), \\ c' + Ac(1 - c)\phi' = 0 \quad \text{in} \quad (-1, 1), \\ \phi(\pm 1) = \tanh \dfrac{\pm 1}{\epsilon}, \\ c(-1) = \sigma_-, \end{cases}$$

where a, b, A and σ_-, the last one positive, are constants, and $\epsilon > 0$ is a small parameter. We will see later in this section that the limit function ϕ satisfies $\phi'(-1) = 0$.

The positivity of ϵ implies the system (6.4) is solvable. In fact, the second equation of (6.4) can be sloved by finding c in terms of ϕ

$$(6.5) \qquad c = \frac{e^{\bar{A} - A\phi}}{1 + e^{\bar{A} - A\phi}},$$

where

$$(6.6) \qquad \bar{A} = \ln \frac{\sigma_-}{1 - \sigma_-} - A\phi(-1).$$

Substituting c into the first equation of (6.4), we obtain a second order equation for ϕ, with boundary values at $x = \pm 1$, which is solvable. We would like to investigate the behavior of a solution $(\phi^\epsilon, c^\epsilon)$ as a small parameter ϵ approaches zero. We are particularly interested in the nature of the interface condition as $\epsilon \to 0$.

6.1. Properties of a solution to the system (6.4). We shall first show the properties of a solution $(c^\epsilon, \phi^\epsilon)$ of the system (6.4). Using the bound $0 < c^\epsilon < 1$, we have

$$(6.7) \qquad -1 - p\epsilon < \phi^\epsilon(x) < 1 + p\epsilon,$$

where $p = 2(|a| + |b|)$. The result is summarized as follows.

Lemma 6.1. *For any $\epsilon > 0$, let ϕ and c be a solution of (6.4) with boundary conditions $\phi(\pm 1) = \pm 1 \pm p\epsilon$. Then $\phi(x)$ is monotonically increasing on $(-1, 1)$ and $c(x)$ is monotonically increasing if $A < 0$ and is monotonically decreasing if $A > 0$.*

Proof. Note from (6.5) that $c(\phi) = \dfrac{e^{\bar{A} - A\phi}}{1 + e^{\bar{A} - A\phi}}$, so that the first equation of (6.4) becomes

$$(6.8) \qquad \epsilon^2 \phi'' + 2\phi(1 - \phi^2) + 2\epsilon\left(a + bc(\phi)\right) = 0,$$

and $\phi(x)$ satisfies boundary conditions $\phi(\pm 1) = \pm 1 \pm p\epsilon$.
Let

$$\phi^r(x) = \phi(x + 1 - r), \quad -1 < x < r < 1, \quad \text{and} \quad \Phi(x) = \phi^r(x) - \phi(x).$$

Then $\Phi(x)$ satisfies

$$(6.9) \quad \epsilon^2 \Phi''(x) + \left[\int_0^1 \left(g'(\tau\phi^r + (1 - \tau)\phi) + \epsilon bc'(\tau\phi^r + (1 - \tau)\phi) \right) d\tau \right] \Phi(x) = 0,$$

in the interval $(-1, r)$ with boundary conditions

$$\Phi(-1) \geq 0 \quad \text{and} \quad \Phi(r) \geq 0,$$

where $-1 < r < 1$ and $g(\phi) = 2\phi(1 - \phi^2)$. Note that the coefficient of Φ in equation (6.9) is bounded in the L^∞-norm and the bound is independent of r. We then have $\Phi(x) > 0$ as $1 + r > 0$ is small due to the maximum principle (In fact, we have that the coefficient of Φ in equation (6.9) is negative as $1 + r$ and ϵ are small. So that the (classical) maximum principle implies $\Phi > 0$ in this case). Now increase r and note that we always have $\Phi(r) \geq 0$ for $-1 < r < 1$, and so by the maximum principle we

80

must continue to have $\Phi(x) > 0$ on $x \in (-1, r)$ for every $r < 1$. This implies that the solution $\phi(x)$ is monotonically increasing. By differentiating equation (6.8) and again by the maximum principle, we find that $\phi'(x) > 0$ in $(-1, 1)$. The monotonicities of the function $c(x)$ follws from (6.5) and $\phi'(x) > 0$.

If the boundary conditions of ϕ are replaced by $\phi(\pm 1) = \tanh \frac{\pm 1}{\epsilon}$, then we do not have the result of the Lemma 6.1. But we can show that any point x where ϕ is not monotonically increasing can occur only when $|\phi(x)| > \tanh \frac{1}{\epsilon}$.

6.2. Comparison with tanh function. For any small $\epsilon > 0$, the property of $\phi(x)$ implies that there exists a unique point $x(\epsilon) \in (-1, 1)$ such that $\phi(x(\epsilon)) = 0$. Now we define a function $\hat{\phi}$ and stretched variable z by

$$
(6.10) \qquad \hat{\phi}(z) \equiv \phi(x), \quad \hat{c}(z) \equiv c(x), \quad z \equiv \frac{x - x(\epsilon)}{\epsilon},
$$

and let

$$
z_- = \frac{-1 - x(\epsilon)}{\epsilon}, \quad z_+ = \frac{1 - x(\epsilon)}{\epsilon}.
$$

Then equation (6.8) may be written as

$$
(6.11) \qquad \hat{\phi}'' + 2(\hat{\phi} - \hat{\phi}^3) + 2\epsilon \left(a + bc(\hat{\phi}) \right) = 0,
$$

with boundary conditions

$$
(6.12) \qquad \hat{\phi}(z_\pm) = \tanh \frac{\pm 1}{\epsilon}.
$$

The following result was proved in [CM].

Lemma 6.2. *Let $\phi(x)$ be a solution of (6.8) and $\hat{\phi}(z)$ be defined as (6.10). Then there exists $\alpha \in (0, 1)$ such that*

$$
(6.13) \qquad |\hat{\phi}(z) - \tanh(z)| \leq C\epsilon^\alpha,
$$

where the constant C is independent of ϵ.

This leads to the following lemma.

Lemma 6.3. *Let $\phi(x)$ be a solution of (6.8). Then there exists an $\alpha \in (0, 1)$ such that*

$$
(6.14) \qquad |\epsilon^2 \phi_x^2(x) - (1 - \phi^2)^2(x)| \leq C\epsilon^\alpha,
$$

where the constant C is independent of ϵ.

Proof. Multiplying (6.8) by ϕ, integrating over $(-1, 1)$ and integrating by parts, we obtain

$$
(6.15) \qquad \int_{-1}^{1} \left(\epsilon^2 \phi_x^2 - (1 - \phi^2)^2 \right) dx
$$

$$
= \epsilon^2 \phi \phi_x \Big|_{x=-1}^{x=1} + \int_{-1}^{1} (3\phi^2 - 1)(1 - \phi^2) dx + \int_{-1}^{1} 2\epsilon(a + bc)\phi \, dx.
$$

By using the estimate $|\phi_x(\pm 1)| \leq C\epsilon^{-1}$, and Lemma 6.2, we have the estimate

$$(6.16) \qquad \int_{-1}^{1} \left(\epsilon^2 \phi_x^2 - (1 - \phi^2)^2 \right) dx \leq C\epsilon^\alpha.$$

On the other hand, by integrating the identity

$$\left(\epsilon^2 \phi_x^2 - (1 - \phi^2)^2 \right)_x = -2\epsilon(a + bc)\phi_x$$

over the interval $(x, r) \subset (-1, 1)$, we obtain

$$(6.17) \qquad \epsilon^2 \phi_x^2(r) - (1 - \phi^2)^2(r) = \epsilon^2 \phi_x^2(x) - (1 - \phi^2)(x) - 2\epsilon \int_x^r (a + bc)\phi_x dx,$$

Integration with respect to x over $(-1, 1)$, and using the estimate

$$\left| \int_x^r (a + bc)\phi_x dx \right| \leq \left| 2a + \frac{b}{A} \ln \frac{1 - c(r)}{1 - c(x)} \right|,$$

one has

$$|\epsilon^2 \phi_x^2(r) - (1 - \phi^2)^2(r)| \leq C\epsilon^\alpha.$$

This proves the lemma.

In order to obtain an interface condition for the function c, we need improve the order of the estimate given in Lemma 6.3. This is given by the following lemma.

Lemma 6.4. Let $\phi(x)$ be a solution of (6.8) and $\hat{\phi}(z)$ be given by (6.10). Then

$$(6.18) \qquad \left| \hat{\phi}_z^2 - (1 - \hat{\phi}^2)^2 \right|_{z=-|\ln \epsilon|}^{z=|\ln \epsilon|} \leq C\epsilon^{1+\alpha},$$

where $\alpha \in (0, 1)$ and the constant C is independent of ϵ.

Proof. Let

$$(6.19) \qquad \psi(z) = \psi^0(z) + \epsilon \psi^1(z) = \left(\eta \psi_i^0 + (1 - \eta)\psi_o^0 \right) + \epsilon \left(\eta \psi_i^1 + (1 - \eta)\psi_o^1 \right),$$

where

$$(6.20) \qquad \psi_i^0(z) = \tanh z,$$

$$(6.21) \qquad \psi_o^0(z) = \begin{cases} 1, & \text{if } z > 0 \\ -1, & \text{if } z < 0. \end{cases}$$

Let $\psi_i^1(z)$ be a solution of the problem

(6.22)
$$\begin{cases} \psi_{izz}^1 + 2\left(1 - 3(\psi_i^0)^2\right)\psi_i^1 + 2(a + b\hat{c}) = 0 \quad \text{in} \quad (z_-, z_+) \\ \psi_i^1(0) = 0, \end{cases}$$

$$\psi_o^1(z) = \frac{1}{2}(a + b\hat{c}),$$

and $\eta(z) \in C^\infty(R^1)$ is a cut-off function such that

$$\eta(z) = \begin{cases} 1, & \text{if } \frac{z_-}{2} + 1 \le z \le \frac{z_+}{2} - 1; \\ 0, & \text{if } z \le \frac{z_-}{2} - 1 \text{ or } z \ge \frac{z_+}{2} + 1, \end{cases}$$

and it is monotonically increasing from 0 to 1 if $z < 0$ and monotonically decreasing from 1 to 0 if $z > 0$.

Notice that the solution $\psi_i^1(z)$ can be expressed by

(6.23) $$\psi_i^1(z) = -A(z) \int_0^z 2B(t)\left(a + b\hat{c}(t)\right) dt - B(z) \int_z^\infty 2A(t)\left(a + b\hat{c}(t)\right) dt,$$

where

(6.24)
$$\begin{cases} A(z) = 1 - \tanh^2 z \\ B(z) = -A(z) \int_0^z \frac{1}{A^2(t)} dt. \end{cases}$$

Thus we obtain

(6.25) $$|\psi_i^1(z)| \le 2\|a + b\hat{c}\|_\infty \left(A(z) \int_0^z |B(t)| dt + |B(z)| \int_z^\infty A(t) dt\right).$$

Similary, by differentiating (6.23) with respect to z, we obtain

(6.26) $$|\psi_{iz}^1(z)| \le 2\|a + b\hat{c}\|_\infty \left(|A'(z)| \int_0^z |B(t)| dt + |B'(z)| \int_z^\infty A(t) dt\right).$$

Therefore, we have the estimate

(6.27) $$\|\psi_i^1(z), \quad \psi_{iz}^1(z)\|_{L^\infty} \le C,$$

where the constant C is independent of ϵ.

To prove Lemma 6.4, we need only to show that

(6.28) $$\left[|\hat{\phi} - \psi^0|^2 + |\hat{\phi}_z - \psi_z^0|^2\right]\Big|_{z=-|\ln \epsilon|}^{z=|\ln \epsilon|} \le C\epsilon^{1+\alpha}.$$

Now consider the difference

(6.29)
$$\Phi(z) = \hat{\phi}(z) - \psi^0(z).$$

Using the identity

$$(\hat{\phi} - \hat{\phi}^3) - (\psi - \psi^3) = \left[1 - 3\psi^2 - 3\psi(\hat{\phi} - \psi) - (\hat{\phi} - \psi)^2\right](\hat{\phi} - \psi),$$

we see that $\Phi(z)$ satisfies

(6.30)
$$\begin{cases} -\Phi_{zz} + 2\Big(3(\psi^0)^2 - 1\Big)\Phi = -2\Big(\Phi^3 + 3\psi^0\Phi^2\Big) + J(z), \\ \Phi(0) = 0, \\ \Phi(z_\pm) = \tanh\dfrac{\pm 1}{\epsilon} - (\pm 1), \end{cases}$$

where

(6.31)
$$J(z) = 2\eta'(\psi_i^0 - \psi_o^0)_z + (\psi_i^0 - \psi_o^0)\Big\{\eta'' + 2\eta(1-\eta)\Big[(1+\eta)(\psi_i^0)^2$$
$$+ (\eta - 2)(\psi_o^0)^2 + (1 - 2\eta)\psi_i^0\psi_o^0\Big]\Big\}.$$

Multiplying Φ on the both sides of the first equation in (6.30) and integrating over (z_-, z_+), we obtain

(6.32)
$$\int_{z_-}^{z_+}\Big(-\Phi\Phi_{zz} + 2[3(\psi^0)^2 - 1]\Phi^2\Big)dz \le -6\int_{z_-}^{z_+}\psi^0\Phi^3 dz + \int_{z_-}^{z_+} J(z)\Phi dz.$$

One has the inequality

(6.33)
$$\int_{z_-}^{z_+}\Big(3(\psi^0)^2 - 1\Big)\Phi^2 dz$$
$$= \Big\{\int_{\{3(\psi^0)^2 - 1 > \delta\}} + \int_{\{3(\psi^0)^2 - 1 < \delta\}}\Big\}\Big[3(\psi^0)^2 - 1\Big]\Phi^2 dz$$
$$\ge \delta\int_{z_-}^{z_+}\Phi^2 dz + (1 + \delta)\int_{|z|\le\sigma}\Phi^2 dz,$$

where $\sigma \le \tanh^{-1}\sqrt{\frac{1+\delta}{3}}$ and $\delta > 0$. Then there exist a (small) positive constant δ and a constant $\lambda > 0$, such that

$$\int_{z_-}^{z_+}\Phi_z^2 dz + 2\int_{z_-}^{z_+}\Big(3(\psi^0)^2 - 1\Big)\Phi^2 dz \ge \lambda\int_{z_-}^{z_+}\Phi_z^2 dz + 2\delta\int_{z_-}^{z_+}\Phi^2 dz,$$

84

and consequently

$$(6.34) \quad \lambda \int_{z_-}^{z_+} \Phi_z^2 dz + 2\delta \int_{z_-}^{z_+} \Phi^2 dz$$

$$\leq 6 \int_{z_-}^{z_+} |\psi^0 \Phi| \Phi^2 dz + \frac{\delta}{2} \int_{z_-}^{z_+} \Phi^2 dz + \frac{1}{2\delta} \int_{z_-}^{z_+} |J|^2 dz + \Phi\Phi_z \Big|_{z_-}^{z_+}.$$

Note also that

$$(6.35) \qquad \|\psi^0\Phi\|_{L^\infty(z_-,z_+)} \leq 2\|\Phi\|_{L^\infty(z_-,z_+)} \leq 2\|\hat\phi - \psi^0\|_{L^\infty(z_-,z_+)} \leq \frac{\delta}{12}$$

as $\epsilon > 0$ is suitable small. We can then obtain the bound

$$(6.36) \qquad \lambda \int_{z_-}^{z_+} \Phi_z^2 dz + \delta \int_{z_-}^{z_+} \Phi^2 dz \leq \frac{1}{2\delta} \int_{z_-}^{z_+} |J|^2 dz + \Phi\Phi_z \Big|_{z_-}^{z_+}.$$

The second term on the right hand side of the above inequality is exponentially small, while the integral $\int_{z_-}^{z_+} |J|^2 dz$ can be estimated by

$$\int_{z_-}^{z_+} |J|^2 dz \leq C\Big\{ \sup_{\eta \neq 0,1} \big[|\psi_i^0 - \psi_o^0|^2 + |\psi_{iz}^0 - \psi_{oz}^0|^2 \big] \Big\},$$

which is also exponentially small. Therefore, from (6.34), we obtain

$$(6.37) \qquad \lambda \int_{z_-}^{z_+} \Phi_z^2 dz + \delta \int_{z_-}^{z_+} \Phi^2 dz \leq C\epsilon^{1+\alpha}.$$

With the above estimate (6.37), we can obtain, by the same method to get (6.37), that

$$(6.38) \qquad \int_{z_-}^{z_+} \Phi_{zz}^2 dz \leq C\epsilon^{1+\alpha}.$$

Hence, the Sobolev imbedding theorem yields $|\Phi(z)|^2$, $|\Phi'(z)|^2 \leq C\epsilon^{1+\alpha}$, where the constant C is independent of ϵ. This proves (6.28), and so Lemma 6.4 follows.

6.3. The condition at crossover point. In light of the estimates established above, we can now consider the limit process as $\epsilon \to 0^+$.

In terms of variable z, we have the system

$$(6.39) \qquad \begin{cases} \dfrac{1}{\epsilon}\big[\hat\phi_{zz} + 2(\hat\phi - \hat\phi^3)\big] + 2(a + b\hat c) = 0, \\[2mm] \hat c_z + A\hat c(1 - \hat c)\hat\phi_z = 0. \end{cases}$$

Note that there exists a $x(0) \in (-1, 1)$ such that (at least for a subsequence ϵ_j) $x(\epsilon) \to x(0)$, and $z_\pm \to \pm\infty$; and we see, from the equation of $\hat{\phi}$, that

$$(6.40) \qquad \frac{1}{\epsilon} \int_{-|\ln \epsilon|}^{|\ln \epsilon|} \left[\hat{\phi}_{zz} \hat{\phi}_z + 2(\hat{\phi} - \hat{\phi}^3) \hat{\phi}_z \right] dz + 2 \int_{-|\ln \epsilon|}^{|\ln \epsilon|} (a + b\hat{c}) \hat{\phi}_z dz = 0.$$

The first integral of (6.40) is of

$$(6.41) \qquad \frac{1}{2\epsilon} \left(\hat{\phi}_z^2 - (1 - \hat{\phi}^2)^2 \right) \Big|_{z=-|\ln \epsilon|}^{z=|\ln \epsilon|} \leq C\epsilon^\alpha;$$

while the second term is

$$(6.42) \qquad 2 \int_{-|\ln \epsilon|}^{|\ln \epsilon|} (a + b\hat{c}) \hat{\phi}_z dz = 2a\hat{\phi} \Big|_{z=-|\ln \epsilon|}^{z=|\ln \epsilon|} + 2b \int_{-|\ln \epsilon|}^{|\ln \epsilon|} \hat{c} \hat{\phi}_z dz$$

$$= 2a\hat{\phi} \Big|_{z=-|\ln \epsilon|}^{z=|\ln \epsilon|} + \frac{2b}{A} \ln(1 - \hat{c}) \Big|_{z=-|\ln \epsilon|}^{z=|\ln \epsilon|}.$$

Letting $\epsilon \to 0^+$, we obtain

$$(6.43) \qquad 4a + \frac{2b}{A} \ln \frac{1 - c^+}{1 - c^-} = 0,$$

where $c^\pm = c^\pm(x(0))$.

As a function $\hat{\phi}^\epsilon(z)$ approaches $\tanh z$, we see that

$$(6.44) \qquad \phi^\epsilon(x) \to -1 + 2H(x - x(0)), \quad \text{as} \quad \epsilon \to 0^+$$

where the function $H(x)$ is the Heaviside function, and

$$(6.45) \qquad c^\epsilon(x) \to \begin{cases} \sigma_-, & \text{if } x < x(0), \\ \sigma_+, & \text{if } x > x(0), \end{cases}$$

where σ_- is given in (6.4) and $\sigma_+ = 1 - (1 - \sigma_-)e^{(-2aA)/b}$.

To summarize, we have

Theorem 6.5. *Let $0 < \sigma_- < 1$ and the couple $(\phi^\epsilon(x), c^\epsilon(x))$ be a solution of the ODE system (6.4) for any $\epsilon > 0$. Then there exists a $x(0) \in (-1, 1)$ such that $(\phi^\epsilon(x), c^\epsilon(x))$ converges to limit functions $(\phi(x), c(x))$ that is given by (6.44) and (6.45) as the parameters $\epsilon \to 0$.*

7. Conclusions

We have proven some of the basic theorems necessary for the mathematical foundations of the phase field alloy model. These include existence, uniqueness and monotonicity of φ and c. These equations also establish new interface conditions as part of a sharp interface model that is a distinguished limit (as interface thickness vanishes) of the phase field alloy equations. In this paper we have established this limit rigorously in some settings. Two important mathematical issues are new to the alloy problem. The first is the presence of two types of layers: one steady state and the other dynamic. The interaction between the formation of these two distinct types of transitions is one of the keys to understanding the role of impurities near an interface. The other new problem arises from the mathematical degeneracy due to the vanishing of the material diffusion constant. We have established the existence of solutions to the fundamental equations in the cases where the interface is fixed. The challenging problems that remain include the study of the full set of equations with an asymptotically vanishing material diffusion coefficient within a general geometric setting. As the interface moves the concentration of impurities that are frozen in will depend strongly upon the local curvature. Consequently a detailed study of this issue will address the question of the distribution of impurities left after solidification.

The introduction of this asymptotically vanishing coefficient also eads to questions involving well-posedness that are sensitive to the nature of the decay of the coefficient as one crosses the interface into the solid.

Both of these new issues that relate to transition layers are intrinsic to the physical problem and are not artifacts of the particular model.

REFERENCES

[C1] G. Caginalp, *Mathematical models of phase boundaries,* Material Instabilities in Continuum Problems and Related Mathematical Problems (J. Ball, ed.), Symposium 1985-1986, Heriot-Watt, Oxford, (1988), 35-52.

[C2] G. Caginalp, *Stefan and Hele-shaw type models as asymptotic limits of phase field equations,* Physics Rev. A, **39** (1989), 887-896.

[CC] G. Caginalp & X. Chen, *Phase field equations in the singular limit of sharp interface problems,* The IMA Volumes in Mathematics and its Applications, **43** (1991), 1-28.

[CF] G. Caginalp & P. Fife, *Dynamics of layered interface arising from phase boundaries,* SIAM J. Appl. Math., **48** (1988), 506-518.

[CJ] G. Caginalp & J. Jones, *A derivation and analysis of phase field models of thermal alloys,* Annals of Physics. **237** (1995) 66-107.

[CM] G. Caginalp & J.B. McLeod, *The interior transition layer for an ordinary differential equation arising from solidification theory,* Quart. Appl. Math., **44** (1986), 155-168.

[CN] G. Caginalp & Y. Nishiura, *The existence of travelling waves for phase field equations and convergence to sharp interface models in the singular limit,* Quart. Appl. Math., **49** (1991), 147-162.

[Co] A.H. Cottrell, *Theoretical Structural Metallurgy,* Camelot Press, London (1955).

[Cr] J. Crank, *Free and moving boundary problems,* Clarendon Press, Oxford (1984).

[CS] G. Caginalp & E. Socolovsky, *Phase field computations of single-needle crystals, crystal growth and motion by mean curvature,* SIAM J. Scientific and Statistical Computing **15** (1994) 106-126.

[CX] G. Caginalp & W. Xie, *Phase-field and sharp-interface alloy models*, Physical Review E, **48** (1993), 1897-1909.

[DHOX] J. N. Dewynne, S. D. Howison, J. R. Ockendon & W. Xie, *Asymptotic Behavior of Solutions to the Stefan Problem with a Kinetic Condition at the Free Boundary*, J. Austral. Math. Soc. ser B, **31** (1989), 81-96.

[ESS] L.C. Evans, M. Soner & P. Souganidis, *Phase transitions and generalized motion by mean curvature*, Comm. Pure Appl. Math., **45** (1992), 1097-1123.

[Fe] J. Fehribach, *Pertubation methods for solid diffusion in a Stefan Problem*, SIAM J. Math. Anal., **19** (1988), 86-99.

[Gl] J. Glimm, *The continuous structure of discontinuities*, in PDE's and continuum models of phase transition (eds. M. Rascle, D. Serre, M. Slemrod) Lecture Notes in Physics, **344** (1989), 177-186, Springer New York.

[HH] P.C. Hohenberg & P.C. Halperin, *Theory of Critical Phenomena*, Rev. Mod. Phys., **49** (1977), 435-480.

[Lu] C. Lupis, *Chemical Thermodynamics of Materials*, North-Holland, New York (1983).

[MS] W. Mullins & R. Sekerka, *Stability of a planar interface during solidfication of a binary alloy*, J. Appl. Phys., **35** (1964), 445-451.

[O] J. Ockendon, *Linear and nonlinear stability of a class of moving boundary problems*, Free Boundary Problems, Pavia 1979, 443-478.

[PXS] M. Primicerio, W. Xie & M. Shu, *On a Problem in Condensed Two-phase Combustion*, Math. Meth. in Appl. Sci., **12** (1990), 519-531.

[S1] R. Sekerka, *Melt growth* in Int. School of Crystallography 7th course: Interfacial aspects of phase transformation, Erice-Trapani, Siccly (Ed. B. Mutaftschiev), D. Reidel Publ (1982), 489-508.

[S2] R. Sekerka, *On the modeling of solid-liquid interface dynamics*, Proc. Darkin Conf. U.S. Steel Res. Lab Monroeville, PA (1976), 301-321.

[WBM1] A. Wheeler, W. Boettinger & G. McFadden, *A phase-field model for isothermal phase trasitions in binary alloys*, Phys. Rev. A, **45** (1992), 7424-7439.

[WBM2] , *A phase field model of solute trapping during solidification*, NIST Preprint.

[Xi1] W. Xie, *The Stefan Problem with a Kinetic Condition at the Free Boundary*, SIAM J. Math. Anal. **21** (1990), 362-373.

[Xi2] W. Xie, *Phase transition in binary systems and the binary alloy problem*, Math. Meth. in Appl. Sci., **15** (1992), 205-221.

Gunduz Caginalp
Department of Mathematics
University of Pittsburgh
Pittsburgh, PA 15260

Weiqing Xie
Department of Mathematics
California State Polytechnic University
Pomona, CA 91768

J W CAHN AND A NOVICK-COHEN

Limiting motion for an Allen–Cahn/Cahn–Hilliard system

Abstract. Long time asymptotics are developed here for an Allen-Cahn/Cahn-Hilliard system derived recently by Cahn & Novick-Cohen [5]. The limiting behavior couples $\mathcal{O}(1)$ motion by mean curvature of antiphase boundaries (APBs) with slow ($\mathcal{O}(\epsilon)$) motion by minus the surface Laplacian of the mean curvature of interphase boundaries (IPBs).

1 Introduction.

Let the Allen-Cahn/Cahn-Hilliard system [5]:

$$(1.1) \qquad u_t = 4\epsilon^2 \nabla \cdot Q(u, v) \nabla \frac{\delta F}{\delta u},$$

$$(1.2) \qquad v_t = -\frac{1}{4} Q(u, v) \frac{\delta F}{\delta v},$$

$$F = \int_\Omega \Big\{ \alpha u(1-u) - \beta v^2 + \frac{\theta}{2}\{(u+v)\ln(u+v) + (u-v)\ln(u-v)$$

$$(1.3) \quad +(1-(u+v))\ln(1-(u+v))+(1-(u-v))\ln(1-(u-v))\}+\frac{1}{2}\epsilon^2\{|\nabla u|^2+|\nabla v|^2\}\Big\}\, dx$$

be defined on a smooth bounded domain $\Omega \subset \mathbf{R}^2$ with Neumann boundary conditions for u where u is a concentration, no-flux boundary for the mass flux $\mathbf{j} = -Q\nabla\mu$ where $Q = Q(u, v)$ is the mobility and $\mu = \frac{\delta F}{\delta u}$ represents the chemical potential, and Neumann boundary conditions for v where v is a non-conserved order parameter. This system can be viewed as a simplest prototype for a system which can exhibit simultaneous ordering and phase separation. See also the discussion in [8],[7].

It is easy to see that the mean mass $\bar{u} = \frac{1}{|\Omega|} \int_\Omega u\, dx$ is conserved under the evolution of the system (1.1)-(1.3) and that the Allen-Cahn/Cahn-Hilliard system corresponds to gradient flow in $H^{-1}(\Omega) \times L^2(\Omega)$. Questions of existence, uniqueness, regularity, and the existence of inertial sets have been addressed for the Allen-Cahn/Cahn-Hilliard system with constant mobility and a quartic polynomial form for the free energy F in Brochet, Hilhorst & Novick-Cohen [1]. One should expect the solution (u, v) of the Allen-Cahn/Cahn-Hilliard system as given in (1.1)-(1.3) to be confined to the domain

$$0 < u + v < 1 \text{ and } 0 < u - v < 1.$$

89

Note that by taking u identically equal to $\frac{1}{2}$ or v identically equal to 0, it is easy to see that the Allen-Cahn/Cahn-Hilliard system encompasses the dynamics of both the Allen-Cahn and the Cahn-Hilliard equations as particular cases.

During late times one can expect the system to be dominated by a finite number of regions throughout which one of the minimizing phases obtainable by a tie-line construction prevail, and these regions can be expected to be separated by thin interfacial regions. Since for $\bar{u} \neq \frac{1}{2}$ and $0 < \alpha < \beta$ the system can contain three different types of minimizers, clearly the possibility of "triple-junctions" or small transitional domains connecting three different types of minimizers arises. In this note formal asymptotics appropriate for the description of possible late time behavior including triple-junction motion are given.

For the Allen-Cahn equation [12], on a time scale $\tau = t$, interfaces between phases move by motion by mean curvature. It has been shown by Cahn, Elliott & Novick-Cohen [3] for the Cahn-Hilliard equation with a mobility proportional to $u(1-u)$ and with a free energy of the form given by (1.3) but with $v \equiv 0$, that on the time scale $\tau = \epsilon^2 t$, formal asymptotics predict that the interface moves by minus the surface Laplacian of the mean curvature. We shall make the physically reasonable assumption that the mobility vanishes in all pure phases. The obvious difficulty which presents itself is that the limiting motion of interest occurs on different time scales in each of the two analogue problems. Thus, under the scaling assumptions employed here, the limiting behavior will indeed be geometric, but the various motions occur on different time scales and it is not obvious how to implement these results as a closed numerical scheme for geometric evolution. Roughly, though, the limiting motion corresponds to interfaces moving by motion by mean curvature coupled at triple-junctions to a pair of interfaces which move by minus the surface Laplacian of the mean curvature. At the triple-junctions, the limiting behavior is governed by a Young's law, a mass flux balance, and a law equating mean curvatures and the chemical potential. Interfaces meet the external boundary at 90°, and a no-flux condition through the external boundary condition arises. Along the interfaces which move by motion by mean curvature, the chemical potential obeys a null surface Laplacian condition. Such motion yields a reasonable description of sintering; in this sense the Allen-Cahn/Cahn-Hilliard system may be considered as a diffuse interface model for sintering. The asymptotics which we consider here also correspond possibly to the Krzanowski instability [10] wherein the disordered phase $(u, v) \approx (\frac{1}{2}, 0)$ constitutes a minor phase, and droplets of the minor phase coagulate along smooth slowly varying antiphase boundaries and detach from the antiphase boundaries at points of high curvature.

2   Preliminaries.

The free energy as given in (1.3) has two minimizers known as ordered phase variants located transcendentally close to the points $(u, v) = (\frac{1}{2}, \pm\frac{1}{2})$ and given implicitly by

(2.1)
$$u = \frac{1}{2}, \quad 2\beta v = \theta\Big\{\ln\Big(\frac{1}{2} + v\Big) + \ln\Big(\frac{1}{2} - v\Big)\Big\},$$

as well as two minima located transcendentally close to the points $(u, v) = (0, 0)$ and $(u, v) = (1, 0)$ known as disordered phases which satisfy

(2.2)
$$-\alpha(1 - 2u) = \theta\{\ln(u) - \ln(1 - u)\}, \quad v = 0.$$

Under the assumption that $0 < \alpha < \beta$, the free energy of the pair of minima given in (2.1) is lower than that of the pair of minima given in (2.2). A tie line construction yields three minima of equal depth consisting of the first pair of minima and one of the two minima from the second set of minima. Without loss of generality taking $\frac{1}{2} < \bar{u} < 1$, these minima correspond to the minima which are located transcendentally close to the points

$$\Big(\frac{1}{2}, -\frac{1}{2}\Big), \ \Big(\frac{1}{2}, +\frac{1}{2}\Big), \ (1, 0).$$

In analogy with taking the mobility to be of the form $M(u) = u(1-u)$ as it appeared in the original derivation of Cahn [4] and Hilliard [9], it is reasonable within the context of the Allen-Cahn/Cahn-Hilliard system to employ a mobility which vanishes in the "pure" phases. For simplicity, let us assume that

$$Q(u, v) = \tilde{Q}\,\Pi_{i=1}^{4}\Big((u + v - w_1^i)^2 + (u - v - w_2^i)^2\Big)^{1/2},$$

where (w_1^i, w_2^i) permute through the values 0 and 1.

Assuming the behavior of interest to occur on a $\tau = \epsilon^3 t$ scale, we write the Allen-Cahn/Cahn-Hilliard system as

(2.3)
$$\epsilon^2 u_t = 4\epsilon\, \nabla \cdot Q(u, v)\nabla\mu,$$

(2.4)
$$\epsilon^3 v_t = -\frac{1}{4}Q(u, v)[F_v(u, v) - \epsilon^2 \Delta v],$$

(2.5)
$$\mu = F_u(u, v) - \epsilon^2 \Delta v.$$

To obtain an intuitive understanding of the use of the $\tau = \epsilon^3 t$ time scale, note that in the analysis of [3], the factor ϵ^2 did not appear in front of the mobility. Additionally, θ is taken to be proportional to ϵ. In order to model the Krzanowski instability, the mean concentration is assumed to be near $\bar{u}$, and hence the disordered phase is a minor phase. For simplicity, let us imagine that the system contains precisely three sub-domains: two large sub-domains corresponding to the order phase variants and a smaller sub-domain corresponding to the disordered phase. The interfacial partitions separating the domains are assumed to have an $\mathcal{O}(\epsilon)$ width. Interfaces between an ordered variant and a disordered phase are known as interphase boundaries (IPB's) and interfaces between two different ordered variants are known as antiphase boundaries (APB's). At an $\mathcal{O}(1)$ distance from the interfaces, u and v are assumed to be transcendentally close to the minimizing values, the flux is assumed to be transcendentally small, and the chemical potential is taken to assume the form of an ϵ-expansion. Since the sub-domains containing the ordered phase variants must be large, it is reasonable to assume that the curvature of the antiphase boundaries is small. More specifically, we assume that

$$(2.6) \qquad \kappa_{interphase} = \kappa_0 + \epsilon \kappa_1 + \mathcal{O}(\epsilon^2).$$

The curvature of the antiphase boundaries is taken to satisfy

$$(2.7) \qquad \kappa_{antiphase} = \epsilon \kappa_1 + \epsilon^2 \kappa_2 + \mathcal{O}(\epsilon^3).$$

We assume the three phase boundaries to meet at a unique triple junction located at some point within the interior of the domain Ω.

3 Outer Solution.

In accordance with the discussion in §2, in the outer region we set

$$(3.1) \qquad u(x, t) = u_i^0 + (\epsilon u^1 + \mathcal{O}(\epsilon^2))e^{-c/\epsilon},$$

$$(3.2) \qquad v(x, t) = v_i^0 + (\epsilon v^1 + \mathcal{O}(\epsilon^2))e^{-c/\epsilon},$$

$$(3.3) \qquad \mu(x, t) = \epsilon \mu^1 + \epsilon^2 \mu^2 + \mathcal{O}(\epsilon^3),$$

$$(3.4) \qquad \mathbf{j} = (\mathbf{j}^0 + \epsilon \mathbf{j}^1 + \mathcal{O}(\epsilon^2))e^{-c/\epsilon}.$$

where the pairs $\{u_i^0, v_i^0\}$ denote the set of (three) minimizers of the constrained free energy and where c is the same coefficient as appears in the expressions for u_i^0 and v_i^0.

Let us assume that $\tilde{\mu} \equiv F_v(u, v) - \epsilon^2 \Delta v$ as well as μ have expansions of the form (3.3). From (3.1) and (3.2) and noting that $Q(u_i^0, v_i^0) = \mathcal{O}(e^{-c/\epsilon})$, it is easy to conclude that the expansions are self-consistent within the framework of equations (2.3)-(2.4). Similarly, the assumptions (3.1)-(3.4) are readily seen to be self-consistent within the context of the definition of j. To check that the assumptions on the expansions are consistent with the definitions of μ and $\tilde{\mu}$, note first that u^0 and v^0 solve the equations

$$F_u(u^0, v^0) = 0 \quad F_v(u^0, v^0) = 0.$$

Linearizing, u^1 and v^1 are given as the solutions to

$$\mu^1 = F_{uu}(u^0, v^0)u^1 + F_{uv}(u^0, v^0)v^1,$$

$$0 = F_{uv}(u^0, v^0)u^1 + F_{vv}(u^0, v^0)v^1.$$

Solving for u^1 and v^1 and noting the form of u^0 and v^0 and the structure of F, it is easy to check that the assumed expansions are self-consistent within this framework of these equations. Higher order self-consistency may readily be demonstrated by induction.

92

4 The inner solution.

4.1 The inner solution for antiphase boundaries.

It is possible to ascertain to leading order the limiting behavior of antiphase boundaries by recalling assumption (2.6) and developing the relevant asymptotic expansions for an "inner solution" within the APB through orders $\mathcal{O}(\epsilon^{-1})$-$\mathcal{O}(\epsilon^2)$. Proceeding in this manner, the concentration and non-conserved order parameter are expanded as

$$U = U^0 + \epsilon U^1 + \epsilon^2 U^2 + \mathcal{O}(\epsilon^3) \qquad V = V^0 + \epsilon V^1 + \epsilon^2 V^2 + \mathcal{O}(\epsilon^3)$$

where U^i and V^i are functions of an inner variable $\rho = r/\epsilon$ which is orthogonal to the interface as well as an arc-length variable s which is normal to ρ. Regular expansion in ϵ are also assumed for the chemical potential and the normal velocity; i.e.,

$$\mu = \epsilon \mu^1 + \epsilon^2 \mu^2 + \mathcal{O}(\epsilon^3) \quad W = W^0 + \epsilon W^1 + \mathcal{O}(\epsilon^2).$$

U^0 and U^1 are found to be symmetric functions of $\rho = r/\epsilon$, and V^0 and V^1 are found to be anti-symmetric functions of the same variable. Moreover, μ^1 satisfies

(4.1)
$$\frac{\partial^2 \mu^1}{\partial s^2} = 0,$$

and the normal velocity of the interface is given to lowest order by

(4.2)
$$W^0 \left[\int_{-\infty}^{\infty} \frac{U^0(\rho) - U^0(-\infty)}{Q(U^0, V^0)}\, d\rho + \int_{-\infty}^{\infty} \frac{(V_\rho^0)^2}{Q(U^0, V^0)}\, d\rho \right] = \frac{\kappa_1}{4} \int_{-\infty}^{\infty} [(U_\rho^0)^2 + (V_\rho^0)^2]\, d\rho.$$

4.2 The inner solution for interphase boundaries.

In order to analyze the motion of interphase boundaries, we set

$$\kappa = \kappa_0 + \epsilon \kappa_1 + \mathcal{O}(\epsilon^2),$$

but otherwise the assumed expansions for the concentration, non-conserved order parameter and chemical potential are taken as in the antiphase analysis. Proceeding through orders $\mathcal{O}(\epsilon^{-1})$-$\mathcal{O}(\epsilon^2)$ yields

(4.3)
$$\mu^1 = \frac{\kappa_0}{[U^0]_-^+} \int_{-\infty}^{\infty} \{(U_\rho^0)^2 + (V_\rho^0)^2\}\, d\rho,$$

and

(4.4)
$$W^0 = 0 \qquad W^1 = -\frac{4\frac{\partial^2 \kappa_0}{\partial s^2}}{([U^0]_-^+)^2} \int_{-\infty}^{\infty} Q(U^0, V^0)\, d\rho \cdot \int_{-\infty}^{\infty} \{(U_\rho^0)^2 + (V_\rho^0)^2\}\, d\rho.$$

5 Triple-junction conditions.

5.1 Young's law.

Following the analysis presented in [2], the stretched variables

$$\eta_i = \frac{\mathbf{x} - \mathbf{m}(t)}{\epsilon}$$

are introduced, where $\mathbf{m}(t)$ denotes the location of the triple-junction and the index i refers to one of the three interfaces Γ_i which meet at the triple-junction. Since the $\mathcal{O}(\epsilon)$ motion of the IPBs is coupled with the $\mathcal{O}(1)$ motion of the APBs at the triple-junction, it is reasonable to assume that $m'(t) = \mathcal{O}(\epsilon)$. For convenience, let Γ_1 denote the antiphase boundary and let Γ_2 and Γ_3 denote the interphase boundaries which meet Γ_1 at the triple-junction. An isosceles triangle T_ϵ with base length proportional to ϵ^β for $0 < \beta < 1$ is constructed at the triple-junction, and its base is taken to be bisected orthogonally by one of the three interfaces. Following the discussion in [2], it is possible to conclude by multiplying equation (2.4)-(2.5) by a suitable factor and integrating over the triangle T_ϵ that

(5.1)
$$\frac{E_1}{\sin \theta_1} = \frac{E_2}{\sin \theta_2} = \frac{E_3}{\sin \theta_3},$$

where

$$E_i = \int_{-\infty}^{\infty} F(U_i^0(\xi), V_i^0(\xi)) \, d\xi$$

denotes the interfacial energy per unit length of Γ_i and θ_i denotes the angle at the triple-junction opposite the ith-interface.

5.2 The balance of fluxes.

Employing the construction presented in §5.1 but taking the base of the isosceles triangle to be orthogonal to the APB, a balance of fluxes condition at the triple-junction is obtainable by integrating equation (2.3) over the triangle T_ϵ. To lowest order, this condition reads

(5.2)
$$0 = M_{APB}^0 \frac{\partial}{\partial s_1} \mu_1^1(0+, t) + M_{IPB}^0 \left[\frac{\partial}{\partial s_2} \mu_2^1(0+, t) + \frac{\partial}{\partial s_3} \mu_3^1(0+, t) \right],$$

where s_i and μ_i^1 refer to the arc-length parametrization and chemical potential of the ith interface, and

$$M_{APB}^0 = \int_{-\infty}^{\infty} Q(U_1^0(\xi), V_1^0(\xi)) \, d\xi$$

$$M_{IPB}^0 = \int_{-\infty}^{\infty} Q(U_2^0(\xi), V_2^0(\xi)) \, d\xi = \int_{-\infty}^{\infty} Q(U_3^0(\xi), V_3^0(\xi)) \, d\xi,$$

where (U_i^0, V_i^0) refers to the inner solution along the interface Γ_i.

5.3 Continuity of the chemical potential at triple-junctions.

Recall by equation (4.1) that for interphase boundaries

$$\mu^1[U^0(+\infty) - U^0(-\infty)] = \kappa_1 \int_{-\infty}^{\infty} \{(U_\rho^0)^2 + (V_\rho^0)^2\}\, d\rho.$$

If the chemical potentials are assumed to be continuous through the triple-junction, i.e.,

$$\mu_1^1(0+, t) = \mu_2^1(0+, t) = \mu_3^1(0+, t),$$

then noting the orientation of the interfaces Γ_2 and Γ_3 and noting that the interfacial energy per unit length along the interfaces Γ_2 and Γ_3 are identical by construction, it follows from (4.1)

(5.3)
$$\sum_{i=2}^{3} \kappa_1^i = 0$$

where κ_1^i denotes the mean curvatures of Γ_i.

6 Conditions at junctions with external boundary.

6.1 The Neumann type condition.

The analysis here follows closely the analysis undertaken in §5.1. See also the discussion in [11]. Let

$$\eta = \frac{\mathbf{x} - \mathbf{m}(t)}{\epsilon},$$

where $\mathbf{m}(t)$ is the point of intersection of an interface Γ (either an APB or an IBP) with the external boundary. Furthermore, let R_ϵ be a rectangle "contained" within Ω with sides proportional to ϵ^β for $0 < \beta < 1$ and with $m(t)$ located at the midpoint of one of its sides. Multiplying by an appropriate factor and integrating equations (2.4)-(2.5) over the rectangle R_ϵ, to lowest order we find that

(6.4)
$$\theta = \frac{1}{2}\pi.$$

6.2 The flux condition.

By reintroducing the R_ϵ-η construction introduced in §6.1, assuming the velocity of the point of contact of the antiphase boundary with the external boundary to be $\mathcal{O}(1)$ and the velocity of the point of contact of interphase boundaries with the external boundary to be $\mathcal{O}(\epsilon)$, and integrating equation (2.3) over the rectangle R_ϵ, it is possible to check that both APB and IPB interfaces satisfy

(6.5)
$$\frac{\partial}{\partial s}\mu^1(0+, t) = 0.$$

For IPB interfaces, this implies that

$$(6.6) \qquad \frac{\partial}{\partial s}\kappa_1(0+,\, t) = 0.$$

7 Conclusions.

In the present note, by looking at a $\tau = \epsilon^3 t$ scale, motion by mean curvature has been coupled with motion by minus the surface Laplacian of the mean curvature but on different time scales. It is intuitively clear that this behavior should lead to an increased differential in the curvatures of the antiphase and interphase boundaries. This observation leads us in a forthcoming paper [6] to examine the behavior of the Allen-Cahn/Cahn-Hilliard system on an even slower scale ($\tau = \epsilon^{7/2}t$) under the assumption that the curvature of APBs is $\mathcal{O}(\epsilon^{3/2})$ and the curvature of IPBs is $\mathcal{O}(\epsilon^{-1})$. Under these assumptions, the two motions: motion by mean curvature and motion by minus the Laplacian of the mean curvature occur on the same time scale. Such assumptions may be appropriate for modelling the Krzanowski instability when the disordered phase constitutes a very minor phase.

Acknowlegement: One of the authors (A. N.-C.) would like to thank Harald Garcke for helpful questions and comments.

References

[1] D. Brochet, D. Hilhorst, and A. Novick-Cohen, Inertial sets for Cahn-Hilliard/Cahn-Allen systems, Appl. Math. Lett. **7** (1994) 83-87.

[2] L. Bronsard and F. Reitich, On singular three-phase boundary motion and the singular limit of a vector-valued Ginzburg-Landau equation, ARMA **124** (1993) 355-379.

[3] J. W. Cahn, C. M. Elliott, and A. Novick-Cohen, The Cahn-Hilliard equation: motion by minus the Laplacian of the mean curvature, accepted: Eur. J. Appl. Math.

[4] J. W. Cahn, On spinodal decomposition, Acta Met. 9 (1961) 795-801.

[5] J. W. Cahn and A. Novick-Cohen, Evolution equations for phase separation and ordering in binary alloys, Journal of Stat. Phys., **76** (1994) 877-909.

[6] A. Novick-Cohen, Triple-junction motion for Allen-Cahn/Cahn-Hilliard systems, preprint.

[7] L.-Q. Chen, Nonequilibrium pattern formation involving both conserved and non-conserved order parameters and effect of long range interactions, Modern Physics Letters, **B 7** (1993) 1857-1881.

[8] T. Eguchi and H. Ninomiya, PC-Visualization of phase transitions in alloys ... kinetics of ordering with phase separation, in *Dynamics of Ordering Processes in Condensed Matter*, S. Komura and H. Furukawa, eds. (Plenum Press, New York, 1988).

[9] J. E. Hilliard, Spinodal decomposition, in Phase Transformations, ASM, Cleveland, 1970, Chapt 12, 497-560.

[10] J. E. Krzanowski and S. M. Allen, Solute-drag effects at migrating diffuse interfaces- I. Theoretical analysis and application to APBs in Fe-Al alloys, Acta Metall. **34** (1986) 1035-1044, and J.E. Krzanowski and S. M. Allen, Solute-drag effects at migrating diffuse interfaces- II. Experimental investigation of APB migration kinetics, Acta Metall. **34** (1986) 1045-1050.

[11] N. C. Owen, J. Rubinstein, P. Sternberg, Minimizers and gradient flows for singularly perturbed bi-stable potentials with a Dirichlet condition, Proc. Roy. Soc. London, Ser. A, **429** (1990) 505-532.

[12] J. Rubinstein, P. Sternberg, J. Keller, Fast reaction, slow diffusion, and curve shortening, SIAM Appl. Math. **49** (1989) 116-133.

A. Novick-Cohen

Department of Mathematics

Technion-IIT

Haifa, Israel 32000

&

J.W. Cahn

Materials Sci. and Eng. Lab.

NIST

Gaithersburg, MD 20899

U.S.A.

A KADOYA, N KATO, N KENMOCHI AND J SHIROHZU

Optimal control on the lateral boundary for a phase separation model with constraint

Abstract. An optimal control problem in which the control space Π is a set of portions Γ of the lateral boundary of a fixed bounded domain Ω in $\mathbb{R}^N (1 \leq N \leq 3)$. In our control problem, each state is described as a non-isothermal phase separation model with constraint which is a coupled system of two nonlinear parabolic PDEs, the energy and mass balance equations, defined on $Q := (0, T) \times \Omega$, and the control action is given through a boundary portion $\Gamma \in \Pi$. In this paper, we shall give an existence result for the optimal control problem and discuss the convergence of regular approximate control problems.

0. Introduction

This paper is concerned with an optimization problem on the lateral boundary $\partial\Omega$ for a non-isothermal diffusive phase separation model formulated in a domain Ω in $\mathbb{R}^N (N = 2 \text{ or } 3)$; for instance, in a binary alloy, our model is described by the (Kelvin) temperature $\theta := \theta(t, x)$ and the local concentration $w = w(t, x)$ of one of the components of the alloy occupying a fixed domain Ω.

Our state problem, referred as $SP(\Gamma)$, is of the form

$$\rho(u)_t + \lambda(w)_t - \Delta u = f \quad \text{in } Q := (0, T) \times \Omega, \tag{0.1}$$

$$w_t - \Delta\{-\mu\Delta w_t - \kappa\Delta w + \xi + g(w) - \lambda'(w)u\} = 0 \quad \text{in } Q, \tag{0.2}$$

$$\xi \in \beta(w) \quad \text{in } Q, \tag{0.3}$$

$$u = h_D \quad \text{on } \Sigma_D := (0, T) \times \Gamma, \tag{0.4}_D$$

$$\frac{\partial u}{\partial n} + n_0 u = h_N \quad \text{on } \Sigma_N := (0.T) \times \Gamma', \ \Gamma' := \partial\Omega \setminus \Gamma, \tag{0.4}_N$$

$$\frac{\partial w}{\partial n} = 0, \ \frac{\partial}{\partial n}\{-\mu\Delta w_t - \kappa\Delta w + \xi + g(w) - \lambda'(w)u\} = 0 \quad \text{on } \Sigma := (0, T) \times \partial\Omega, \tag{0.5}$$

$$u(0, \cdot) = u_0, w(0, \cdot) = w_0 \quad \text{in } \Omega, \tag{0.6}$$

where ρ is a continuous increasing function; β is a maximal monotone graph in $\mathbb{R} \times \mathbb{R}$; λ and g are smooth functions, and λ' is the derivative of λ; μ, κ and n_0 are positive constants; f, h_D, h_N, u_0 and w_0 are prescribed data.

General modeling for phase separation phenomena is referred, for instance, to [1, 7]; in [1] the existence of weak solutions was discussed, but the uniqueness question has been open for such a wide class of models. Recently, system $SP(\Gamma)$ was introduced in [4, 5] as a system which is a reasonable approximation to some physical cases and uniquely solved in the variational sense.

From some industrial motivation we are interested in the control of the behaviour of w, which is possibly done by taking not only f, h_D and h_N as driving parameters, but also the location of a boundary portion Γ' (or equivalently Γ) on which flux condition $(0.4)_N$ (or Dirichlet condition $(0.4)_D$) is imposed. In this paper we especially pay our attention to the efficiency of location of Γ' (or Γ) as a driving parameter for the control of w; therefore we consider a model optimization problem in which f, h_D and h_N are fixed and each state problem is given by $SP(\Gamma)$ for Γ in a class of boundary portions.

In optimization problems the choice of cost functional J is of course important, too; in particular, we are interested in the case when a term of the form $|w_\Gamma - w_d|^2_{C(\overline{Q})}$ is included in the expression of J, for example,

$$J(\Gamma) := A \int_Q |u_\Gamma - u_d|^2 dxdt + B|w_\Gamma - w_d|^2_{C(\overline{Q})} + C \int_{\Sigma(\Gamma')} |h_d|^2 d\sigma dt \qquad (0.7)$$

where A, B, C are positive constants, u_d, w_d, h_d are given in $L^2(Q)$, $C(\overline{Q})$, $L^2(\Sigma)$, respectively, and $\{u_\Gamma, w_\Gamma\}$ is the solution of state problem $SP(\Gamma)$; $d\sigma$ stands for the surface element on $\partial\Omega$.

Throughout this paper, we use the following notation. For a general (real) Banach space Y, we denote by $|\cdot|_Y$ the norm in Y and by Y^* the dual of Y. Also, for a positive finite number T, we denote by $C_w([0,T];Y)$ the space of all weakly continuous functions $u : [0,T] \to Y$, and by definition "$u_n \to u$ in $C_w([0,T];Y)$ as $n \to +\infty$" means that for each $z^* \in Y^*$, $\langle z^*, u_n(t) \rangle_{Y^*,Y}$ converges to $\langle z^*, u(t) \rangle_{Y^*,Y}$ uniformly in $t \in [0,T]$ as $n \to +\infty$, where $\langle \cdot, \cdot \rangle_{Y^*,Y}$ is the duality pairing between Y^* and Y.

For simplicity we put

$$H := L^2(\Omega), \ V := H^1(\Omega), \ H_0 := \{v \in H; \int_\Omega zdx = 0\}, V_0 := V \cap H_0,$$

and

$$\Pi := \{\Gamma \subset \partial\Omega; \ \Gamma \text{ is compact in } \partial\Omega, \ \sigma(\Gamma) > 0\}.$$

For each $\Gamma \in \Pi$, we put

$$V(\Gamma) := \{z \in V; z = 0 \text{ a.e. on } \Gamma\}$$

which is a closed subspace of V, and

$$(v,w) := \int_\Omega vwdx \qquad\qquad \text{for } v, w \in H,$$

$$(v,w)_{\partial\Omega} := \int_{\partial\Omega} vwd\sigma \qquad\qquad \text{for } v, w \in L^2(\partial\Omega),$$

$$a(v,w) := \int_\Omega \nabla v \cdot \nabla wdx \qquad\qquad \text{for } v, w \in V,$$

$\pi :=$ the projection from H onto H_0, i.e.

$$\pi(z)(x) := z(x) - \frac{1}{|\Omega|} \int_\Omega z(y) dy,$$

where $|\Omega|$ is the volume of Ω. We note that H_0 and V_0 are Hilbert spaces with $|z|_{H_0} := |z|_H$ and $|z|_{V_0} := |\nabla z|_H$, respectively, as well as $V(\Gamma)$ with $|z|_{V(\Gamma)} := \{a(z,z) + n_0 \int_{\partial\Omega} |z|^2 d\sigma\}^{1/2}$; we use symbol $(\cdot,\cdot)_0$ for the inner product in H_0, $\langle\cdot,\cdot\rangle_0$ for the duality pairing between V_0^* and V_0, $\langle\cdot,\cdot\rangle_\Gamma$ for the duality pairing between $V(\Gamma)^*$ and $V(\Gamma)$ for each $\Gamma \in \Pi$.

As usual, identifying H and H_0 with their duals, we have

$$V_0 \subset H_0 \subset V_0^*, \quad V \subset H \subset V^*$$

and

$$V(\Gamma) \subset H \subset V(\Gamma)^* \quad \text{for any } \Gamma \in \Pi$$

with dense and compact embeddings. Also, we denote by F_0 the duality mapping from V_0 onto V_0^*, and by F_Γ the duality mapping from $V(\Gamma)$ onto $V(\Gamma)^*$, which are respectively defined by

$$\langle F_0 z, \eta \rangle := a(z,\eta) \quad \text{for all } z, \eta \in V_0,$$

and

$$\langle F_\Gamma z, \eta \rangle_\Gamma := a(z,\eta) + n_0 \int_{\partial\Omega} z\eta d\sigma \quad \text{for all } z, \eta \in V(\Gamma).$$

In general, given a subset E of $\overline{\Omega}$, χ_E denotes the characteristic function of E defined on $\overline{\Omega}$.

We now introduce a notion of convergence in Π. By definition, a sequence $\{\Gamma_n\} \subset \Pi$ converges to $\Gamma \in \Pi$, denoted by $\Gamma_n \to \Gamma$ in Π as $n \to +\infty$, if the following conditions (C1) – (C3) are satisfied:

(C1) If $\{n_k\}$ is a subsequence of $\{n\}$, $z_k \in V(\Gamma_{n_k})$ and $z_k \to z$ weakly in V as $k \to +\infty$, then $z \in V(\Gamma)$.

(C2) For any $z \in V(\Gamma)$, there is a sequence $\{z_n\} \subset V$ such that $z_n \in V(\Gamma_n)$, $n = 1, 2, \cdots$, and $z_n \to z$ in V as $n \to +\infty$.

(C3) $\chi_{\Gamma_n} \to \chi_\Gamma$ in $L^1(\partial\Omega)$ as $n \to +\infty$.

Also, a subset Π' of Π is said to have property (C), if Π' is compact in the sense of (C1) – (C3), namely, any sequence $\{\Gamma_n\}$ of Π' contains a subsequence convergent to a certain $\Gamma \in \Pi'$.

1. Formulation of an optimization problem

First of all, we mention precise assumptions on the data ρ, λ, β, g, f, h_D, h_N, u_0 and w_0; Ω is a bounded domain in $\mathbb{R}^N$, $N = 2$ or 3, with smooth boundary $\partial\Omega$ and T is a fixed positive number.

(H1) ρ is a maximal monotone graph in $\mathbb{R} \times \mathbb{R}$ whose domain $D(\rho)$ and range $R(\rho)$ are open in $\mathbb{R}$, and it is locally bi-Lipschitz continuous as a function from $D(\rho)$ onto $R(\rho)$, and there are constants $A_0 > 0$ and α with $1 \le \alpha < 2$ such that

$$|\rho(r_1) - \rho(r_2)| \ge \frac{A_0|r_1 - r_2|}{|r_1 r_2|^\alpha + 1} \quad \text{for all } r_1, r_2 \in D(\rho). \tag{1.1}$$

(H2) β is a maximal monotone graph in $\mathbb{R} \times \mathbb{R}$ such that $\overline{D(\beta)} = [\sigma_*, \sigma^*]$ for constants σ_*, σ^* with $-\infty < \sigma_* < \sigma^* < +\infty$.

(H3) λ is a C^2-function from $\mathbb{R}$ into itself and g is a C^1-function from $\mathbb{R}$ into itself; λ' is the derivative of λ.

(H4) (i) $f \in W^{1,2}(0, T; H)$;

(ii) $h_D \in W^{1,2}(0, T; H^{1/2}(\partial\Omega))$ such that there is a function $\tilde{h}_D \in W^{1,2}(0, T; V)$ with $\rho(\tilde{h}_D) \in W^{1,2}(0, T; V)$;

(iii) $h_N \in W^{1,2}(0, T; L^2(\partial\Omega)) \cap L^\infty(\Sigma)$ such that

$$n_0 \inf D(\rho) \le h_N(t, x) \le n_0 \sup D(\rho) \quad \text{for a.e. } (t, x) \in \Sigma \tag{1.2}$$

and there are positive constants A_1 and A_1' such that

$$\rho(r)(n_0 r - h_N(t, x)) \ge -A_1|r| - A_1' \quad \text{for all } r \in D(\rho) \text{ and a.e. } (t, x) \in \Sigma. \tag{1.3}$$

(H5) (i) $u_0 \in V$ such that $\rho(u_0) \in H$ and $u_0 = h_D(0, \cdot)$ a.e. on $\partial\Omega$;

(ii) $w_0 \in H^2(\Omega)$ such that

$$\sigma_* < \frac{1}{|\Omega|} \int_\Omega w_0 dx =: m < \sigma^* \tag{1.4}$$

and $\dfrac{\partial w_0}{\partial n} = 0$ a.e. on $\partial\Omega$ and there is $\xi_0 \in H$ satisfying

$$\xi_0 \in \beta(w_0) \quad \text{a.e. in } \Omega, \ -\kappa\Delta w_0 + \xi_0 \in V.$$

Corresponding to functions h_D, h_N and $\Gamma \in \Pi$, we consider the function $h_\Gamma : [0, T] \to V$ given by

$$\begin{cases} h_\Gamma(t) = h_D(t) \quad \text{a.e. on } \Gamma, \\ a(h_\Gamma(t), z) + (n_0 h_\Gamma(t) - h_N(t), z)_{\partial\Omega} = 0 \quad \text{for all } z \in V(\Gamma); \end{cases} \tag{1.5}$$

note under condition (H4) and $\sigma(\Gamma) \ge \sigma_0$ for a positive constant σ_0 that such a function h_Γ exists in $W^{1,2}(0, T; V)$ and $|h_\Gamma|_{W^{1,2}(0,T;V)} \le K$ for a certain constant K depending only on quantities in (H4) and σ_0. Moreover, if $\Gamma_n \to \Gamma$ in Π as $n \to +\infty$, then $h_{\Gamma_n} \to h_\Gamma$ in $C([0,T];V)$ as $n \to +\infty$ (cf. [6]).

We now give the weak formulation for state problem $SP(\Gamma)$ for each $\Gamma \in \Pi$.

Definition 1.1. A couple $\{u, w\}$ of functions $u : [0,T] \to V$ and $w : [0,T] \to H^2(\Omega)$ is called a (weak) solution of $SP(\Gamma)$, if the following properties (w1) – (w4) are fulfilled:

(w1) $u - h_\Gamma \in C_w([0,T];V(\Gamma))$, $\rho(u) \in C_w([0,T];H)$, $\rho(u)' \in L^2(0,T;V(\Gamma)^*)$,

$w \in C_w([0,T];H^2(\Omega))$ with $\dfrac{\partial w(t)}{\partial n} = 0$ a.e. on $\partial\Omega$ for all $t \in [0,T]$, and $w' \in L^2(0,T;H)$.

(w2) $u(0) = u_0$ and $w(0) = w_0$.

(w3) For all $z \in V(\Gamma)$ and a.e. $t \in [0,T]$,

$$\frac{d}{dt}(\rho(u)(t) + \lambda(w)(t), z) + a(u(t), z) + n_0(u(t) - h_\Gamma(t), z)_{\partial\Omega} = (f(t), z). \tag{1.6}$$

(w4) There exists a function $\xi \in L^2(0,T;H)$ such that $\xi \in \beta(w)$ a.e. in Q and

$$\frac{d}{dt}(w(t), \eta - \mu\Delta\eta) + \kappa(\Delta w(t), \Delta\eta) - (g(w(t)) + \xi(t) - \lambda'(w(t))u(t), \Delta\eta) = 0 \tag{1.7}$$
$$\text{for all } \eta \in H^2(\Omega) \text{ with } \frac{\partial\eta}{\partial n} = 0 \text{ a.e. on } \partial\Omega \text{ and a.e. } t \in [0,T].$$

Remark 1.1. Let $\{u, w\}$ be the solution of $SP(\Gamma)$ in the above sense. Then it follows from (1.7) in (w4) that

$$\frac{d}{dt}(w(t), 1) = 0, \text{ i.e. } \frac{d}{dt}\int_\Omega w(t, x)dx = 0,$$

so that $\dfrac{1}{|\Omega|}\displaystyle\int_\Omega w(t, x)dx = m$ for all $t \in [0,T]$ and $w(t) - m \in V_0$ for all $t \in [0,T]$.

According to a result [5, Theorem 2.2], problem $SP(\Gamma)$ has an unique solution $\{u, w\}$ for each $\Gamma \in \Pi$; more precisely, in [5] the case of $\Gamma = \phi$ was treated, but the existence, uniqueness and regularity results remain valid for the case of $\sigma(\Gamma) > 0$ and they can be proved by slight modification of those in [5]; the solution $\{u, w\}$ of $SP(\Gamma)$ has further regularity $w' \in L^2(0,T;V_0) \cap L^\infty(0,T;H_0)$.

Based on the solvability of $SP(\Gamma)$, we now propose an optimization problem.

For a given non-empty subset Π_c of Π having property (C), our optimization problem, denoted by $P(\Pi_c)$, is to find a set $\Gamma_* \in \Pi_c$ such that

$$J(\Gamma_*) = \inf_{\Gamma \in \Pi_c} J(\Gamma), \tag{1.8}$$

where $J(\Gamma)$ is defined by (0.7).

102

Our main results are stated as follows.

Theorem 1.1. *Let Π_c be a non-empty subset of Π having property(C). Then, optimization problem $P(\Pi_c)$ has at least one solution $\Gamma_* \in \Pi_c$.*

The above existence result is obtained from the following theorem on the continuous dependence of the solution $\{u_\Gamma, w_\Gamma\}$ of $SP(\Gamma)$ upon $\Gamma \in \Pi$.

Theorem 1.2. *Let $\{\Gamma_n\}$ be a sequence in Π such that $\Gamma_n \to \Gamma$ in Π as $n \to +\infty$, and $\{u_n, w_n\}$ and $\{u, w\}$ be the solutions of $SP(\Gamma_n)$ and $SP(\Gamma)$, respectively. Then*

$$u_n \to u \text{ in } C_w([0,T]; V), \quad w_n \to w \text{ in } C_w([0,T]; H^2(\Omega)) \tag{1.9}$$

as $n \to +\infty$.

It is easily seen from Theorem 1.2 that any minimizing sequence $\{\Gamma_n\} \subset \Pi_c$ of the cost functional $J(\cdot)$ on Π_c contains a subsequence convergent to a solution of $P(\Pi_c)$.

We shall sketch the proofs of the theorems in sections 2 and 3.

2. Uniform estimates with respect to Γ

In proving the theorems a crucial step is to give uniform estimates of solutions of $SP(\Gamma)$ with respect to Γ. In this section conditions (H1) – (H5) are always supposed and Π_c is a non-empty subset of Π having property (C).

We denote by $\{u_\Gamma, w_\Gamma\}$ the solution of $SP(\Gamma)$ for each $\Gamma \in \Pi$, and by ξ_Γ the function ξ in condition (w4) of Definition 1.1.

In terms of the duality mappings $F_\Gamma : V(\Gamma) \to V(\Gamma)^*$ and $F_0 : V_0 \to V_0^*$ the variational identities (1.6) and (1.7) are respectively written in the equivalent forms

$$\rho(u_\Gamma)'(t) + \lambda(w_\Gamma)'(t) + F_\Gamma(u_\Gamma(t) - h_\Gamma(t)) = f(t) \text{ in } V(\Gamma)^* \text{ for a.e. } t \in [0,T], \tag{2.1}$$

and

$$F_0^{-1}w_\Gamma'(t) + \mu w_\Gamma'(t) + \kappa F_0(\pi w_\Gamma(t)) + \pi[\xi_\Gamma(t) + g(w_\Gamma(t)) - \lambda'(w_\Gamma(t))u_\Gamma(t)] = 0$$
$$\text{in } H_0 \text{ for a.e. } t \in [0,T], \tag{2.2}$$

In the following three lemmas uniform estimates for solutions of $SP(\Gamma)$ with respect to Γ are given.

First, compute (at least formally) the following (1) – (3):

(1) Multiply both sides of (2.1) by $u_\Gamma(t) - h_\Gamma(t)$ and integrate in time.
(2) Multiply both sides of (2.2) by $w_\Gamma'(t)$ and integrate in time.
(3) Multiply both sides of (2.1) by $\rho(u_\Gamma(t)) - \rho(\tilde{h}_D(t))$ and integrate in time.

Then we get:

Lemma 2.1. *There is a positive constant M_1, depending only on the quantities in (H1) –(H5) and Π_c, such that*

$$|u_\Gamma - h_\Gamma|_{L^2(0,T;V(\Gamma))} + |\rho(u_\Gamma)|_{L^\infty(0,T;H)} + |\rho^*(\rho(u_\Gamma))|_{L^\infty(0,T;L^1(\Omega))}$$

$$+|w_\Gamma|_{L^\infty(0,T;V)} + |w'_\Gamma|_{L^2(0,T;V^*)} + |\widehat{\beta}(w_\Gamma)|_{L^\infty(0,T;L^1(\Omega))} \tag{2.3}$$

$$\leq M_1 \quad \text{for all } \Gamma \in \Pi_c,$$

where ρ^* is a primitive of ρ^{-1} and $\widehat{\beta}$ is a lower-semicontinuous convex function on $\mathbb{R}$ such that $\beta = \partial\widehat{\beta}$ (= subdifferential of $\widehat{\beta}$ in $\mathbb{R}$).

Secondary, compute (formally) the following (4) – (5):
(4) Multiply both sides of (2.1) by $u'_\Gamma(t) - h'_\Gamma(t)$ and integrate in time.
(5) Multiply both sides of $\dfrac{d}{dt}(2.2)$ by $w'_\Gamma(t)$ and integrate in time.
Then, with the help of uniform estimate (2.3), we have:

Lemma 2.2. *There is a positive constant M_2, depending only on the quantities in (H1) –(H5) and Π_c, such that*

$$|u_\Gamma - h_\Gamma|_{L^\infty(0,T;V(\Gamma))} + |w'_\Gamma|_{L^\infty(0,T;H_0)} + |w'_\Gamma|_{L^2(0,T;V_0)} \leq M_2 \quad \text{for all } \Gamma \in \Pi_c, \tag{2.4}$$

The above constant M_2 essentially depends upon a given constant $\mu > 0$.
Next, we obtain uniform estimates for w_Γ in $L^\infty(0,T;H^2(\Omega))$ and ξ_Γ in $L^\infty(0,T;H)$ from estimates (2.3) and (2.4) by applying the theory on semilinear elliptic PDEs including maximal monotone perturbation β (cf. [2]).

Lemma 2.3. *There is a positive constant M_3, depending only on the quantities in (H1) –(H5) and Π_c, such that*

$$|w_\Gamma|_{L^\infty(0,T;H^2(\Omega))} + |\xi_\Gamma|_{L^\infty(0,T;H)} \leq M_3 \quad \text{for all } \Gamma \in \Pi_c. \tag{2.5}$$

The rigorous computations corresponding (1) –(5) can be done for an adequate approximate problem to $SP(\Gamma)$, and estimates (2.3) – (2.5) can be obtained as the limits of those for the approximate solutions. In detail, see a forthcoming paper [3].

3. Proofs of Theorems 1.1 and 1.2

First we give a proof of Theorem 1.2, using the same notation as in the statement of Theorem 1.2. According to estimates (2.3) – (2.5), we may assume, by taking a subsequence of $\{n\}$ if necessary, that

$$w_n \to \overline{w} \quad \text{in } C([0,T];V) \text{ and } C_w([0,T];H^2(\Omega)) \tag{3.1}$$

$$w_n' \to \overline{w}' \quad \text{weakly in } L^2(0,T;V_0) \text{ and weakly* in } L^\infty(0,T;H_0) \tag{3.2}$$

$$\xi_n \to \overline{\xi} \quad \text{weakly* in } L^\infty(0,T;H) \tag{3.3}$$

$$u_n \to \overline{u} \quad \text{weakly* in } L^\infty(0,T;V) \text{ with } \overline{u} - h_\Gamma \in L^\infty(0,T;V(\Gamma)) \tag{3.4}$$

and

$$\rho(u_n) \to \overline{\rho} \quad \text{in } C_w([0,T];H) \tag{3.5}$$

for some functions $\overline{u}$, $\overline{w}$ and $\overline{\xi}$. In fact, (3.1) – (3.4) are directly obtained from (2.4) and (2.5); by the way it follows from (3.1) and (3.3) that

$$\overline{\xi} \in \beta(\overline{w}) \quad \text{a.e. in } Q. \tag{3.6}$$

Also, (3.5) is inferred as follows. We observe from (2.4) and the relation $\rho(u_n)' = -F_n(u_n - h_n) - \lambda(w_n)' + f$ that $\{\rho(u_n)\}$ and $\{\rho(u_n)'\}$ are bounded in $L^\infty(0,T;H)$ and $L^\infty(0,T;H^{-1}(\Omega))$, respectively, so that $\rho(u_n) \to \overline{\rho}$ in $C([0,T];H^{-1}(\Omega))$ and hence in $C_w([0,T];H)$, where $F_n := F_{\Gamma_n}$ and $h_n := h_{\Gamma_n}$.

Next, by integrating (1.6) for $\{u_n, w_n\}$ in time we have that

$$(\rho(u_n(s)), z) = (\rho(u_0), z) + (\lambda(w_n(s)) - \lambda(w_0), z)$$

$$-a\left(\int_0^s (u_n - h_n)dt, z\right) - n_0\left(\int_0^s (u_n - h_n)dt, z\right)_{\partial\Omega} + \left(\int_0^s f dt, z\right) \tag{3.7}$$

$$\text{for all } z \in V(\Gamma_n) \text{ and all } s \in [0,T].$$

Now, let z be any function in $V(\Gamma)$, and take a sequence $\{z_n\} \subset V$ such that $z_n \to z$ in V and $z_n \in V(\Gamma_n)$. Then, substituting z_n as z in (3.7) and passing to the limit as $n \to +\infty$, we get from (3.1), (3.4) and (3.5) that

$$(\overline{\rho}(s), z) = (\rho(u_0), z) + (\lambda(\overline{w}(s)) - \lambda(w_0), z)$$

$$-a\left(\int_0^s (\overline{u} - h_\Gamma)dt, z\right) - n_0\left(\int_0^s (\overline{u} - h_\Gamma)dt, z\right)_{\partial\Omega} - \left(\int_0^s f dt, z\right) \tag{3.8}$$

$$\text{for all } z \in V(\Gamma) \text{ and all } s \in [0,T].$$

Lemma 3.1. $\overline{\rho} = \rho(\overline{u})$ and $\overline{u} - h_\Gamma \in C_w([0,T];V(\Gamma))$.

Proof. Take $u_n(s) - h_n(s)$ as z in (3.7) and integrate in s on $[0,T]$. Then

$$\int_0^T (\rho(u_n), u_n - h_n)ds$$

$$= \int_0^T (\rho(u_0), u_n - h_n)ds + \int_0^T (\lambda(w_n) - \lambda(w_0), u_n - h_n)ds$$

$$-\frac{1}{2}|\nabla \int_0^T (u_n - h_n)dt|_H^2 - \frac{n_0}{2}|\int_0^T (u_n - h_n)dt|_{L^2(\partial\Omega)}^2 + \int_0^T \left(\int_0^s f dt, u_n - h_n\right)ds,$$

since

$$a(\int_0^s (u_n - h_n)dt, u_n(s) - h_n(s)) = \frac{1}{2}\frac{d}{dt}|\nabla \int_0^T (u_n - h_n)dt|_H^2$$

and

$$(\int_0^s (u_n - h_n)dt, u_n(s) - h_n(s))_{\partial\Omega} = \frac{1}{2}\frac{d}{dt}|\int_0^T (u_n - h_n)dt|_{L^2(\partial\Omega)}^2.$$

Therefore

$$\limsup_{n\to+\infty}\int_0^T (\rho(u_n), u_n - h_n)ds$$

$$\leq \int_0^T (\rho(u_0), \bar{u} - h_\Gamma)ds + \int_0^T (\lambda(\bar{w}) - \lambda(w_0), \bar{u} - h_\Gamma)ds$$

$$-\frac{1}{2}|\nabla \int_0^T (\bar{u} - h_\Gamma)dt|_H^2 - \frac{n_0}{2}|\int_0^T (\bar{u} - h_\Gamma)dt|_{L^2(\partial\Omega)}^2 + \int_0^T (\int_0^s f\,dt, \bar{u} - h_\Gamma)ds.$$

By (3.8), the right hand side of the above inequality is equal to $\int_0^T (\bar{\rho}, \bar{u} - h_\Gamma)ds$, whence

$$\limsup_{n\to+\infty}\int_0^T (\rho(u_n), u_n - h_n)ds \leq \int_0^T (\bar{\rho}, \bar{u} - h_\Gamma)ds.$$

This shows that

$$\limsup_{n\to+\infty}\int_0^T (\rho(u_n), u_n)ds \leq \int_0^T (\bar{\rho}, \bar{u})ds, \tag{3.9}$$

since $h_n \to h_\Gamma$ in $C([0,T]; V)$. Inequality (3.9) together with the maximal monotonicity of ρ in $L^2(0, T; H)$ implies that $\bar{\rho} = \rho(\bar{u})$. Finally by Remark 1.3 in [8] it follows from $\rho(\bar{u}) \in C_w([0, T]; H)$ and $\bar{u} \in L^\infty(0, T; V)$ that $\bar{u} \in C_w([0, T]; V)$ and hence $\bar{u} - h_\Gamma \in C_w([0, T]; V(\Gamma))$. $\square$

Lemma 3.2. $u_n \to \bar{u}$ in $C_w([0, T]; V)$ as $n \to +\infty$.

Proof. If not, then there would exist a positive number ε_0, an element z^* in V^*, a subsequence $\{n_k\}$ of $\{n\}$ and a sequence $\{t_k\} \subset [0, T]$ such that

$$\langle z^*, u_{n_k}(t_k) - \bar{u}(t_k)\rangle \geq 3\varepsilon_0 \quad \text{for all } k = 1, 2, \cdots, \tag{3.10}$$

where $\langle\cdot,\cdot\rangle$ stands for the duality pairing between V^* and V. By taking further a subsequence of $\{n_k\}$ if necessary, we may assume that $t_k \to t_0 \in [0, T]$ for a certain t_0 and $u_{n_k}(t_k) \to \bar{z}$ weakly in V as $k \to +\infty$. In this case we have $\rho(u_{n_k}(t_k)) \to \rho(\bar{z})$ weakly in H as $k \to +\infty$ and

$$|\langle z^*, u_{n_k}(t_k) - \bar{z}\rangle| \leq \varepsilon_0, \quad |\langle z^*, \bar{u}(t_k) - \bar{u}(t_0)\rangle| \leq \varepsilon_0$$

for all sufficiently large k. Combining this with (3.10) we derive that

$$\langle z^*, \bar{z} - \bar{u}(t_0)\rangle \geq \varepsilon_0. \tag{3.11}$$

On the other hand, it follows from (3.5) and Lemma 3.1 that $\rho(u_{n_k}(t_k)) \to \rho(\bar{u}(t_0))$ weakly in H, so that $\rho(\bar{z}) = \rho(\bar{u}(t_0))$, i.e. $\bar{z} = \bar{u}(t_0)$. This contradicts (3.11). $\square$

It is easy now to see from convergences (3.1) – (3.5) with (3.6) and Lemmas 3.1, 3.2, that $\{\bar{u}, \bar{w}\}$ is the unique solution of $SP(\Gamma)$, i.e. $\{\bar{u}, \bar{w}\} = \{u, w\}$. Thus the proof of Theorem 1.2 is complete.

Proof of Theorem 1.1: Let $\{\Gamma_n\}$ be a minimizing sequence in Π_c of the functional J on Π_c, i.e.

$$J(\Gamma_n) \to J^* := \inf_{\Gamma \in \Pi_c} J(\Gamma).$$

Then, by property (C), we may assume that $\Gamma_n \to \Gamma_*$ in Π for some $\Gamma_* \in \Pi_c$. Now, denoting by $\{u_n, w_n\}$ and $\{u_*, w_*\}$ the solutions of $SP(\Gamma_n)$ and $SP(\Gamma_*)$, respectively, we see from Theorem 1.2 that

$$u_n \to u_* \text{ in } C_w([0, T]; V), \quad w_n \to w_* \text{ in } C_w([0, T]; H^2(\Omega)),$$

which guarantees that $J(\Gamma_n) \to J(\Gamma_*)$, whence $J_* = J(\Gamma_*)$. Thus Γ_* is a solution of $P(\Pi_c)$.

$\square$

4. Regular approximation for $P(\Pi_c)$

In this section, from the numerical point of view we discuss regular approximation of $SP(\Gamma)$ and $P(\Pi_c)$.

At first, we introduce the approximation ρ^ν, β^ε and χ_Γ^τ for ρ, β and χ_Γ, respectively, which are defined below.

(a) Let $D(\rho) := (r_*, r^*)$ for $-\infty \leq r_* < r^* \leq +\infty$, and choose two families $\{a_\nu; 0 < \nu \leq 1\}$ and $\{b_\nu; 0 < \nu \leq 1\}$ in $D(\rho)$ such that

$$r_* < a_\nu < a_{\nu'} < a_1 < b_1 < b_{\nu'} < b_\nu < r^* \text{ if } 0 < \nu < \nu' < 1$$

and

$$a_\nu \downarrow r_*, \ b_\nu \uparrow r^* \text{ as } \nu \downarrow 0.$$

Then, $\rho^\nu : \mathbb{R} \to \mathbb{R}$ is defined for each $\nu \in (0, 1]$ by

$$\rho^\nu(r) := \begin{cases} \rho(b_\nu) + r - b_\nu & \text{for } r > b_\nu, \\ \rho(r) & \text{for } a_\nu \leq r \leq b_\nu, \\ \rho(a_\nu) + r - a_\nu & \text{for } r < a_\nu. \end{cases}$$

(b) For each $0 < \varepsilon \leq 1$, β^ε is the Yosida-approximation of β, namely,

$$\beta^\varepsilon(r) := \frac{r - (I + \varepsilon\beta)^{-1}r}{\varepsilon}, \quad r \in \mathbb{R}.$$

(c) Let $\{\chi_\Gamma^\tau\} := \{\chi_\Gamma^\tau; 0 < \tau \le 1, \Gamma \in \Pi_c\}$ be a family of smooth functions on $\partial\Omega$ and suppose that it satisfies the following properties $(\chi 1) - (\chi 3)$:

$(\chi 1)$ $0 \le \chi_\Gamma \le \chi_\Gamma^\tau \le 1$; $\mathrm{supp}(\chi_\Gamma^\tau) \subset \{x \in \partial\Omega; dist(x, \Gamma) \le \tau\}$ for all $\tau \in (0, 1]$ and $\Gamma \in \Pi_c$.

$(\chi 2)$ For each $\tau \in (0, 1]$, $\{\chi_\Gamma^\tau; \Gamma \in \Pi_c\}$ is compact in $L^1(\partial\Omega)$.

$(\chi 3)$ Let $V(\tau, \Gamma) := \{z \in V; \chi_\Gamma^\tau z = 0 \text{ a.e. on } \Gamma\}$ for each $\tau \in (0, 1]$ and $\Gamma \in \Pi_c$. If $\tau_n \downarrow 0$ and $\Gamma_n \in \Pi_c$, then there are a subsequence $\{n_k\}$ of $\{n\}$ and $\Gamma \in \Pi_c$ such that $\chi_{\Gamma_{n_k}}^{\tau_{n_k}} \to \chi_\Gamma$ in $L^1(\partial\Omega)$ as $k \to \infty$, and $V(\tau_{n_k}, \Gamma_{n_k}) \to V(\Gamma)$ in V as $k \to \infty$ in the sense of Mosco [6].

Now we propose a regular approximation for $SP(\Gamma)$, referred to as $SP(\Gamma)^{\nu\varepsilon\tau\delta}$, where $\nu, \varepsilon, \tau, \delta \in (0, 1]$, by the penalty method:

$$\rho^\nu(u)_t + \lambda(w)_t - \Delta u = f \quad \text{in } Q, \tag{4.1}$$

$$w_t - \Delta(-\mu\Delta w_t - \kappa\Delta w + \beta^\varepsilon(w) + g(w) - \lambda'(w)u) = 0 \quad \text{in } Q, \tag{4.2}$$

$$\frac{\partial u}{\partial n} = -\frac{\chi_\Gamma^\tau}{\delta}(u - h_D) + (1 - \chi_\Gamma^\tau)(h_N - n_0 u) \quad \text{on } \Sigma, \tag{4.3}$$

$$\frac{\partial w}{\partial n} = 0, \; \frac{\partial}{\partial n}(-\mu\Delta w_t - \kappa\Delta w + \beta^\varepsilon(w) + g(w) - \lambda'(w)u) = 0 \quad \text{on } \Sigma, \tag{4.4}$$

$$u(0) = u_{0\nu} := \min\{\max\{u_0, a_\nu\}, b_\nu\}, \; w(0) = w_0 \quad \text{in } \Omega. \tag{4.5}$$

The notion of a weak solution of $SP(\Gamma)^{\nu\varepsilon\tau\delta}$ is given below.

Definition 4.1. A couple $\{u, w\}$ of functions $u : [0, T] \to V$ and $w : [0, T] \to H^2(\Omega)$ is called a solution of $SP(\Gamma)^{\nu\varepsilon\tau\delta}$, if the following conditions (w1)' – (w4)' are satisfied:

(w1)' $u \in W^{1,2}(0, T; H) \cap C([0, T]; V)$,

$w \in W^{1,2}(0, T; H) \cap C_w([0, T]; H^2(\Omega))$ with $\dfrac{\partial w(t)}{\partial n} = 0$ a.e. on $\partial\Omega$ for all $t \in [0, T]$.

(w2)' $u(0) = u_{0\nu}$, $w(0) = w_0$.

(w3)' For all $z \in V$ and a.e. $t \in [0, T]$,

$$(\rho^\nu(u)'(t) + \lambda(w)'(t), z) + a(u(t), z)$$

$$+ (\frac{\chi_\Gamma^\tau}{\delta}(u(t) - h_D(t)) - (1 - \chi_\Gamma^\tau)(h_N(t) - n_0 u(t)), z)_{\partial\Omega} = (f(t), z).$$

(w4)' For all $\eta \in H^2(\Omega)$ with $\dfrac{\partial\eta}{\partial n} = 0$ a.e. on $\partial\Omega$ and a.e. $t \in [0, T]$,

$$(w'(t), \eta - \mu\Delta\eta) + \kappa(\Delta w(t), \Delta\eta) - (g(w(t)) + \beta^\varepsilon(w(t)) - \lambda'(w(t))u(t), \Delta\eta) = 0.$$

According to a result in [4], $SP(\Gamma)^{\nu\varepsilon\tau\delta}$ has a unique solution $\{u, w\}$. Our regular approximate optimization problem $P(\Pi_c)^{\nu\varepsilon\tau\delta}$ is to find $\Gamma_*^{\nu\varepsilon\tau\delta} \in \Pi_c$ such that

$$J^{\nu\varepsilon\tau\delta}(\Gamma_*^{\nu\varepsilon\tau\delta}) = \inf_{\Gamma \in \Pi_c} J^{\nu\varepsilon\tau\delta}(\Gamma),$$

where

$$J^{\nu\varepsilon\tau\delta}(\Gamma) := A\int_Q |u - u_d|^2 dx dt + B|w - w_d|^2_{C(\overline{Q})} + C\int_\Sigma (1 - \chi_\Gamma^\tau)|h_d|^2 d\sigma dt,$$

$\{u, w\}$ being the solution of $SP(\Gamma)^{\nu\varepsilon\tau\delta}$.

Finally we mention a convergence result.

Theorem 4.1. *Let Π_c, $\{\rho^\nu\}$, $\{\beta^\varepsilon\}$, $\{\chi_\Gamma^\tau\}$ be as above. Then:*
(1) For $\nu, \varepsilon, \tau, \delta \in (0, 1]$, $P(\Pi_c)^{\nu\varepsilon\tau\delta}$ has at least one solution $\Gamma_*^{\nu\varepsilon\tau\delta} \in \Pi_c$.
(2) Let $\{\nu_n\}$, $\{\varepsilon_n\}$, $\{\tau_n\}$ and $\{\delta_n\}$ be any null sequences and let $\{\Gamma_n := \Gamma_*^{\nu_n\varepsilon_n\tau_n\delta_n}\}$ be a sequence of solutions of $P(\Pi_c)^{\nu_n\varepsilon_n\tau_n\delta_n}$. Then, $\{\Gamma_n\}$ contains a subsequence convergent in Π and any limit Γ_* is a solution of $P(\Pi_c)$.

For a detailed proof, see a forthcoming paper [3].

References

[1] H. W. Alt and I. Pawlow, Existence of solutions for non-isothermal phase separation, Adv. Math. Soc. Appl. 1(1992), 319 - 409.

[2] H. Brézis, M. Crandall and A. Pazy, Perturbations of nonlinear maximal monotone sets, Comm. Pure Appl. Math. 23(1970), 123 - 144.

[3] A. Kadoya, N. Kato, N. Kenmochi and J. Shirohzu, Optimal control for a class of phase separation models, in preparation.

[4] N. Kenmochi and M. Niezgódka, Nonlinear System for non-isothermal diffusive phase Separation, J. Math Anal. Appl. 188(1994), 651 - 679.

[5] N. Kenmochi and M. Niezgódka, Viscosity approach for modelling non-isothermal diffusive phase separation, Japan J. Industrial Appl. Math., (1996), to appear.

[6] U. Mosco, Convergence of convex sets and of solutions of variational inequalities, Advances Math., 3(1969), 510-585.

[7] O. Penrose and P. C. Fife, Thermodynamically consistent models of phase-field type for the kinetics of phase transitions, Physica D, 13(1990), 44-62.

[8] J. Shirohzu, N. Sato and N. Kenmochi, Asymptotic convergence in models for phase change problems, in Nonlinear Analysis and Applications, pp.361-385, GAKUTO Inter. Ser. Math. Sci. Appl., Vol. 7, Gakkōtosho, Tokyo, 1995.

A. Kadoya
Dept. Management Sci., Fac. Commercial Sci.

Hiroshima Shudo University
Asaminami-ku, Hiroshima, 731-31 Japan

N. Kato
Information Sci. Dept.
Taisei Corporation
Shinjuku-ku, Tokyo, 169 Japan

N. Kenmochi
Dept. Math., Fac. Education
Chiba Univ.
Chiba, 263 Japan

J. Shirohzu
Dept. Math., Graduate School Sci. Tech.
Chiba Univ.
Chiba, 263 Japan

N KENMOCHI

Global attractor of the semigroup associated to a phase-field model with constraint

Abstract. This paper is concerned with large-time behaviour of solutions of a phase-field (solid-liquid phase transition) model with constraint. The model is a system of two nonlinear parabolic PDEs satisfied by the temperature $\theta = \theta(t, x)$ and the order parameter $w = w(t, x)$; in our model the so-called double obstacles are imposed on the order parameter, i.e., $\sigma_* \leq w \leq \sigma^*$, where σ_* and σ^* are the threshold values for w corresponding to pure solid and pure liquid, respectively. The objective of this paper is to generate a semigroup $\{S(t)\}$ associated to our phase-field model and construct its global attractor.

1. Introduction

Let us consider the following autonomous system of PDEs, referred as (PFC):

$$[\rho(u) + \lambda(w)]_t - \Delta u + \nu\rho(u) = f(x) \quad \text{in } Q := (0, +\infty) \times \Omega, \tag{1.1}$$

$$w_t - \kappa\Delta w + \xi + g(w) = \lambda'(w)u \quad \text{in } Q, \tag{1.2}$$

$$\xi \in \beta(w) \quad \text{in } Q, \tag{1.3}$$

$$\frac{\partial u}{\partial n} + n_o u = h(x), \quad \frac{\partial w}{\partial n} = 0 \quad \text{on } \Sigma := (0, +\infty) \times \Gamma, \tag{1.4}$$

$$u(0, \cdot) = u_o, \quad w(0, \cdot) = w_o \quad \text{in } \Omega, \tag{1.5}$$

where Ω is a bounded domain in $\mathbf{R}^N$ ($1 \leq N \leq 3$) with smooth boundary $\Gamma := \partial\Omega$; $\rho(\cdot)$ is an increasing function, possibly having a singularity, such as $\rho(u) = -\frac{1}{u}$ for $-\infty < u < 0$; $\lambda(\cdot)$ and $g(\cdot)$ are smooth functions on $\mathbf{R}$, and $\lambda'(\cdot)$ is the derivative of λ; β is a maximal monotone graph in $\mathbf{R} \times \mathbf{R}$ with bounded damain $D(\beta)$; κ, n_o and ν are positive constants; f, h, u_o and w_o are prescribed data.

This system is motivated by solid-liquid phase transition processes. In such a context, $\theta = \rho(u)$ is the (absolute) temperature and w is the (non-conserved) order parameter which has two threshold values σ_* and σ^* with $-\infty < \sigma_* < \sigma^* < +\infty$, namely $\sigma_* \leq w(t, x) \leq \sigma^*$ a.e. on Q, and $w = \sigma_*$, $\sigma_* < w < \sigma^*$ and $w = \sigma^*$ correspond respectively to pure solid, mixture and pure liquid. Within the Landau-Ginzburg theory for thermodynamical phase transition, (1.1) and (1.2) can be interpreted as kinetic equations for the internal energy $e := \rho(u) + \lambda(w)$ and order papameter w. We refer to [12] for physical interpretation of system (1.1)-(1.2).

Many theoretical results have been established, for instance [2, 4, 5, 6, 7, 9, 11, 13] since the earlier papers [14, 16] not including maximal monotone constraint β.

111

In [9], problem (PFC) with $\nu = 0$ was treated and the existence, uniqueness and large-time behaviour of the solution were discussed, and it is easily checked that all the results remain valid in the case of $\nu > 0$, too. Uniqueness was shown there only in the case when initial data are good enough, more precisely $u_o \in H^1(\Omega)$ and $w_o \in H^2(\Omega)$ satisfying appropriate compatibility conditions. But this uniqueness result is not enough, because it seems natural from some features of equations (1.1) and (1.2) that the semigroup associated to (PFC) is generated on a subset of the product space $\rho^{-1}(L^2(\Omega)) \times H^1(\Omega)$, where $\rho^{-1}(L^2(\Omega)) = \{v; \rho(v) \in L^2(\Omega)\}$.

Recently, in [8] a new uniqueness result was established in a very wide class, which is adequate for the objective of this paper, under the additional assumption that

$$\lambda \text{ is convex on } \overline{D(\beta)} \text{ and } D(\rho) \subset (-\infty, 0]. \tag{1.6}$$

In this paper, assuming (1.6) and taking advantage of this uniqueness result, we shall generate a semigroup $\{S(t)\}_{t \geq 0}$, consisting of operators $S(t)$ which assigns to each initial value $[u_o, w_o]$ the element $[u(t), w(t)]$, on a subset of $\rho^{-1}(L^2(\Omega)) \times H^1(\Omega)$. In case $\rho : \mathbf{R} \to \mathbf{R}$ is bi-Lipschitz continuous (hence $\rho^{-1}(L^2(\Omega)) = L^2(\Omega)$), the semigroup $\{S(t)\}$ has the standard continuity property in $L^2(\Omega) \times H^1(\Omega)$, so the general theory on attractors (cf. [15]) can be applied to the construction of its global attractor; see [1] and [10] for the related works. In the singular case of ρ, a global attractor will be constructed still in the space $L^2(\Omega) \times H^1(\Omega)$, but unfortunately our semigroup $\{S(t)\}$ lacks the continuity at $t = 0$ in $L^2(\Omega) \times H^1(\Omega)$ in the usual sense. This means that we can not directly apply the general theory to our case. However, so far as the construction of the global attractor is concerned, some modified versions of [15] are available to our case; for this purpose, a Lyapunov-like functional will be introduced. In particular, the term $\nu\rho(u)$ in (1.1) is very important in order to get an absorbing set.

Notation. In general, for a (real) Banach space Y the norm is denoted by $|\cdot|_Y$ and the dual space by Y^*. For a given compact interval $[t_o, t_1]$, $C_w([t_o, t_1]; Y)$ is the space of all weakly continuous functions from $[t_o, t_1]$ into Y, and "$u_n \to u$ in $C_w([t_o, t_1]; Y)$ (as $n \to +\infty$)" means that for each $y^* \in Y^*$, $\langle y^*, u_n(t) - u(t)\rangle_{Y^*,Y} \to 0$ uniformly in $t \in [t_o, t_1]$, where $\langle \cdot, \cdot \rangle_{Y^*,Y}$ denotes the duality pairing between Y^* and Y.

Throughout this paper, let Ω be a bounded domain in $\mathbf{R}^N$, $1 \leq N \leq 3$, with smooth boundary $\Gamma := \partial\Omega$. As the norm of $H^1(\Omega)$ we use

$$|z|_{H^1(\Omega)} := \left\{ \int_\Omega |\nabla z|^2 dx + n_o \int_\Gamma |z|^2 d\Gamma \right\}^{\frac{1}{2}},$$

which is equivalent to the standard norm of $H^1(\Omega)$. As usual, we have

$$H^1(\Omega) \subset L^2(\Omega) \subset H^1(\Omega)^*$$

with dense and compact embeddings. We use the following notation:
    $(\cdot, \cdot)$: the inner product in $L^2(\Omega)$;

112

$(\cdot, \cdot)_\Gamma$: the inner product in $L^2(\Gamma)$;

$a(v, z) := \displaystyle\int_\Omega \nabla z \cdot \nabla v dx$ for $v, z \in H^1(\Omega)$;

$\langle \cdot, \cdot \rangle$ the duality pairing between $H^1(\Omega)^*$ and $H^1(\Omega)$;
F: the duality mapping from $H^1(\Omega)$ onto $H^1(\Omega)^*$ defined by

$$\langle Fv, z \rangle = a(v, z) + n_o(v, z)_\Gamma, \quad v, z \in H^1(\Omega);$$

F_o: the operator from $H^1(\Omega)$ into $H^1(\Omega)^*$ defined by

$$\langle F_o v, z \rangle = a(v, z), \quad v, z \in H^1(\Omega).$$

We see that if $\ell_1 \in L^2(\Omega)$, $\ell_2 \in L^2(\Gamma)$ and $\ell \in H^1(\Omega)^*$ is given by the formula

$$\langle \ell, z \rangle := (\ell_1, z) + (\ell_2, z)_\Gamma, \quad z \in H^1(\Omega),$$

then $\ell = Fv$ (resp. $\ell = F_N v$) is formally equivalent to

$$-\Delta v = \ell_1 \text{ in } \Omega, \quad \frac{\partial v}{\partial n} + n_o v = \ell_2 \text{ (resp. } \frac{\partial v}{\partial n} = \ell_2) \text{ on } \Gamma; \tag{1.7}$$

in particular, if $\ell_2 \in H^{\frac{1}{2}}(\Gamma)$, then $v \in H^2(\Omega)$ and hence (1.7) is rigorously verified.

For a function $k : [t_o, t_1] \to \mathbf{R}$ we say that k is a function of bounded variation with absolutely continuous decreasing part on $[t_o, t_1]$ if there is an absolutely continuous function k_o on $[t_o, t_1]$ such that $k(t) + k_o(t)$ is non-decreasing on $[t_o, t_1]$; in this case, k is differentiable almost everywhere, its derivative k' is integrable on $[t_o, t_1]$ and

$$k(t) - k(s) \leq \int_s^t k'(\tau) d\tau \quad \text{for all } t_o \leq s \leq t \leq t_1.$$

This notion will be used sometimes in the expression of energy inequalities.

2. Existence and uniqueness results

We begin with specifying all the data, except initial data u_o, w_o, in our system.

(H1) ρ is a maximal monotone graph in $\mathbf{R} \times \mathbf{R}$ with open domain $D(\rho)$ and open range $R(\rho)$ in $\mathbf{R}$, and it is increasing and locally bi-Lipschitz continuous as a function from $D(\rho)$ onto $R(\rho)$; we denote by ρ^{-1} the inverse of ρ and fix proper l.s.c. convex functions ρ^* and $\hat{\rho}$ on $\mathbf{R}$ whose subdifferential coincides with ρ^{-1} and ρ in $\mathbf{R}$, respectively.

(H2) β is a maximal monotone graph in $\mathbf{R} \times \mathbf{R}$ with bounded domain $D(\beta)$ having non-empty interior in $\mathbf{R}$, say $\overline{D(\beta)} = [\sigma_*, \sigma^*]$ for some constants σ_*, σ^* with $-\infty < \sigma_* < \sigma^* < +\infty$; we fix a non-negative proper l.s.c. convex function $\hat{\beta}$ on $\mathbf{R}$ whose subdifferential coincides with β in $\mathbf{R}$.

(H3) λ and its derivative λ' are Lipschitz continuous functions on $[\sigma_*, \sigma^*]$, and

$$\lambda''(w)u \leq 0 \quad \text{for a.e. } w \in [\sigma_*, \sigma^*] \text{ and all } u \in D(\rho). \tag{2.1}$$

(H4) g is a Lipschitz continuous function on $[\sigma_*, \sigma^*]$; we fix a primitive $\hat{g}$ of g which is non-negative on $[\sigma_*, \sigma^*]$.

(H5) κ, n_o and ν are positive constants.

(H6) $f \in L^2(\Omega)$ and $h \in L^\infty(\Gamma)$ such that

$$n_o \sup D(\rho) \geq h(x) \geq n_o \inf D(\rho) \quad \text{for a.e. } x \in \Gamma$$

and there are positive constants A_1 and A_1' satisfying

$$\rho(r)(n_o r - h(x)) \geq -A_1|r| - A_1' \quad \text{for all } r \in D(\rho) \text{ and a.e. } x \in \Gamma.$$

Moreover, as the set of possible initial data we consider

$$D := \{[v, z]; \rho(v) \in L^2(\Omega) \text{ with } \rho^*(\rho(v)) \in L^1(\Omega), \ z \in H^1(\Omega) \text{ with } \hat{\beta}(z) \in L^1(\Omega)\},$$

and as the set of good initial data

$$D_o := \{[v, z] \in D; v \in H^1(\Omega), \ z \in H^2(\Omega) \text{ with } \frac{\partial z}{\partial n} = 0 \text{ a.e. on } \Gamma,$$

$$\exists \tilde{z} \in L^2(\Omega) \text{ such that } \tilde{z} \in \beta(z) \text{ a.e. on } \Omega\}.$$

Initial data $[u_o, w_o]$ are given in D, and we write (PFC;u_o, w_o) for (PFC), if the initial data u_o, w_o are explicitly indicated.

Now, the weak formulation for (PFC) is given as follows.

Definition 2.1. Given initial data $[u_o, w_o] \in D$ and a finite number $T > 0$, a couple of functions $u : [0, T] \to H^1(\Omega)$ and $w : [0, T] \to H^1(\Omega)$ is called a (weak) solution of (PFC;u_o, w_o]) on $[0, T]$, if the following conditions (w1) - (w4) are fulfilled:

(w1) $u \in L^2(0, T; H^1(\Omega))$, $\rho(u) \in C_w([0, T]; L^2(\Omega)) \cap W^{1,2}(0, T; H^1(\Omega)^*)$,
$w \in C([0, T]; H^1(\Omega)) \cap W^{1,2}(0, T; L^2(\Omega))$.

(w2) $\rho(u)(0) = \rho(u_o)$ and $w(0) = w_o$.

(w3) For all $z \in H^1(\Omega)$ and a.e. $t \in [0, T]$

$$\langle \rho(u)'(t) + \lambda(w)'(t), z \rangle + a(u(t), z) + \nu(\rho(u(t)), z) + (n_o u(t) - h, z)_\Gamma = (f, z). \tag{2.2}$$

(w4) There is $\xi \in L^2(0, T; L^2(\Omega))$ such that $\xi \in \beta(w)$ a.e. on $Q_T := (0, T) \times \Omega$ and

$$(w'(t), z) + \kappa a(w(t), z) + (\xi(t) + g(w(t)) - \lambda'(w(t))u(t), z) = 0$$

for all $z \in H^1(\Omega)$ and a.e. $t \in [0, T]$.

A couple of functions $\{u, w\}$ is called a global solution of (PFC), if it is a solution of (PFC) on $[0, T]$ for every finite $T > 0$.

By "$\xi \in \beta(w)$ (hence $\sigma_* \leq w \leq \sigma^*$) a.e. on Q_T" in condition (w4) of Definition 2.1 the above weak formulation is independent of the behaviour of functions λ and g on the outside of $[\sigma_*, \sigma^*]$. Therefore, without loss of generality we may assume that

$$\text{the support of } g \text{ is compact in } \mathbf{R} \text{ and } \lambda \text{ is linear outside } [\sigma_*, \sigma^*]. \tag{2.4}$$

and

$$\hat{g} \text{ is still nonnegative on } \mathbf{R}. \tag{2.5}$$

For the function $f \in L^2(\Omega)$ and $h \in L^\infty(\Gamma)$ there exists a function $h_o \in H^1(\Omega)$ with $\rho(h_o) \in L^2(\Omega)$ such that

$$a(h_o, z) + (n_o h_o - h, z)_\Gamma + \nu(\rho(h_o), z) = (f, z) \quad \text{for all } z \in H^1(\Omega).$$

With this function h_o and the duality mapping $F : H^1(\Omega) \to H^1(\Omega)^*$, the variational identity (2.2) is equivalently written in the form

$$\rho(u)' + \lambda(w)' + F(u - h_o) + \nu(\rho(u) - \rho(h_o)) = 0 \quad \text{in } H^1(\Omega)^*, \text{ a.e. } t \in [0, T]. \tag{2.6}$$

Also, with the mapping $F_o : H^1(\Omega) \to H^1(\Omega)^*$ the variational identity (2.3) is equivalently written in the form

$$w' + \kappa F_o w + \xi + g(w) = \lambda'(w)u \quad \text{in } L^2(\Omega), \text{ a.e. } t \in [0, T]. \tag{2.7}$$

We first recall a result which guarantees the uniqueness of a solution of (PFC).

Theorem 2.1. *Assume that (H1) - (H6) hold. Let $\{u_i, w_i\}$, $i = 1, 2$, be any solutions of (PFC) on $[0, T]$ associated to initial data $[u_{oi}, w_{oi}] \in D$. Then, with notation $e_i := \rho(u_i) + \lambda(w_i)$ for $i = 1, 2$,*

$$|e_1(t) - e_2(t)|^2_{H^1(\Omega)^*} + |w_1(t) - w_2(t)|^2_{L^2(\Omega)}$$

$$\leq e^{R_o(t-s)}(|e_1(s) - e_2(s)|^2_{H^1(\Omega)^*} + |w_1(s) - w_2(s)|^2_{L^2(\Omega)}) \tag{2.8}$$

$$\text{for all } 0 \leq s \leq t \leq T,$$

where R_o is a positive constant depending only on κ, n_o and the Lipschitz constants of λ and g.

In [8; Theorem 2.1] the same inequality as (2.8) was obtained for problem (PFC) with $\nu = 0$ and the proof is available to the case of $\nu > 0$ with a slight modification. Hypothesis (2.1) in (H3) is essential for the proof of (2.8).

Next, let us introduce functionals J_1 and J_2 which are respectively defined by

$$J_1(v) := \int_\Omega \rho^*(\rho(v)) dx - (\rho(v), h_o) + C_1$$

for all $v \in \rho^{-1}(L^2(\Omega)) := \{v; \rho(v) \in L^2(\Omega)\}$, and

$$J_2(z) := \frac{\kappa}{2}|\nabla z|^2_{L^2(\Omega)} + \int_\Omega (\hat{\beta}(z) + \hat{g}(z))dx - (\lambda(z), h_o) + C_2$$

for all $z \in H^1(\Omega)$, where C_1 and C_2 are positive constants so that J_1 and J_2 are non-negative. In fact, such constants exist, since

$$\rho^*(r) - rh_o(x) \geq \rho^*(\rho(h_o(x))) - \rho(h_o(x))h_o(x) \quad \text{for all } r \in \mathbf{R} \text{ and a.e. } x \in \Omega, \quad (2.9)$$

and $|(\lambda(z), h_o)|$ is bounded by a positive constant independent of z with $\sigma_* \leq z \leq \sigma^*$; note here that $J_1(v)$ (resp. $J_2(z)$) is equal to $+\infty$, if $\int_\Omega \rho^*(\rho(v))dx = +\infty$ (resp. $\int_\Omega \hat{\beta}(z)dx = +\infty$). We put

$$J(v, z) := J_1(v) + J_2(z) \quad \text{for } [v, z] \in \rho^{-1}(L^2(\Omega)) \times H^1(\Omega);$$

by definition, J is finite and non-negative on D.

Theorem 2.2. *Assume that (H1) - (H6) hold, and let $[u_o, w_o] \in D$. Then $(PFC;u_o, w_o)$ has a (unique) global solution $\{u, w\}$ such that*

$$\begin{cases} u \in C_w([\delta, T]; H^1(\Omega)), \ w \in C_w([\delta, T]; H^2(\Omega)), \\ w' \in L^\infty(\delta, T; L^2(\Omega)) \cap L^2(\delta, T; H^1(\Omega)), \\ \xi \in L^\infty(\delta, T; L^2(\Omega)), \end{cases} \quad (2.10)$$

for every $0 < \delta < T < +\infty$, where ξ is the function as in (w4) of Definition 2.1. Moreover, the following inequalities (i) - (iii) hold.

(i) *$J(u, w)$ is of bounded variation with absolutely continuous decreasing part on $\mathbf{R}_+$ and*

$$\frac{d}{dt}J(u(t), w(t)) + |u(t) - h_o|^2_{H^1(\Omega)} + |w'(t)|^2_{L^2(\Omega)}$$

$$+\nu(\rho(u(t)) - \rho(h_o), u(t) - h_o) = 0 \quad (2.11)$$

for a.e. $t \geq 0$.

(ii) *$|\rho(u)|^2_{L^2(\Omega)}$ is of bounded variation with absolutely continuous decreasing part on $\mathbf{R}_+$ and there is a positive constant M_1 independent of initial data $[u_o, w_o] \in D$ such that*

$$\frac{d}{dt}|\rho(u(t))|^2_{L^2(\Omega)} + \nu|\rho(u(t))|^2_{L^2(\Omega)} \leq M_1\{|u(t) - h_o|^2_{H^1(\Omega)} + |w'(t)|^2_{L^2(\Omega)} + 1\} \quad (2.12)$$

for a.e. $t \geq 0$.

116

(iii) There is a positive constant M_2 independent of initial data $[u_o, w_o] \in D$ such that

$$(s - t_o) \left\{ |u(s) - h_o|_{H^1(\Omega)}^2 + |w'(s)|_{L^2(\Omega)}^2 + \nu \left| \int_\Omega \hat{\rho}(u(s)) dx \right| \right\}$$

$$+ \kappa \int_{t_o}^s (t - t_o) |\nabla w'(t)|_{L^2(\Omega)}^2 dt \tag{2.13}$$

$$\leq M_2 \{ J(u(t_o), w(t_o)) + |\rho(u(t_o))|_{L^2(\Omega)}^2 + 1 \},$$

for all $t_o \geq 0$ and a.e. $s \in [t_o, t_o + 2]$.

Remark 2.1. From (2.13) of Theorem 2.2 we further derive an estimate of the form

$$(s - t_o)\{ |w(s)|_{H^2(\Omega)}^2 + |\xi(s)|_{L^2(\Omega)}^2 \}$$

$$\leq M_3 \{ J(u(t_o), w(t_o)) + |\rho(u(t_o))|_{L^2(\Omega)}^2 + 1 \} \tag{2.14}$$

for all $t_o \geq 0$ and a.e. $s \in (t_o, t_o + 2]$,

where M_3 is a positive constant independent of initial data $[u_o, w_o] \in D$. In fact, since

$$\kappa F_o w(t) + \xi(t) = -w'(t) - g(w(t)) + \lambda'(w(t))u(t) =: \ell(t) \text{ in } L^2(\Omega) \tag{2.15}$$

for a.e. $t \geq 0$,

it follows from a regularity result in [3] that

$$|w(t)|_{H^2(\Omega)}^2 + |\xi(t)|_{L^2(\Omega)}^2 \leq M_3'(|\ell(t)|_{L^2(\Omega)}^2 + 1) \qquad \text{for a.e. } t \geq 0,$$

with a constant M_3' independent of solutions $\{u, w\}$. Combining this with (2.13), we obtain an inequality of the form (2.14).

Remark 2.2. Let $\{u, w\}$ be the global solution of (PFC) obtained by Theorem 2.1. Then, since $w(t)$ satisfies (2.15) for a.e. $t \geq 0$ and w is weakly continuous from $(0, +\infty)$ into $H^2(\Omega)$, it follows automatically from (2.15) that $\xi(t)$ is defined for all $t > 0$ and $\frac{\partial w(t)}{\partial n} = 0$ a.e. on Γ for all $t > 0$, so that $[u(t), w(t)] \in D_o$ for all $t > 0$. By the way, (2.14) holds true for all $t_o \geq 0$ and all $s \in (t_o, t_o + 2]$.

We shall give a brief proof of Theorem 2.2 in the next section.

3. Proof of Theorem 2.2

A. The case $[u_o, w_o] \in D_o$. First, consider the case of good initial data, i.e. $[u_o, w_o] \in D_o$. In this case, it was proved in [13; Theorem 1.1] that problem (PFC) with $\nu = 0$ has one and only one global solution $\{u, w\}$ having regularity (2.10) with $\delta = 0$. A slight modification of that proof is available to the case of $\nu > 0$, so in the regular case of initial data we omit the detalis of the existence proof.

Let us now derive inequalities (2.11) - (2.13) by means of formal calculations; the rigorous calculations should be done in the step of regular approximations for (PFC), as the construction of a solution shows in [13].

(a) Derivation of (2.11): Multiplying (2.6) by $u(t) - h_o$, we have (formally)

$$(\rho(u)'(t), u(t)) - \frac{d}{dt}(\rho(u(t)) + \lambda(w(t)), h_o) + |u(t) - h_o|^2_{H^1(\Omega)}$$

$$+\nu(\rho(u(t)) - \rho(h_o), u(t) - h_o) = -(\lambda(w)'(t), u(t)) \qquad (3.1)$$

$$\text{for a.e. } t \geq 0.$$

Since $(\rho(u)', u) = \frac{d}{dt}\int_\Omega \rho^*(\rho(u))dx$, it follows from (3.1) that

$$\frac{d}{dt}\{J_1(u(t)) - (\lambda(w(t)), h_o)\} + |u(t) - h_o|^2_{H^1(\Omega)} + \nu(\rho(u(t)) - \rho(h_o), u(t) - h_o) \qquad (3.2)$$

$$= -(\lambda(w)'(t), u(t)), \quad \text{a.e. } t \geq 0.$$

Secondly, multiplying (2.7) by $w'(t)$, we have

$$|w'(t)|^2_{L^2(\Omega)} + \frac{d}{dt}\left\{\frac{\kappa}{2}|\nabla w(t)|^2_{L^2(\Omega)} + \int_\Omega(\hat{\beta}(w(t)) + \hat{g}(w(t)))dx\right\} \qquad (3.3)$$

$$= (\lambda(w)'(t), u(t)), \quad \text{a.e. } t \geq 0.$$

Adding (3.2) and (3.3) immediately yields (2.11).

(b) Derivation of (2.12): Take $\rho(u(t))$ as a (formal) test function z in (2.2) to get

$$\frac{1}{2}\frac{d}{dt}|\rho(u(t))|^2_{L^2(\Omega)} + \nu|\rho(u(t))|^2_{L^2(\Omega)} + a(u(t), \rho(u(t))) + (n_o u(t) - h, \rho(u(t)))_\Gamma \qquad (3.4)$$

$$= (f, \rho(u(t))) - (\lambda(w)'(t), \rho(u(t))), \quad \text{a.e. } t \geq 0.$$

Since $a(u(t), \rho(u(t))) \geq 0$ (cf. the monotonicity of ρ and

$$(n_o u(t) - h, \rho(u(t))) \geq -A_1|u(t)|_{L^1(\Omega)} - A_1'|\Omega| \qquad \text{(cf. condition (H6))},$$

it follows from (3.4) that

$$\frac{1}{2}\frac{d}{dt}|\rho(u(t))|^2_{L^2(\Omega)} + \nu|\rho(u(t))|^2_{L^2(\Omega)} \qquad (3.5)$$

$$\leq A_1|u(t)|_{L^1(\Omega)} + |f|_{L^2(\Omega)}|\rho(u(t))|_{L^2(\Omega)} + L_{\lambda'}|w'(t)|_{L^2(\Omega)}|\rho(u(t))|_{L^2(\Omega)},$$

where $L_{\lambda'}$ is the Lipschitz constant of λ'. It is atraightforward to obtain (2.12) from (3.5).

(c) Derivation of (2.13): We compute the following items (1), (2), (3):

(1) Multiply (2.6) by u' and $\frac{d}{dt}$(2.7) by w'.

(2) Add the two results of (1).

118

(3) Multiply the result of (2) by $(t - t_o)$ and integrate in time over $[t_o, s]$. Now we formally compute (1), and have

$$(\rho(u)'(t), u'(t)) + \frac{d}{dt}\left\{\frac{1}{2}|u(t) - h_o|^2_{H^1(\Omega)} + \nu \int_\Omega \hat{\rho}(u(t))dx - \nu(\rho(h_o), u(t))\right\}$$

$$= -(\lambda(w)'(t), u'(t)), \quad \text{a.e. } t \geq 0, \tag{3.6}$$

and

$$\frac{1}{2}\frac{d}{dt}|w'(t)|^2_{L^2(\Omega)} + \kappa|\nabla w'(t)|^2_{L^2(\Omega)} + (\beta(w)'(t), w'(t)) + (g(w)'(t), w'(t))$$

$$= (\lambda(w)'(t), u'(t)) + \int_\Omega \lambda''(w(t))|w'(t)|^2 u(t)dx, \quad \text{a.e. } t \geq 0. \tag{3.7}$$

Here, we note that

$$(\rho(u)'(t), u'(t)) \geq 0, \quad (\beta(w)'(t), w'(t)) \geq 0 \quad \text{(cf. the monotonicity of } \rho, \beta)$$

and

$$\int_\Omega \lambda''(w)|w'(t)|^2 u(t)dx \leq 0 \quad \text{(cf. (2.1) in (H3)).}$$

Therefore, the addition of (3.6) and (3.7) yields that

$$\frac{d}{dt}\left\{\frac{1}{2}|u(t) - h_o|^2_{H^1(\Omega)} + \frac{1}{2}|w'(t)|^2_{L^2(\Omega)} + \nu \int_\Omega \hat{\rho}(u(t))dx - \nu(\rho(h_o), u(t))\right\}$$

$$+\kappa|\nabla w'(t)|^2_{L^2(\Omega)} \leq L_g|w'(t)|^2_{L^2(\Omega)}, \quad \text{a.e. } t \geq 0, \tag{3.8}$$

where L_g is the Lipschitz constant of g.

Besides, multiplying (3.8) by $(t - t_o)$, $t \geq t_o$, we have after simple arrangement of coefficients

$$\frac{d}{dt}\left\{(t - t_o)\left(|u(t) - h_o|^2_{H^1(\Omega)} + |w'(t)|^2_{L^2(\Omega)} + 2\nu \int_\Omega \hat{\rho}(u(t))dx - 2\nu(\rho(h_o), u(t))\right)\right\}$$

$$+2\kappa(t - t_o)|\nabla w'(t)|^2_{L^2(\Omega)} \leq 2L_g(t - t_o)|w'(t)|^2_{L^2(\Omega)} \tag{3.9}$$

$$+|u(t) - h_o|^2_{H^1(\Omega)} + |w'(t)|^2_{L^2(\Omega)} + 2\nu \int_\Omega \hat{\rho}(u(t))dx - 2\nu(\rho(h_o), u(t)), \quad \text{a.e. } t \geq t_o.$$

Now, integrate (3.9) in t over $[t_o, s]$ with $s \leq t_o + 2$. Then

$$(s - t_o)\left\{|u(s) - h_o|^2_{H^1(\Omega)} + |w'(s)|^2_{L^2(\Omega)} + 2\nu \int_\Omega \hat{\rho}(u(s))dx - 2\nu(\rho(h_o, u(s)))\right\}$$

$$+2\kappa \int_{t_o}^s (t - t_o)|\nabla w'|^2_{L^2(\Omega)}dt \leq 2L_g \int_{t_o}^s (t - t_o)|w'|^2_{L^2(\Omega)}dt \tag{3.10}$$

$$+\int_{t_o}^s \left\{|u - h_o|^2_{H^1(\Omega)} + |w'|^2_{L^2(\Omega)} + 2\nu \int_\Omega \hat{\rho}(u)dx - 2\nu(\rho(h_o), u)\right\}dt.$$

119

It is easy to arrive at an inequality of the form (2.13) from (3.10) with the help of (2.11), (2.12) and the fact that $\hat{\rho}(u) \leq \hat{\rho}(h_o) + \rho(u)(u - h_o)$.

B. *The case* $[u_o, w_o] \in D$

Given $[u_o, w_o] \in D$, choose a sequence of initial data $\{[u_{on}, w_{on}]\}$ in D_o such that

$$\rho(u_{on}) \to \rho(u_o) \text{ in } L^2(\Omega), \qquad \rho^*(\rho(u_{on})) \to \rho^*(\rho(u_o)) \text{ in } L^1(\Omega),$$

$$w_{on} \to w_o \text{ in } H^1(\Omega), \qquad \hat{\beta}(w_{on}) \to \hat{\beta}(w_o) \text{ in } L^1(\Omega)$$

(as $n \to +\infty$); hence $J(u_{on}, w_{on}) \to J(u_o, w_o)$. For each n, let $\{u_n, w_n\}$ be the global solution of (PFC;u_{on}, w_{on}). Then, by uniform estimates derived from (2.11) - (2.14) we may assume that

$$u_n \to u \quad \text{weakly in } L^2(0, T; H^1(\Omega)), \tag{3.11}$$

$$t^{\frac{1}{2}} u_n \to t^{\frac{1}{2}} u \quad \text{weakly* in } L^\infty(0, T; H^1(\Omega)), \tag{3.12}$$

$$\rho(u_n) \to \bar{\rho} \quad \text{weakly* in } L^\infty(0, T; L^2(\Omega)) \tag{3.13}$$

$$\begin{cases} w_n \to w \text{ weakly in } W^{1,2}(0, T; L^2(\Omega)), \text{ in } C([0, T]; L^2(\Omega)) \\ \qquad \text{and in } C_w([0, T]; H^1(\Omega)), \end{cases} \tag{3.14}$$

$$t^{\frac{1}{2}} w_n \to t^{\frac{1}{2}} w \quad \text{weakly* in } L^\infty(0, T; H^2(\Omega)), \tag{3.15}$$

$$t^{\frac{1}{2}} w_n' \to t^{\frac{1}{2}} w' \quad \text{weakly in } L^2(0, T; H^1(\Omega)) \text{ and weakly* in } L^\infty(0, T; L^2(\Omega)), \tag{3.16}$$

$$\xi_n \to \xi \text{ weakly in } L^2(0, T; L^2(\Omega)), \quad t^{\frac{1}{2}} \xi_n \to t^{\frac{1}{2}} \xi \text{ weakly* in } L^\infty(0, T; L^2(\Omega)) \tag{3.17}$$

for some functions u, $\bar{\rho}$, w and ξ; note further that (3.14) and (3.17) imply that

$$\xi \in \beta(w) \quad \text{a.e. on } Q_T, \tag{3.18}$$

and (3.11) - (3.13) imply that

$$\begin{cases} \rho(u_n)' \to \bar{\rho}' = -F(u - h_o) - \lambda(w)' - \nu(\bar{\rho} - \rho(h_o)) \\ \qquad\qquad\qquad \text{weakly in } L^2(0, T; H^1(\Omega)^*), \\ t^{\frac{1}{2}} \rho(u_n)' \to t^{\frac{1}{2}} \bar{\rho}' \quad \text{weakly* in } L^\infty(0, T; H^1(\Omega)^*), \end{cases} \tag{3.19}$$

and hence

$$\rho(u_n) \to \bar{\rho} \quad \text{in } C([0, T]; H^1(\Omega)^*) \text{ and } C_w([0, T]; L^2(\Omega)).$$

We observe that

$$\limsup_{n \to +\infty} \int_0^T (\rho(u_n) - \bar{\rho}, u_n) dt$$

$$= \limsup_{n \to +\infty} \int_0^T \langle \rho(u_n) - \bar{\rho}, u_n \rangle dt$$

$$\leq \limsup_{n \to +\infty} |\rho(u_n) - \bar{\rho}|_{C([0,T];H^1(\Omega)^*)} \int_0^T |u_n|_{H^1(\Omega)} dt$$

$$= 0,$$

120

namely

$$\limsup_{n \to +\infty} \int_0^T (\rho(u_n), u_n)dt \le \int_0^T (\bar\rho, u)dt,$$

which shows by the maximal montonicity of ρ in $L^2(0, T; L^2(\Omega))$ that

$$\bar\rho = \rho(u) \quad \text{a.e. on } Q_T. \tag{3.20}$$

Convergences (3.11) - (3.17) and (3.19) with (3.18) and (3.20) are enough to see that the limit $\{u, w\}$ is a solution of (PFC;u_o, w_o) having regularity (2.10), and satisfying inequalities (2.11) - (2.13).

4. The semigroup and its global attractor

Based on Theorem 2.2 we generate a semigroup $\{S(t)\}_{t\ge0}$ on D associated to (PFC). As usual, for each $t \ge 0$, $S(t) : D \to D$ is defined by

$$S(t)[u_o, w_o] := [u(t), w(t)], \quad [u_o, w_o] \in D, \tag{4.1}$$

where $\{u, w\}$ is the global solution of (PFC;u_o, w_o). It is clear that $\{S(t)\}$ forms a semigroup on D.

Theorem 4.1. *Assume that (H1) - (H6) hold, and let $\{S(t)\}$ be the semigroup on D given by (4.1). Then the following statements hold.*

(a) Let $0 < \delta < T < +\infty$. Then $S(\cdot)[v, z] \in C_w([\delta, T]; H^1(\Omega) \times H^2(\Omega))$ for any $[v, z] \in D$. Furthermore, if $[v_n, z_n] \in D$, $n = 1, 2, \cdots$, $[v, z] \in D$, $\{J(v_n, z_n)\}$ is bounded, $\rho(v_n) \to \rho(v_o)$ weakly in $L^2(\Omega)$ and $z_n \to z$ weakly in $H^1(\Omega)$ as $n \to +\infty$, then

$$S(\cdot)[v_n, z_n] \to S(\cdot)[v, z] \quad \text{in } C_w([\delta, T]; H^1(\Omega) \times H^2(\Omega))$$

as $n \to +\infty$.

(b) $S(t)D \subset D_o$ for all $t > 0$, and for each $R > 0$ the set $D_R := \{[v, z] \in D; J(v, z) \le R\}$ is positively invariant for $\{S(t)\}$, i.e. $S(t)D_R \subset D_R$ for all $t \ge 0$.

Assertion (b) of Theorem 4.1 is a direct consequence of Theorem 2.2 and Remarks 2.1, 2.2. Assertion (a) can be proved in the same way as that in (B) of the previous section.

Theorem 4.2. *Assume that (H1) - (H6) hold, and let $\{S(t)\}$ be the semigroup on D given by (4.1). Then there exists a subset A of D_o such that*

(i) A is compact and connected in $L^2(\Omega) \times H^1(\Omega)$, and is bounded in $H^1(\Omega) \times H^2(\Omega)$;

(ii) A is invariant for $\{S(t)\}$, i.e. $A = S(t)A$ for all $t \ge 0$;

(iii) for each subset B of D with $\sup_{[v,z] \in B}\{J(v, z) + |\rho(v)|^2_{L^2(\Omega)}\} < +\infty$, and for each $\varepsilon > 0$ there exists a finite time $T_{B,\varepsilon} > 0$ such that

$$\text{dist}_{L^2(\Omega) \times H^1(\Omega)}(S(t)[v, z], A) < \varepsilon \quad \text{for all } [v, z] \in B \text{ and } t \geq T_{B,\varepsilon},$$

where $\text{dist}_{L^2(\Omega) \times H^1(\Omega)}$ stands for the distance in $L^2(\Omega) \times H^1(\Omega)$.

By Theorem 4.2 we can say that A is a global attractor of $\{S(t)\}$. The key for the construction of the global attractor is to find an absorbing set for the semigroup $\{S(t)\}$.

Lemma 4.1. *There are positive constants ε_o, ε_1 and N_o such that*

$$\frac{d}{dt}\{J(u(t), w(t)) + \varepsilon_o|\rho(u(t))|^2_{L^2(\Omega)}\} + \varepsilon_1\{J(u(t), w(t)) + \varepsilon_o|\rho(u(t))|^2_{L^2(\Omega)}\} \leq N_o, \quad (4.2)$$

$$a.e. \ t \geq 0$$

for all global solutions $\{u, w\}$ with initial data $[u_o, w_o] \in D$.

Proof. By the definition of subdifferential ρ^{-1} of ρ^* in $\mathbf{R}$ we observe that

$$(\rho(u(t)) - \rho(h_o), u(t) - h_o)$$
$$\geq \int_\Omega \rho^*(\rho(u(t)))dx - (\rho(u(t)), h_o) + (\rho(h_o), h_o) - \int_\Omega \rho^*(\rho(h_o))dx \qquad (4.3)$$
$$= J_1(v) + (\rho(h_o), h_o) - \int_\Omega \rho^*(\rho(h_o))dx$$

for all $t \geq 0$. Also, multiply (2.7) by $w(t) - r_o$ with $\beta(r_o) \ni 0$, we have

$$(w'(t), w(t) - r_o) + \kappa|\nabla w(t)|^2_{L^2(\Omega)} + \int_\Omega \hat{\beta}(w(t))dx$$

$$\leq -(g(w(t)), w(t) - r_o) + (\lambda'(w(t))u(t), w(t) - r_o)$$

which yields

$$\kappa|\nabla w(t)|^2_{L^2(\Omega)} + \int \hat{\beta}(w(t))dx \leq R_1\{|w'(t)|_{L^2(\Omega)} + |u(t)|_{L^2(\Omega)} + 1\} \qquad \text{for a.e. } t \geq 0,$$
$$(4.4)$$

where R_1 is a positive constant independent of all solutions $\{u, w\}$ of (PFC). It is derived from (4.4) that

$$J_2(w(t)) \leq R_2\{|w'(t)|_{L^2(\Omega)} + |u(t) - h_o|_{H^1(\Omega)} + 1\} \quad \text{for a.e. } t \geq 0, \qquad (4.5)$$

where R_2 is a positive constant independent of all solutions $\{u, w\}$ of (PFC). Now, noting (4.3) and (4.5), and computing "(2.11) + $\varepsilon_o \times$ (2.12)" with a sufficiently small positive number ε_o, we get an inequality of the form (4.2) for certain positive constants ε_1 and N_o. $\Diamond$

In what follows, J_o is the functional defined by

$$J_o(v, z) := J(v, z) + \varepsilon_o |\rho(v)|^2_{L^2(\Omega)}, \qquad [v, z] \in \rho^{-1}(L^2(\Omega)) \times H^1(\Omega),$$

where ε_o is the same as in Lemma 4.1.

Lemma 4.2. *There exists a subset B_o of D such that*

(1) $\sup\limits_{[v,z] \in B_o} J_o(v, z) \leq +\infty$,

(2) for any subset B with $\sup_{[v,z] \in B} J_o(v, z) < +\infty$ there exists a finite time $t_B > 0$ satisfying

$$S(t)B \subset B_o \qquad \text{for all } t \geq t_B.$$

Proof. From (4.2) of Lemma 4.1 it follows that

$$J_o(S(t)[v, z]) \leq e^{-\varepsilon_1 t} J_o(v, z) + \frac{N_o}{\varepsilon_1} \qquad \text{for all } t \geq 0 \text{ and } [v, z] \in D.$$

Therefore, the set $B_o := \{[v, z] \in D; J_o(v, z) \leq 1 + \frac{N_o}{\varepsilon_1}\}$ is a required one. $\Diamond$

Lemma 4.3. *Let B be any subset of D with $\sup_{[v,z] \in B} J_o(v, z) < +\infty$ and δ be any positive number. Then $B_\delta := \text{cl}\{\cup_{t \geq \delta} S(t)B\}$ is in D_o, compact in $L^2(\Omega) \times H^1(\Omega)$, bounded in $H^1(\Omega) \times H^2(\Omega)$ and $\sup_{[v,z] \in B_\delta} J_o(v, z) < +\infty$, where $\text{cl}(\cdot)$ stands for the closure of $(\cdot)$ in $L^2(\Omega) \times H^1(\Omega)$.*

It is easy to get the above lemma from Theorem 2.2.

Proof of Theorem 4.2: The construction of A is standard. As usual, we are going to show that

$$A := \bigcap_{s>0} \text{cl}\{\bigcup_{t \geq s} S(t)B_o\} \tag{4.6}$$

satisfies the required properties, where B_o is the absorbing set obtained by Lemma 4.2.

From the above definition (4.6) of A it is clear that

$$\begin{cases} X := [v, z] \in A \text{ if and only if there are sequences } \{t_n\} \text{ with } t_n \to +\infty \\ (\text{as } n \to +\infty) \text{ and } \{X_n := [v_n, z_n]\} \text{ in } B_o \text{ such that} \\ \qquad S(t_n)X_n \to X \text{ in } L^2(\Omega) \times H^1(\Omega). \end{cases} \tag{4.7}$$

Moreover, on account of Lemmas 4.2, 4.3, it is easily seen that $A \subset D_o \cap B_o$, and in (4.7) the sequences $\{t_n\}$ and $\{X_n\}$ can be chosen so as to satisfy further that for some $\tilde{X} := [\tilde{v}, \tilde{z}] \in D_o$

$$\begin{cases} X_n \to \tilde{X}, \quad S(t_n)X_n \to X \quad \text{weakly in } H^1(\Omega) \times H^2(\Omega), \\ \sup_{n \geq 1} J_o(X_n) < +\infty, \quad \sup_{n \geq 1} J_o(S(t_n)X_n) < +\infty. \end{cases} \tag{4.8}$$

Now, the invariance of A is shown as follows. Let $t > 0$ and $X := [v, z]$ be any element of A. Then there are sequences $\{t_n\}$ with $t_n \to +\infty$ and $\{X_n\} \subset B_o$ satisfying (4.7) with (4.8). Then, by (a) of Theorem 4.1,

$$S(t_n + t)X_n = S(t)S(t_n)X_n \to S(t)X \quad \text{weakly in } H^1(\Omega) \times H^2(\Omega).$$

Hence $S(t)X \in A$ by characterization (4.7), and $S(t)A \subset A$.

Conversely, we show $A \subset S(t)A$ for $t > 0$. Let X and $\{X_n\}$ be as above. Then, $\{Y_n := S(t_n - t)X_n; n \geq n_o\}$ is bounded in $H^1(\Omega) \times H^2(\Omega)$ and $\sup_{n \geq n_o} J_o(Y_n) < +\infty$, if n_o is sufficiently large, so we may assume that $Y_n \to Y$ weakly in $H^1(\Omega) \times H^2(\Omega)$ for some $Y \in A$. By (a) of Theorem 4.1 again, it follows that

$$S(t)Y_n \to S(t)Y \quad \text{weakly in } H^1(\Omega) \times H^2(\Omega).$$

This implies that $X = S(t)Y (\subset S(t)A)$, since $S(t)Y_n = S(t_n)X_n \to X$ weakly in $H^1(\Omega) \times H^2(\Omega)$. Hence $A \subset S(t)A$. Thus (ii) has been obtained. Assertion (i) is standard; in particular, the connectedness of A can be drived from the convexity of D with the help of characterization (4.7) and (4.8) for A and the convergence result (a) of Theorem 4.1. We omit its detailed proof. Assertion (iii) immediately follows from the definition (4.6) of A. $\Diamond$

Finally, we mention an interesting remark concerning the generation of another semigroup associated to (PFC). Inequality (2.8) of Theorem 2.1 suggests us that there is a semigroup $\{E(t)\}_{t \geq 0}$ on a subset D_* of $H^1(\Omega)^* \times L^2(\Omega)$ such that

(E1) D_* is the closure of $D_E := \{[\rho(v) + \lambda(z)]; [v, z] \in D\} (\subset L^2(\Omega) \times H^1(\Omega))$ in $H^1(\Omega)^* \times L^2(\Omega)$;

(E2) $E(t)[e_o, w_o] := [e(t), w(t)]$ for each $[e_o, w_o] \in D_E$, where $e(t) = \rho(u) + \lambda(w)$ and $\{u, w\}$ is the global solution of (PFC;u_o, w_o) with $u_o := \rho^{-1}(e_o - \lambda(w_o))$.

In fact, on account of (2.8) of Theorem 2.1, $\{E(t)\}$ is uniquely determined by (E1) and (E2). Now, many interesting questions, which are still open, are proposed. For instance we have the following questions.

- Is it true that (smoothing effect) $E(t)[e_o, w_o] \in D_E$ for $t > 0$ and $[e_o, w_o] \in D_*$?

- Does there exist a global attractor of $\{E(t)\}$ in $H^1(\Omega)^* \times L^2(\Omega)$?

References

1. P. W. Bates and S. Zheng, Inertial manifolds and inertial sets for the phase-field equations, J. Dynamical Diff. Equations, **4** (1992), 375-398.

2. J. F. Blowey and C. M. Elliott, A phase-field model with a double obstacle potential, in *Motion by Mean Curvature*, pp. 1-22, Walter de Gruyter, Berlin-New York, 1994.

3. H. Brézis, M. G. Crandall and A. Pazy, Perturbations of nonlinear maximal monotone sets, Comm. Pure Appl. Math., **23** (1970), 123-144.

4. P. Colli and J. Sprekels, On a Penrose-Fife Model with zero interfacial energy leading to a phase-field system of relaxed Stefan type, preprint, **No.104**, IAAS, Berlin, 1994.

5. A. Damlamian, N. Kenmochi and N. Sato, Phase field equations with constraint, in *Nonlinear Mathematical Problems in Industry II*, pp. 391-404, GAKUTO Inter. Ser. Math. Sci. Appl., **Vol.2**, Gakkōtosho, Tokyo, 1993.

6. W. Horn, J. Sprekels and S. Zheng, Global existence of smooth solutions to the Penrose-Fife model for Ising ferromagnets, Adv. Math. Sci. Appl., to appear.

7. A. Ito and N. Kenmochi, Asymptotic behaviour of solutions to phase field models with constraints, Funk. Ekvac., to appear, Tech. Rep. Math. Sci., Chiba Univ., **Vol.9**(1993).

8. N. Kenmochi, Uniqueness of the solution to a nonlinear system arising in phase transition, in *Nonlinear Analysis and Applications*, pp. 261-271, GAKUTO Inter. Ser. Math. Sci. Appl., **Vol.7**, Gakkōtosho, Tokyo, 1995.

9. N. Kenmochi and M. Niezgódka, Systems of nonlinear parabolic equations for phase change problems, Adv. Math. Sci. Appl., **3**(1993/94), 89-117.

10. N. Kenmochi, M. Niezgódka and S. Zheng, Global attractor of a non-isothermal model for phase separation, in *Curvature Flows and Related Topics*, pp. 129-143, GAKUTO Inter. Ser. Math. Sci. Appl., **Vol.5**, Gakkōtosho, Tokyo, 1995.

11. Ph. Laurençot, Solutions to a Penrose-Fife model of phase-field type, J. Math. Anal. Appl., **185**(1994), 262-274.

12. O. Penrose and P. C. Fife, Thermodynamically consistent models of phase-field type for the kinetics of phase transitions, Physica D, **43**(1990), 44-62.

13. J. Shirohzu, N. Sato and N. Kenmochi, Asymptotic convergence in models for phase change problems, in *Nonlinear Analysis and Applications*, pp. 361-385, GAKUTO Inter. Ser. Math. Sci. Appl., **Vol.7**, Gakkōtosho, Tokyo, 1995.

14. J. Sprekels and S. Zheng, Global smooth solutions to a thermodynamically consistent model of phase-field type in higher space dimensions, J. Math. Anal. Appl. **176**(1993), 200-223.

15. R. Temam, *Infinite Dimensional Dynamical Systems in Mechanics and Physics*, Springer-Verlag, Berlin, 1988.

16. S. Zheng, Global existence for a thermodynamically consistent model of phase field type, Differential Integral Equations., **5**(1992), 241-253.

Nobuyuki Kenmochi: Department of Mathematics, Faculty of Education,
Chiba University, 1-33 Yayoi-chō, Inage-ku, Chiba, 263 Japan

T KOYAMA

On a heat equation with hysteresis in the source term

1. Introduction.

The aim of this note is to construct a unique solution of the following initial boundary value problem of the system of two parabolic equations:

$$(P) \quad \begin{cases} u_t - \Delta u + w = 0 & \text{in } Q := \Omega \times (0, \infty), \\ w_t - \Delta w + \partial I(u, \cdot)w \ni 0 & \text{in } Q, \\ u|_\Gamma = g_1, \ w|_\Gamma = g_2, & \text{on } \Sigma := \Gamma \times (0, \infty), \\ u(\cdot, 0) = u_0, \ w(\cdot, 0) = w_0 & \text{in } \Omega. \end{cases}$$

Here Ω is a bounded domain of $\mathbf{R}^N$ with smooth boundary Γ. I is a functional on $\mathbf{R}^2$ defined by

$$I(\xi, \eta) := \begin{cases} 0 & \text{if } f_a(\xi) \leq \eta \leq f_d(\xi), \\ +\infty & \text{otherwise,} \end{cases}$$

where f_a and f_d are given monotone nondecreasing functions such that

$$(f.1) \qquad\qquad f_a(\xi) \leq f_d(\xi) \qquad \text{for all } \xi \in \mathbf{R}.$$

For any function $\xi(t)$, $\eta \mapsto I(\xi(t), \eta)$ is a time dependent proper lower semicontinuous and convex functional on $\mathbf{R}$, and its subdifferential operator is

$$\partial I(\xi(t), \cdot)\eta := \begin{cases} [0, \infty) & \text{if } f_a(\xi(t)) < \eta = f_d(\xi(t)), \\ \{0\} & \text{if } f_a(\xi(t)) < \eta < f_d(\xi(t)), \\ (-\infty, 0] & \text{if } f_a(\xi(t)) = \eta < f_d(\xi(t)), \\ \mathbf{R} & \text{if } f_a(\xi(t)) = \eta = f_d(\xi(t)), \end{cases}$$

with domain $D(\partial I(\xi(t), \cdot)) = \{\eta \in \mathbf{R}; f_a(\xi(t)) \leq \eta \leq f_d(\xi(t))\}$. For each $\xi(t) \in AC([0, T])$ and $\eta_0 \in \mathbf{R}$ with $f_a(\xi(0)) \leq \eta_0 \leq f_d(\xi(0))$, the initial value problem for O.D.E.

$$\begin{cases} \eta_t + \partial I(\xi(t), \cdot)\eta \ni 0 & \text{for } t > 0, \\ \eta(0) = \eta_0, \end{cases}$$

has a unique solution $\eta(t)$. The operator $\xi(\cdot) \mapsto \eta(\cdot)$ involves hysteresis type memory effect and is called *Hysteron operator*. For this, see [K-P] and [H]. Therefore, if u and w are regarded as temperature and intensity of heat source respectively, (P) is a model

126

for heat conduction with feed back controle, which controller has memory property and spatial diffusion property.

In the case w does not cause diffution, 2nd. equation of (P) is replaced by

$$w_t + \partial I(u, \cdot)w \ni 0 \quad \text{in } Q,$$

and this is essentially O. D. E.. In such case, existence, uniqueness and asymptotic stability results are obtained in [V], [K-K-V] [K-V] and [K].

This time, besides the memory effect, we introduce spatial diffution effect for w, and the 2nd. equation of (P) comes to *the heat equation with moving obstacles*. To get spatial regularity of w, we need stronger assumptions on initial data and functions f_a and f_d, than diffusion-less case.

2. Problem and Result.

Besides $(f.1)$, we make the following assumptions:

$(f.2)$
$$f_a, f_d \in C_b^2(\mathbf{R}) \text{ and}$$
$$|f_a'|_{L^\infty(\mathbf{R})}, |f_d'|_{L^\infty(\mathbf{R})} \leq L, \ |f_a''|_{L^\infty(\mathbf{R})}, |f_d''|_{L^\infty(\mathbf{R})} \leq M,$$

(BV)
$$g_1, g_2 \in C(\Gamma),$$

(IV)
$$u_0, w_0 \in W^{2,2}(\Omega) \cap L^\infty(\Omega),$$
$$u_0|_\Gamma = g_1, \quad w_0|_\Gamma = g_2,$$

and that

(C)
$$f_a(u_0) \leq w_0 \leq f_d(u_0), \quad \text{a.e. on } \Omega,$$
$$f_a(g_1) \leq g_2 \leq f_d(g_1), \quad \text{a.e. on } \Gamma.$$

Then the following theorem holds.

Theorem. *Under the assumptions $(f.1)$, $(f.2)$, (BV), (IV) and (C), the problem (P) has, for each $T > 0$, a unique solution (u, w) in*

$$L^\infty(0, T; W^{1,2}(\Omega)) \cap W^{1,2}(0, T; L^2(\Omega)) \cap L^\infty(Q_T),$$

3. Sketch of proof.

First, split the problem into

$(P)_u$
$$\begin{cases} u_t - \Delta u + w = 0 & \text{in } Q, \\ u|_\Gamma = g_1 & \text{on } \Sigma, \\ u(\cdot, 0) = u_0 & \text{in } \Omega, \end{cases}$$

and

$$(P)_w \quad \begin{cases} w_t - \Delta w + \partial I(u(t), \cdot)w(t) \ni 0 & \text{in } Q, \\ w|_\Gamma = g_2 & \text{on } \Sigma, \\ w(\cdot, 0) = w_0 & \text{in } \Omega, \end{cases}$$

Next, approximate them by time-discretization with time unit $\tau > 0$, and Yosida approximation with parameter $\lambda > 0$ as follows:

$$(P)_{u,\tau} \quad \begin{cases} \tau^{-1}(u_i - u_{i-1}) - \Delta u_i + w_i = 0 & \text{in } \Omega, \quad i = 1, \cdots, n, \\ u_i|_\Gamma = g_1 & \text{on } \Sigma, \quad i = 1, \cdots, n, \end{cases}$$

and

$$(P)_{w,\tau,\lambda} \quad \begin{cases} \tau^{-1}(w_i - w_{i-1}) - \Delta w_i + \partial I(u_i, \cdot)_\lambda w_i = 0 \\ \qquad\qquad\qquad\qquad \text{in } \Omega, \quad i = 1, \cdots, n, \\ w_i|_\Gamma = g_2 & \text{on } \Sigma, \quad i = 1, \cdots, n. \end{cases}$$

Here $\partial I(u_i, \cdot)_\lambda$ is the Yosida approximation for $\partial I(u_i, \cdot)$ defined by

$$\partial I(u_i(x), \cdot)_\lambda \eta = \frac{1}{\lambda}[(\eta - f_d(u_i(x)))^+ - (\eta - f_a(u_i(x)))^-],$$

for each $x \in \Omega$.

The idea of the proof is illustrated as follows: Problems $(P)_{u,\tau}$ and $(P)_{w,\tau,\lambda}$ define operators $\{w_i\} \mapsto \{u_i\}$ and $\{u_i\} \mapsto \{w_i\}$, and composite operators of these two operators have unique fixed points $\{u_{i,\lambda}\}$ and $\{w_{i,\lambda}\}$, which are approximate solution of (P). Letting τ, $\lambda \downarrow 0$ gives a solution (u, w) of (P).

For the simplicity, put

$$\delta u_i := \frac{1}{\tau}(u_i - u_{i-1}), \quad \delta w_i := \frac{1}{\tau}(w_i - w_{i-1}), \quad i = 1, 2, \cdots,$$

$$F(\xi, \eta) := (\eta - f_d(\xi))^+ - (\eta - f_a(\xi))^-,$$

and

$$\varphi(u) := \frac{1}{2}|\nabla u|_{L^2(\Omega)}^2.$$

(Step 1).

Lemma 1. (i) *For each fixed* $\{w_i\}_{i=1,\cdots,n} \subset W^{1,2}(\Omega) \cap L^\infty(\Omega)$, $(P)_{u,\tau}$ *has a unique solution* $\{u_i\}_{i=1,\cdots,n} \subset W^{1,2}(\Omega) \cap L^\infty(\Omega)$ *which satisfies*

$$(1) \qquad \frac{1}{2}\sum_{k=1}^n \tau|\delta u_k|_{L^2(\Omega)}^2 + \varphi(u_n) \le \varphi(u_0) + \frac{1}{2}\sum_{k=1}^n \tau|w_k|_{L^2(\Omega)}^2,$$

and

$$(2) \qquad |u_i|_{L^\infty(\Omega)} \le |u_0|_{L^\infty(\Omega)} + \sum_{k=1}^i \tau |w_k|_{L^\infty(\Omega)}, \quad i = 1, \cdots, n.$$

(ii) *Let us denote the solution* $\{u_i\}$ *of* $(P)_{u,\tau}$ *by* $\mathcal{K}_\tau\{w_i\}$, *and put* $\{u_{i,j}\} = \mathcal{K}_\tau\{w_{i,j}\}$, $j = 1, 2$. *Then*

$$(3) \qquad |u_{i,1} - u_{i,2}|_{L^\infty(\Omega)} \le \sum_{k=1}^i \tau |w_{k,1} - w_{k,2}|_{L^\infty(\Omega)}, \quad i = 1, \cdots, n.$$

Proof of Lemma 1. Existence and energy estimate (1) are well known. L^∞ estimate (2) is obtained by the maximum principle. To show (3), first note that

$$\frac{1}{\tau}((u_{i,1} - u_{i,2}) - (u_{i-1,1} - u_{i-1,2})) - \Delta(u_{i,1} - u_{i,2}) + (w_{i,1} - w_{i,2}) = 0.$$

Suppose $\max_\Omega(u_{i,1} - u_{i,2})$ is obtained at $x_0 \in \mathbf{R}$. Then

$$
\begin{aligned}
(u_{i,1} - u_{i,2})(x_0) &\le (u_{i-1,1} - u_{i-1,2})(x_0) - \tau(w_{i,1} - w_{i,2})(x_0) \\
&\le \max_\Omega(u_{i-1,1} - u_{i-1,2}) - \tau \min_\Omega(w_{i,1} - w_{i,2}),
\end{aligned}
$$

because $\Delta(u_{i,1} - u_{i,2})(x_0) \le 0$. In the similar way we have

$$\min_\Omega(u_{i,1} - u_{i,2}) \ge \min_\Omega(u_{i-1,1} - u_{i-1,2}) - \tau \max_\Omega(w_{i,1} - w_{i,2}).$$

Sum them up in i, we have (3). $\square$

Lemma 2. (i) *For each fixed* $\{u_i\}_{i=1,\cdots,n} \subset W^{1,2}(\Omega) \cap L^\infty(\Omega)$, $(P)_{w,\tau,\lambda}$ *has a unique solution* $\{w_i\}_{i=1,\cdots,n} \subset W^{1,2}(\Omega) \cap L^\infty(\Omega)$ *which satisfies*

$$
(4) \qquad
\begin{aligned}
\sum_{k=1}^n \tau |\delta w_k|^2_{L^2(\Omega)} &+ \varphi(w_n) + \frac{1}{2\lambda}|F(u_n, w_n)|^2_{L^2(\Omega)} \\
&\le \varphi(u_0) + C(u) \sum_{k=1}^n L\tau |\delta u_k|_{L^\infty(\Omega)},
\end{aligned}
$$

where

$$C(u) := \max_{i=1,\cdots,n} (M\varphi(u_i) + L|\Delta u_i|_{L^1(\Omega)} + L|\delta u_i|_{L^1(\Omega)}),$$

and

$$
(5) \qquad
\begin{aligned}
|w_i|_{L^\infty(\Omega)} &\le |w_0|_{L^\infty(\Omega)} + \max_{k=1,\cdots,i}(|f_a(u_k)|_{L^\infty(\Omega)} + |f_d(u_k)|_{L^\infty(\Omega)}), \\
& \qquad\qquad\qquad\qquad\qquad\qquad\qquad i = 1, \cdots, n.
\end{aligned}
$$

(ii) *Let us denote the solution $\{w_i\}$ of $(P)_{w,\tau,\lambda}$ by $\mathcal{F}_{\tau,\lambda}\{u_i\}$, and put $\{w_{i,j}\} = \mathcal{F}_{\tau,\lambda}\{u_{i,j}\}$, $j = 1, 2$. Then*

(6)
$$|w_{i,1} - w_{i,2}|_{L^\infty(\Omega)} \le L \max_{k=1,\cdots,i} |u_{k,1} - u_{k,2}|_{L^\infty(\Omega)} \quad i = 1, \cdots, n.$$

Proof of Lemma 2. Because $\eta \mapsto \partial I(u_i(x), \cdot)_\lambda \eta$ is Lipschitz continuous for each fixed i and $x \in \Omega$, existence is proved by the standard argument. To derive energy estimate (4), muliply

(7)
$$\delta w_i - \Delta w_i + \frac{1}{\lambda} F(u_i, w_i) = 0$$

by δw_i and note

$$(-\Delta w_i, \delta w_i) \ge \frac{1}{\tau}(\varphi(w_i) - \varphi(w_{i-1}))$$

and

$$\left(\frac{1}{\lambda} F(u_i, w_i), \delta w_i\right) \ge \frac{1}{\lambda\tau}(F(u_i, w_i), F(u_i, w_i) - F(u_{i-1}, w_{i-1}))$$
$$- \frac{1}{\lambda\tau}|F(u_i, w_i)| \max(|f_d(u_i) - f_d(u_{i-1})|, |f_d(u_i) - f_d(u_{i-1})|).$$

This leads to

$$\tau|\delta w_i|^2_{L^2(\Omega)} + \varphi(w_i) + \frac{1}{2\lambda}|F(u_i, w_i)|^2_{L^2(\Omega)} \le \varphi(w_{i-1}) + \frac{1}{2\lambda}|F(u_{i-1}, w_{i-1})|^2_{L^2(\Omega)}$$
$$+ \frac{1}{\lambda}|F(u_i, w_i)|_{L^1(\Omega)} L|u_i - u_{i-1}|_{L^\infty(\Omega)}.$$

Summing this up for $i = 1, \cdots, n$ gives

$$\sum_{k=1}^{n} \tau|\delta w_k|^2_{L^2(\Omega)} + \varphi(w_n) + \frac{1}{\lambda}|F(u_n, w_n)|^2_{L^2(\Omega)} \le \varphi(u_0) + \frac{L\tau}{\lambda}\sum_{k=1}^{n}|F(u_k, w_k)|_{L^1(\Omega)}|\delta u_k|_{L^\infty(\Omega)}.$$

Next, multiply (7) by $\mathrm{sgn}(F(u_i, w_i))$ and note

$$(\delta w_i, \mathrm{sgn}(F(u_i, w_i))) \ge \frac{1}{\tau}(F(u_i, w_i) - F(u_{i-1}, w_{i-1}), \mathrm{sgn}(F(u_i, w_i)))$$
$$- \frac{1}{\tau}\max(|f_d(u_i) - f_d(u_{i-1})|_{L^1(\Omega)}, |f_a(u_i) - f_a(u_{i-1})|_{L^1(\Omega)}),$$
$$\ge \frac{1}{\tau}|F(u_i, w_i)|_{L^1(\Omega)} - \frac{1}{\tau}|F(u_{i-1}, w_{i-1})|_{L^1(\Omega)} - L|\delta u_i|_{L^1(\Omega)},$$

$$(-\Delta w_i, \mathrm{sgn}\, F(u_i, w_i)) \ge - \int_{\partial\{w_i > f_d(u_i)\}} \frac{\partial}{\partial n}(w_i - f_d(u_i))\, dS$$
$$+ \int_{\partial\{w_i < f_a(u_i)\}} \frac{\partial}{\partial n}(w_i - f_a(u_i))\, dS$$
$$- \int_{\partial\{w_i > f_d(u_i)\}} \Delta f_d(u_i)\, dx + \int_{\partial\{w_i < f_a(u_i)\}} \Delta f_a(u_i)\, dx$$
$$\ge -2M\varphi(u_i) - L|\Delta u_i|_{L^1(\Omega)} =: K_i$$

130

and

$$\left(\frac{1}{\lambda}F(u_i, w_i), \mathrm{sgn}(F(u_i, w_i))\right) = \frac{1}{\lambda}|F(u_i, w_i)|_{L^1(\Omega)}.$$

Then we have

$$\left(1 + \frac{\tau}{\lambda}\right)|F(u_i, w_i)|_{L^1(\Omega)} \le |F(u_{i-1}, w_{i-1})|_{L^1(\Omega)} + \tau(L|\delta u_i|_{L^1(\Omega)} + K_i).$$

Multiply this by $\left(1 + \frac{\tau}{\lambda}\right)^{i-1}$ and sum up in i, we have

$$\left(1 + \frac{\tau}{\lambda}\right)^{i}|F(u_i, w_i)|_{L^1(\Omega)} \le \tau \max_{k=1,\cdots,i}(L|\delta u_i|_{L^1(\Omega)} + K_i)\sum_{k=0}^{i-1}\left(1 + \frac{\tau}{\lambda}\right)^{k}.$$

This means

$$|\partial I(u_i, \cdot)_\lambda w_i|_{L^1(\Omega)} = \frac{1}{\lambda}|F(u_i, w_i)|_{L^1(\Omega)} \le \max_{k=1,\cdots,i}(L|\delta u_i|_{L^1(\Omega)} + K_i),$$

and (4). To show (5), compare w_i with solutions $w_{i,*}$ $(* = a, d)$ of

$$\begin{cases} \delta w_{i,*} - \Delta w_{i,*} + \frac{1}{\lambda}(w_{i,*} - f_*(u_i)) = 0 \text{ in } \Omega, \quad i = 1, \cdots, n, \\ w_{i,*}|_\Gamma = g_2 \text{ on } \Sigma, \quad i = 1, \cdots, n, \end{cases}$$

and apply maximum princilple to $w_{i,*}$. To show (6), it is enough to show

$$|w_{i,1} - w_{i,2}|_{L^\infty(\Omega)} \le \max\left(\begin{array}{c} |w_{i-1,1} - w_{i-1,2}|_{L^\infty(\Omega)}, \\ |f_a(u_{i,1}) - f_a(u_{i,2})|_{L^\infty(\Omega)}, \\ |f_d(u_{i,1}) - f_d(u_{i,2})|_{L^\infty(\Omega)} \end{array}\right).$$

First, note that

$$w_{i,1} - w_{i,2} - \tau\Delta(w_{i,1} - w_{i,2}) + \frac{\tau}{\lambda}(F(u_{i,1}, w_{i,1}) - F(u_{i,2}, w_{i,2}))$$
$$= w_{i-1,1} - w_{i-1,2},$$

and put

$$\alpha := -\tau\Delta(w_{i,1} - w_{i,2}) + \frac{\tau}{\lambda}(F(u_{i,1}, w_{i,1}) - F(u_{i,2}, w_{i,2})).$$

Suppose that $\max_\Omega(w_{i,1} - w_{i,2})$ is attained at $x_0 \in \Omega$, and $\alpha(x_0) < 0$. Then, because $\Delta(w_{i,1} - w_{i,2})(x_0) \le 0$, we have

$$F(u_{i,1}, w_{i,1}) - F(u_{i,2}, w_{i,2}) < 0.$$

And this leads to

$$(f_a(u_{i,1}) - f_a(u_{i,2}))(x_0) > (w_{i,1} - w_{i,2})(x_0), \text{ or}$$
$$(f_d(u_{i,1}) - f_d(u_{i,2}))(x_0) > (w_{i,1} - w_{i,2})(x_0).$$

□

(Step 2). By (3) and (6), composite operator

$$\mathcal{K}_\tau \circ \mathcal{F}_{\tau,\lambda} : (W^{1,2}(\Omega) \cap L^\infty(\Omega))^n \to (W^{1,2}(\Omega) \cap L^\infty(\Omega))^n$$

has a unique fixed point $\{u_{i,\lambda}\}_{i=1,\cdots,n}$ which gives a solution of

$$(P)_{\tau,\lambda} \quad \begin{cases} \delta u_{i,\lambda} - \Delta u_{i,\lambda} + w_{i,\lambda} = 0 & \text{in } \Omega, \quad i = 1, \cdots, n, \\ \delta w_{i,\lambda} - \Delta w_{i,\lambda} + \partial I(u_{i,\lambda}, \cdot)_\lambda w_{i,\lambda} = 0 & \text{in } \Omega, \quad i = 1, \cdots, n, \\ u_{i,\lambda}|_\Gamma = g_1, \quad w_{i,\lambda}|_\Gamma = g_2, & \text{on } \Sigma, \quad i = 1, \cdots, n, \\ u_{0,\lambda} = u_0, \quad w_{0,\lambda} = w_0, & \text{on } \Omega, \end{cases}$$

together with $\{w_{i,\lambda}\} := \mathcal{F}_{\tau,\lambda}\{u_{i,\lambda}\}$.

Moreover, (3) and (6) gives

$$|\delta u_{i,\lambda}|_{L^\infty(\Omega)} \leq |\delta u_{1,\lambda}|_{L^\infty(\Omega)} + \sum_{k=2}^{i} \tau L |\delta u_{k,\lambda}|_{L^\infty(\Omega)},$$

and therefore

$$(8) \qquad |\delta u_{i,\lambda}|_{L^\infty(\Omega)} \leq |\delta u_{1,\lambda}|_{L^\infty(\Omega)} (1 - L\tau)^{-i+1}.$$

Now put $i = [t/\tau]$ for $0 < t \leq T$. Then $(u_{[t/\tau],\lambda}, w_{[t/\tau],\lambda})$ gives an approximate solution of (P), and (8) leads to

$$(9) \qquad |\delta u_{[t/\tau],\lambda}|_{L^\infty(\Omega)} \leq e^{3Lt/2} |\delta u_{1,\lambda}|_{L^\infty(\Omega)} \qquad \text{for } 0 < \tau \leq \frac{1}{2L}.$$

On the other hand, from (2) and (5), we have

$$|u_{i,\lambda}|_{L^\infty(\Omega)}, \ |w_{i,\lambda}|_{L^\infty(\Omega)} \leq A(|u_0|_{L^\infty(\Omega)} + |w_0|_{L^\infty(\Omega)}) + B,$$

with some constants A and B which are independent of i, τ and λ. Therefore,

$$|\delta u_{1,\lambda}|_{L^\infty(\Omega)} = \left| \frac{1}{\tau}((1 - \tau\Delta)^{-1}(u_0 - \tau w_{1,\lambda}) - u_0) \right|_{L^\infty(\Omega)}$$
$$\leq |\Delta u_0|_{L^\infty(\Omega)} + C'|w_{1,\lambda}|_{L^\infty(\Omega)} + C' \leq C,$$

and

$$|\delta u_{[t/\tau],\lambda}|_{L^\infty(\Omega)} \leq C e^{3LT/2}$$

where the constants C' and C depend only on u_0 and w_0.

By Lemmas 1 and 2, the quantities

$$\frac{1}{2} \sum_{k=1}^{[t/\tau]} \tau |\delta u_{k,\lambda}|^2_{L^2(\Omega)} + \varphi(u_{[t/\tau],\lambda}),$$

$$|u_{[t/\tau],\lambda}|_{L^\infty(\Omega)},$$

$$\sum_{k=1}^{[t/\tau]} \tau |\delta w_{k,\lambda}|^2_{L^2(\Omega)} + \varphi(w_{[t/\tau],\lambda}) + \frac{1}{2\lambda} |F(u_{[t/\tau],\lambda}, w_{[t/\tau],\lambda})|^2_{L^2(\Omega)},$$

$$|w_{[t/\tau],\lambda}|_{L^\infty(\Omega)},$$

remain bounded as $\tau, \lambda \downarrow 0$. By standard arguments,

$$u(t) := \lim_{\tau,\lambda\downarrow 0} u_{[t/\tau],\lambda}, \quad \text{and} \quad w(t) := \lim_{\tau,\lambda\downarrow 0} w_{[t/\tau],\lambda}$$

exist and give the solution of (P) in

$$L^\infty(0,T;W^{1,2}(\Omega)) \cap W^{1,2}(0,T;L^2(\Omega)) \cap L^\infty(Q_T).$$

□

References

[H] M. Hilpert, On uniqueness for evolution problems with hysteresis, Mathematical Models for Phase Change Problems, ed. J. F. Rodrigues, Intern. Ser. Numer. Math. Vol. 88 ,377-388, Birkhäuser, Basel-Boston-Berlin, (1989).

[K] T. Koyama, Asymptotic stability for a class of variational inequalities with hysteresis. Advances in Mathematical Sciences and Applications, **2**, 369-389 (1993).

[K-K-V] N. Kenmochi, T. Koyama and A. Visintin, On a class of variational inequalities with memory terms, PROGRESS IN PARTIAL DIFFERENTIAL EQUATIONS: ELLIPTIC AND PARABOLIC PROBLEMS, C. Bandle et al.(editors), Pitman research note in mathematics series, 266, Longman, Essex UK. 164-175 (1992).

[K-V] N. Kenmochi and A. Visintin, Asymptotic stability for parabolic variational inequalities with hysteresis, MODELS OF HYSTERESIS, A. Visintin (editor), Pitman research note in mathematics series, 286, Longman, Essex UK. 59-70 (1993).

[K-P] M. A. Krasnoselskiĭ and A. V. Pokrovskiĭ, *Systems with hysteresis* (Russian), Nauka, Moskow, 1983. English translation: Springer, Berlin, 1989.

[V] A. Visintin, Evolution problems with hysteresis in source term, SIAM J. Math. Anal. **17**, 1113-1138 (1986).

Tetsuya Koyama
Department of Mathematics,
Hiroshima Institute of Technology,
Miyake, Saiki-ku, Hiroshima, 731-51 Japan.
tkoyama@ds5.cc.it-hiroshima.ac.jp

Ph LAURENÇOT

Degenerate Cahn–Hilliard equation as limit of the phase-field equation with non-constant thermal conductivity[1]

Abstract: In this note, we study the relationship between the degenerate Cahn-Hilliard equation and the classical phase-field model with non-constant thermal conductivity. We consider the phase-field model with logarithmic free energy, and prove that weak solutions to this problem constructed by a suitable regularization process converge to a weak solution to the degenerate Cahn-Hilliard equation.

1   Introduction

In this note, we study the behaviour of weak solutions (ϕ^α, u^α) to

$$(1.1) \qquad \alpha\, \phi_t^\alpha + \ln\left(\frac{1+\phi^\alpha}{1-\phi^\alpha}\right) \;=\; \xi^2\, \Delta\phi^\alpha + 2\, \phi^\alpha + u^\alpha \ \text{ in } \Omega \times (0, +\infty),$$

$$(1.2) \qquad \alpha\, u_t^\alpha + \phi_t^\alpha \;=\; div\left(B(\phi^\alpha)\, \nabla u^\alpha\right) \ \text{ in } \Omega \times (0, +\infty),$$

$$(1.3) \qquad \frac{\partial\phi^\alpha}{\partial n} \;=\; 0, \quad \frac{\partial u^\alpha}{\partial n} = 0 \ \text{ on } \Gamma \times (0, +\infty),$$

$$(1.4) \qquad \phi^\alpha(0) \;=\; \phi_0^\alpha, \ \ u^\alpha(0) = u_0^\alpha \ \text{ in } \Omega,$$

as α decreases to zero. Here, Ω is a bounded open subset of $\mathbb{R}^N$ ($N \geq 1$), with smooth boundary Γ, α and ξ are positive real numbers, and B denotes the thermal conductivity ; we assume here that B is given by :

$$(1.5) \qquad\qquad B(r) = \begin{cases} 1 - r^2 & \text{if } |r| \leq 1, \\ 0 & \text{otherwise.} \end{cases}$$

The problem $(1.1) - (1.4)$ arises from the modelling of nonisothermal phase transitions in binary systems using the phase-field approach. Here, ϕ^α denotes the order parameter (which is the state variable characterizing the different phases) and u^α, the temperature (see [Ca], [CT]). Let us notice that, since $B(\phi)$ vanishes for $|\phi| \geq 1$, (1.2) is a degenerate parabolic equation.

Formally, if we set $\alpha = 0$ in $(1.1) - (1.4)$, we find that the limit of (ϕ^α, u^α) (if it exists) should satisfy

$$(1.6) \qquad\qquad u \;=\; -\xi^2\, \Delta\phi + \ln\left(\frac{1+\phi}{1-\phi}\right) - 2\, \phi \ \text{ in } \Omega \times (0, +\infty),$$

$$(1.7) \qquad\qquad \phi_t \;=\; div\left(B(\phi)\, \nabla u\right) \ \text{ in } \Omega \times (0, +\infty),$$

$$(1.8) \qquad\qquad \frac{\partial\phi}{\partial n} \;=\; 0, \quad \frac{\partial u}{\partial n} = 0 \ \text{ on } \Gamma \times (0, +\infty),$$

$$(1.9) \qquad\qquad \phi(0) \;=\; \phi_0 \ \text{ in } \Omega.$$

---

[1]Part of this work was done while the author was a Ph.D. student at the Equipe de Mathématiques, CNRS URA 741, Université de Franche-Comté in Besançon, France

The problem $(1.6) - (1.9)$ is a degenerate Cahn-Hilliard equation, which has been studied by C.M. Elliott and H. Garcke in [EG], where existence of weak solutions is proved.

Concerning $(1.1) - (1.4)$, existence of weak solutions has been investigated in [La] : these weak solutions arise as limits of solutions of a regularized version of $(1.1) - (1.4)$. It is the purpose of this work to prove rigorously that weak solutions to $(1.1) - (1.4)$ constructed by a suitable regularization process converges as α decreases to zero to a weak solution of $(1.6) - (1.9)$, which belongs to the same class as that of C.M. Elliott and H. Garcke ([EG]). Of course, the lack of uniqueness of weak solutions to $(1.1) - (1.4)$ and $(1.6) - (1.9)$ prevents us from getting more precise results.

A similar result has been proved by B. Stoth when B is a positive constant, and for a smooth double well potential ([St]). However, the method does not seem to apply here because of the degeneracy of B.

Let us finally mention that the method we use in this paper relies strongly on the particular choice of B and the logarithmic free energy, and does not seem to extend to the general case.

2   Convergence to the degenerate Cahn-Hilliard equation

We put

$$(2.1) \qquad \beta(r) = \ln\left(\frac{1+r}{1-r}\right), \quad r \in (-1,1),$$

and denote by $\hat{\beta}$, the convex function vanishing at $r = 0$ whose subdifferential in $\mathbb{R}$ is β, namely

$$(2.2) \qquad \hat{\beta}(r) = (1+r)\,\ln(1+r) + (1-r)\,\ln(1-r), \quad r \in [-1,1].$$

We then put $F = \hat{\beta} + F_0$, where

$$(2.3) \qquad F_0(r) = 1 - r^2, \quad r \in [-1,1].$$

We now introduce some notations and assumptions on the initial data $(\phi_0^\alpha, u_0^\alpha)$ in (1.4). We denote by V', the dual space of $H^1(\Omega)$, and by $\langle .,. \rangle_{V',V}$, the duality pairing between $H^1(\Omega)$ and V'. We also put, for $T > 0$,

$$Q_T = \Omega \times (0,T), \quad \Sigma_T = \Gamma \times (0,T).$$

The assumptions on the initial data are as follows :
(A) for each $\alpha \in (0,1)$, $(\phi_0^\alpha, u_0^\alpha) \in H^1(\Omega) \times L^2(\Omega)$ and satisfy :

$$(2.4) \qquad \hat{\beta}(\phi_0^\alpha) \in L^1(\Omega),$$

$$(2.5) \qquad \frac{\xi^2}{2}\,|\phi_0^\alpha|_{H^1(\Omega)}^2 + \int_\Omega F(\phi_0^\alpha)\,dx + \frac{\alpha}{2}\,|u_0^\alpha|_{L^2(\Omega)}^2 \leq C_0,$$

for some constant $C_0 > 0$, and the sequence (ϕ_0^α) converges strongly in $L^2(\Omega)$ to some function $\phi_0 \in H^1(\Omega)$ as $\alpha \to 0$.

We now give the definition of a weak solution to $(1.1) - (1.4)$ we use in the sequel.

Definition 2.1 *Let $T > 0$ and $(\phi_0^\alpha, u_0^\alpha) \in H^1(\Omega) \times L^2(\Omega)$ be such that (2.4) holds. A pair of functions (ϕ^α, u^α) is a weak solution to (1.1) − (1.4) on $(0, T)$ if there exist functions (ζ^α, J^α) satisfying :*

(i) $\phi^\alpha \in W^{1,2}(0, T, L^2(\Omega)) \cap L^\infty(0, T, H^1(\Omega)) \cap L^2(0, T, H^2(\Omega))$, $\phi^\alpha(0) = \phi_0^\alpha$,

(ii) $\zeta^\alpha \in L^2(Q_T)$, $\zeta^\alpha = \beta(\phi^\alpha)$ *a.e. in Q_T (which yields that $\phi^\alpha \in (-1, 1)$ a.e. in Q_T),*

(iii) $u^\alpha \in W^{1,2}(0, T, V') \cap L^\infty(0, T, L^2(\Omega))$, $u^\alpha(0) = u_0^\alpha$,

(iv) $J^\alpha \in L^2(Q_T, \mathbb{R}^N)$, $J^\alpha = \nabla\left(B(\phi^\alpha)\, u^\alpha\right) - u^\alpha\, \nabla B(\phi^\alpha)$,

and such that,

$$(\textbf{2.6}) \qquad \alpha\, \phi_t^\alpha + \zeta^\alpha \;=\; \xi^2\, \Delta\phi^\alpha + 2\, \phi^\alpha + u^\alpha, \text{ a.e. in } Q_T,$$

$$(\textbf{2.7}) \qquad \frac{\partial \phi^\alpha}{\partial n} \;=\; 0, \text{ a.e. on } \Sigma_T,$$

$$(\textbf{2.8}) \qquad \alpha \int_0^T \langle u_t^\alpha, \eta \rangle_{V',V}\, ds + \int_0^T \int_\Omega \phi_t^\alpha\, \eta\, dx ds + \int_0^T \int_\Omega J^\alpha . \nabla \eta\, dx ds = 0,$$

for any $\eta \in L^2(0, T, H^1(\Omega))$.

  A pair of functions (ϕ^α, u^α) is called a weak solution to (1.1) − (1.4) on $(0, +\infty)$ if it is a weak solution to (1.1) − (1.4) on $(0, T)$ for each $T > 0$.

  It follows from [La, Thm. 2.1] that there exists at least one weak solution to (1.1) − (1.4) on $(0, +\infty)$ for any $(\phi_0^\alpha, u_0^\alpha)$ in $H^1(\Omega) \times L^2(\Omega)$ such that (2.4) holds.
  In order to state our convergence result, we need to specify how we construct the weak solution we shall deal with in the sequel : from now on, $(\phi_0^\alpha, u_0^\alpha)_{\alpha \in (0,1)}$ is a given family of functions satisfying assumption **(A)**. We also fix $T > 0$.
For $\lambda \in (0, 1)$, we put $B_\lambda = B + \lambda$.
We infer from [La, Thm. 2.1] the following result :

Proposition 2.2 *For each $(\alpha, \lambda) \in (0, 1)^2$, there exist functions $(\phi^{\alpha,\lambda}, \zeta^{\alpha,\lambda}, u^{\alpha,\lambda})$ satisfying*

(i) $\phi^{\alpha,\lambda} \in W^{1,2}(0, T, L^2(\Omega)) \cap L^\infty(0, T, H^1(\Omega)) \cap L^2(0, T, H^2(\Omega))$, $\phi^{\alpha,\lambda}(0) = \phi_0^\alpha$,

(ii) $\zeta^{\alpha,\lambda} \in L^2(Q_T)$, $\zeta^{\alpha,\lambda} = \beta(\phi^{\alpha,\lambda})$ *a.e. in Q_T ($\phi^{\alpha,\lambda} \in (-1, 1)$ a.e. in Q_T),*

(iii) $u^{\alpha,\lambda} \in W^{1,2}(0, T, V') \cap L^\infty(0, T, L^2(\Omega)) \cap L^2(0, T, H^1(\Omega))$, $u^{\alpha,\lambda}(0) = u_0^\alpha$,

and such that,

$$(\textbf{2.9}) \qquad \alpha\, \phi_t^{\alpha,\lambda} + \zeta^{\alpha,\lambda} \;=\; \xi^2\, \Delta\phi^{\alpha,\lambda} + 2\, \phi^{\alpha,\lambda} + u^{\alpha,\lambda}, \text{ a.e. in } Q_T,$$

$$(\textbf{2.10}) \qquad \frac{\partial \phi^{\alpha,\lambda}}{\partial n} \;=\; 0, \text{ a.e. on } \Sigma_T,$$

$$\alpha \int_0^T \langle u_t^{\alpha,\lambda}, \eta \rangle_{V',V} \, ds + \int_0^T \int_\Omega \phi_t^{\alpha,\lambda} \, \eta \, dx ds + \int_0^T \int_\Omega B_\lambda(\phi^{\alpha,\lambda}) \, \nabla u^{\alpha,\lambda} . \nabla \eta \, dx ds = 0,$$

for any $\eta \in L^2(0, T, H^1(\Omega))$. Moreover, it holds :

$$\int_0^t \int_\Omega \left(\alpha \, |\phi_t^{\alpha,\lambda}|^2 + B_\lambda(\phi^{\alpha,\lambda}) \, |\nabla u^{\alpha,\lambda}|^2 \right) \, dx ds + \mathcal{L}_\alpha(\phi^{\alpha,\lambda}(t), u^{\alpha,\lambda}(t)) \leq \mathcal{L}_\alpha(\phi_0^\alpha, u_0^\alpha),$$

where

$$\mathcal{L}_\alpha(\psi, v) = \int_\Omega \left(\frac{\xi^2}{2} \, |\nabla \psi|^2 + F(\psi) + \frac{\alpha}{2} \, |v|^2 \right) \, dx.$$

Now, a proof similar to that of [La, Thm. 2.1] (to which we refer) yields :

Proposition 2.3 *For any $\alpha \in (0,1)$, there is a subsequence of $(\phi^{\alpha,\lambda}, u^{\alpha,\lambda})_{\lambda \in (0,1)}$ that converges to a weak solution (ϕ^α, u^α) to $(1.1) - (1.4)$ on $(0,T)$ as λ decreases to zero. More precisely, $(\phi^{\alpha,\lambda})$ converges to ϕ^α strongly in $L^2(0, T, H^1(\Omega))$ and weakly in $L^2(0, T, H^2(\Omega))$ and $W^{1,2}(0, T, L^2(\Omega))$, $(\zeta^{\alpha,\lambda})$ converges weakly to ζ^α in $L^2(Q_T)$, $(u^{\alpha,\lambda})$ converges to u^α strongly in $\mathcal{C}(0, T, V')$ and weakly-$*$ in $L^\infty(0, T, L^2(\Omega))$, and $(B_\lambda(\phi^{\alpha,\lambda}) \, \nabla u^{\alpha,\lambda})$ converges weakly to J^α in $L^2(Q_T)$. Moreover, $(B_\lambda(\phi^{\alpha,\lambda})^{1/2} \, \nabla u^{\alpha,\lambda})$ converges weakly to $\tilde{J}^\alpha$ in $L^2(Q_T)$ with $J^\alpha = B(\phi^\alpha)^{1/2} \, \tilde{J}^\alpha$, and it holds :*

$$(\textbf{2.11}) \qquad \int_0^t \int_\Omega \left(\alpha \, |\phi_t^\alpha|^2 + |\tilde{J}^\alpha|^2 \right) \, dx ds + \mathcal{L}_\alpha(\phi^\alpha(t), u^\alpha(t)) \leq \mathcal{L}_\alpha(\phi_0^\alpha, u_0^\alpha).$$

Note that Proposition 2.3 gives a weak solution to $(1.1) - (1.4)$ on $(0,T)$, which satisfy in addition the Liapunov estimate (2.14). From now on, (ϕ^α, u^α) denotes the weak solution to $(1.1) - (1.4)$ on $(0,T)$ we obtained in Proposition 2.3.

We now state our main result :

Theorem 2.4 *There is a subsequence of (ϕ^α, J^α) which converges to (ϕ, J), where*

(i) $\phi \in W^{1,2}(0, T, V') \cap \mathcal{C}([0,T], L^2(\Omega)) \cap L^\infty(0, T, H^1(\Omega)) \cap L^2(0, T, H^2(\Omega)),$

(ii) $\phi(0) = \phi_0, \; \dfrac{\partial \phi}{\partial n} = 0$ *a.e. on Σ_T and $\phi \in [-1,1]$ a.e. in Q_T,*

(iii) $J \in L^2(Q_T, \mathbb{R}^N),$

and (ϕ, J) satisfies :

$$(\textbf{2.12}) \qquad\qquad\qquad \phi_t = div(J) \quad in \; L^2(0, T, V'),$$

$$\int_0^T \int_\Omega J.\eta \, dx ds = \xi^2 \int_0^T \int_\Omega div\,(B(\phi)\,\eta) \; \Delta\phi \, dx ds$$
$$(\textbf{2.13}) \qquad\qquad\qquad\qquad + \int_0^T \int_\Omega (2 - 2\,B(\phi)) \; \eta.\nabla\phi \, dx ds,$$

for any $\eta \in L^2(0, T, H^1(\Omega, \mathbb{R}^N)) \cap L^\infty(Q_T, \mathbb{R}^N)$ such that $\eta.n = 0$ on Σ_T.

138

3 Proof of Theorem 2.4

3.1 Preliminary results

Hereafter, we state the main lemma that will allow us to derive sufficient estimates on (ϕ^α, u^α) to be able to pass to the limit as α decreases to zero.

Lemma 3.1 *We consider $\psi_0 \in H^1(\Omega)$, $f \in L^2(Q_T)$ and (τ, κ), two positive real numbers and assume that*

(**3.1**) $$\hat{\beta}(\psi_0) \in L^1(\Omega),$$

where $\hat{\beta}$ is given by (2.2). Then, the solution ψ to

(**3.2**) $$\tau \, \psi_t - \kappa \, \Delta\psi + \beta(\psi) - 2 \, \psi \; = \; f \; in \, Q_T,$$

(**3.3**) $$\frac{\partial\psi}{\partial n} \; = \; 0 \; on \, \Sigma_T,$$

(**3.4**) $$\psi(0) \; = \; \psi_0 \; in \, \Omega,$$

where β is given by (2.1) satisfies :

(**3.5**) $$\int_0^T \int_\Omega f \, div \, (B(\psi) \, \eta) \, dxds \; = \; \int_0^T \int_\Omega div \, (B(\psi) \, \eta) \, (\tau \, \psi_t - \kappa \, \Delta\psi) \, dxds$$
$$- \int_0^T \int_\Omega (2 - 2 \, B(\psi)) \, \eta . \nabla\psi \, dxds,$$

for any $T > 0$ and $\eta \in L^2(0, T, H^1(\Omega, I\!\!R^N)) \cap L^\infty(Q_T, I\!\!R^N)$ such that $\eta.n = 0$ on Σ_T. If, in addition, f belongs to $L^2(0, T, H^1(\Omega))$, it holds :

(**3.6**) $$\kappa \int_0^T \int_\Omega |\Delta\psi|^2 \, dxds \; \leq \; \frac{\tau}{2} \int_\Omega |\nabla\psi_0|^2 \, dx + 2 \int_0^T \int_\Omega |\nabla\psi|^2 \, dxds$$
$$+ \frac{1}{4} \int_0^T \int_\Omega (1 - \psi^2) \, |\nabla f|^2 \, dxds,$$

for any $T > 0$.

Proof of Lemma 3.1

For $\epsilon \in (0, 1)$, we consider the Yosida approximation β_ϵ of β, which is given by :

(**3.7**) $$\beta_\epsilon = \frac{Id - \gamma_\epsilon}{\epsilon}, \quad \gamma_\epsilon = (Id + \epsilon \, \beta)^{-1}.$$

The function γ_ϵ is a contraction from $I\!\!R$ to $I\!\!R$ with $\gamma_\epsilon(I\!\!R) \subset (-1, 1)$, while β_ϵ is a Lipschitz continuous function with Lipschitz constant ϵ^{-1} (see e.g. [Br]). Furthermore,

(**3.8**) $$|\gamma_\epsilon(r) - r| \; \leq \; \epsilon \, |\beta(r)|, \quad r \in (-1, 1),$$

(**3.9**) $$\beta_\epsilon'(r) \; = \; \frac{2}{1 + 2\epsilon - (\gamma_\epsilon(r))^2}, \quad r \in I\!\!R.$$

We denote by ψ^ϵ the solution to

$$(3.10) \qquad \tau\,\psi_t^\epsilon - \kappa\,\Delta\psi^\epsilon + \beta_\epsilon(\psi^\epsilon) - 2\,\psi^\epsilon \;=\; f \;\text{ in } Q_T,$$

$$(3.11) \qquad \frac{\partial\psi^\epsilon}{\partial n} \;=\; 0 \;\text{ on } \Sigma_T,$$

$$(3.12) \qquad \psi^\epsilon(0) \;=\; \psi_0 \;\text{ in } \Omega.$$

Let $T > 0$. It follows from standard arguments (see e.g. [Br]) that (ψ^ϵ) converges to ψ for the weak topologies of $L^2(0,T,H^2(\Omega))$ and of $W^{1,2}(0,T,L^2(\Omega))$ and for the strong topology of $L^2(0,T,H^1(\Omega))$ as ϵ decreases to zero.

Since (ψ^ϵ) converges strongly to ψ in $L^2(0,T,H^1(\Omega))$, we may assume that (ψ^ϵ) converges to ψ almost everywhere in Q_T. Since $\psi(x,t)$ belongs to $(-1,1)$ for almost every (x,t) in Q_T, we infer from (3.8) that

$$\begin{aligned}
|\gamma_\epsilon(\psi^\epsilon(x,t)) - \psi(x,t)| &\;\leq\; |\psi^\epsilon(x,t) - \psi(x,t)| + |\gamma_\epsilon(\psi(x,t)) - \psi(x,t)| \\
&\;\leq\; |\psi^\epsilon(x,t) - \psi(x,t)| + \epsilon\,|\beta(\psi(x,t))|.
\end{aligned}$$

Thus, $(\gamma_\epsilon(\psi^\epsilon))$ converges to ψ almost everywhere in Q_T, and (3.9) yields that

$$(3.13) \qquad (\beta_\epsilon'(\psi^\epsilon)) \text{ converges to } 2(1-\psi^2)^{-1} \text{ a.e. in } Q_T.$$

We first prove (3.5). We consider $\eta \in L^2(0,T,H^1(\Omega,\mathbb{R}^N)) \cap L^\infty(Q_T,\mathbb{R}^N)$ such that $\eta.n = 0$ on Σ_T. We take the scalar product in $L^2(Q_T)$ of (3.10) with $div\,(B(\psi^\epsilon)\,\eta)$ and get

$$\begin{aligned}
\int_0^T\!\!\int_\Omega f\,div\,(B(\psi^\epsilon)\,\eta)\;dxds \;=\; & \int_0^T\!\!\int_\Omega (\tau\,\psi_t^\epsilon - \kappa\,\Delta\psi^\epsilon)\,div\,(B(\psi^\epsilon)\,\eta)\;dxds \\
& - \int_0^T\!\!\int_\Omega \beta_\epsilon'(\psi^\epsilon)\,B(\psi^\epsilon)\,\nabla\psi^\epsilon.\eta\;dxds \\
(3.14) \qquad & + 2\int_0^T\!\!\int_\Omega B(\psi^\epsilon)\,\nabla\psi^\epsilon.\eta\;dxds.
\end{aligned}$$

Since (ψ^ϵ) converges strongly to ψ in $L^2(0,T,H^1(\Omega))$, and since B is Lipschitz continuous, the sequence $(B(\psi^\epsilon))$ converges strongly to $B(\psi)$ in $L^2(0,T,H^1(\Omega))$. Therefore,

$$div\,(B(\psi^\epsilon)\,\eta) \to div\,(B(\psi)\,\eta) \;\text{ in } L^2(Q_T).$$

Next, it follows from (3.13) that $(\beta_\epsilon'(\psi^\epsilon)\,B(\psi^\epsilon))$ converges to 2 a.e. in Q_T. Furthermore, since

$$|\gamma_\epsilon(r)| \leq |r|, \quad r \in \mathbb{R},$$

and B vanishes outside $(-1,1)$, it holds :

$$|\beta_\epsilon'(\psi^\epsilon)\,B(\psi^\epsilon)| \leq 2.$$

It then follows from Lebesgue dominated convergence theorem that $(\beta_\epsilon'(\psi^\epsilon)\,B(\psi^\epsilon))$ converges to 2 in $L^2(Q_T)$. We may then pass to the limit in (3.14) and obtain (3.5).

We now prove (3.6).

Since β_ϵ is Lipschitz continuous, $\beta_\epsilon(\psi^\epsilon)$ belongs to $L^2(0, T, H^1(\Omega))$. We take the scalar product in $L^2(Q_T)$ of (3.10) with $(-\Delta\psi^\epsilon)$ and integrate by parts ; since $\beta'_\epsilon > 0$, Young inequality yields :

$$\kappa \int_0^T \int_\Omega |\Delta\psi^\epsilon|^2 \, dxds \ + \ \int_0^T \int_\Omega \beta'_\epsilon(\psi^\epsilon) \, |\nabla\psi^\epsilon|^2 \, dxds$$

$$\leq \frac{\tau}{2} \int_\Omega |\nabla\psi_0|^2 \, dx \ + \ 2 \int_0^T \int_\Omega |\nabla\psi^\epsilon|^2 \, dxds$$

$$(3.15) \quad + \frac{1}{2} \int_0^T \int_\Omega \beta'_\epsilon(\psi^\epsilon) \, |\nabla\psi^\epsilon|^2 \, dxds \ + \ \frac{1}{2} \int_0^T \int_\Omega \beta'_\epsilon(\psi^\epsilon)^{-1} \, |\nabla f|^2 \, dxds.$$

We infer from (3.9) and (3.13) that $(2\beta'_\epsilon(\psi^\epsilon)^{-1} \, |\nabla f|^2)$ converges to $(1 - \psi^2) \, |\nabla f|^2$ a.e. in Q_T and that

$$0 \leq 2\beta'_\epsilon(\psi^\epsilon)^{-1} \, |\nabla f|^2 \leq 4 \, |\nabla f|^2 \in L^1(Q_T).$$

It then follows from Lebesgue dominated convergence theorem that $(2\beta'_\epsilon(\psi^\epsilon)^{-1} \, |\nabla f|^2)$ converges to $(1 - \psi^2) \, |\nabla f|^2$ in $L^1(Q_T)$. Since $\beta'_\epsilon > 0$, we may then pass to the limit in (3.15) and obtain (3.6). $\qquad\qquad\square$

3.2 Proof of Theorem 2.4

In the following, we denote by C any positive constant depending only on Ω, N, ξ, C_0 in (2.5) and T.

Lemma 3.2 *For any $\alpha \in (0, 1)$, it holds :*

$$(3.16) |\phi^\alpha|_{L^\infty(0,T,H^1(\Omega))} + |J^\alpha|_{L^2(Q_T)} + \alpha^{1/2} \left(|\phi^\alpha_t|_{L^2(Q_T)} + |u^\alpha|_{L^\infty(0,T,L^2(\Omega))} \right) \ \leq \ C,$$

$$(3.17) \qquad\qquad\qquad\qquad\qquad\qquad\qquad\qquad |\phi^\alpha|_{L^2(0,T,H^2(\Omega))} \ \leq \ C.$$

Proof of Lemma 3.2

It follows at once from (2.5), (2.14), the positivity of F and the relationship between J^α and $\tilde{J}^\alpha$ that

$$(3.18) |\nabla\phi^\alpha|_{L^\infty(0,T,L^2(\Omega))} + \alpha^{1/2} \, |\phi^\alpha_t|_{L^2(Q_T)} + \alpha^{1/2} \, |u^\alpha|_{L^\infty(0,T,L^2(\Omega))} + |J^\alpha|_{L^2(Q_T)} \leq C.$$

Next, we take $\eta = 1$ in (2.8) and use (2.5) and (3.18) to get

$$(3.19) \qquad\qquad\qquad\qquad \left| \int_\Omega \phi^\alpha(t) \, dx \right| \leq C.$$

Then, (3.16) follows from $(3.18) - (3.19)$ and Poincaré inequality.

Now, for any $\lambda \in (0, 1)$, $u^{\alpha,\lambda}$ belongs to $L^2(0, T, H^1(\Omega))$. On the one hand, we infer from Lemma 3.1 that it holds

$$\xi^2 \int_0^T \int_\Omega |\Delta\phi^{\alpha,\lambda}|^2 \, dxds \ \leq \ \frac{\alpha}{2} \int_\Omega |\nabla\phi^\alpha_0|^2 \, dx + 2 \int_0^T \int_\Omega |\nabla\phi^{\alpha,\lambda}|^2 \, dxds$$

$$(3.20) \qquad\qquad\qquad\qquad + \frac{1}{4} \int_0^T \int_\Omega (1 - (\phi^{\alpha,\lambda})^2) \, |\nabla u^{\alpha,\lambda}|^2 \, dxds.$$

On the other hand, we infer from (2.5) and (2.12) that

$$\frac{\xi^2}{2} \, |\nabla \phi^{\alpha,\lambda}|^2_{L^\infty(0,T,L^2(\Omega))} + \int_0^T \int_\Omega B_\lambda(\phi^{\alpha,\lambda}) \, |\nabla u^{\alpha,\lambda}|^2 \, dxds \le C.$$

It follows from the above estimate that

(3.21) $$\int_0^T \int_\Omega \left(2 \, |\nabla \phi^{\alpha,\lambda}|^2 + \frac{1}{2} \, (1 - (\phi^{\alpha,\lambda})^2) \, |\nabla u^{\alpha,\lambda}|^2 \right) dxds \le C.$$

It now follows from (2.5), (3.20) and (3.21) that

$$\int_0^T \int_\Omega |\Delta \phi^{\alpha,\lambda}|^2 \, dxds \le C.$$

Then (3.17) follows from the previous estimate, the weak convergence of $(\phi^{\alpha,\lambda})$ to ϕ^α in $L^2(0,T,H^2(\Omega))$ (see Prop. 2.3) and standard elliptic arguments. $\qquad \square$

In the next lemma, we derive further estimates.

Lemma 3.3 *For any $\alpha \in (0,1)$ and $h \in (0,T)$, one has :*

(3.22) $$|e^\alpha|_{L^\infty(0,T,L^2(\Omega))} + |e^\alpha_t|_{L^2(0,T,V')} \;\le\; C,$$

(3.23) $$|\phi^\alpha(t+h) - \phi^\alpha(t)|_{V'} \;\le\; C \, h^{1/4}, \;\; t \in [0, T-h],$$

where $e^\alpha = \alpha \, u^\alpha + \phi^\alpha$.

Proof of Lemma 3.3
First, (3.22) is a straightforward consequence of (3.16) and (2.8).

Next, let $h \in (0,T)$ and $t \in [0, T-h]$. On the one hand, it follows from (3.22) that

$$|e^\alpha(t+h) - e^\alpha(t)|_{V'} \le C \, h^{1/2},$$

hence

(3.24) $$|\phi^\alpha(t+h) - \phi^\alpha(t)|_{V'} \le \alpha \, |u^\alpha(t+h) - u^\alpha(t)|_{V'} + C \, h^{1/2}.$$

On the other hand, we infer from (3.16) and (2.8) that

$$\alpha^{3/2} \, |u^\alpha_t|_{L^2(0,T,V')} \le C,$$

hence

(3.25) $$\alpha^{3/2} \, |u^\alpha(t+h) - u^\alpha(t)|_{V'} \le C \, h^{1/2}.$$

It also follows from (3.16) that

(3.26) $$\alpha^{1/2} \, |u^\alpha(t+h) - u^\alpha(t)|_{V'} \le C.$$

Combining (3.25) − (3.26) yields

$$\alpha \, |u^\alpha(t+h) - u^\alpha(t)|_{V'} \le C \, h^{1/4}.$$

This last estimate, together with (3.24) yields (3.23). □

Since the sequence (ϕ^α) is bounded in $L^\infty(0, T, H^1(\Omega))$, in $L^2(0, T, H^2(\Omega))$ and satisfies (3.23), we infer from [Si, Thm. 5] that

(3.27)
$$(\phi^\alpha) \text{ is relatively compact in } \mathcal{C}([0, T], L^2(\Omega))$$
$$\text{and in } L^2(0, T, H^1(\Omega)).$$

Next, it follows from (3.22) and [Si, Cor. 4] that

(3.28)
$$(e^\alpha) \text{ is relatively compact in } \mathcal{C}([0, T], V').$$

We infer from (3.16) − (3.17), (3.22) and (3.27) − (3.28) that there exist

$$\phi \in \mathcal{C}([0, T], L^2(\Omega)) \cap L^\infty(0, T, H^1(\Omega)) \cap L^2(0, T, H^2(\Omega)),$$
$$e \in W^{1,2}(0, T, V') \cap L^\infty(0, T, L^2(\Omega)), \quad J \in L^2(Q_T),$$

and a subsequence of $(\phi^\alpha, e^\alpha, J^\alpha)$ (which we still denote by $(\phi^\alpha, e^\alpha, J^\alpha)$) such that

$$\phi^\alpha \to \phi \text{ in } \mathcal{C}([0, T], L^2(\Omega)), \text{ in } L^2(0, T, H^1(\Omega)),$$
$$\text{and a.e. in } Q_T,$$
$$\phi^\alpha \rightharpoonup \phi \text{ in } L^2(0, T, H^2(\Omega)),$$

(3.29)
$$e^\alpha \to e \text{ in } \mathcal{C}([0, T], V'),$$

$$e_t^\alpha \rightharpoonup e_t \text{ in } L^2(0, T, V'),$$

$$J^\alpha \rightharpoonup J \text{ in } L^2(Q_T).$$

It also follows from (3.16) that

(3.30)
$$\alpha \, \phi_t^\alpha \to 0 \text{ in } L^2(Q_T),$$

$$\alpha \, u^\alpha \to 0 \text{ in } L^2(Q_T).$$

Furthermore, since B is Lipschitz continuous, it follows from (3.29) that

(3.31)
$$B(\phi^\alpha) \to B(\phi) \text{ in } L^2(0, T, H^1(\Omega)).$$

We first infer from (3.29) − (3.30) that $e = \phi$. We then pass to the limit in (2.8) and get (2.15).

It remains to identify J.

We consider $\eta \in L^2(0, T, H^1(\Omega, \mathbb{R}^N)) \cap L^\infty(Q_T, \mathbb{R}^N)$ such that $\eta.n = 0$ on Σ_T. On the one hand, we infer from Lemma 3.1 that

$$\int_0^T \int_\Omega u^\alpha \, div \, (B(\phi^\alpha) \, \eta) \, dx ds \; = \; \int_0^T \int_\Omega div \, (B(\phi^\alpha) \, \eta) \, \left(\alpha \, \phi_t^\alpha - \xi^2 \, \Delta \phi^\alpha \right) \, dx ds$$
$$- \int_0^T \int_\Omega (2 - 2 \, B(\phi^\alpha)) \, \eta.\nabla \phi^\alpha \, dx ds,$$

143

which yields, thanks to $(3.29) - (3.31)$,

$$\lim_{\alpha \to 0} \int_0^T \int_\Omega u^\alpha \; div \, (B(\phi^\alpha) \, \eta) \;\; dx ds \;\; = \;\; -\xi^2 \int_0^T \int_\Omega div \, (B(\phi) \, \eta) \;\; \Delta\phi \; dx ds$$

$$(3.32) \qquad\qquad\qquad - \int_0^T \int_\Omega (2 - 2 \, B(\phi)) \; \eta . \nabla \phi \; dx ds.$$

On the other hand,

$$\int_0^T \int_\Omega J^\alpha . \eta \; dx ds = - \int_0^T \int_\Omega u^\alpha \; div \, (B(\phi^\alpha) \, \eta) \;\; dx ds,$$

hence

$$(3.33) \qquad \lim_{\alpha \to 0} \int_0^T \int_\Omega u^\alpha \; div \, (B(\phi^\alpha) \, \eta) \;\; dx ds = - \int_0^T \int_\Omega J . \eta \; dx ds.$$

Then, (2.16) follows from (3.32) and (3.33). The proof of Theorem 2.4 is thus complete.
□

References

[Br] H. BREZIS, *Opérateurs maximaux monotones et semi-groupes de contractions dans les espaces de Hilbert*, North Holland, Amsterdam, 1973.

[Ca] G. CAGINALP, An analysis of a phase field model of a free boundary, *Arch. Rat. Mech. Anal.* **92** (1986), 205-245.

[CT] J.W. CAHN, J.E. TAYLOR, Surface motion by surface diffusion, *Acta metall. mater.* **42** (1994), 1045-1063.

[EG] C.M. ELLIOTT, H. GARCKE, On the Cahn-Hilliard equation with degenerate mobility, *SIAM J. Math. Anal.*, to appear.

[La] Ph. LAURENÇOT, Weak solutions to a phase-field model with non-constant thermal conductivity, *Quart. Appl. Math.*, to appear.

[Si] J. SIMON, Compact sets in the space $L^p(0, T; B)$, *Ann. Mat. Pura Appl.* **146** (1987), 65-96.

[St] B.E.E. STOTH, The Cahn-Hilliard equation as limit of the phase-field equations, SFB 256 - Report **304**, Bonn, 1993.

Ph. Laurençot
CNRS UMR 9973 & INRIA-Lorraine *Projet Numath*
Laboratoire de Mathématiques, Université de Nancy I
BP 239, F-54506 Vandoeuvre les Nancy Cedex, France

N SATO

Periodic solutions of phase field equations with constraint

Abstract. A model for solid-liquid phase transitions, which is a coupled system of second order nonlinear parabolic PDEs(kinetic equations for the internal energy and the order parameter), is considered. It is quite natural to expect that if the forcing term in the kinetic equation for the internal energy is periodic in time, then the large-time behavior of the solution is asymptotically periodic as $t \to +\infty$. In this paper, as the first step of the reserch on this subject, we shall show the existence of a periodic solution of our model.

1. Introduction

We consider a nonlinear system of the following form (1.1)-(1.2):

(1.1)
$$\frac{\partial \rho(u)}{\partial t} + \frac{\partial w}{\partial t} - \triangle u = f(t, x) \qquad \text{in } Q := (0, T_0) \times \Omega,$$

(1.2)
$$\nu \frac{\partial w}{\partial t} - \kappa \triangle w + \beta(w) + g(w) \ni u \quad \text{in } Q$$

with lateral boundary conditions:

(1.3 − 1)
$$\frac{\partial u}{\partial n} + \alpha_N(x)u = h_N(t, x) \quad \text{on } \Sigma := (0, T_0) \times \Gamma,$$

(1.3 − 2)
$$\frac{\partial w}{\partial n} = 0 \quad \text{on } \Sigma,$$

and T_0-periodic conditions:

(1.4)
$$u(T_0, \cdot) = u(0, \cdot), \quad w(T_0, \cdot) = w(0, \cdot).$$

Here Ω is a bounded domain in R^N ($N \geq 1$) with smooth boundary $\Gamma := \partial\Omega$; ρ is a monotone increasing and bi-Lipschitz continuous function on R; ν and κ are positive constants; β is a maximal monotone graph in $R \times R$; g is a smooth function defined on R; α_N is a positive, bounded and measurable function on Γ; T_0 is a positive constant; f, h_N are given data. For simplicity system (1.1)-(1.3) is refered as (P) and (P) with periodic condition (1.4) as (PP).

It is the purpose of this paper to discuss (PP), which is called the phase-field model (cf.[6,2,11,5,3,9]) with constraint, from a view-point of the theory (cf.[1,7,8]) of time-dependent subdifferentials of convex functions in a Hilbert space. Let us consider the space $V := H^1(\Omega)$ with norm $|z|_V := \{|\nabla w|^2_{L^2(\Omega)} + \int_\Gamma \alpha_N |z|^2 d\Gamma\}^{\frac{1}{2}}$ and its dual space V^* which is a Hilbert space with inner product $(v, z)_* :=_{V^*} \langle v, F^{-1}z \rangle_V$ where $_{V^*}\langle \cdot, \cdot \rangle_V$ denotes the duality pairing between V^* and V and F is the duality mapping from V onto V^*. Then, $X := V^* \times L^2(\Omega)$ becomes a Hilbert space with inner product $([e_1, w_1], [e_2, w_2])_X = (e_1, e_2)_* + \nu(w_1, w_2)$, where $(\cdot, \cdot)$ stands for the inner product in $L^2(\Omega)$.

Now, choose $h : [0, T_0] \to V$ such that for each $t \in [0, T_0]$

$$(1.5) \qquad \int_\Omega \nabla h(t) \cdot \nabla z\, dx + \int_\Gamma \alpha_N h(t) z\, d\Gamma = \int_\Gamma h_N(t) z\, d\Gamma \quad \text{for all } z \in V.$$

Note that if $h_N \in W^{1,2}(0, T_0; L^2(\Gamma))$, then $h \in W^{1,2}(0, T_0; V)$.

Next, denote by $\hat{\beta}$ a proper l.s.c. non-negative convex function on R such that the subdifferential $\partial\hat{\beta}$ of $\hat{\beta}$ coincides with β in R. Using these h and $\hat{\beta}$, for each $t \in [0, T_0]$ we introduce the following proper l.s.c. convex function φ^t_κ on X:

$$\varphi^t_\kappa(U) = \begin{cases} \int_\Omega \rho^*(e - w)dx + \dfrac{\kappa}{2}|\nabla w|^2_{L^2(\Omega)} + \int_\Omega \hat{\beta}(w)dx - (h(t), e) \\ \qquad \text{if } U = [e, w] \in L^2(\Omega) \times H^1(\Omega) \text{ with } \hat{\beta}(w) \in L^1(\Omega), \\[2mm] +\infty \quad \text{otherwise,} \end{cases}$$

where ρ^* is a non-negative primitive of ρ^{-1}.

According to [9;Section 3] (or [4;Theorem 2.1]), (PP) can be reformulated as an evolution equation in X of the following form:

$$(1.6) \qquad U'(t) + \partial\varphi^t_\kappa(U(t)) \ni [f(t), -\frac{1}{\nu}g(w(t))] \quad \text{in } X, \quad 0 \le t \le T_0,$$

$$U(0) = U(T_0),$$

where $U(t) = [\rho(u(t)) + w(t), w(t)]$ and $\partial\varphi^t_\kappa$ is the subdifferential of φ^t_κ in X.

For the Cauchy problem of (1.6) we refer to [3] or [9].

2. The case when g is independent of w

In this section, we consider system (P) with $g \equiv \ell(t, x)$, denoted by (P_ℓ), with initial condition

$$(2.1) \qquad u(0, \cdot) = u_0, \quad w(0, \cdot) = w_0 \quad \text{in } \Omega$$

or with T_0-periodic condition (1.4). For simplicity, we denote (P_ℓ) with (2.1) by (CP_ℓ) and (P_ℓ) with (1.4) by (PP_ℓ).

These problems are discussed in Damlamian-Kenmochi-Sato [4]. Following it we recall some results.

We make the following assumptions (A1)-(A5):

(A1) $\rho : R \to R$ is a increasing and bi-Lipschitz continuous function.

(A2) β is a maximal monotone graph in $R \times R$ with non-empty interior of $D(\beta)$ in R. There is a positive constant C_1 such that

$$\hat{\beta}(r) \geq C_1 r^2 \quad \text{for any } r \in R;$$

(A3) $h_N \in W^{1,2}(0, T_0; L^2(\Gamma))$ with $h_N(0, \cdot) = h_N(T_0, \cdot)$;

(A4) $f \in L^2(0, T_0; L^2(\Omega))$;

(A5) $\ell \in L^2(0, T_0; L^2(\Omega))$.

Note here that under (A3) the function h given by (1.5) belongs to $W^{1,2}(0, T_0; V)$ and $h(0) = h(T_0)$.

Definition 2.1. (1) Let $0 < T < +\infty$. Then a couple $\{u, w\}$ of functions $u : [0, T] \to V^*$ and $w : [0, T] \to L^2(\Omega)$ is called a (weak) solution of (P_ℓ) on $[0, T]$, if the following conditions (w1)-(w3) are fulfilled:

(w1) $\rho(u) \in C([0, T]; V^*) \cap W_{loc}^{1,2}((0, T]; V^*) \cap L^2(0, T; L^2(\Omega))$, $u \in L_{loc}^2((0, T]; H^1(\Omega))$, $w \in C([0, T]; L^2(\Omega)) \cap W_{loc}^{1,2}((0, T]; L^2(\Omega)) \cap L^2(0, T; H^1(\Omega))$ and $\hat{\beta}(w) \in L^1(0, T; L^1(\Omega))$;

(w2) For all $z \in V$ and a.e. $t \in [0, T]$ we have

$$_{V^*}\langle \rho(u)'(t) + w'(t), z \rangle_V + \int_\Omega \nabla(u(t) - h(t)) \cdot \nabla z \, dx \; + \; \int_\Gamma \alpha_N(u(t) - h(t)) z \, d\Gamma$$
$$= \; (f(t), z),$$

where $\rho(u)'$ and w' denote the derivatives of $\rho(u)$ and w in time t, respectively;

(w3) there exists $\xi \in L_{loc}^2((0, T]; L^2(\Omega))$ such that $\xi \in \beta(w)$ a.e. in $Q_T := (0, T) \times \Omega$ and

$$\nu(w'(t), z) + \kappa \int_\Omega \nabla w(t) \cdot \nabla z \, dx + (\xi(t), z) = (u(t) - \ell(t), z)$$

for all $z \in H^1(\Omega)$ and a.e. $t \in [0, T]$.

(2) Let $J = [0, T]$ with $0 < T < +\infty$, and $u_0 \in V^*$, $w_0 \in L^2(\Omega)$. Then a couple $\{u, w\}$ of functions $u : J \to V^*$ and $w : J \to L^2(\Omega)$ is called a (weak) solution of (CP_ℓ) on J, if the couple $\{u, w\}$ is a solution of (P_ℓ) on J such that

$$u(0) = u_0 \quad \text{and} \quad w(0) = w_0.$$

(3) A couple $\{u, w\}$ of functions $u : [0, T_0] \to V^*$ and $w : [0, T_0] \to L^2(\Omega)$ is called a (weak) solution of (PP_ℓ), if the couple $\{u, w\}$ is a solution of (P_ℓ) on $[0, T_0]$ and satisfies the T_0-periodicity conditions, i.e.

$$u(0) = u(T_0) \text{ in } V^*, \quad w(0) = w(T_0) \text{ in } L^2(\Omega).$$

According to [9;Section 3] (or [4;Theorem 2.1]) we see that (P_ℓ) can be reformulated as an evolution equation in X of the following form:

$$U'(t) + \partial \varphi_\kappa^t(U(t)) \ni [f(t), -\frac{1}{\nu}\ell(t)] \quad \text{in } X, \ 0 \le t \le T_0.$$

From the general theory [8], we have:

Theorem 2.1. *(Damlamian-Kenmochi-Sato [4]) Assume that (A1)-(A5) hold. Then we have the following statements:*

(1) If the initial data satisfy

$$u_0 \in V^*, \ w_0 \in L^2(\Omega) \text{ with } w_0 \in \overline{D(\beta)} \text{ a.e. on } \Omega,$$

then (CP_ℓ) admits one and only one solution $\{u, w\}$ on $[0, T]$ such that

$$t^{\frac{1}{2}} u' \in L^2(0, T_0; V^*), \quad t^{\frac{1}{2}} u \in L^2(0, T_0; H^1(\Omega)),$$

$$tu' \in L^2(0, T_0; L^2(\Omega)), \quad tu \in L^\infty(0, T_0; H^1(\Omega)),$$

$$tw' \in L^2(0, T_0; L^2(\Omega)), \quad t^{\frac{1}{2}} w \in L^\infty(0, T_0; H^1(\Omega)),$$

$$t\hat\beta(w) \in L^\infty(0, T_0), \quad t^{\frac{1}{2}}\xi \in L^2(0, T_0; L^2(\Omega)),$$

where ξ is the function in condition (w3).

(2) If the initial data satisfy that

$$u_0 - h(0) \in V,$$

$$w_0 \in H^1(\Omega) \text{ with } \hat\beta(w) \in L^1(\Omega),$$

then (CP_ℓ) admits one and only one solution $\{u, w\}$ on $[0, T_0]$ such that

(2.2) $$\rho(u)' \in L^2(0, T_0; L^2(\Omega)), \quad u \in L^\infty(0, T_0; V),$$

(2.3) $$w' \in L^2(0, T_0; L^2(\Omega)), \quad \hat\beta(w) \in L^\infty(0, T_0; L^1(\Omega)),$$

(2.4) $$\xi \in L^2(0, T_0; L^2(\Omega)), \quad w \in L^\infty(0, T_0; H^1(\Omega))$$

where ξ is the function in condition (w3).

Note that for the above $\{u(t), w(t)\}$, $U(t) := [\rho(u(t)) + w(t), w(t)]$ is a solution of Cauchy problem of equation (1.6) with $g(w(t)) = \ell(t)$. Considering the mapping S from $\overline{D(\varphi^0)}$ to $\overline{D(\varphi^{T_0})}$ which assings each $U_0 := [\rho(u_0) + w_0, w_0]$ to $U(T_0)$, we can show that S has a fixed point in $\overline{D(\varphi^0)}$. Therefore we have the following theorem:

Theorem 2.2. *(Damlamian-Kenmochi-Sato [4]) (PP$_\ell$) admits at least one solution. In particular, β is strictly monotone, then the solution of (PP$_\ell$) is unique.*

Remark 2.1. By Theorem 2.1, any solution $\{u, w\}$ of (PP$_\ell$) has the regularity properties (2.2)-(2.4).

Remark 2.2. Since $h \in W^{1,2}(0, T_0; V)$, we can formally write (PP$_\ell$) in the form

$$(2.5) \qquad \rho(u)'(t) + w'(t) - \Delta(u(t) - h(t)) = f(t) \quad \text{in } Q,$$

$$(2.6) \qquad \nu w'(t) - \kappa \Delta w(t) + \beta(w(t)) \ni u(t) - \ell(t) \quad \text{in } Q,$$

$$\frac{\partial(u(t) - h(t))}{\partial n} + \alpha_N(x)(u(t) - h(t)) = 0 \quad \text{on } \Sigma,$$

$$\frac{\partial w(t)}{\partial n} = 0 \quad \text{on } \Sigma$$

and

$$u(T_0) = u(0), \quad w(T_0) = w(0) \quad \text{in } L^2(\Omega).$$

We give uniform estimates of solutions of (PP$_\ell$) with respect to f, h and ℓ.

Lemma 2.1. *Let $\{u, w\}$ be a solution of (PP$_\ell$), $\ell \in L^\infty(0, T_0; L^2(\Omega))$ and $r_0 \in D(\hat{\beta})$ where $\hat{\beta}$ is a positive primitive of β. Then, the following estimates hold:*

(1) There is a positive constant C_2 independent of ℓ and β such that

$$(2.7) \qquad \frac{\nu}{2}|w'|^2_{L^2(0,T_0;L^2(\Omega))} + \frac{1}{2}|u - h|^2_{L^2(0,T_0;V)} \le C_2 \left(|\ell|^2_{L^2(0,T_0;L^2(\Omega))} + R_0 \right)$$

where

$$R_0 := \int_0^{T_0} \{|f(t)|^2_{L^2(\Omega)} + |h(t)|^2_{L^2(\Omega)} + |h'(t)|^2_{L^2(\Omega)} + 1\}dt.$$

(2) There is a positive constant C_3 independent of ℓ and β such that

$$(2.8) \qquad \begin{aligned} & \kappa|\nabla w|^2_{L^2(0,T_0;L^2(\Omega))} + \left\| \int_\Omega \hat{\beta}(w)dx \right\|_{L^1(0,T_0)} \\ & \le C_3 \left(|\ell|^2_{L^2(0,T_0;L^2(\Omega))} + r_0^2 + \hat{\beta}(r_0) + R_0 \right). \end{aligned}$$

(3) There is a positive constant C_4 independent of ℓ and β such that

$$(2.9) \qquad |u' - h'|^2_{L^2(0,T_0;L^2(\Omega))} \le C_4 \left(|\ell|^2_{L^2(0,T_0;L^2(\Omega))} + r_0^2 + \hat{\beta}(r_0) + R_0 \right).$$

(4) There is a positive constant C_5 independent of ℓ and β such that
$$(2.10)$$
$$|u(t) - h(t)|^2_V \le C_5 \left(|\ell|^2_{L^2(0,T_0;L^2(\Omega))} + r_0^2 + \hat{\beta}(r_0) + R_0 \right) \quad \text{for all } t \in [0, T_0].$$

(5) There is a positive constant C_6 independent of ℓ and β such that

$$|w(t)|^2_{L^2(\Omega)} + \kappa |\nabla w(t)|^2_{L^2(\Omega)}$$
$$(2.11)$$
$$\le \; C_6 \left(\ell|^2_{L^2(0,T_0;L^2(\Omega))} + r_0^2 + \hat{\beta}(r_0) + R_0 \right) \quad \text{for all } t \in [0, T_0].$$

Proof. Multiplying (2.5) by $u(t) - h(t)$ and (2.6) by $w'(t)$, we get

$$\frac{d}{dt} \left\{ \int_\Omega \rho^*(\rho(u))(t)dx - (\rho(u)(t), h(t)) - \int_\Omega \hat{\rho}(h)(t)dx \right\} + |u(t) - h(t)|^2_V$$
$$= \; -(w'(t), u(t)) + (w'(t), h(t)) + (f(t), u(t) - h(t)) - (\rho(u)(t) - \rho(h)(t), h'(t))$$

and

$$(2.12) \qquad \nu |w'(t)|^2_{L^2(\Omega)} + \frac{d}{dt} \left\{ \frac{\kappa}{2} |\nabla w(t)|^2_{L^2(\Omega)} + \int_\Omega \hat{\beta}(w(t))dx \right\}$$
$$= \; (u(t), w'(t)) - (\ell(t), w'(t))$$

for a.e. $t \in [0, T_0]$ where $\hat{\rho}$ is a positive primitive of ρ and ρ^* is a positive primitive of the inverse ρ^{-1} of ρ.

Adding these two equalities, after some arrangements we arrive at an inequality of the form

$$\frac{\nu}{2} |w'(t)|^2_{L^2(\Omega)} +$$

$$+ \frac{d}{dt} \left\{ \frac{\kappa}{2} |\nabla w(t)|^2_{L^2(\Omega)} + \int_\Omega \hat{\beta}(w(t))dx + \right.$$

$$+ \int_\Omega \rho^*(\rho(u))(t)dx - (\rho(u)(t), h(t)) - \int_\Omega \hat{\rho}(h)(t)dx \right\}$$

$$+ \frac{1}{2} |u(t) - h(t)|^2_V$$

$$\le \; C_7 \left\{ |f(t)|^2_{L^2(\Omega)} + |h(t)|^2_{L^2(\Omega)} + |h'(t)|^2_{L^2(\Omega)} + |\ell(t)|^2_{L^2(\Omega)} + 1 \right\}$$

for a.e. $t \in [0, T_0]$, where C_7 is a positive constant depending only on ν and ρ. Here, integrate the above inequality over $[0, T_0]$ to get

$$\frac{\nu}{2}|w'|^2_{L^2(0,T_0;L^2(\Omega))} + \frac{1}{2}|u - h|^2_{L^2(0,T_0;V)} \le C_7 \left(R_0 + |\ell|^2_{L^2(0,T_0;L^2(\Omega))} \right);$$

we used above the periodicity $w(0) = w(T_0)$ and $u(0) = u(T_0)$. Hence, we get (2.7). Next, multiplying (2.5) by $u'(t) - h'(t)$, we get

$$(2.13) \quad C_8|u'(t) - h'(t)|^2_{L^2(\Omega)} + \frac{d}{dt}|u(t) - h(t)|^2_V \le C_9 \left(|f(t) - w'(t)|^2_{L^2(\Omega)} + |h'(t)|^2_{L^2(\Omega)} \right)$$

for a.e. $t \in [0, T_0]$ where C_8 and C_9 are positive constants depending only on ρ. Indeed,

$$(\rho(u)'(t), u'(t) - h'(t))$$

$$= \quad (\rho'(u)(t)u'(t) - \rho'(u)(t)h'(t) + \rho'(u)(t)h'(t), u'(t) - h'(t))$$

$$= \quad \int_\Omega \rho'(u)(t)|u'(t) - h'(t)|^2 dx + (\rho'(u)(t)h'(t), h'(t))$$

for a.e. $t \in [0, T_0]$, where ρ' is the detrivative of ρ and ρ' is bounded. Integrating the above inequality over $[0, T_0]$ we get (2.9). Next, multiplying (2.6) by $w(t) - r_0$ with $r_0 \in D(\hat{\beta})$ we get

$$(2.14) \quad \begin{aligned} &\frac{\nu}{2}\frac{d}{dt}|w(t) - r_0|^2_{L^2(\Omega)} + \kappa|\nabla w(t)|^2_{L^2(\Omega)} + \int_\Omega \hat{\beta}(w(t))dx \\ &\le \quad (u(t) - \ell(t), w(t) - r_0) + \int_\Omega \hat{\beta}(r_0)dx \end{aligned}$$

for a.e. $t \in [0, T_0]$. Therefore we get an inequality of the form

$$\frac{\nu}{2}\frac{d}{dt}|w(t) - r_0|^2_{L^2(\Omega)} + \kappa|\nabla w(t)|^2_{L^2(\Omega)} + \frac{C_1}{2}|w(t)|^2_{L^2(\Omega)}$$

$$\le \quad C_{10}(|u(t) - h(t)|^2_{L^2(\Omega)} + |h(t)|^2_{L^2(\Omega)} + |\ell(t)|^2_{L^2(\Omega)} + r_0^2 + |\Omega|\hat{\beta}(r_0) + 1)$$

where C_{10} is a positive constant depending only on C_1. Integrate the above inequality over $[0, T_0]$ to get

$$(2.15) \quad \begin{aligned} &\kappa|\nabla w|^2_{L^2(0,T_0;L^2(\Omega))} + \frac{C_1}{2}|w|^2_{L^2(0,T_0;L^2(\Omega))} \\ &\le \quad C_{10}(|u - h|^2_{L^2(0,T_0;L^2(\Omega))} + |\ell|^2_{L^2(0,T_0;L^2(\Omega))} + T_0 r_0^2 + T_0|\Omega|\hat{\beta}(r_0) + R_0). \end{aligned}$$

Here, integrating the inequality (2.14) over $[0, T_0]$ we get

$$(2.16) \quad \begin{aligned} &\kappa|\nabla w|^2_{L^2(0,T_0;L^2(\Omega))} + \left| \int_\Omega \hat{\beta}(w)dx \right|_{L^1(0,T_0)} \\ &\le \quad C_{11}(|u - h|^2_{L^2(0,T_0;L^2(\Omega))} + |w|^2_{L^2(0,T_0;L^2(\Omega))} + \\ &\qquad + |\ell|^2_{L^2(0,T_0;L^2(\Omega))} + T_0 r_0^2 + T_0|\Omega|\hat{\beta}(r_0) + R_0), \end{aligned}$$

where C_{11} is a positive constant independent of ℓ. From this inequality with estimates (2.7) and (2.15), we obtain (2.8). By the way, integrating the inequality (2.13) over $[s, t]$ with $0 \leq s < t \leq T_0$, we have

$$\frac{1}{2}|u(t) - h(t)|_V^2$$

(2.17) $\quad \leq \quad \frac{1}{2}|u(s) - h(s)|_V^2$

$$+ 2C_9\left(|w'|_{L^2(0,T_0;L^2(\Omega))}^2 + |\ell|_{L^2(0,T_0;L^2(\Omega))}^2 + T_0 r_0^2 + T_0|\Omega|\hat{\beta}(r_0) + R_0\right).$$

Moreover, integrating the above inequality over $[0, t]$ with respect to s, we have

$$\frac{t}{2}|u(t) - h(t)|_V^2 \leq \frac{1}{2}|u - h|_{L^2(0,T_0;V)}^2 + 2T_0 C_9\left(|w'|_{L^2(0,T_0;L^2(\Omega))}^2\right.$$
$$\left. + |\ell|_{L^2(0,T_0;L^2(\Omega))}^2 + T_0 r_0^2 + T_0|\Omega|\hat{\beta}(r_0) + R_0\right).$$

for all $t \in [0, T_0]$. In particular,

$$\frac{1}{2}|u(T_0) - h(T_0)|_V^2 \left(= \frac{1}{2}|u(0) - h(0)|_V^2\right) \leq \frac{1}{2T_0}|u - h|_{L^2(0,T_0;V)}^2$$
$$+ 2C_9\left(|w'|_{L^2(0,T_0;L^2(\Omega))}^2 + |\ell|_{L^2(0,T_0;L^2(\Omega))}^2 + T_0 r_0^2 + T_0|\Omega|\hat{\beta}(r_0) + R_0\right).$$

This and (2.17) with $s = 0$ imply that (2.10) holds. Next, from (2.12) it follows that

$$\frac{\kappa}{2}|\nabla w(t)|_{L^2(\Omega)}^2 + \int_\Omega \hat{\beta}(w(t))dx$$

$$\leq \quad \frac{\kappa}{2}|\nabla w(s)|_{L^2(\Omega)}^2 + \int_\Omega \hat{\beta}(w(s))dx+$$

$$+ C_{12}(|u - h|_{L^2(0,T_0;L^2(\Omega))}^2 + |\ell|_{L^2(0,T_0;L^2(\Omega))}^2 + r_0^2 + |\Omega|\hat{\beta}(r_0) + R_0)$$

for all $s, t \in [0, T_0]$ with $s \leq t$, where C_{12} is a positive constant independent of ℓ and β. Moreover, integrating the above inequality over $[0, t]$ with respect to s, we have

$$\frac{\kappa t}{2}|\nabla w(t)|_{L^2(\Omega)}^2 + t\int_\Omega \hat{\beta}(w(t))dx$$

(2.18) $\quad \leq \quad \frac{\kappa}{2}|\nabla w|_{L^2(0,T_0;L^2(\Omega))}^2 + \left|\int_\Omega \hat{\beta}(w)dx\right|_{L^1(0,T_0)}$

$$+ C_{12}T_0(|u - h|_{L^2(0,T_0;L^2(\Omega))}^2 + |\ell|_{L^2(0,T_0;L^2(\Omega))}^2 + r_0^2 + \hat{\beta}(r_0) + R_0)$$

for all $t \in [0, T_0]$. In particular,

$$\frac{\kappa}{2}|\nabla w(T_0)|_{L^2(\Omega)}^2 + \int_\Omega \hat{\beta}(w(T_0))dx \quad \left(= \frac{\kappa}{2}|\nabla w(0)|_{L^2(\Omega)}^2 + \int_\Omega \hat{\beta}(w(0))dx\right)$$

(2.19) $\quad \leq \quad \frac{\kappa}{2T_0}|\nabla w|_{L^2(0,T_0;L^2(\Omega))}^2 + \frac{1}{T_0}\left|\int_\Omega \hat{\beta}(w)dx\right|_{L^1(0,T_0)} +$

$$+ C_{12}(|u - h|_{L^2(0,T_0;L^2(\Omega))}^2 + |\ell|_{L^2(0,T_0;L^2(\Omega))}^2 + r_0^2 + \hat{\beta}(r_0) + R_0).$$

152

From (2.18) and (2.19) with estimates (2.7) and (2.8), just as the proof for (2.10), we see that

$$\frac{\kappa}{2}|\nabla w(t)|^2_{L^2(\Omega)} + \int_\Omega \hat{\beta}(w(t))dx \leq C_{13}(|\ell|^2_{L^2(0,T_0;L^2(\Omega))} + r_0^2 + \hat{\beta}(r_0) + R_0)$$

for all $t \in [0, T_0]$ where C_{13} is a positive constant independent of ℓ and β, so that, from the assumption (A2), we get (2.11). □

Corollary 2.1. *There is a positive constant C_{14} independent of ℓ and β such that*

$$|w|^2_{L^2(0,T_0;L^2(\Omega))} \leq C_{14}\left(|\ell|^2_{L^2(0,T_0;L^2(\Omega))} + r_0^2 + \hat{\beta}(r_0) + R_0\right).$$

Moreover, in Damlamian-Kenmochi-Sato [4], the structure of solutions of (PP$_\ell$) was investigated.

3. Main Result

In addition to assumptions (A1)-(A4), suppose that

(A5) $g : R \to R$ is a locally Lipschitz continuous function on R and there is a positive constant L_g such that $L_g < C_1$ and

$$(g(r_1) - g(r_2))(r_1 - r_2) \geq -L_g|r_1 - r_2|^2$$

for any $r_1, r_2 \in R$.

Moreover, we assume that there is a positive primitive $\hat{g}$ of g.

We first give the weak formulation of (PP).

Definition 3.1. A couple $\{u, w\}$ of functions $u : [0, T_0] \to V^*$ and $w : [0, T_0] \to L^2(\Omega)$ is called a (weak) solution of (PP) if the following conditions (w1)'-(w4)' are fulfilled:

(w1)' $\rho(u) \in C([0, T_0]; V^*) \cap W^{1,2}([0, T_0]; V^*) \cap L^2(0, T_0; L^2(\Omega))$, $u \in L^2([0, T_0]; H^1(\Omega))$, $w \in C([0, T_0]; L^2(\Omega)) \cap W^{1,2}([0, T_0]; L^2(\Omega)) \cap L^2(0, T_0; H^1(\Omega))$, and $\hat{\beta}(w) + \hat{g}(w) \in L^1(0, T_0; L^1(\Omega))$;

(w2)' = (w2) in Definition 2.1;

(w3)' there exists a function ξ such that $\xi \in \beta(w)$ a.e. in Q, $\xi + g(w)$ belongs to the space $L^2([0, T_0]; L^2(\Omega))$ and

$$\nu(w'(t), z) + \kappa \int_\Omega \nabla w(t) \cdot \nabla z dx + (\xi(t) + g(w(t)), z) = (u(t), z)$$

for all $z \in H^1(\Omega)$ and a.e. $t \in [0, T_0]$.

(w4)' $u(T_0) = u(0), w(T_0) = w(0)$ in $L^2(\Omega)$.

Remark 3.1. From assumption (A6), we see that $g(r) + L_g r (r \in R)$ is monotone increasing continuous function. Therefore, since $\tilde{\beta}(r) := \beta(r) + g(r) + L_g r$ is a maximal monotone graph in $R \times R$, (1.2) can be written in the following equation

$$\nu \frac{\partial w}{\partial t} - \kappa \triangle w + \tilde{\beta}(w) - L_g w \ni u \quad \text{in } Q.$$

Hence we may assume that g is uniformly Lipschitz continuous function on R with Lipschitz constant L_g without loss of generality.

Our main result is mentioned as follows:

Theorem 3.1. *Under conditions (A1)-(A4) and (A6), (PP) admits at least one solution.*

Proof of Theorem 3.1. Step 1. In this step, we assume that β is strictly monotone and $\overline{D(\beta)}$ is a compact interval i.e. $\overline{D(\beta)} = [\sigma_*, \sigma^*]$ for some constants σ_* and σ^* with $-\infty < \sigma_* < \sigma^* < +\infty$.

Consider a set $Y \subset L^2(0, T; L^2(\Omega))$ defined by

$$Y := \{\tilde{w}; \sigma_* \leq \tilde{w}(t, x) \leq \sigma^* \text{ a.e. in } Q\}.$$

Next, for each $\tilde{w} \in Y$, solve the problem (PP$_\ell$) with $\ell \equiv g(\tilde{w}) \in L^2(0, T_0; L^2(\Omega))$; let $\{u, w\}$ be its unique solution. It is clear that $w \in Y$. From Lemma 2.1, we have

$$\frac{\nu}{2} |w'|^2_{L^2(0, T_0; L^2(\Omega))} \leq C_2 \left(T_0 |\Omega| \sup_{r \in [\sigma_*, \sigma^*]} |g(r)|^2 + R_0 \right) (=: C_2'),$$

$$\kappa |w|^2_{L^2(0, T_0; H^1(\Omega))} \leq C_3 \left(T_0 |\Omega| \sup_{r \in [\sigma_*, \sigma^*]} |g(r)|^2 + r_0^2 + \hat{\beta}(r_0) \right) (=: C_3').$$

Now, introduce a subset Y_1 of Y as follows:

$$Y_1 := \{\tilde{w} \in Y; \frac{\nu}{2} |\tilde{w}'|^2_{L^2(0, T_0; L^2(\Omega))} \leq C_2', \kappa |\tilde{w}|^2_{L^2(0, T_0; H^1(\Omega))} \leq C_3'\}.$$

Clearly, Y_1 is a non-empty compact convex subset of $L^2(0, T_0; L^2(\Omega))$. Next, consider the mapping S from Y_1 into Y_1 which assigns each $\tilde{w} \in Y_1$ to the solution $w \in Y_1$ of (PP$_\ell$) with $\ell \equiv g(\tilde{w})$. S is continuous in Y_1 with respect to the topology of $L^2(0, T_0; L^2(\Omega))$. In fact, let $\{\tilde{w}_n\}$ be any convergent sequence in Y_1 with limit function $\tilde{w} \in Y_1$ with respect to the topology of $L^2(0, T_0; L^2(\Omega))$ as $n \to +\infty$ and $\{u_n, w_n\}$ be the unique solution of (PP$_\ell$) with $\ell \equiv g(\tilde{w}_n)$. Since $|g(\tilde{w}_n)|_{L^2(0, T_0; L^2(\Omega))} \leq T_0 |\Omega| \sup_{r \in [\sigma_*, \sigma^*]} |g(r)|^2$ for any n, by Lemma 2.1 we can find a subsequence $\{n_k\}$ with $n_k \to +\infty$ and u, w such that

$$w'_{n_k} \to w' \quad \text{weakly in } L^2(0, T_0; L^2(\Omega)),$$

$$w_{n_k} \to w \quad \text{weakly in } L^2(0, T_0; H^1(\Omega)),$$

154

$$u_{n_k} \to u \quad \text{weakly in } L^2(0, T_0; V),$$
$$u'_{n_k} \to u' \quad \text{weakly in } L^2(0, T_0; L^2(\Omega)).$$

From the Aubin's compactness theorem with estimate (4)-(5) of Lemma 2.1 it follows that

$$u_{n_k} \to u \quad \text{in } C([0, T_0]; L^2(\Omega))$$

and

$$w_{n_k} \to w \quad \text{in } C([0, T_0]; L^2(\Omega)),$$

therefore

$$\rho(u_{n_k}) \to \rho(u) \quad \text{in } C([0, T_0]; L^2(\Omega))$$
$$u_{n_k}(0) = u_{n_k}(T_0) \to u(0) = u(T_0) \text{ in } L^2(\Omega),$$

and

$$w_{n_k}(0) = w_{n_k}(T_0) \to w(0) = w(T_0) \text{ in } L^2(\Omega).$$

Since $g(\tilde{w}_{n_k}) \to g(\tilde{w})$ in $L^2(0, T_0; L^2(\Omega))$ as $k \to +\infty$, from a convergence theorem [8;Theorem 2.7.1], we see that $\{u, w\}$ is a solution of (PP$_\ell$) with $\ell \equiv g(\tilde{w})$. This implies that S is continuous with respect to the topology of $L^2(0, T_0; L^2(\Omega))$. Hence it follows from the Schauder's fixed point theorem that S has at least one fixed point in Y_1, which gives a solution of (PP). $\qquad \square$

Lemma 3.1. *Under the same assumptions as in Step 1 of the proof of Theorem 3.1, letting $\{u, w\}$ be a solution of (PP) and $r_0 \in (\sigma_*, \sigma^*)$, we have the following inequalities:*

(1)   *There is a positive constant C_{15} independent of β such that*

(3.1)
$$\frac{\nu}{2}|w'|^2_{L^2(0,T_0;L^2(\Omega))} + \frac{1}{2}|u - h|^2_{L^2(0,T_0;V)} \le C_{15} R_0.$$

(2)   *There is a positive constant C_{16} independent of β such that*

(3.2)
$$\kappa|\nabla w|^2_{L^2(0,T_0;L^2(\Omega))} + \left|\int_\Omega \hat{\beta}(w)dx\right|_{L^1(0,T_0)} \le C_{16}\left(r_0^2 + \hat{\beta}(r_0) + R_0\right).$$

(3)   *There is a positive constant C_{17} independent of β such that*

(3.3)
$$|u' - h'|^2_{L^2(0,T_0;L^2(\Omega))} \le C_{17}\left(r_0^2 + \hat{\beta}(r_0) + R_0\right).$$

(4)   *There is a positive constant C_{18} independent of β such that*

(3.4)
$$|u(t) - h(t)|^2_V \le C_{18}\left(r_0^2 + \hat{\beta}(r_0) + R_0\right) \quad \text{for all } t \in [0, T_0].$$

(5)   *There is a positive constant C_{19} independent of β such that*

(3.5)
$$|w(t)|^2_{L^2(\Omega)} + \kappa|\nabla w(t)|^2_{L^2(\Omega)} \le C_{19}\left(r_0^2 + \hat{\beta}(r_0) + R_0\right) \quad \text{for all } t \in [0, T_0].$$

Proof. Noting that $(g(w(t)), w'(t)) = \dfrac{d}{dt} \int_\Omega \hat{g}(w(t))dx$ for all $x \in [0, T_0]$, we can prove the above lemma quite similarly to Lemma 2.1, where $\hat{g}$ is a positive primitive of g. $\square$

Corollary 3.1. *Let* $r_0 \in (\sigma_*, \sigma^*)$. *There is a positive constant* C_{23} *independent of* β *such that*

$$|w|_{L^2(0,T_0;L^2(\Omega))}^2 \leq C_{23} \left(R_0 + \int_\Omega \hat{\beta}(r_0)dx \right).$$

Proof of Theorem 3.1. Step 2. Thanks for Remark 3.1 it is enough to treat the case when g is a Lipschitz continuous function with Lipschitz constant $L_g < C_1$. Also, consider the general case of β; let $\sigma_* = \inf D(\beta)$ and $\sigma^* = \sup D(\beta)$. Now, choose two sequences $\{a_n\}$ and $\{b_n\}$ in R so that $a_n \searrow \sigma_*$ and $b_n \nearrow \sigma^*$ (as $n \to +\infty$), and approximate $\hat{\beta}$ by

$$\hat{\beta}_n(r) := \hat{\beta}(r) + I_{[a_n,b_n]}(r) + \frac{1}{2n}|r|^2,$$

where $I_{[a_n,b_n]}$ is the indicator function of $[a_n, b_n]$. Put $\beta_n := \partial \hat{\beta}_n$ and consider the problems (PP) with β_n for each $n \in N$. Let $\{u_n, w_n\}$ be its unique solution. Note that there is a constant $r_0 \in D(\hat{\beta})$ such that $\hat{\beta}_n(r_0) \leq \hat{\beta}(r_0) + 1$ for large n. Hence, Lemma 3.1 and Corollary 3.1 may give uniformly estimates respect to n. By Lemma 3.1 and Corollary 3.1 we can find a sequence $\{n_k\}$ with $n_k \to +\infty$, and u, w such that

$$w_{n_k}' \to w' \quad \text{weakly in } L^2(0, T_0; L^2(\Omega)),$$

$$w_{n_k} \to w \quad \text{weakly in } L^2(0, T_0; H^1(\Omega)),$$

$$u_{n_k} \to u \quad \text{weakly in } L^2(0, T_0; V),$$

$$u_{n_k}' \to u' \quad \text{weakly in } L^2(0, T_0; L^2(\Omega)).$$

From the Aubin's compactness theorem with estimate (4)-(5) of Lemma 3.1 it follows that

$$u_{n_k} \to u \quad \text{in } C([0, T_0]; L^2(\Omega))$$

and

$$w_{n_k} \to w \quad \text{in } C([0, T_0]; L^2(\Omega)),$$

therefore

$$\rho(u_{n_k}) \to \rho(u) \quad \text{in } C([0, T_0]; L^2(\Omega)),$$

$$u_{n_k}(0) = u_{n_k}(T_0) \to u(0) = u(T_0) \text{ in } L^2(\Omega),$$

$$w_{n_k}(0) = w_{n_k}(T_0) \to w(0) = w(T_0) \text{ in } L^2(\Omega)$$

and

$$g(w_{n_k}) \to g(w) \quad \text{in } L^2(0, T_0; L^2(\Omega)).$$

Moreover, since $\hat{\beta}_{n_k} \to \hat{\beta}$ on R in the sense of Mosco (cf. [10]), we see easily that

$$\varphi_{n_k}^t \to \varphi^t \text{ on } L^2(\Omega) \text{ in the sense of Mosco as } k \to +\infty,$$

156

where φ^t is as in section 1 and $\varphi^t_{n_k}$ is φ^t corresponding to β_{n_k}. Therefore, from a convergence theorem [8;Theorem 2.7.1], we conclude that $\{u, w\}$ is a solution of (PP). $\square$

References

[1] H. Attouch and A. Damlamian, Problémes d'évolution dans les Hilbert et applications, J. Math. Pures Appl., **54** (1975), 53–74.

[2] G. Caginalp, An analysis of a phase field model of a free boundary, Arch. Rat. Mech. Anal., **92** (1986), 205–245.

[3] A. Damlamian, N. Kenmochi and N. Sato, Phase field equations with constraints, *"Nonlinear Mathematical Problems in Industry"*, pp. 391-404, Gakuto. Inter. Ser. Math. Sci. Appl. Vol. **2**, Gakkōtosho, Tokyo, 1993.

[4] A. Damlamian, N. Kenmochi and N. Sato, Subdifferential Operator Approach to a Class of Nonlinear Systems for Stefan Problems with Phase Relaxation, Nonlinear Anal. TMA., **23** (1994), 115–142.

[5] C. M. Elliott and S. Zheng, Global existence and stability of solutions to the phase field equations, *Free Boundary Problems*; pp. 48–58, Intern. Ser. Numer. Math. Vol. 95, Birkhäurer, Basel, 1990.

[6] G. J. Fix, Phase field methods for free boundary problems, *Free Boundary Problems: Theory and Applications,* pp.580–589, Pitman Reserch Notes in Math. Ser. Vol. 79, 1983.

[7] N. Kenmochi, Some nonlinear parabolic variational inequalities, Israel J. Math., **22** (1975), 304–331.

[8] N. Kenmochi, Solvability of nonlinear evolution equations with time-dependent constraints and application, Bull. Fac. Education, Chiba Univ., **30** (1981), 1–87.

[9] N. Kenmochi, Systens of nonlinear PDEs arising from dynamical phase transitions, *Phase Transitions and Hysteresis*; pp. 39–86, Lecture Notes in Math. 1584, Springer-Verlag, Berlin, 1993.

[10] U. Mosco, Convergence of convex sets and of solutions variational inequalities, Advances Math. **3**(1968), 510–585.

[11] A. Visintin, Stefan problems with phase relaxation, IMA J. Appl. Math., **34** (1985), 225–245.

Naoki SATO

Nagaoka National College of Technology, Depatment of Mathematics

888 Nishikatakai-machi, Nagaoka City, Niigata, 940 JAPAN

P STRZELECKI

Quasilinear elliptic systems of Ginsburg–Landau type

1. Introduction

Let $\Omega \subset \mathbb{R}^n$ be an open, bounded, simply connected domain with smooth boundary. Consider the energy functional E_ε of Ginzburg–Landau type given by

$$(1) \qquad E_\varepsilon(u) := \frac{1}{n} \int_\Omega |\nabla u|^n \, dx + \frac{1}{4\varepsilon^n} \int_\Omega (1 - |u|^2)^2 \, dx \,.$$

E_ε is defined for maps $u \in W^{1,n}(\Omega, \mathbb{R}^n)$, i.e., $u = (u_1, \ldots, u_n)$ with each u_i belonging to the standard Sobolev space $W^{1,n}(\Omega)$. For a fixed smooth boundary condition $g : \partial\Omega \to S^{n-1}$, set

$$W_g^{1,n} := \{ u \in W^{1,n}(\Omega, \mathbb{R}^n) \mid u = g \text{ on } \partial\Omega \} \,.$$

The minimum
$$(2) \qquad \underset{u \in W_g^{1,n}}{\text{Min}} \; E_\varepsilon(u)$$

is achieved by some (in general non-unique, see examples in Section 2) map u_ε which solves (in the distributional sense) the Euler–Lagrange system

$$(3) \qquad - \text{div}\left(|\nabla u_\varepsilon|^{n-2} \nabla u_\varepsilon^i \right) = \frac{1}{\varepsilon^n} u_\varepsilon^i (1 - |u_\varepsilon|^2), \qquad i = 1, \ldots, n.$$

In this note, we consider the following

PROBLEM. Study the asymptotic behaviour of u_ε as $\varepsilon \to 0$, for $n \geq 3$.

We assume in addition that the boundary condition $g : \partial\Omega \to S^{n-1}$ is topologically trivial, i.e. $\deg g = 0$.

The two dimensional case. For $n = 2$, the functional (1) and its analogues are often considered in various vortex models of phase transitions. In applications, the complex-valued function u_ε is related to

- the density of superconducting electrons in type II superconductors; $|u_\varepsilon| \approx 1$ corresponds to the superconducting state and $|u_\varepsilon| \approx 0$ corrensponds to the normal state;

- the two-dimensional magnetization vector in magnets and the condenstate wavefunction in superfluids.

The parameter ε is usually *very* small in relevant physical applications. Thefeore, it is of great mathematical interest to analyze the asymptotic behaviour of u_ε as $\varepsilon \to 0$ (even if the limiting problem has no physical meaning). Bethuel, Brezis and Hélein have recently proved, among other things, the following results concerning the asymptotic behaviour of u_ε.

"Easy" case. Assume that $g : \partial\Omega \to S^1$ is smooth, and $\deg g = 0$. The asymptotic behaviour of the whole family $\{u_\varepsilon \; : \; \varepsilon > 0\}$ is then fully described by the following.

Theorem 1.1 (Bethuel, Brezis, Hélein [1]). *Let $n = 2$ and let u_ε be a minimizer of E_ε in $W_g^{1,2}(\Omega, \mathbb{R}^2)$. Denote by u_0 the unique S^1-valued harmonic mapping with $u_0\big|_{\partial\Omega} = g$. As $\varepsilon \to 0$, u_ε tends to u_0 in $C^{1,\alpha}(\overline{\Omega})$ for any $\alpha < 1$, and in $C^k(K)$ for any natural k and any compact $K \subset \Omega$.*

Tough case. If the boundary condition $g : \partial\Omega \to S^1$ has nonzero degree, $\deg g = d > 0$, then the situation is more complicated. Main difficulty stems from the fact that $E_\varepsilon(u_\varepsilon)$ blows like $|\log\varepsilon|$ for $\varepsilon \to 0$.

Theorem 1.2 (Bethuel, Brezis, Hélein [2]). *Assume that $\Omega \subset \mathbb{R}^2 \equiv \mathbb{C}$ is starhaped. Then, one can find a sequence $\varepsilon_k \to 0$, precisely d points a_1, a_2, ..., a_d in Ω, and a smooth harmonic map $u_* : \Omega \setminus \{a_1, \ldots, a_d\} \to S^1$ satisfying $u_* = g$ on $\partial\Omega$ such that*

$$u_{\varepsilon_k} \to u_* \qquad \text{in } C^{1,\alpha}(\overline{\Omega} \setminus \{a_1, \ldots, a_d\}) \text{ for any } \alpha \in (0,1),$$
$$\text{and in } C_{\text{loc}}^k(\Omega \setminus \{a_1, \ldots, a_d\}) \text{ for any } k.$$

Moreover, u_* coincides with the so-called canonical harmonic map, i.e. is of the form

$$u_*(z) = \prod_{j=1}^d \left(\frac{z - a_j}{|z - a_j|} \right) \cdot \exp(i\varphi),$$

where $\varphi : \Omega \to \mathbb{R}$ is a harmonic function with $\exp(i\varphi) = \prod_{j=1}^d \left(\frac{|z-a_j|}{z-a_j} \right) g$ on $\partial\Omega$.

Another theorem from [2] states that the configuration $\{a_1, \ldots, a_d\}$ of singularities of u_* minimizes a function of d complex variables, $W : \Omega \times \cdots \times \Omega \to \mathbb{R}$, the so-called *renormalized energy*. The beautiful analysis of [2] heavily depends on the linearity of the Laplace operator and on the nice structure of complex numbers that we can use in two dimensions. To deal with more realistic problems arising in the modelling of phase transitions in magnetic and superconducting media, it seems necessary to gain first some insight into mathematical phenomena ocurring for more general variational problems, with more complicated nonlinearities in their Euler–Lagrange equations. Some of relevant open problems are listed in [2]; one of them is to describe the asymptotic behaviour of u_ε in the general case, for $n \geq 3$.

The results. We begin with necessary definitions. Consider the n-Dirichlet integral

$$\mathbf{I}_n(u) = \int_\Omega |\nabla u|^n \, dx.$$

A map $u \in W_g^{1,n}(\Omega, S^{n-1})$ is *(weakly) n-harmonic* iff it is a critical point of $\mathbf{I}_n$ in the class $W_g^{1,n}(\Omega, S^{n-1})$ with respect to variations in the range. It is well known that n-harmonic maps are precisely the weak solutions of the nonlinear degenerate elliptic system

$$(4) \qquad\qquad - \operatorname{div}\left(|\nabla u|^{n-2}\nabla u\right) = u|\nabla u|^n .$$

By a *minimizing n-harmonic map* we mean here any minimizer of $\mathbf{I}_n$ in the class $W_g^{1,n}(\Omega, S^{n-1})$. For a fixed $g : \partial\Omega \to S^{n-1}$ with $\deg g = 0$, denote by $\mathcal{M}_g$ the set of all minimizing n-harmonic maps u with $u = g$ on $\partial\Omega$.

Proposition 1.3. *If $\{u_\varepsilon \mid \varepsilon > 0\}$ is a family of minimizers for the problem (2), then one can choose a sequence $\varepsilon_k \overset{k=\infty}{\longrightarrow} 0$, and a minimizing n-harmonic map $u_* \in \mathcal{M}_g$ such that $u_{\varepsilon_k} \overset{k=\infty}{\longrightarrow} u_*$ strongly in $W^{1,n}(\Omega, \mathbb{R}^n)$ and a.e.*

Proof. Pick a map $w \in \mathcal{M}_g$. (This is the only place where the assumption $\deg g = 0$ is used: it implies that $W_g^{1,n}(\Omega, S^{n-1})$ is nonempty!) Since $|w| = 1$ a.e., we have

$$(5) \qquad\qquad E_\varepsilon(u_\varepsilon) \le E_\varepsilon(w) = \frac{1}{n}\int_\Omega |\nabla w|^n\, dx .$$

Hence, the family $\{u_\varepsilon : \varepsilon > 0\}$ is bounded in $W^{1,n}(\Omega, \mathbb{R}^n)$. Thus, one can find a sequence $\varepsilon_k \to 0$ and a map $u_* \in W^{1,n}(\Omega, \mathbb{R}^n)$ such that

$$(6) \qquad \begin{cases} \nabla u_{\varepsilon_k} \rightharpoonup \nabla u_* & \text{weakly in } L^n(\Omega, \mathbb{R}^{n^2}), \\ u_{\varepsilon_k} \to u_* & \text{strongly in } L^n(\Omega, \mathbb{R}^n) \text{ and a.e..} \end{cases}$$

Using Fatou's Lemma, (5), and (6), one checks that $u_* \in \mathcal{M}_g$ is also a minimizing n-harmonic map. Now, recall a classical theorem of functional analysis (weak convergence in L^p plus convergence of norms implies, for $1 < p < \infty$, strong convergence in L^p) to conclude that $u_{\varepsilon_k} \to u_*$ strongly in $W^{1,n}$. $\qquad\square$

Remark. In general, it may happen that, for a fixed boundary condition g, different sequences of minimizers u_ε converge to different limits (see Section 2.) Such nonunique asymptotic behaviour (in the "easy" case of topologically trivial boundary condition) is excluded for $n = 2$.

The next theorem is our main result.

Theorem 1.4. *Let u_{ε_k} be, as in Proposition 1.3, a sequence of minimizers of E_{ε_k} which converges to some minimizing n-harmonic map u_* in the $W^{1,n}$-norm. Then, for any $\alpha \in (0,1)$ and any compact subset K of Ω, u_{ε_k} tends to u_* in the space $C^\alpha(K)$.*

A rough sketch of the proof is given in Section 3 (the interested reader is referred to [9] for more details).

Remark. For the tougher case $\deg g = d \ne 0$, Min Chun Hong [8] has proved that a subsequence of u_ε converges weakly in $W^{1,n}(G, \mathbb{R}^n)$, where $G = (\Omega$ except $|d|$ distinct points), to some map u_* which is n-harmonic on G.

A very interesting recent paper [5] of Z. Han and Y. Li contains a result more general than that of Hong. Namely, Han and Li prove that u_{ε_k} converges to u_* strongly in $C^0_{\text{loc}}(\overline{\Omega} \setminus \{a_1, \ldots, a_{|d|}\})$. Their method of proof is different from ours; in particular, differentiation of systems (which we use to obtain gradient bounds) is replaced by a new regularity theorem, obtained via pertubartion arguments, for p-harmonic systems with Hölder continuos coefficients.

The notation throughout the rest of this paper is either standard or selfexplanatory. By B_r or $B(a, r)$ we denote the Euclidean ball in $\mathbb{R}^n$ of radius r, centered at a. The length of the gradient is defined by the formula $|\nabla u|^2 := \sum_{1 \leq i, j \leq n} \left(\frac{\partial u^i}{\partial x_j}\right)^2$. Finally, C denotes a general constant which may change from one line to another; $C(n)$ denotes a constant depending *only* on n.

2. An example of nonuniqueness: planar n-harmonic maps

In this Section, we show an example of nonuniqueness of minimizing n-harmonic maps from B^n into S^{n-1}. The general idea follows the papers of Hardt and Kinderlehrer [6], and Hardt, Kinderlehrer, and Lin [7]. Some technicalities are simplified here.

For sake of simplicity, let $n = 3$ and write, for some $q > 0$ to be specified later,

$$(7) \qquad g(x) = (\cos qx_3, \sin qx_3, 0) \qquad \text{for } (x_1, x_2, x_3) \in \partial B^3 \equiv S^2.$$

Then, define $v : B^3 \to S^2$ by the same formula. By a straightforward calculation one verifies that, for every positive q, v is a smooth S^2 valued 3-harmonic map with $|\nabla v| \equiv q$ and $\mathbf{I}_3(v) = 4\pi q^3/3$. We shall show that v is not *minimizing* in the class $W_g^{1,3}(B^3, S^2)$, and that a minimizing $u = (u^1, u^2, u^3) \in W_g^{1,3}(B^3, S^2)$ cannot be planar, i.e. its coordinate u^3 must be nonzero on a set of nonzero Lebesgue measure.

To this end, pick a cutoff function $\eta \in C_0^\infty(B^3)$ with $\eta \equiv \frac{\pi}{2}$ on a smaller ball $B(0, 1 - \mu)$ and $|\nabla \eta| \leq 2/\mu$. Put

$$w(x) = (\cos \eta(x) \cos qx_3, \cos \eta(x) \sin qx_3, \sin \eta(x)).$$

Then, $|\nabla w|^2 = \cos^2(\eta)|\nabla v|^2 + |\nabla \eta|^2$. Therefore,

$$\begin{aligned}
\mathbf{I}_3(w) &= \int_{B^3} |\nabla w|^3 \, dx \leq \sqrt{2} \int_{B^3} (\cos^3(\eta)|\nabla v|^3 + |\nabla \eta|^3) \, dx \\
&\leq \frac{4\pi\sqrt{2}}{3}(1 - (1 - \mu)^3) \left(q^3 + \frac{8}{\mu^3}\right) < 200(\mu q^3 + \mu^{-2}).
\end{aligned}$$

Obviously, for μ small and q large, the last expression does not exceed $\mathbf{I}_3(v) = 4\pi q^3/3$ (it is enough to take e.g. $q = \mu^{-1} = 100$). Therefore v is not minimizing.

Suppose now that $w = (w^1, w^2, 0) \in W_g^{1,3}(B^3, S^2)$ were a minimizing 3-harmonic map. By a theorem of Bethuel and Zheng [3, Lemma 1] one could write $w^1 = \cos(\theta)$,

$w^2 = \sin(\theta)$ for some real-valued function θ of class $W^{1,3}$. Since w satisfies (4) for $n = 3$, one can easily check that θ solves the non-constrained 3-harmonic equation, $\mathrm{div}(|\nabla\theta|\nabla\theta) = 0$. Moreover, the boundary condition implies

$$\theta(x_1, x_2, x_3) - qx_3 \in \{2k\pi \mid k \in \mathbb{Z}\} \qquad \text{for } (x_1, x_2, x_3) \in S^2.$$

But if the trace of a Sobolev function has all its values in a discrete subset of $\mathbb{R}$, then it is constant. Hence, without loss of generality we can assume that $\theta(x) = qx_3$ on S^2. By the monotonicity of non-constrained 3-harmonic operator (with Dirichlet boundary conditions) we conclude that $\theta(x) = qx_3$ in B^3. Hence, $w \equiv v$, a contradiction. Therefore, no minimizer can be planar in our case.

Take now a minimizer $\psi = (\psi^1, \psi^2, \psi^3)$. The above reasoning clearly implies that $W_g^{1,3}$ contains two *distinct minimizers* for $\mathbf{I}_3$, i.e. $(\psi^1, \psi^2, +|\psi^3|)$ and $(\psi^1, \psi^2, -|\psi^3|)$, and one nonminimizing planar 3-harmonic map. Moreover, if one minimizes E_ε in the class $W_g^{1,3}$, there are, for small values of ε, at least two distinct minimizers $(u_\varepsilon^1, u_\varepsilon^2, \pm|u_\varepsilon^3|)$. Since minimizers of $\mathbf{I}_3$ are not planar, subsequences of $(u_\varepsilon^1, u_\varepsilon^2, +|u_\varepsilon^3|)$ and $(u_\varepsilon^1, u_\varepsilon^2, -|u_\varepsilon^3|)$ don't converge to the same limit.

The above example raises obvious questions. Can the cardinality of $\mathcal{M}_g$ be arbitrary? Can $\mathcal{M}_g$ be an infinite set? (This is possible for harmonic maps $u : B^3 \to S^2$, see [7].) How to characterize those $u \in \mathcal{M}_g$ which can be limits of subsequences of u_ε? These problems shall be object of further study.

3. Sketch of proof of Theorem 1.4

Step 1. Maximum principle. From (3), using $\psi_m = u_i \min(m, (|u_\varepsilon|^2 - 1)_+)$, $m = 1, 2, \ldots$, as a testing function, we obtain after a brief calculation the inequality

$$\frac{1}{2}\int_{A_m} |\nabla u_\varepsilon|^{n-2}|\nabla(|u_\varepsilon|^2 - 1)_+|^2 \, dx + \frac{1}{\varepsilon^n}\int_{A_m} (|u_\varepsilon|^2 - 1)_+^2 \, dx \le 0,$$

where $A_m = \{x \in \Omega \mid (|u_\varepsilon|^2 - 1)_+(x) < m\}$. Since m is arbitrary, this implies

(8) $$|u_\varepsilon(x)| \le 1 \qquad \text{for a.e. } x \in \Omega.$$

Step 2. Cacciopoli estimate and its consequences. Considering difference quotients, check first that $|\nabla u_\varepsilon|^{(n-2)/2}\nabla u_\varepsilon$ has first order distributional partial derivatives locally in L^2. Then, differentiate both sides of (3) with respect to x_j and use appropriate test functions to obtain, after a standard but tedious computation, the following

Lemma 3.1. *If u_ε is a weak solution to (3), $\beta \ge 0$, $n \ge 3$, and $w_\varepsilon := |\nabla u_\varepsilon|^2$, then, for any $\zeta \in C_0^\infty(\Omega)$,*

$$\int_\Omega |\nabla(w_\varepsilon^{\frac{n+\beta}{4}})|^2\zeta^2 \, dx \le \frac{2(n+\beta)}{\varepsilon^n}\int_\Omega (1 - |u_\varepsilon|^2)w_\varepsilon^{\frac{\beta+2}{2}}\zeta^2 \, dx + 9n^2\int_\Omega w_\varepsilon^{\frac{n+\beta}{2}}|\nabla\zeta|^2 \, dx.$$

Applying Moser and De Giorgi–Stampacchia iteration techniques in a way mimicking the arguments of DiBenedetto and Friedman [4], and using a scaling argument, one shows that for u_ε with $\int_\Omega |\nabla u_\varepsilon|^n \, dx \leq C$ we have $|\nabla u_\varepsilon| \leq C/\varepsilon$ on any compact subset of Ω. As a consequence, one proves that $|u_{\varepsilon_k}| \to 1$ uniformly on compact subsets of Ω.

Step 3. Getting rid of ε. Here, we modify a trick which, in the case $n = 2$, is due to Bethuel, Brezis and Hélein [1, steps A3 and A4].

Since $|u_\varepsilon| \approx 1$ for small ε, one can rewrite the Euler–Lagrange system in the form

$$\frac{1 - |u_\varepsilon|^2}{\varepsilon^n} = - \sum_{1 \leq i \leq n} \operatorname{div}\left(|\nabla u_\varepsilon|^{n-2}\nabla u_\varepsilon^i\right)\frac{u_\varepsilon^i}{|u_\varepsilon|^2}.$$

We combine this equality with Lemma 3.1, compute the divergence and integrate by parts the term with the Laplacian Δu_ε^k. Applying several times Cauchy–Schwarz inequality in a standard way and absorbing appropriate terms, we obtain the estimate

$$(9) \qquad \int_\Omega |\nabla(w_\varepsilon^{\frac{n+\beta}{4}})|^2 \zeta^2 \, dx \leq C(n, \beta)\left(\int_\Omega w_\varepsilon^{\frac{n+\beta+2}{2}} \zeta^2 \, dx + \int_\Omega w_\varepsilon^{\frac{n+\beta}{2}} |\nabla \zeta|^2 \, dx\right).$$

The main concern now is to estimate the first integral on the right hand side. To this end, set $p = 2n/(n + 2)$ so that the Sobolev conjugate exponent of p is $p^* = np/(n-p) = 2$. By Sobolev imbedding theorem and Hölder inequality with exponents $\frac{n+2}{n}$ and $\frac{n+2}{2}$, we have

$$
\begin{aligned}
\int_\Omega w_\varepsilon^{\frac{n+\beta+2}{2}} \zeta^2 \, dx \;\leq\;& C(n)\left(\int_\Omega |\nabla(w_\varepsilon^{\frac{n+\beta}{4}})|^2 \zeta^2 \, dx\right) \cdot \left(\int_{\{\zeta \neq 0\}} w_\varepsilon^{\frac{n}{2}} \, dx\right)^{\frac{2}{n}} \\
(10) \qquad & + C(n)\left(\int_\Omega w_\varepsilon^{\frac{n+\beta}{2}} |\nabla \zeta|^2 \, dx\right) \cdot \left(\int_{\{\zeta \neq 0\}} w_\varepsilon^{\frac{n}{2}} \, dx\right)^{\frac{2}{n}}.
\end{aligned}
$$

Here, as before, ζ is a smooth cutoff function, $w_\varepsilon = |\nabla u_\varepsilon|^2$, and $\beta \geq 0$.

Step 4. Final argument: gradient bounds independent of ε. By Proposition 1.3, ∇u_{ε_k} converges strongly in L^n. Therefore, if the support of ζ is small enough, the integral

$$(11) \qquad I(\varepsilon_k) := \int_{\{\zeta \neq 0\}} w_\varepsilon^{\frac{n}{2}} \, dx \equiv \int_{B_R} |\nabla u_\varepsilon|^n \, dx$$

is (uniformly) small for all ε_k. One can combine (9) with (10) and then use (11) to rewrite the resulting inequality in the form

$$\int_\Omega |\nabla(w_{\varepsilon_k}^{\frac{s}{2}})|^2 \zeta^2 \, dx \leq C(n)s^2 \int_\Omega w_{\varepsilon_k}^s |\nabla \zeta|^2 \, dx, \qquad s \in [n/2, M),$$

where $M > n/2$ is a fixed number depending only on the diameter of the set $\{\zeta \neq 0\}$. Next, we start a Moser iteration and obtain the following

Lemma 3.2. *For any compact set $K \subset \Omega$ and any $q \in (1, \infty)$ there exists a constant C, depending only on n, q, K, and the boundary data g, such that*

$$(12) \qquad \|(w_{\varepsilon_k})^q\|_{W^{1,2}(K)} \leq C \qquad \text{for } k = 1, 2, \ldots$$

Here, ε_k stands for the sequence selected in Proposition 1.3.

To finish the whole proof, we apply a diagonal procedure, Rellich–Kondrachov compactness theorem, and Sobolev imbedding theorem.

Remark. The assumption that u_ε minimizes E_ε was used only to establish the convergence $u_{\varepsilon_k} \to u_*$ in $W^{1,n}$.

4. Concluding remarks

Theorem 1.4 is of course not a final result. We expect that it is possible to obtain at least local $C^{1,\alpha}$ convergence, and C^α convergence of u_{ε_k} on $\overline{\Omega}$, also in the case of variable boundary data g_ε sufficiently close to a fixed map $g_0 : \partial\Omega \to S^{n-1}$. This would allow to improve the results obtained by Han–Li [5] and Hong [8] for the nonzero degree case.

The search for the renormalized energy (which would allow to predict the position of singularities a_i of the limiting map u_* in the case $\deg g \neq 0$) still remains a challenging and interesting open problem.

Acknowledgement. This research has been started while the author stayed at the University of Paris VI as an ESF/FBP fellow. He is grateful to Professor Haïm Brezis for his valuable help and encouragement.

The author is also partially supported by KBN grant no. 2 PO3A 034 08.

References

[1] F. Bethuel, H. Brezis, and F. Hélein. Asymptotics for the minimization of a Ginzburg–Landau functional. *Calculus of Variations and PDE*, 1:123–148, 1993.

[2] F. Bethuel, H. Brezis, and F. Hélein. *Ginzburg–Landau vortices*, volume 13 of *Progress in Nonlinear Differential Equations and Their Applications*. Birkhauser, Boston, 1994.

[3] F. Bethuel and X. Zheng. Density of smooth functions between two manifolds in Sobolev spaces. *J. of Funct. Anal.*, 80:60–75, 1988.

[4] E. DiBenedetto and A. Friedman. Regularity of solutions of nonlinear degenerate parabolic systems. *J. für die Reine und Angew. Math.*, 349:83–128, 1984.

[5] Z. Han and Y. Li. Degenerate elliptic systems and applications to Ginzburg–Landau type equations, I. Preprint, Rutgers University, 1995 (to appear in *Calculus of Variations and PDE*).

[6] R. Hardt and D. Kinderlehrer. Mathematical questions of liquid crystlas theory. In *College de France Seminar IV (3)*, 1987.

[7] R. Hardt, D. Kinderlehrer, and F. H. Lin. The variety of configurations of static liquid crystlas. In H. Berestycki, J.-M. Coron, and I. Ekeland, editors, *Variational methods*. Birkhauser, 1990.

[8] M. C. Hong. Asymptotic behavior for minimizers of a Ginzburg–Landau functional in higher dimensions associated with n-harmonic maps. Preprint, 1995.

[9] P. Strzelecki. Asymptotics for the minimization of a Ginzburg–Landau energy in n dimensions. *Colloquium Math.*, in press.

Paweł Strzelecki
Institute of Mathematics and ICM, Warsaw University
ul. Banacha 2
00–097 Warszawa, Polska (Poland)
e-mail: `pawelst@mimuw.edu.pl`

A VISINTIN
Hysteresis and free boundary problems

Abstract: In this note we review the definition of hysteresis opererator and some classes of free boundary problems with hysteresis, and present a collection of references.

1. Hysteresis

Hysteresis appears in several phenomena, in physics, engineering, chemistry, biology, economics, and so on. Typical examples are plasticity, ferromagnetism, ferroelectricity. A systematic investigation of the mathematical properties of hysteresis only began in the 1970s, but is now in full development.

A Hysteresis Loop. As an example let us consider the hysteresis loop outlined in Fig. 1. If u increases from u_A to u_C the pair (u, w) moves along the curve ABC; conversely, if u decreases from u_C to u_A, (u, w) moves along the path CDA. Moreover if u inverts its motion when $u_A < u(t) < u_C$, then (u, w) moves into the interior of $\mathcal{L}$ (the region bounded by $ABCDA$) along a curve which must be prescribed by the specific hysteresis model. One may assume that any interior point of $\mathcal{L}$ can be attained by a suitable choice of the input function u.

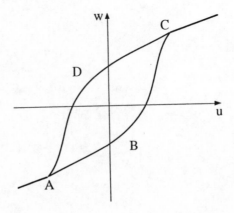

Figure 1. Continuous *hysteresis loop.*

At any instant t, $w(t)$ depends on the previous evolution of u (*memory effect*), and on the initial state of the system. In the most simple setting one assumes a dependence of the form

$$w(t) = [\mathcal{F}(u, w^0)](t) \qquad \forall t \in [0, T], \tag{1.1}$$

where $[\mathcal{F}(u, w^0)](0) = w^0$, and $(u(0), w^0) \in \mathcal{L}$. Here $\mathcal{F}$ represents an operator acting in an appropriate space of time dependent functions, e.g. $C^0([0, T])$, for any fixed w^0. Obviously $\mathcal{F}$ must be *causal*: the output $w(t)$ may not depend on $u_{|]t,T]}$.

Here it is implicitly assumed that the pair $(u(0), w^0)$ characterizes the initial state of the system. However in several cases the state depends on *inner variables*, whose initial value must be then specified.

Rate Independence. We also require the path of the pair $(u(t), w(t))$ to be invariant with respect to any increasing C^∞-diffeomorphism $\varphi : [0, T] \to [0, T]$:

$$\mathcal{F}(u \circ \varphi, w^0) = \mathcal{F}(u, w^0) \circ \varphi \qquad \text{in } [0, T] \tag{1.2}$$

(i.e., if $u \mapsto w$ then $u \circ \varphi \mapsto w \circ \varphi$). We regard this invariance property as the characteristic feature of hysteresis, and define hysteresis as *rate independent memory*. Any rate independent causal operator will be named a *hysteresis operator*.

This definition excludes any viscous-type memory, such as those represented by time convolution.

P.D.E.s with Hysteresis. O.D.E.s and P.D.E.s can be coupled with a hysteresis law. For instance if Ω is a domain of $\mathbf{R}^N$ and A an elliptic operator, one can couple (1.1) (set at each space point) with either

$$\frac{\partial}{\partial t}(u + w) + Au = f \qquad \text{in } Q := \Omega \times]0, T[, \tag{1.3}$$

or

$$\frac{\partial u}{\partial t} + Au + w = f \qquad \text{in } Q, \tag{1.4}$$

and with suitable initial and boundary conditions; here f is a given function. For $A = \text{curl}^2$ ($A = -\frac{d^2}{dx^2}$ is $N = 1$), (1.3) and (1.1) are a (simplified) model of the evolution of a ferromagnetic system. (1.3) can be derived by coupling the Maxwell equations with the Ohm law and neglecting the displacement current; here u represents the magnetic field, and w the magnetization. (1.4) with $A = -\Delta$ and (1.1) may represent heat diffusion in presence of a distributions of thermostats. These examples can be generalized in several ways; for instance in some applications the hysteresis operator acts on the boundary of Ω.

Hysteresis operators and P.D.E.s of this sort have been studied in [10], for instance.

2. Stefan Problem with Hysteresis

Hysteresis in Phase Transitions. A conference (Trento 1989) and a school (Montecatini 1993, see [1]) have been devoted to the interplay between hysteresis and free boundary problems. Here we just review some examples.

The classical *Landau model* of phase transitions illustrates that *nonconvexity, hysteresis* and *phase transitions* are strictly related. Let us consider a nonconvex potential $F : \mathbf{R} \to \mathbf{R}$, e.g. $F(y) = (y^2 - 1)^2$ for any $y \in \mathbf{R}$. Its derivative F' is nonmonotone, and its inverse is multivalued; in evolution discontinuous hysteresis can then occur.

In space distributed systems hysteresis must be *sustained* by space interaction, which often appears as *surface tension*. If the latter vanishes F must be replaced by the *convexified* potential F^{**}, whose derivative $(F^{**})'$ is derived from F' through the classical *Maxwell lever rule*. The inverse of $(F^{**})'$ is also discontinuous. In space distributed systems discontinuities in the constitutive relations typically correspond to the occurrence of free boundaries. For instance the weak formulation of the standard Stefan problem can be derived through such a convexification.

In Fig. 1 we outlined a continuous hysteresis operator. One can also deal with *discontinuous* operators; an example is easily obtained modifying the sign function, replacing the critical value 0 by two thresholds ρ_1, ρ_2 ($\rho_1 < \rho_2$) for downward and upward jumps (respectively), cf. Fig. 2(a).

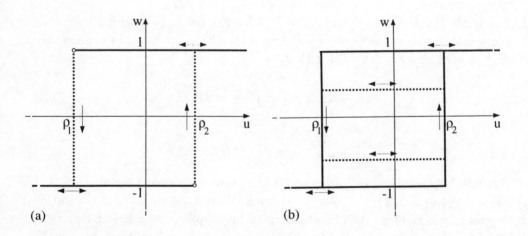

(a)   (b)

Figure 2. *Delayed relay operator (ranging in $\{-1, 1\}$), and corresponding completion (allowing for all $(u, w) \in [\rho_1, \rho_2] \times [-1, 1]$).*

This operator h_ρ (where $\rho := (\rho_1, \rho_2)$) is named *(delayed) relay*. It tranforms any input funtion $u \in C^0([0, T])$ and any initial output $\xi \in \{-1, 1\}$ (such that $\xi = -1$

if $u(0) < \rho_1$, and $\xi = 1$ if $u(0) > \rho_2$) into the output w as follows. $w(t)$ attains the value -1 as long as $u(t) > \rho_2$; as u overtakes the values ρ_2, w jumps to 1. There it stays until w descends below ρ_1; at that time w jumps to -1, and so on.

Dealing with P.D.E.s one is induced to extend the graph of the relay by allowing w to attain intermediate values between -1 and 1, cf. Fig. 2(b), to get a multivalued operator k_ρ having a closed graph in appropriate function spaces; see [10; Chap. VI]. This completion procedure is analogous to that of replacing the sign function (with $\text{sign}_0(0) = 0$) by the sign graph (with $\text{sign}(0) = [-1, 1]$).

One can then couple either the P.D.E. (1.3) or (1.4) with the constitutive relation

$$w \in k_\rho(u, w^0) \qquad \text{in } Q. \tag{2.1}$$

This yields a free boundary problem which is sometimes named *Stefan problem with hysteresis*. This can be regarded as a rough model of undercooling and superheating in phase transitions, or for $N = 1$ as a simplified model of ferromagnetic evolution. See Little and Showalter [25], Verdi and V. [33], V. [36, 10]. A different hysteresis relation has been considered by Meirmanov and Shemetov [28].

3. Other Free Boundary Problems with Hysteresis

Several phenomena of applicative interest lead to free boundary problems containing hysteresis nonlinearities. Here are just some (simplified) examples.

(1) In some models hysteresis appears in the source or boundary terms. For instance in a two-phase system a distribution of thermostats can be used to control phase transitions. One can then consider the following Stefan-type problem:

$$\frac{\partial}{\partial t}[u + H(u)] - \Delta u + \mathcal{F}(u) = f \qquad \text{in } \mathcal{D}'(Q). \tag{3.1}$$

Results are known for $\mathcal{F}$ *continuous* hysteresis operator. See Friedman and Hoffmann [13], Götz, Hoffmann and Meirmanov [14], Hoffmann and Kenmochi [15], Hoffmann, Kornstädt and Sprekels[16], Hoffmann and Sprekels [17], Hoffmann, Sprekels and Niezgódka [18], Sprekels [30].

(2) Heat diffusion in presence of thermostats is represented by an equation of the form

$$\frac{\partial u}{\partial t} - \Delta u + \mathcal{F}(u) = f \qquad \text{in } \mathcal{D}'(Q). \tag{3.2}$$

If $\mathcal{F}$ is discontinuous, e.g. equal to the relay operator k_ρ, this also is a free boundary problem. See Alt [11], Hornung and Showalter [21], Stoth [31], Verdi and V. [33], V. [10; Chap. XI].

(3) Hysteresis also occurs in biology. For instance, growth of bacteria in presence of nutrients has been represented by Hoppensteadt and Jäger by a model which

includes a parabolic P.D.E. similar to (1.4), with the notable difference that u is a vector. See Chiu, Hoppensteadt and Jäger [12], Hoppensteadt and Jäger [19], Hoppensteadt, Jäger and Pöppe [20], Jäger [22, 23], V. [10; Chap. XI].

This phenomenon is analogous to that of *Liesegang rings* in chemistry.

(4) *Ferromagnetism* is a typical hysteresis phenomenon. According to a classical model (see e.g. Chikazumi and Charap [52]), a ferromagnetic body splits up into several (unknown) uniformly magnetized domains. These are separated by thin transition layers, which can also be represented as free boundaries, cf. Anzellotti, Baldo and V. [50]. The mathematical properties of this model have not yet been studied systematically.

An interesting macroscopic model of scalar ferromagnetic hysteresis was proposed by Preisach [56] fifty years ago, and has also been applied to other phenomena; see Mayergoyz [8]. Attempts have been done to extend this model to vectors; see e.g. Damlamian and V. [53], Mayergoyz [55], V. [10; Sect. VI.3]. However none of them seems to describe ferromagnetism satisfactorily. Therefore the problem of formulating a physically acceptable vectorial model of ferromagnetic hysteresis is still open, let alone that of coupling such a model with the Maxwell equations.

Hysteresis also occurs in ferroelectricity, and coupling with the Maxwell equations yields nonlinear P.D.E.s. Apparently this setting has not yet been studied in the framework of function spaces.

(5) Hysteresis is quite evident in continuum mechanics. *Plasticity* is an obvious example. In an elasto-plastic body the elastic and yielding phases are separated by a free boundary. This problem has been satisfactorily treated in the framework of *convex analysis,* however a formulation in terms of hysteresis operators might lead to new results.

Shape memory alloys have been developed in the last years, and are accounted for elsewhere in these proceedings.

Damage and *fatigue* are essentially rate independent, and so can also be labbelled as hysteresis phenomena. Analytical properties of models of fatigue have been studied by Brokate, Dressler and Krejčí [51], Krejčí [54]; but related P.D.E.s have not yet been addressed.

(6) *Saturated-unsaturated fluid flow* through porous media leads to the system

$$\begin{cases} \varphi \dfrac{\partial s}{\partial t} - \nabla \cdot k \left(\dfrac{\nabla p}{\rho g} + \vec{\xi} \right) = 0 & \text{in } \mathcal{D}'(Q) \\ k = \mathcal{F}_1(s), \ s = \mathcal{F}_2(p) & \text{in } Q, \end{cases} \tag{3.3}$$

coupled with appropriate initial and boundary conditions, including a so-called *seepage condition,* of Signorini-type. Here φ is the porosity, g the gravity acceleration, ρ the density, $\vec{\xi}$ a vertical upward vector. The saturation s, the pressure p, the hydraulic conductivity k are unknown. $\mathcal{F}_1$ and $\mathcal{F}_2$ are continuous hysteresis operators

170

(of *Preisach-type,* for instance). See Alt, Luckhaus and V. [49] for the analogous model without hysteresis.

A large technical literature has been devoted to hysteresis in porous media filtration, see e.g. [41 – 48], but its mathematical aspects have not yet been studied.

Hysteresis occurs in a multitude of physical phenomena; for instance it is quite evident in *adsorption.* Its mathematical properties have not yet been investigated.

Finally we note that so far numerical analysis of P.D.E.s including hysteresis has received little attention. See Verdi and V. [33, 34].

4. References

(i) **Basic References on Hysteresis.** Some monographs are (or are going to be) at disposal of the reader interested to the mathematics of hysteresis: [5], [6], [7], [8], [10]. More than 300 references can be found in the latter volume.

1. A. Bossavit, C. Emson, I.D. Mayergoyz: Géométrie différentielle, éléments finis, modèles d' hystérésis. Eyrolles, Paris 1991
2. M. Brokate: Optimale Steuerung von gewöhnlichen Differentialgleichungen mit Nichtlinearitäten vom Hysteresis-Typ. Lang, Frankfurt am Main (1987). English translation: Optimal control of ordinary differential equations with nonlinearities of hysteresis type. In: Automation and Remote Control, 52 (1991), 53 (1992)
3. M. Brokate, K. Dressler, P. Krejčí, T.I. Seidman, L. Tavernini, A. Visintin: Contributions to the session on "Problems in hysteresis". In: Proceedings of World Congress of Nonlinear Analysts, Tampa, August 1992 (to appear)
4. M. Brokate, N. Kenmochi, I. Müller, J.F. Rodrigues, C. Verdi (A. Visintin, ed.): Phase Transitions and Hysteresis. Lecture Notes in Mathematics, vol. 1584. Springer, Berlin 1994
5. M. Brokate, J. Sprekels: Hysteresis phenomena in phase transitions (Monograph in preparation)
6. M.A. Krasnosel'skiĭ, A.V. Pokrovskiĭ: Systems with hysteresis. Springer, Berlin 1989. (Russian edition: Nauka, Moscow 1983)
7. P. Krejčí: Convexity, hysteresis and dissipation in hyperbolic equations. (Monograph in preparation)
8. I.D. Mayergoyz: *Mathematical models of hysteresis.* Springer, New York 1991
9. A. Visintin (ed.): Models of hysteresis. Longman, Harlow 1993
10. A. Visintin: Differential models of hysteresis. Springer, Berlin 1994

(ii) **Free Boundary Problems with Hysteresis.** The following is just an incomplete list of papers in which hysteresis and free boundary problems appear as the main issue. (For instance papers on shape memory alloys are discussed and quoted in Sprekels's contribution.)

11. H.W. Alt: On the thermostat problem. Control and Cybernetics **14** (1985) 171–193

12. C. Chiu, F.C. Hoppensteadt, W. Jäger: Analysis and computer simulation of acretion patterns in bacterial cultures. J. Math. Biol. **32** (1994) 809–834

13. A. Friedman, K.-H. Hoffmann: Control of free boundary problems with hysteresis. S.I.A.M. J. Control Opt. **26** (1988) 42–55

14. I.G. Götz, K.-H. Hoffmann, A.M. Meirmanov: Periodic solutions of the Stefan problem with hysteresis-type boundary conditions. Manuscripta Math. **78** (1993) 179–199

15. K.-H. Hoffmann, N. Kenmochi: Two-phase Stefan problems with feedback controls. In: Mathematical Models for Phase Change Problems (J. F. Rodrigues, ed.). Birkhäuser, Basel 1989, pp. 239–260

16. K.-H. Hoffmann, H.-J. Kornstädt, J. Sprekels: Automatische Steuerung freier Ränder bei Stefan-Problemen durch Thermostatkontrollen am festen Rand. Z. Anal. Anwend. **4** (1985) 225–233

17. K.-H. Hoffmann, J. Sprekels: Real-time control in free boundary problems connected with the continuous casting of steel. In: Optimal control of partial differential equations (K.-H. Hoffmann, W. Krabs, eds.), Birkhäuser, Basel 1984, pp. 127–143

18. K.-H. Hoffmann, J. Sprekels, M. Niezgódka: Feedback control via thermostats of multidimensional two-phase Stefan problem. Nonlinear Analysis, T.M.A. **15** (1990) 955–976

19. F.C. Hoppensteadt, W. Jäger: Pattern formation by bacteria. In: Lecture Notes in Biomathematics (W. Jäger, H. Rost, P. Tautu, eds.), vol. 38. Springer, Berlin 1979, pp. 68–81

20. F.C. Hoppensteadt, W. Jäger, C. Pöppe: A hysteresis model for bacterial growth patterns. In: Lecture Notes in Biomathematics (W. Jäger, J.D. Murray, eds.), vol. 55. Springer, Berlin 1984, pp. 123–134

21. U. Hornung, R.E. Showalter: PDE-models with hysteresis on the boundary. In: Models of hysteresis (A. Visintin, ed.). Longman, Harlow 1993, pp. 30–38

22. W. Jäger: A diffusion-reaction system modelling spatial patterns. In: Equadiff. **5**, Bratislava, 1981. Teubner, Leipzig 1982, pp. 151–158

23. W. Jäger: Mathematical modelling in chemistry and biology. Interd. Sci. Rev. **11** (1986) 181–188

24. P. Krejčí: Hysteresis operators – a new approach to evolution differential inequalities. Comment. Math. Univ. Carolinae **30** (1989) 525–536

25. T. Little, R. Showalter: The super-Stefan problem. International Journal of Engineering Science

26. T. Little, R. Showalter: Semilinear parabolic equations with Preisach hysteresis. Differential and Integral Equations **7** (1994) 1021–1040

27. T. Little, R. Showalter: Vector hysteresis and semilinear parabolic equations. Preprint 1994

28. A. Meirmanov, N. Shemetov: The Stefan problem with hysteresis-type phase function. Advances in Mathematical Sciences and Applications (to appear)

29. J.C.W. Rogers: The unique solution of a diffusion consumption problem with hysteresis. Control and Cybernetics **15** (1987) 483–498

30. J. Sprekels: On the automatic control of the free boundary in a one-phase Stefan problem. In: Applied nonlinear functional analisis (R. Gorenflo, K.-H. Hoffmann, eds.). Peter Lang, Frankfurt 1983, pp. 301–310

31. B. Stoth: On periodic solutions of linear thermostat problems. Diplom thesis, Bonn 1987

32. C. Verdi: Numerical aspects of free boundary and hysteresis problems. In: Modelling and analysis of phase transition and hysteresis phenomena (A. Visintin, ed.). Proceedings of a C.I.M.E. course, Montecatini 1993. Lecture Notes in Mathematics, vol. 1584. Springer, Heidelberg 1994 (to appear)

33. C. Verdi, A. Visintin: Numerical approximation of hysteresis problems. I.M.A. J. Numer. Anal. **5** (1985) 447–463

34. C. Verdi, A. Visintin: Numerical approximation of the Preisach model for hysteresis. Math. Model. and Numer. Anal. **23** (1989) 335–356

35. A. Visintin: A model for hysteresis of distributed systems. Ann. Mat. Pura Appl. **131** (1982) 203–231

36. A. Visintin: A phase transition problem with delay. Control and Cybernetics **11** (1982) 5–18

37. A. Visintin: Evolution problems with hysteresis in the source term. S.I.A.M. J. Math. Anal. **17** (1986) 1113–1138

38. A. Visintin: On the Preisach model for hysteresis. Nonlinear Analysis, T.M.A. **9** (1984) 977–996

39. A. Visintin: On hysteresis in phase transitions. Control and Cybernetics **14** (1985) 297–307

40. A. Visintin: Hysteresis and semigroups. In: Models of hysteresis (A. Visintin, ed.). Longman, Harlow 1993, pp. 192–206

(iii) Hysteresis in Porous Media Filtration. Here is just a short selection of the many engineering papers dealing with this topic.

41. Y. Mualem: Modified approach to capillary hysteresis based on a similarity hypothesis. Water Resour. Res. **9** (1973) 1324–1331

42. Y. Mualem: A conceptual model of hysteresis. Water Resour. Res. **10** (1974) 514–520

43. Y. Mualem: Theory of universal hysteretical properties of unsaturated porous media. Proc. Fort Collins Fluid Intern. Hydrol. Symp. 1977, Water Resour. Res. Public. 1979

44. A. Poulovassilis: The effect of the entrapped air on the hysteresis curves of a porous body and on its hydraulic conductivity. Soil Sci. **109** (1970) 154–162

45. A. Poulovassilis, E.C. Childs: The hysteresis of pure water: the non-independence of domains. Soil Sci. **112** (1970) 301–312

46. A. Poulovassilis, W.M. El-Ghamry: The dependent domain theory applied to scanning curves of any order in hysteretic soil water relationships. Soil Sci. **126** (1978) 1–8

47. A. Poulovassilis, E. Tzimas: The hysteresis in the relationship between hydraulic conductivity and suction. Soil Sci. **117** (1974) 250–256

48. A. Poulovassilis, E. Tzimas: The hysteresis in the relationship between hydraulic conductivity and soil water content. Soil Sci. **120** (1975) 327–331

(iv) Other References.

49. H.W. Alt, S. Luckhaus, A. Visintin: On nonstationary flow through porous media. Ann. Matem. Pura Appl. **136** (1984) 303–316
50. G. Anzellotti, S. Baldo, A. Visintin: Asymptotic behavior of the Landau-Lifshitz model of ferromagnetism. Appl. Math. Optim. **23** (1991) 171–192
51. M. Brokate, K. Dressler, P. Krejčí: Dissipation of energy, fatigue and damage evaluation (to appear)
52. S. Chikazumi, S.H. Charap: Physics of magnetism. Wiley, New York 1964
53. A. Damlamian, A. Visintin: Une généralisation vectorielle du modèle de Preisach pour l'hystérésis. C.R. Acad. Sci. Paris, Série I **297** (1983) 437–440
54. P. Krejčí: Modelling of singularities in elastoplastic materials with fatigue. Apl. Mat. **39** (1994) 137–160
55. I.D. Mayergoyz: Vector Preisach model of hysteresis. J. Appl. Phys. **63** (1988) 2995–3000
56. F. Preisach: Über die magnetische Nachwirkung. Z. Physik **94** (1935) 277–302

Augusto Visintin
Universita' degli Studi di Trento
Dipartimento di Matematica
via Sommarive 14, 38050 Povo (Trento) - Italy

Part 3.

Mathematical miscellania related to free boundary problems

D BLANCHARD

Renormalized solutions for parabolic problems with $L1$ data

Abstract: In the present paper, the notion and properties of renormalized solutions are developed for two model parabolic problems with L^1 data. In contrast with the framework of weak solutions, this class of solutions (that are not in general weak solutions) allows to obtain existence and uniqueness results.

1   Introduction

To illustrate the notion of renormalized solutions we consider here the following two model problems :

PROBLEM 1

$$\frac{\partial u}{\partial t} - \mathrm{div}\,a(Du) = f \text{ in } \Omega \times (0,T),$$
$$u(t=0) = u_0 \text{ in } \Omega,$$
$$u = 0 \text{ on } \partial\Omega \times (0,T).$$

PROBLEM 2

$$\frac{\partial b(u)}{\partial t} - \Delta u + \mathrm{div}\,\phi(u) = f \text{ in } \Omega \times (0,T),$$
$$b(u)(t=0) = b(u_0) \text{ in } \Omega,$$
$$u = 0 \text{ on } \partial\Omega \times (0,T).$$

In both Problems 1 and 2, Ω is a bounded domain of $\mathbb{R}^N(N \geq 1)$, $T > 0$ and $f \in L^1(\Omega \times (0,T))$.

In Problem 1, a is a monotone operator with $(p-1)$-growth at infinity $(p > 1)$ and $u_0 \in L^1(\Omega)$. Remark that a is not assumed to be strictly monotone.

In Problem 2, b is a C^1-function defined on an open interval $]\underline{a}, \overline{a}[$ of $\mathbb{R}$ (possibly $\underline{a} = -\infty$ and/or $\overline{a} = +\infty$), such that b^{-1} is continuous on $\mathbb{R}$ (see assumption B2 for details). The $\mathbb{R}^N$-valued function ϕ is assumed to be continuous on $\mathbb{R}$, and $b(u_0) \in L^1(\Omega)$.

As far as Problem 1 is concerned, existence of a weak solution (in the sense that each term makes sense as a distribution) may be established for $p > \frac{2N+1}{N+1}$ (see Blanchard [1], thm 3) but, to our knowledge, the uniqueness of such a weak solution is not established.

When Problem 2 is investigated a first difficulty is due to the fact that the function ϕ is not restricted by any growth condition at infinity. It follows that deriving an estimate on $\phi(u)$ seems to be an arduous task. This lack of estimate is indeed a serious obstacle in proving existence of a weak solution for Problem 2. A second difficulty is due to the nonlinear term $\frac{\partial b(u)}{\partial t}$ since the function b may admit a so-called "infinite bareer" for a finite value $\bar{a}$ (or $\underline{a}$) of the variable u (i.e. $b(u) \to +\infty$ as $u \to \bar{a}$). In the case where $\phi = 0$ and for a nonlinear operator $-\operatorname{div}a(Du)$ in place of Δu this last question is investigated in Blanchard-Francfort [2] (thm 2) where an existence result is proved under the extra assumption that $f \in L^1(0,T; L^p(\Omega))$ ($p > 1$ being related to growth of $a(Du)$ at infinity). Indeed the proof given in the above mentioned paper does not work when f belongs to $L^1(\Omega \times (0,T))$. Another interesting question related to Problems of type 1 deals with the uniqueness of a solution. As far as weak solutions are concerned, this question is still an open problem even in the case $\phi = 0$.

For all these reasons and in order to prove an existence and uniqueness result for Problems 1 and 2, we propose to use the notion of renormalized solutions introduced by Di Perna and Lions ([7]) as far as Boltzmann equations are concerned. This notion of solutions was then adapted to a few elliptic versions of Problem 2 in Boccardo, Diaz, Giachetti and Murat [6], in Lions and Murat [8], and Murat [9], [10].

2 Definition of renormalized solutions and comments

The assumptions on the data for Problems 1 and 2 are the following :
ASSUMPTIONS FOR PROBLEM 1

(A1) $f \in L^1(\Omega \times (0,T))$, $u_0 \in L^1(\Omega)$

(A2) $(a(\xi) - a(\xi'), \xi - \xi')_N \geq 0$, $\forall \xi, \xi' \in \mathbb{R}^N$,

(A3) $a(0) = 0$

(A4) $\exists p > 1, \exists \alpha > 0, \exists c_1, c_2 \geq 0$ such that

$$(a(\xi), \xi)_N \geq \alpha \|\xi\|^p_N, \quad \forall \xi \in \mathbb{R}^N,$$
$$\|a(\xi)\|_N \leq c_1 + c_2 \|\xi\|^{p-1}_N, \quad \forall \xi \in \mathbb{R}^N.$$

ASSUMPTIONS FOR PROBLEM 2

(B1) $f \in L^1(\Omega \times (0,T))$ and u_0 is a measurable function such that $b(u_0) \in L^1(\Omega)$.

(B2) b is a strictly increasing C^1-function defined on an open interval $(\underline{a}, \bar{a})$ of $\mathbb{R}$ such that $b(0) = 0$, $\lim_{r \to \bar{a}} b(r) = +\infty$ and $\lim_{r \to \underline{a}} b(r) = -\infty$ and such that b^{-1} is continuous on $\mathbb{R}$;

(B3) ϕ is a continuous function defined on $\mathbb{R}$ with value in $\mathbb{R}^N$; i.e. $\phi = (\phi_1, \cdots, \phi_N)$ with $\phi_i \in C^0(\mathbb{R})$.

178

Let us denote by $T_K(r)$ the truncation function at height K $(K \geq 0)$ i.e.

$$T_K(r) = \min(K, \max(r, -K)).$$

The following Definitions 1 and 2 are nonlinear parabolic versions of the definition of renormalized solutions for the elliptic problems investigated in [6], [8], [9], [10].

Definition 1 *A real valued function u defined on $\Omega \times (0, T)$ is a renormalized solution of Problem 1 if*

(1) $$u \in C^0([0, T]; L^1(\Omega));$$

(2) $$T_K(u) \in L^p(0, T; W_0^{1,p}(\Omega)), \text{ for any } K \geq 0;$$

(3) $$\int_{n \leq |u| \leq n+1} |Du|^p dxdt \to 0 \text{ as } n \to +\infty;$$

for any $C^\infty(\mathbb{R})$-function S such that S' has a compact support (i.e. $S' \in C_0^\infty(\mathbb{R})$), u solves the equation

(4) $$\frac{\partial S(u)}{\partial t} - \operatorname{div}[S'(u)a(Du)] + S''(u)a(Du)Du = fS'(u) \text{ in } \Omega \times (0, T);$$

(5) $$u(t = 0) = u_0 \text{ in } \Omega.$$

Definition 2 *A measurable function u defined on $\Omega \times (0, T)$ is a renormalized solution of Problem 2 if*

(6) $$b(u) \in L^\infty(0, T; L^1(\Omega));$$

(7) $$T_K(u) \in L^2(0, T; H_0^1(\Omega)) \text{ for any } K \geq 0;$$

(8) $$\int_{\{n \leq |b(u)| \leq n+1\}} b'(u)|Du|^2 dxdt \to 0 \text{ as } n \to \infty;$$

for any $C^\infty(\mathbb{R})$-function S such that S' has a compact support (i.e. $S' \in C_0^\infty(\mathbb{R})$) u solves the equation

$$\frac{\partial S(b(u))}{\partial t} - \operatorname{div}[S'(b(u))Du] + S''(b(u))b'(u)|Du|^2$$

(9) $$+ \operatorname{div}[S'(b(u))\phi(u)] - S''(b(u))b'(u)\phi(u)Du = fS'(b(u)) \text{ in } \Omega \times (0, T);$$

and $b(u)$ satisfies the initial condition

$$S(b(u))(t = 0) = S(b(u_0)) \text{ in } \Omega.$$

Remark 1 Loosely speaking equation (4) is derived through the formal pointwise multiplication of the equation of Problem 1 by the function $S'(u)$.

Note that in Definition 1 Du is not defined even as a distribution but due to (2) each term in (4) has a meaning in $\mathcal{D}'(\Omega \times (0, T))$. Indeed if M is such that $\operatorname{supp} S' \subset [-M, M]$, $S'(u)a(Du)$ coincides with $S''(u)a(DT_M(u))$ while $S''(u)a(Du)Du$ coincides with $S'(u)a(DT_M(u))DT_M(u)$. In view of the growth assumption (A4) and (2) we deduce that $S'(u)a(Du)$ and $S''(u)a(Du)Du$ identify respectively with an element of $L^{p'}(\Omega \times (0,T))$ and of $L^1(\Omega \times (0,T))$ $\left(\frac{1}{p'} + \frac{1}{p} = 1\right)$. Indeed since $fS'(u) \in L^1(\Omega \times (0,T))$ (by **2**), it follows that equation (4) takes place in $L^1(\Omega \times (0,T)) + L^{p'}\left(0,T; W^{-1,p'}(\Omega)\right)$.

Remark 2 Equation (9) is formally derived through multiplication of the equation of Problem 2 by $S'(b(u))$.

As in definition 1, Du is not defined as a distribution in Definition 2, but due to (7) each term in (9) has a meaning in $L^1(Q) + L^2(0,T; H^{-1}(\Omega))$. Indeed if M is again such that $\operatorname{supp} S' \subset [-M, M]$, the properties of b stated in B2 show that $|u| \leq K_M$ if $|b(u)| \leq M$ where K_M is defined by $K_M = \max\{-b^{-1}(M), b^{-1}(M)\}$. This allows to replace in (9) u and $b(u)$ respectively by $T_{K_M}(u)$ and $T_M(b(u))$. Since $b'(u)$ and $\phi(u)$ are bounded on the set $\{u; |b(u)| \leq M\}$, the regularity (7) of $T_{K_M}(u)$ together with B1 imply that each term of (9) (except the first one) belongs to either $L^1(Q)$ or $L^2(0,T; H^{-1}(\Omega))$. As an exemple one has

$$S''(b(u))b'(u)|Du|^2 = S''(b(u))b'(u)|DT_{K_M}(u)|^2 \in L^1(\Omega \times (O,T)),$$

since

$$|S''(b(u))b'(u)| \leq \|S''\|_{L^\infty(\ )} \max_{r \in [b^{-1}(-M), b^{-1}(M)]} |b'(r)| \ .$$

The above considerations show that equation (9) takes place in $\mathcal{D}'(\Omega \times (0,T))$ and that $\frac{\partial S(b(u))}{\partial t}$ belongs to $L^2\left(0,T; H^{-1}(\Omega)\right) + L^1(\Omega \times (0,T))$. It follows that $S(b(u)) \in C^0\left([0,T]; W^{-1,s}(\Omega)\right)$ for $s < \inf\left(2, \frac{N}{N-1}\right)$ so that the initial condition on $S(b(u))$ in (9) makes sense (see e.g. Simon [11]).

3 Existence and uniqueness results

The following Theorems are concerned with existence results and comparison principles for renormalized solutions of Problems 1 and 2.

Theorem 1 *Under assumptions (A1)-(A4), there exists a unique renormalized solution of Problem 1 (in the sense of Definition 1). If u_1 and u_2 are the renormalized solutions for the respective data (f_1, u_0^1) and (f_2, u_0^2) with*

$$f_1 \leq f_2 \ a.e. \ in \ \Omega \times (0,T),$$

and

$$u_0^1 \leq u_0^2 \ a.e. \ in \ \Omega,$$

then

$$u_1 \leq u_2 \ a.e. \ in \ \Omega \times (0,T).$$

Theorem 2 * *Existence result : Under the assumptions (B1)-(B3), there exists a renormalized solution of Problem 2 (in the sense of Definition 2).*
  * *Comparison principle and uniqueness result : in addition to assumptions (B1)-(B3), the functions ϕ and b' are assumed to be locally Lipchitz. Moreover assume that for any compact subset C of $(\underline{a}, \overline{a})$, there exists $\alpha(C) > 0$ such that*

$$b'(r) \geq \alpha(C) \; ; \; \forall r \in C \; .$$

Then, if u_1 and u_2 are two renormalized solutions for the respective data (f_1, u_0^1) and (f_2, u_0^2) with

$$f_1 \leq f_2 \; a.e. \; in \; \Omega \times (0, T),$$

and

$$u_0^1 \leq u_0^2 \; a.e. \; in \; \Omega,$$

one has

$$u_1 \leq u_2 \; a.e. \; in \; \Omega \times (0, T).$$

The proof of Theorems 1 and 2 are respectively detailed in Blanchard and Murat [3] and in Blanchard and Redwane[5] (see also Blanchard and Redwane [4]). Let us just mention the main ideas and estimates that permit to establish the above results.

Sketch of proof of Theorem 1: Assume that u^ε is a solution of Problem 1 where f and u_0 are replaced by smooth approximations f^ε and u_0^ε (when $\varepsilon \to 0$). The main difficulty in passing to the limit in equation (9) for u^ε when $\varepsilon \to 0$ concerns the term $S''(u^\varepsilon)a(Du^\varepsilon)Du^\varepsilon$ which identifies with $S''(u^\varepsilon)a(DT_M(u^\varepsilon))DT_M(u^\varepsilon)$.

Due to assumptions (A1)-(A4), using $T_M(u^\varepsilon)$ as a test function in Problem 1 for u^ε easily shows that

(10) $$DT_M(u^\varepsilon) \text{ is bounded in } L^p(\Omega \times (0, T)) \; ,$$

and

(11) $$a(DT_M(u^\varepsilon)) \text{ is bounded in } L^{p'}(\Omega \times (0, T)),$$

for any $M \geq 0$.

Since indeed u^ε is a compact sequence converging to u in $C^0([0, T]; L^1(\Omega))$ (by standard arguments) and because S'' is a smooth function, the key point is then to prove that

(12) $$a(DT_M(u^\varepsilon))DT_M(u^\varepsilon) \rightharpoonup a(DT_M(u)DT_M(u) \text{ in } L^1(\Omega \times (0, T)) - \text{weak}.$$

Actually the above result is an easy consequence of the following estimate

(13) $$\lim_{\varepsilon \to 0} \lim_{\eta \to 0} \int_{\Omega \times (0,T)} (a(DT_M(u^\varepsilon)) - a(DT_M(u^\eta)))(DT_M(u^\varepsilon) - DT_M(u^\eta)) \, dx dt = 0.$$

which is established in Blanchard [1] (and used to prove existence of a weak solution when $p > \frac{2N+1}{N+1}$).

The proof of the comparison principle in Theorem 1 relies on the following technique. Let S_n be a smooth approximation of the function T_{n+1} such that $\text{supp} S_n' \subset [-(n+1), -n] \cup [n, n+1]$.

Then take the difference of equations (4) with $S = S_n$ for u_1 and u_2, use the test function $T_K^+(S_n(u_1) - S_n(u_2))$ in the resulting equation ($K \geq 0$ and $r^+ = \sup(r, 0)$) and let n goes to infinity. In this limit process the two terms of the type

$$\int_{\Omega \times (0,T)} S_n''(u)a(Du)DuT_K^+(S_n(u_1) - S_n(u_2))dxdt$$

go to zero, because of (3) and $\text{supp} S_n'' \subset [-(n+1), -n] \cup [n, n+1]$.

Using (A2)-(A3) and (3), the following estimate is then derived

$$\lim_{n \to +\infty} \int_{\Omega \times (0,T)} (S_n'(u_1)a(Du_1) - S_n'(u_2)a(Du_2))D\left(T_K^+(S_n(u_1) - S_n(u_2))\right) dxdt \geq 0 .$$

In view of the assumptions on the data (f_1, u_0^1) and (f_2, u_0^2) the contribution of the parabolic term then gives

$$\int_0^{(u_1 - u_2)^+(x,t)} T_K(s)ds \leq 0 \text{ a.e. in } \Omega \times (0,T),$$

for any $K \geq 0$, which leads to the result.

Sketch of proof of Theorem 2 Let us denote by u^ε a solution of Problem 2 where b, ϕ, f and $b(u_0)$ are replaced by smooth approximations b_ε, ϕ_ε, f^ε and $b^\varepsilon(u_0^\varepsilon)$. The main difficulty in passing to the limit in equation (9) for u^ε concerns the behaviour of the term $S''(b_\varepsilon(u^\varepsilon)) b_\varepsilon'(u^\varepsilon)|Du^\varepsilon|^2$ which identifies $S''(b_\varepsilon(u^\varepsilon))T_M'(b(u^\varepsilon))b_\varepsilon'(u^\varepsilon)|Du^\varepsilon|^2$ (recall that $\text{supp} S' \subset [-M, M]$).

Using $T_K(u^\varepsilon)$ and $T_M(b_\varepsilon(u^\varepsilon))$ as test functions in Problem 2 for u^ε shows that

(14) $\qquad\qquad DT_K(u^\varepsilon)$ is bounded in $L^2(\Omega \times (0,T))$,

(15) $\qquad\qquad T_M'(b_\varepsilon(u^\varepsilon))(b_\varepsilon'(u^\varepsilon))^{\frac{1}{2}} Du^\varepsilon$ is bounded in $L^2(\Omega \times (0,T))$,

for any $K \geq 0$ and $M \geq 0$ (remark that in these processes the term involving $\phi(u)$ has a zero contribution by Stoke's formula and the boundary condition $u^\varepsilon = 0$). From the estimate on $T_K(u^\varepsilon)$ and the properties of b (see B2), we deduce that

$$u^\varepsilon \to u \quad \text{a.e. in } \Omega \times (0,T)$$
$$b_\varepsilon(u^\varepsilon) \to b(u) \quad \text{a.e. in } \Omega \times (0,T)$$

where u is a measurable function defined on $\Omega \times (0,T)$.

182

Due to these pointwise convergences and to the smooth character of S'', the key point is then to show that

$$(16) \qquad T'_M(b_\varepsilon(u^\varepsilon))b'_\varepsilon(u^\varepsilon)|Du^\varepsilon|^2 \to T'_M(b(u))b'(u)|Du|^2 \text{ in } L^1(\Omega \times (0,T))$$

as ε goes to zero.

The proof of the above result is rather technical (again see [5] for details) but let us mention that we use a decomposition of the type

$$\left| (b'_\varepsilon(u^\varepsilon))^{\frac{1}{2}} Du^\varepsilon - \left(b'_\eta(u^\eta)\right)^{\frac{1}{2}} Du^\eta \right|^2$$
$$= (Db_\varepsilon(u^\varepsilon) - Db_\eta(u^\eta)) D(u^\varepsilon - u^\eta) + \left| (b'_\varepsilon(u^\varepsilon))^{\frac{1}{2}} - \left(b'_\eta(u^\eta)\right)^{\frac{1}{2}} \right|^2 Du^\varepsilon Du^\eta$$

and then the continuous character of b' on $(\underline{a}, \overline{a})$ (which is essential at this step and percludes the consideration of Stefan's type problem).

To establish the comparison result of Theorem 2, we use the test function

$$sg^+_\sigma \left(S_n(b(u_1)) - S_n(b(u_2)) \right)$$

in the difference of equations (9) with $S = S_n$ for u_1 and u_2 (sg^+_σ is a standard approximation of sg^+ and S_n is defined in the sketchy proof of Theorem 1). Then we let $\sigma \to 0$ and $n \to +\infty$. In this limit process the contribution of the terms of the type $S''_n(b(u))b'(u)|Du|^2$ vanishes because of (8). The local Lipschitz-character of ϕ permits to control the terms involving ϕ, because $S'(b(u))$ vanishes is $|b(u)| > n+1$.

Indeed the worst term to control when $\sigma \to 0$ is

$$\int_{\Omega \times (0,T)} (S'_n(b(u_1))Du_1 - S'_n(b(u_2))Du_2) D \left(sg^+_\sigma(S_n(b(u_1)) - S_n(b(u_2))) \right) dxdt ,$$

and we show that its limit is positive when $\sigma \to 0$ by using a decomposition of type (10), the assumption $b' \geq \alpha(c)$ and the Lipchitz-character of b'.

The contribution of the parabolic term (when $n \to \infty$) together with the assumptions on the data allow to conclude that $b(u_1) \leq b(u_2)$ a.e. in $\Omega \times (0,T)$, which in turn implies $u_1 \leq u_2$ a.e. .

4 Concluding Remarks

Considering renormalized solutions (instead of weak solution) allows to solve a few parabolic problems involving nonlinearities without any growth condition at infinity (e.g. Problem 2) and to obtain uniqueness results for L^1 data (which are still open questions as far as weak solutions are concerned, even for the standard Problem 1). Moreover proving existence of such solutions generally raises a few "natural" questions pertaining to the behaviour of the approximated fields. Namely, in view of estimates

(10) and (11) the convergence result (12) arises as a "natural" question and as it is mentioned in the skecht of proof of Theorem 1, passing to the limit in (4) mainly relies on (12). As far as Problem 2 is concerned, passing to the limit in (9) is strongly linked to the convergence result (16) which is the answer to a "natural" question on the behaviour of u^ε in view of estimate (15).

References

[1] BLANCHARD D. Truncation and monotonicity methods for parabolic equations. *Nonlinear Anal. TMA*, 21(10):725–743, 1993.

[2] BLANCHARD D., FRANCFORT G. A few results on a class of degenerate parabolic equations. *Ann. Scu. Norm. Sup. Pisa*, 8(2):213–249, 1991.

[3] BLANCHARD D., MURAT F. Renormalized solutions of nonlinear parabolic problems with L^1 data : existence and uniqueness. (to appear).

[4] BLANCHARD D., REDWANE H. Renormalized solutions for a class of nonlinear evolution problem.

[5] BLANCHARD D., REDWANE H. Solutions renormalisées d'équations paraboliques à deux nonlinéarités. *C. R. Acad. Sci.*, 319:831–835, 1994.

[6] BOCCARDO L., DIAZ J.I., GIACHETTI D., MURAT F. Existence of a solution for a weaker form of a nonlinear elliptic equation. In *Recent advances in nonlinear elliptic and parabolic problems*, number 208 in Pitman Res. Notes. Math. Ser., pages 229–246, Harlow, 1988. Longman Sci. Tech.

[7] DI PERNA R.J., LIONS P.L. On the Cauchy problem for Boltzmann equations : global existence and weak stability. *Ann. of Math.*, 130(2):321–366, 1989.

[8] LIONS P.L., MURAT F. Solutions renormalisées d'équations elliptiques. (to appear).

[9] MURAT F. Solutions renormalizadas de EDP elipticas non lineares. Technical Report R93023, Laboratoire d'Analyse Numérique, Paris VI, 1993.

[10] MURAT F. Equations elliptiques non linéaires avec second membre L^1 ou mesure. In *Comptes Rendus du 26ème Congrès d'Analyse Numérique*, les Karellis, France, 1994.

[11] SIMON J. Compact sets in $L_p(0,T;b)$. *Ann. Mat. Pura. Appl.*, 146(4):65–96, 1987.

Dominique Blanchard
URA-CNRS 1378 Analyse et Modèles Stochastiques
Université de Rouen
76821 Mont Saint Aignan Cedex, France

and

Laboratoire d'Analyse Numérique
Tour 55-65, Université Pierre et Marie Curie
4, Place Jussieu
75252 Paris Cedex 05, France.

M H GIGA AND Y GIGA
Consistency in evolutions by crystalline curvature

Dedicated to Professor Kôji Kubota on his sixtieth birthday

Abstract. Motion of curves by crystalline energy is often considered for "admissible" piecewise linear curves. This is because the evolution of such curves can be described by a simple system of ordinary differential equations. Recently, a generalized notion of solutions based on comparison principle is introduced by the authors. In this note we show that a classical admissible solution is always a generalized solution in our sense.

1. Introduction. Motion by crystalline evergy or crystalline curvature is interpreted as a typical example of geometric evolutions by nonsmooth interfacial energy. Let Γ_t denote an embedded curve in the plane depending on time t. Let $\mathbf{n}$ be the unit normal vector field of Γ_t determining the orientation of Γ_t and let V denote the normal velocity in the direction of $\mathbf{n}$. We consider the equation of Γ_t of the form

$$(1) \qquad V = -\frac{1}{\beta(\mathbf{n})} \left(\sum_{i=1}^{2} \frac{\partial}{\partial x_i}((\partial_i \gamma)(\mathbf{n})) + C(t) \right) \quad \text{on} \quad \Gamma_t.$$

Here $\gamma : \mathbf{R}^2 \to \mathbf{R}$ is of the form

$$\gamma(q) = |q|\gamma_0(q/|q|)$$

and γ_0, β are given positive functions defined on the unit circle and $\partial_i \gamma$ denotes the partial derivative $\partial \gamma/\partial q_i$ as a function on $\mathbf{R}^2$. The function $C(t)$ is a given continuous function. An interfacial energy γ_0 is called *crystalline* if its Frank diagram

$$\text{Frank}(\gamma_0) = \{(q_1, q_2) \in \mathbf{R}^2 ; \gamma(q) = 1\}$$

is a convex polygon. In this case because of jumps of first derivatives of γ, (1) is no longer a usual partial differential equation of coordinate representations of Γ_t.

Taylor [T1] proposed an evolution governed by (1) when γ_0 is crystalline by restricting Γ_t as "admissible" polygon. A system of ordinary differential equation is derived by a variational principle when $\beta = \text{const}.\gamma^{-1}, C \equiv 0$. Independently, Angenent and Gurtin [AG] derived the same system for general β and C from the balance of forces and the second law of thermodynamics.

We shall recall their equation when Γ_t is given as a graph of a function. Such a version is given in [GirK 1] for $C \equiv 0$. Let Γ_t be given as a graph of a function $y = u(t, x), x \in \mathbf{R}$. Then (1) becomes

$$(2) \qquad u_t = a(u_x)\big[(W'(u_x))_x - C(t)\big]$$

with

$$(3) \qquad \begin{cases} a(p) = (1 + p^2)^{1/2} M(p), \\[2mm] \dfrac{1}{M(p)} = \beta\Big(-\dfrac{p}{(1 + p^2)^{1/2}}, \dfrac{1}{(1 + p^2)^{1/2}}\Big), \\[2mm] W(p) = \gamma(-p, 1) \end{cases}$$

provided that $\mathbf{n}$ is taken upward [GMHG]. If γ_0 is crystalline, then W' is a piecewise constant nondecreasing function whose jump discontinuities consists of finitely many points $p_1 < p_2 < \cdots < p_m$.

We say a function v on $\mathbf{R}$ is *admissible crystal* if (i) v is a piecewise linear continuous function with slopes belong to $P = \{p_i\}_{i=1}^m$; (ii) let p_i be a slope of v in an interval (a_1, a_2) where v_x has jump at a_1 and a_2. Then the slopes of $v(x)$ for $x < a_1$ (near a_1) or for $x > a_2$ (near a_2) are either p_{i+1} or p_{i-1} with $i + 1 \le m, i - 1 \ge 1$. The graph of such a function is called a Wulff curve in [EGS].

We say $u = u(t, x)$ is an *admissible evolving crystal* on a time interval J if $u(t, \cdot)$ is admissible crystal and jumps of u_x move smoothly in time for $t \in J$. (This definition is consistent with that in [GG].) For an admissible evolving crystal an evolution equation corresponding to (2) is derived in [T1] and [AG]. It is of the form

$$(4) \qquad u_t = a(u_x)\Big(\dfrac{\chi\Delta}{L} - C(t)\Big) \quad \text{on} \quad x_j(t) < x < x_{j+1}(t).$$

Here for fixed t, $\{x_j(t)\}$ is a discrete set and it consists of jumps of $u_x(t, \cdot)$ and $u(t, \cdot)$ is linear on $(x_j(t), x_{j+1}(t))$; j runs in either a finite set $\{1, 2, \cdots, d\}$, the set of natural numbers $\mathbf{N} = \{1, 2, \cdots\}$ (or $-\mathbf{N} = \{-1, -2, \cdots\}$) or the set $\mathbf{Z}$ of integers; in the first two cases we use convention that $x_1 = -\infty$ ($x_{-1} = +\infty$), and in the first case $x_d = +\infty$. The quantity L denotes the length of (x_j, x_{j+1}) i.e.,

$$L = x_{j+1}(t) - x_j(t).$$

If (x_j, x_{j+1}) is an infinite interval, we interpret $1/L$ as zero. The quantity Δ is defined by

$$\Delta = W'(p_k + 0) - W'(p_k - 0)$$

where $u_x = p_k$ on (x_j, x_{j+1}). The quantity χ is called a transition number. It takes the value 1 (resp. -1) if $u(t, \cdot)$ is convex (resp. concave) around (x_j, x_{j+1}); otherwise $\chi = 0$. The quantity $\chi\Delta/L$ is often called a crystalline curvature or weighted curvature [T1,2]. Note that on an admissible evolving crystal the value of a outside P is irrelevant to define the equation (4).

The equation (4) yields a system of ordinary differential equations (ODE) at least for x_j's ([AG], [T1,2], [GirK1,2]). We shall postpone to present this equation. If $\{x_j\}$ is a finite set or $u(t, \cdot)$ is periodic in x, the number of unknown is finite so that a local existence theorem for ODE applies. For example we observe that if initial data is an admissible crystal and periodic, then there is a unique admissible evolving crystal satisfying (4) at least for a short time. However, x_j may agree with x_{j+1} in a finite time. Fortunately, if $x_j(t_0) = x_{j+1}(t_0)$ for some time, then $\chi = 0$ for $(x_j(t), x_{j+1}(t))$, $0 < t < t_0$. In other words a facet with $\chi = \pm 1$ does not disappear. Moreover, at most three consecutive x_j's may agree at one time. Even at such a time t_0 where some facet disappears, $u(t_0 - 0, \cdot)$ is an admissible crystal. There observations are given in [GirK1] with $C \equiv 0$ but it extends to $C \not\equiv 0$ provided that C is continuous in t.

Note that even if admissible crystalline evolution loses facet at $t = t_0$, one can restart with $u(t_0, \cdot)$ and solve (4) again. By this process a global solution of (4) is obtained. To be precise we say $u = u(t, x)$ $(0 < t < T)$ is weakly admissible evolving crystal if u is an admissible evolving crystal on $(t_\ell, t_{\ell+1})$, $\ell = -1, 0, \cdots, k$ for some $0 = t_{-1} < t_0 < t_1 < \cdots < t_k < t_{k+1} = T$ and u is continuous across t_ℓ, $\ell = 0, \cdots, k$. (This definition is consistent with that in [GG].) Using this terminology, we obtain for example that there is a unique weakly admissible evolving crystal satisfying (4) globally-in-time if initial data is an admissible.

The main goal of this note is to show that weakly admissible evolving crystal is indeed a generalized solution introduced by the authors [GMHG] (under some conditin for β if $C \not\equiv 0$.) This means our notion of solutions is a natural extension. In [EGS] analogous statement is proved for generalized solutions based on the nonlinear semigroup theory [FG]. However, since the generalized solution in [FG] is only defined for $C \equiv 0$, their statement is restricted for $C \equiv 0$.

Recently, motion by crystalline energy is studied extensively. Instead of mentioning all related articles we only list review articles [T2], [GirK2] and [GMHG].

We conclude this section by deriving a system of ODEs from (4). For simplicity we assume that an admissible evolving crystal $u(t, \cdot)$ $(0 \le t < T)$ is periodic in x (with period ω) so that for some d, $x_{i+d} = x_i + \omega$ for all i in $\mathbf{Z}$. Let $L_j(t)$ denote the length of $(x_{j-1}(t), x_j(t))$, i.e.

$$(5) \qquad L_j(t) = x_j(t) - x_{j-1}(t), \quad i = 1, \cdots, d.$$

Let $R_j(t)$ denote $[x_{j-1}(t), x_j(t)]$ so that the interior $R_j(t)^0 = (x_{j-1}(t), x_j(t))$. Let $(u_t)_j = (u_t)_j(t)$ denote $u_t(t, x)$ for $x \in R_j(t)^0$. Let $(u_x)_j = (u_x)_j(t)$ denote the slope of $u(t, \cdot)$ on $R_j(t)^0$. i.e.,

$$(6) \qquad (u_x)_j = u_x(t, x) \quad \text{for} \quad x_{j-1}(t) < x < x_j(t).$$

Since u is continuous,

$$(7) \qquad dL_j(t)/dt = \rho_j^0 (u_t)_j + \rho_j^{-1}(u_t)_{j-1} + \rho_j^1 (u_t)_{j+1}, \quad j = 1, \cdots, d$$

with

$$\rho_j^0 = ((u_x)_j - (u_x)_{j-1})^{-1} + ((u_x)_{j+1} - (u_x)_j)^{-1}$$
$$\rho_j^{-1} = -((u_x)_j - (u_x)_{j-1})^{-1}$$
$$\rho_j^1 = -((u_x)_{j+1} - (u_x)_j)^{-1}.$$

This follows from the elementary geometry and does not depend on the special evolution equation (4). We now invoke (4) which is rewritten as

(8)
$$(u_t)_j = a((u_x)_j)(\Lambda_j - C(t)), \quad j = 1, \cdots, d$$
$$\Lambda_j = \chi_j \Delta / L_j$$

with $\Delta = W'((u_x)_j + 0) - W'((u_x)_j - 0)$, where χ_j is the transition number on $R_j^0 = (x_{j-1}, x_j)$. Since $(u_x)_j$ and χ_j are determined initially, the equations (7) and (8) yield a system of ODEs for L_j $(j = 1, \cdots, d)$; note that $(u_0)_t = (u_d)_t$ in (7) so the system (7)-(8) is closed. As in [GirK1] differentiating $u(x_j(t), t)$ in t we get

(9)
$$dx_j/dt = -((u_t)_{j+1} - (u_t)_j)/((u_x)_{j+1} - (u_x)_j)$$

for $j = 1, \cdots, d$. The function $(u_t)_j$ is computable from (7), (8) so the evolution of x_j is determined by (9).

Derivation of (7), (8), (9) is found in [GirK1], where $C \equiv 0$ and $a(p) = (1 + p^2)^{1/2}$ is assumed but it extends to our setting with no essential change. The systems (7), (8) is found in [AG] in a little bit different form ; L_j is replaced by the length of the graph $y = u(t, x)$ on (x_{j-1}, x_j) and $(u_t)_j$ is replaced by the normal velocity. For later convenience by j-th facet we mean the graph of $y = u(t, \cdot)$ on $R_j(t)$. The point $(x_j(t), u(t, x_j(t)))$ is called a corner.

2. Generalized solutions.

We recall from [GMHG] our definition of generalized solution when W' is a piecewise constant function with jumps on P. Note that our definition and results are still valid for more general W as in [GMHG].

Definition(P-faceted). Let Ω be an open interval. A function ϕ in $C(\Omega)$ is called P-faceted at x_0 in Ω if ϕ fulfills the following conditions.

There are a closed nontrivial finite interval $I(\subset \Omega)$ containing x_0 and p in P such that ϕ agrees with an affine function

$$\ell_p(x) = p(x - x_0) + \phi(x_0)$$

in I and $\phi(x) \neq \ell_p(x)$ for all $x \in J \setminus I$ with some neighborhood $J(\subset \Omega)$ of I. The interval I is called a faced region of ϕ containing x_0 and is denoted by $R(\phi, x_0)$. The value p is called the slope of the facet.

Definition(Weighted curvature with driving term). Let c be a constant and x_0 be a point in Ω. For $\phi \in C(\Omega)$ we set the value

$$\Lambda_W(\phi, x_0, ; c) = W''(\phi'(x_0))\phi''(x_0) + c(= 0 + c \text{ since } W'' = 0 \text{ outside } P)$$

if ϕ is second differentiable at x_0 and $\phi'(x_0) \notin P$ and

$$\Lambda_W(\phi, x_0; c) = \frac{\chi}{L}\Delta_i + c$$

if ϕ is P-faceted at x_0 in Ω with slope p_i, where $\Delta_i = W'(p_i + 0) - W'(p_i - 0)$. Here $L = L(\phi, x_0)$ is the length of the faceted region I containing x_0 and $\chi = \chi(\phi, x_0)$ is the transition number defined by

$$\begin{aligned}
\chi &= +1 \quad \text{if} \quad \phi \geq \ell_{p_i} \quad \text{in} \quad J, \\
\chi &= -1 \quad \text{if} \quad \phi \leq \ell_{p_i} \quad \text{in} \quad J, \\
\chi &= 0 \quad\quad \text{otherwise}
\end{aligned}$$

for some neighborhood J of the facet region I.

Definition. A function $\phi \in C^2(\Omega)$ belongs to a class $C_P^2(\Omega)$ if ϕ is P-faceted at x_0 in Ω whenever $\phi'(x_0)$ belongs to P. For $\phi \in C_P^2$, $\Lambda_W(\phi, x; c)$ is defined for all $x \in \Omega$ so we often write it by $\Lambda_W(\phi; c)(x)$.

Definition (Space of admissible functions). For $Q = (0, T) \times \Omega$ let $A_P(Q)$ be the set of functions on Q of the form

$$\phi(x) + g(t), \quad \phi \in C_P^2(\Omega), \quad g \in C^1(0, T).$$

An element of $A_P(Q)$ is called an admissible function.

We are now in position to define our generalized solution in the viscosity sense.

Definition (Generalized solution). A real-valued function u on Q is a (*viscosity*) *subsolution* of

(E) $$u_t - a(u_x)\Lambda_W(u; -C) = 0$$

if the upper semicontinuous envelope $u^* < \infty$ in $\overline{Q}$ and

(*) $$\psi_t(\hat{t}, \hat{x}) - a(\psi_x(\hat{t}, \hat{x}))\Lambda_W(\psi(\hat{t}, \cdot); -C(\hat{t}))(\hat{x}) \leq 0$$

whenever $(\psi, (\hat{t}, \hat{x})) \in A_P(Q) \times Q$ fulfills

$$\max_Q(u^* - \psi) = (u^* - \psi)(\hat{t}, \hat{x}).$$

A (*viscosity*) *supersolution* is defined by replacing $u^*(< \infty)$ by the lower semicontinuous envelope $u_*(> -\infty)$, max by min and the inequality in (*) by the opposite one. If u is sub- and supersolution, u is called a *viscosity solution* or a *generalized solution*. Hereafter we avoid to use the word viscosity because this is nothing to do with viscosity in the sense of physics corresponding to (E). A key result in [GMHG] is

190

Fundamental Comparison Theorem. *Let u and v be a sub- and supersolution of (E), respectively, where Ω is a bounded open interval. Assume that $C \in C[0, T)$ and $a \in C(\mathbf{R})$ with $a \geq 0$. If $u^* \leq v_*$ on the parabolic boundary $(= [0, T) \times \partial\Omega \cup \{0\} \times \bar{\Omega})$ of Q, then $u^* \leq v_*$ in Q.*

Existence Theorem. *Suppose that $u_0 \in C(\mathbf{R})$ is periodic with period ω. Then there is a unique global generalized solution $u \in C([0, \infty) \times \mathbf{R})$ of (E) with $u(0, x) = u_0(x)$ and $u(t, x + \omega) = u(t, x)$.*

3. Consistency.
We always assume that a is a nonnegative continuous function and that $C \in C[0, \infty)$. Our goal is to prove :

Theorem. *Assume that u is a weakly admissible evolving crystal satisfying (4) defined in $(0, \infty) \times \mathbf{R}$. If $C \not\equiv 0$, assume that a satisfies*

$$a(p) = \theta a(p_k) + (1 - \theta)a(p_{k+1})$$
$$\text{for} \quad p = \theta p_k + (1 - \theta)p_{k+1}, 0 \leq \theta \leq 1, 1 \leq k \leq m - 1$$

where $P = \{p_1 < \cdots < p_m\}$. Then u is a generalized solution of (E) (in $(0, \infty) \times \mathbf{R}$).

We shall mainly give a proof for $C \equiv 0$ and point out how to alter the proof for general C.

We shall only prove that u is a subsolution of (E) in $Q = (0, \infty) \times \Omega$ with $\Omega = \mathbf{R}$ since the proof for supersolution is the same. Since $u \in C(\overline{Q})$, the property $u^* < \infty$ in $\overline{Q}$ is trivially satisfied. Let $(\hat{t}, \hat{x}) \in Q$ and $\psi \in A_P(Q)$ satisfy

$$\max_Q(u - \psi) = (u - \psi)(\hat{t}, \hat{x}).$$

Our goal is to show $(*)$. As usual we may assume $(u - \psi)(\hat{t}, \hat{x}) = 0$. By the definition of $A_P(Q)$, ψ is of the form

$$\psi(t, x) = \phi(x) + g(t), \quad \phi \in C_P^2(\Omega), \quad g \in C^1(0, \infty).$$

Note that at $\hat{t}$ some facet may disappear. However, $x_j(t)$ together with its time derivative $\dot{x}_j(t)$ is always continuous on $(\hat{t} - \delta, \hat{t}]$ for sufficiently small $\delta > 0$.

Lemma 1. *Assume that $C = 0$. Assume that $\hat{x}$ is the abscissa of a corner of $u(\hat{t}, \cdot)$ i.e. $\hat{x} = x_i(\hat{t} - 0)$ for some integer i. Let j be the largest integer such that $\hat{x} = x_j(\hat{t} - 0)$. Let $p_{k-1} \in P$ be $(u_x)_{j+1}$, the slope of $u(t, \cdot)$ on $R_{j+1}(t)^\circ = (x_j(t), x_{j+1}(t))$ for t close to $\hat{t}$ $(t < \hat{t})$; $(u_x)_{j+1}$ is independent of t.*
(i) $p_{k-1} \leq \phi'(\hat{x}) \leq p_k$.
(ii) If $x_{j-1}(\hat{t} - 0) < x_j(\hat{t} - 0)$ then $(u_x)_j = p_k$. Moreover, $(u_t)_{j+1}(\hat{t} - 0) \leq 0$ and $(u_t)_j(\hat{t} - 0) \leq 0$.

191

(iii) If $x_{j-1}(\hat{t} - 0) = x_j(\hat{t} - 0)$ and $(u_x)_j(t) = p_k$ for t close to $\hat{t}$ and $t < \hat{t}$, then $(u_t)_j(\hat{t} - 0) = 0$ and $(u_t)_{j+1}(\hat{t} - 0) \leq 0$.

(iv) It holds that $g'(\hat{t}) \leq 0$ provided that either assumption of (ii) or (iii) holds.

Remark. The assumption $C = 0$ is invoked only to prove the statements for u_t and g'.

Proof. (i) Since $u(\hat{t}, \cdot)$ is an admissible crystal and $u(\hat{t}, \cdot) - \phi$ takes its maximum in Ω at $\hat{x}$, we see $p_{k-1} \leq \phi'(\hat{x}) \leq p_k$.

(ii) The first assertion is trivial. Since the transition numbers χ_{j+1} and χ_j cannot equal one, from (8) it follows that

$$(u_t)_{j+1}(t) \leq -a(p_{k-1})C(t) = 0,$$
$$(u_t)_j(t) \leq -a(p_k)C(t) = 0$$

for t sufficiently close to $\hat{t}$ and $t < \hat{t}$, where we have used $C = 0$.

(iii) In this case j-th facet on $(x_{j-1}(t), x_j(t))$ vanishes at $\hat{t}$. Its transition number χ_j must be zero as pointed out in Section 1. Since $C = 0$, the equation (8) yields $(u_t)_j(\hat{t} - 0) = 0$. Since $(u_x)_j = p_k$, the transition number χ_{j+1} cannot be one so $(u_t)_{j+1}(\hat{t} - 0) \leq 0$ follows.

(iv) We take y such that

$$x_j(t) < y < x_{j+1}(t) \quad \text{for} \quad \hat{t} - \delta < t \leq \hat{t}$$

by taking $\delta > 0$ sufficiently small. Our assumption

$$\max_Q(u - \psi) = (u - \psi)(\hat{t}, \hat{x})$$

implies that

$$u(t, x_j(t)) - u(\hat{t}, \hat{x}) \leq \psi(t, x_j(t)) - \psi(\hat{t}, \hat{x}), \quad \hat{t} - \delta < t \leq \hat{t}.$$

The left hand side equals

$$u(t, y) + p_{k-1}(x_j(t) - y) - u(\hat{t}, y) - p_{k-1}(\hat{x} - y)$$
$$= u(t, y) - u(\hat{t}, y) + p_{k-1}(x_j(t) - \hat{x})$$

for $t \in (\hat{t} - \delta, \hat{t})$. Multiplying the preceeding inequality with $-(t - \hat{t})^{-1}$ and sending t to $\hat{t}$ with $t < \hat{t}$ yields

$$-u_t(\hat{t} - 0, y) - p_{k-1}\dot{x}_j(\hat{t} - 0) \leq -\phi'(\hat{x})\dot{x}_j(\hat{t} - 0) - g'(\hat{t})$$

or

$$g'(t) \leq (u_t)_{j+1}(\hat{t} - 0) + \dot{x}_j(\hat{t} - 0)(p_{k-1} - \phi'(\hat{x})).$$

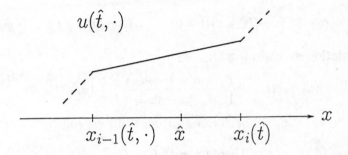

Fig.1. $\hat{x}$ belongs to interior of i-th faceted region $R_i(\hat{t})^\circ$.

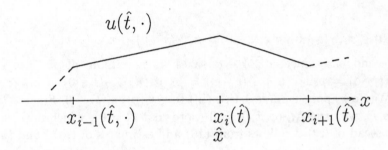

Fig.2. $\hat{x}$ coincides with $x_i(\hat{t})$.

By (9) we see

$$\dot{x}_j(\hat{t} - 0) = -\{(u_t)_{j+1}(\hat{t} - 0) - (u_t)_j(\hat{t} - 0)\}\Big/(p_{k-1} - p_k).$$

Using this relation we end up with

$$(10) \qquad g'(\hat{t}) \le (u_t)_{j+1}(\hat{t} - 0)\left\{1 - \frac{\phi'(\hat{x}) - p_{k-1}}{p_k - p_{k-1}}\right\} + (u_t)_j(\hat{t} - 0)\frac{\phi'(\hat{x}) - p_{k-1}}{p_k - p_{k-1}}.$$

Since $p_{k-1} \le \phi'(\hat{x}) \le p_k$, in both cases (ii) and (iii) we conclude $g'(\hat{t}) \le 0$. ■

We continue to prove our Theorem with $C = 0$.

Step 1. If $\phi'(\hat{x})$ does not belong to P, then $\hat{x}$ must be the abscissa of a corner of $u(\hat{t}, \cdot)$. As pointed out in Section 1, at most three consecutive $x_j(\hat{t} - 0)$'s may agree with $\hat{x}$. If exactly two $x_j(\hat{t} - 0)$'s agree with $\hat{x}$, then $u_x(\hat{t}, \cdot)$ is constant near $\hat{x}$ which leads $\phi'(\hat{x}) \in P$. Thus it only happens either

$$x_{i-1}(\hat{t} - 0) < x_i(\hat{t} - 0) = \hat{x} < x_{i+1}(\hat{t} - 0) \qquad \text{(see Fig.2)}$$

or

$$x_{i-2}(\hat{t} - 0) < x_{i-1}(\hat{t} - 0) = x_i(\hat{t} - 0) = x_{i+1}(\hat{t} - 0) < x_{i+2}(\hat{t} - 0) \quad \text{(see Fig.4 and 5).}$$

In the second case since $u(\hat{t}, \cdot) - \phi$ takes its maximum at $\hat{x}$, $(u_x)_{i+1}(\hat{t} - 0) = p_k$, $(u_x)_i(\hat{t} - 0) = p_{k-1}$, $(u_x)_{i-1}(\hat{t} - 0) = p_k$ if $(u_x)_{i+2}(\hat{t} - 0) = p_{k-1}$; Fig.5 is excluded. We now apply Lemma 1 (ii), (iii) to get (iv) $g'(\hat{t}) \le 0$. Since $\phi'(\hat{x})$ does not belong to P, we have $\Lambda_W(\phi, \hat{x}; 0) = 0$ (note that $C = 0$), which now yields (∗). (If $C \not\equiv 0$, instead of (iv) it follows from (10) and estimates of $(u_t)_j$ and $(u_t)_{j+1}$ that

$$g'(\hat{t}) \le -C(\hat{t})a(\phi'(\hat{x}))$$

provided that a satisfies the assumption in our Main Theorem. Note that (10) is still valid for $C \not\equiv 0$.)

Step 2. Assume that ϕ is faceted at $\hat{x}$. Then the situation is divided into four cases.

Case A (No facet disappearing near $\hat{x}$. See Fig.1).
$$x_{i-1}(\hat{t} - 0) < x_i(\hat{t} - 0) < \hat{x} < x_{i+1}(\hat{t} - 0) < x_{i+2}(\hat{t} - 0) \qquad \text{(for some integer } i\text{).}$$

Case B (No facet disappearing near $\hat{x}$. See Fig.2).
$$x_{i-1}(\hat{t} - 0) < \hat{x} = x_i(\hat{t} - 0) < x_{i+1}(\hat{t} - 0) \qquad \text{(for some integer } i\text{).}$$

Case C (Annihilation of one facet near $\hat{x}$. See Fig.3).
$$x_{i-2}(\hat{t} - 0) < x_{i-1}(\hat{t} - 0) = x_i(\hat{t} - 0) < x_{i+1}(\hat{t} - 0)$$
and $x_{i-2}(\hat{t} - 0) < \hat{x} < x_{i+1}(\hat{t} - 0)$ \qquad (for some integer i).

Case D (Annihilation of two facets near $\hat{x}$. See Fig.4 and 5).
$$x_{i-1}(\hat{t} - 0) = x_i(\hat{t} - 0) = x_{i+1}(\hat{t} - 0),$$
$$x_{i-2}(\hat{t} - 0) < x_{i-1}(\hat{t} - 0), \ x_{i+1}(\hat{t} - 0) < x_{i+2}(\hat{t} - 0)$$
and $x_{i-2}(\hat{t} - 0) < \hat{x} < x_{i+2}(\hat{t} - 0)$ \qquad (for some integer i).

We need to compare Λ_W of ϕ and $u(\hat{t}, \cdot)$ near $\hat{x}$. We recall a variant of the maximum principle in [GG].

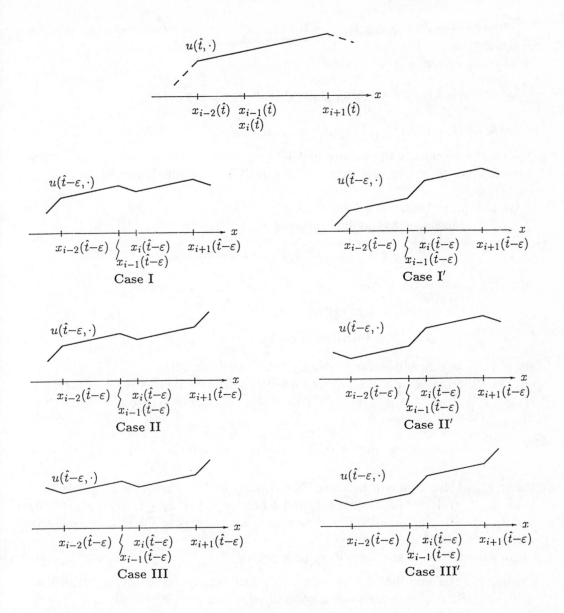

Fig.3. There are 6 cases of i-th facet disappearing without its adjacent neighbors disappearing.

Lemma 2. *Let f be in $C_P^2(\mathbf{R})$. Let h be an admissible crystal, and let c be a real number. Assume that $f \geq h$ on $\mathbf{R}$.*
(i) *If $f = h$ around a point $\overline{x}$, then*
$$\Lambda_W(f, \overline{x}; c) \geq \Lambda_W(h, \overline{x}; c).$$
(ii) *If $R(f, \overline{x}) \cap R(h, x') = \{\overline{x}\}$ for some x' and $f(\overline{x}) = h(\overline{x})$ then*
$$\Lambda_W(f, \overline{x}; c) \geq \Lambda_W(h, x'; c)$$
provided that f' on $R(f, \overline{x})$ equals h' on $R(h, x')^\circ$.

The proof is essentially the same as in [GG] so is omitted. The second part of Lemma 2 corresponds to *proper edge-edge touching* in[GG]. We shall abbreviate $\Lambda_W(f, x; 0)$ by $\Lambda(f, x)$.

We now prove $(*)$ in each cases.

Case A. Since $u - \psi$ takes its maximum at $(\hat{t}, \hat{x})$ and u is smooth near $(\hat{t}, \hat{x})$,

$$\phi'(\hat{x}) = (u_x)_i(\hat{t}) \in P, \quad g'(\hat{t}) = (u_t)_i(\hat{t}).$$

Applying Lemma 2 we obtain

$$g'(\hat{t}) - a(\phi'(\hat{x}))\Lambda(\phi, \hat{x})$$
$$\leq (u_t)_i(\hat{t}) - a((u_x)_i(\hat{t}))\Lambda(u(\hat{t}, \cdot), \hat{x}) = 0,$$

since u is a weakly admissible evolving crystal satisfying (4).

Case B. From Lemma 1 it follows that $g'(\hat{t}) \leq 0$. This implies $(*)$ if $\Lambda(\phi, \hat{x}) \geq 0$. If $\Lambda(\phi, \hat{x}) < 0$ so that $\chi(\phi, \hat{x}) = -1$, then there is $\overline{x} \neq \hat{x}$ in $R(\phi, \hat{x})^\circ \cap (x_{i-1}(\hat{t} - 0), x_{i+1}(\hat{t} - 0))$ such that

$$(u - \psi)(\hat{t}, \overline{x}) = \max_Q(u - \psi) = 0.$$

For $(\hat{t}, \overline{x})$ the situation now becomes Case A, C or D. If $(*)$ holds for $(\hat{t}, \overline{x})$, then $(*)$ holds for $(\hat{t}, \hat{x})$ since $\Lambda(\phi, \overline{x}) = \Lambda(\phi, \hat{x})$ and $\phi'(\hat{x}) = \phi'(\overline{x})$. If $x_{i-2}(\hat{t}-0) < x_{i-1}(\hat{t}-0)$ and $x_{i+1}(\hat{t} - 0) < x_{i+2}(\hat{t}-0)$, it is clear that Case A occurs so the proof is complete without studying Case C and D.

Combining Case A and Case B we have proved :

Corollary. *Assume that u is an admissible evolving crystal satisfying (4) defined in $[0, \infty) \times \mathbf{R}$. If $C \equiv 0$, then u is a generalized solution of (E).*

We continue to study cases C and D.

Case C. As pointed out in Section 1 we know that on disappearing facets the transition number equals zero [GirK1]. It turns out that there are only six possible pictures (cases (I)-(III), (I')-(III') in Fig.3) of the graph of u at the time $t - \varepsilon$ just before $\hat{t}$. We set $p_k = u_x(\hat{t}, \hat{x})$ and observe that

$$(u_x)_{i-1}(\hat{t} - 0) = (u_x)_{i+1}(\hat{t} - 0) = p_k.$$

In Propositions 1 - 3 we consider Case C.

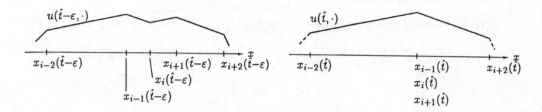

Fig.4. Case(i): i-th and $i+1$-th facets disappear without their adjacent neighbors disappearing.

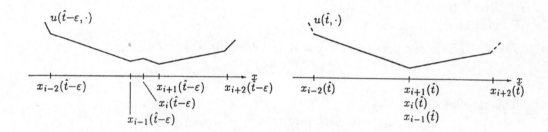

Fig.5. Case(ii): i-th and $i+1$-th facets disappear without their adjacent neighbors disappearing.

Proposition 1.

$$g'(t) \leq \min\{u_t(\hat{t} - 0, y); y \in R(\phi, \hat{x})^o \cap R(u(\hat{t}, \cdot), \hat{x})^o \quad \text{with} \quad u_x(\hat{t} - 0, y) = p_k\}.$$

Proof. Since $u - \psi$ takes its maximum at $(\hat{t}, y)$ with $y \in R(\phi, \hat{x})^o \cap R(u(\hat{t}, \cdot), \hat{x})^o$ and $u_x(\hat{t} - 0, y) = p_k$,

$$g'(\hat{t}) \leq u_t(\hat{t} - 0, y).$$

∎

We shall always use this property to estimate $g'(t)$.

Proposition 2. *Assume that j-th facet satisfies*

$$\chi_j(\hat{t} - 0) = -1, \quad (u_x)_j(\hat{t} - 0) = p_k,$$
$$R_j(\hat{t} - 0) \subset R(u(\hat{t}, \cdot), \hat{x}) \quad \text{and} \quad R_j(\hat{t} - 0) \neq R(u(\hat{t}, \cdot), \hat{x}).$$

Then,
(i) $R_j(\hat{t} - 0)^o \cap R(\phi, \hat{x})^o = \phi,$
(ii) $\chi(\phi, \hat{x}) \neq -1.$

Proof. (i) Suppose that there is $y \in R_j(\hat{t} - 0)^o \cap R(\phi, \hat{x})^o$. Clearly

$$\max_Q (u - \psi) = (u - \psi)(\hat{t}, y).$$

This implies

$$u_t(\hat{t} + 0, y) \leq (u_t)_j(\hat{t} - 0) = a(p_k)\Lambda_j(\hat{t} - 0)$$

where $\Lambda_j(t) = \chi_j(t)\Delta/L_j(t)$ with $\Delta = W'(p_k + 0) - W'(p_k - 0)$. Since

$$u_t(\hat{t} + 0, y) = a(p_k)\Lambda(u(\hat{t} + 0, \cdot), y) = a(p_k)\Lambda(u(\hat{t}, \cdot), y),$$

we now obtain

$$\Lambda(u(\hat{t}, \cdot), y) \leq \Lambda_j(\hat{t} - 0).$$

Since $\chi_j(\hat{t} - 0) = -1$, so that $\Lambda_j(\hat{t} - 0) < 0$, this implies $\chi(u(\hat{t}, \cdot), y) = -1$. Dividing both sides of $\Lambda \leq \Lambda_j$ by Δ and χ_j, we obtain

$$L(u(\hat{t}, \cdot), \hat{x}) = L(u(\hat{t}, \cdot), y) \leq L_j(\hat{t} - 0).$$

This contradicts the inclusion relation of $R_j(\hat{t} - 0)$ and $R(u(\hat{t}, \cdot), \hat{x})$.
(ii) Suppose that $\chi(\phi, \hat{x}) = -1$. Since $u - \psi$ attains its maximum zero at $(\hat{t}, \hat{x})$, $u(\hat{t}, \cdot) \leq \phi$ in $\mathbf{R}$. This implies $\chi(u(\hat{t}, \cdot), \hat{x}) = -1$ and $R(u(\hat{t}, \cdot), \hat{x}) \subset R(\phi, \hat{x})$. Since $R_j(\hat{t} - 0) \subset R(u(\hat{t}, \cdot), \hat{x})$ this contradicts (i) so we have proved $\chi(\phi, \hat{x}) \neq -1$. ∎

Let ϕ be faceted at x_0 with slope p, and let $\bar{x}$ be the right (resp. left) boundary point of $R(\phi, x_0)$. The value $\chi^+(\phi, x_0)$ (resp. $\chi^-(\phi, x_0)$) is assigned to one if $\phi \geq \ell_p$ near $\bar{x}$ and to minus one otherwise, where ℓ_p is a linear function of slope p defined in Section 2. For u at $x_j(t)$ we set

$$\chi_j^c(t) = \begin{cases} 1 & \text{if} \quad (u_x)_{j+1} > (u_x)_j, \\ -1 & \text{if} \quad (u_x)_{j+1} < (u_x)_j. \end{cases}$$

Proposition 3. *(i) Assume that $x_i(\hat{t}-0) \leq \hat{x} < x_{i+1}(\hat{t}-0)$ and $\chi_{i+1}^c(\hat{t}-0) = 1$.*
Then $\chi^+(\phi,\hat{x}) = 1$.
(ii) Assume that $x_{i-2}(\hat{t}-0) < \hat{x} \leq x_{i-1}(\hat{t}-0)$ and $\chi_{i-2}^c(\hat{t}-0) = 1$. Then $\chi^-(\phi,\hat{x}) = 1$.

Proof. The proof of (ii) parallels that of (i) so we only present the proof of (i). Suppose that $\chi^+(\phi,\hat{x}) = -1$. Then

$$R(\phi,\hat{x}) \supset \{x > \hat{x}; x \in R(u(\hat{t},\cdot),\hat{x})\}.$$

Since $u(\hat{t},\cdot) \leq \phi$ in $\mathbf{R}$, the $i+2$-th facet disappears at $t = \hat{t}$, so that $(u_x)_{i+3} = p_k$ ($=u_x(\hat{t},\hat{x})$). Since $\chi_{i+1}^c(\hat{t}-0) = 1$ and $\chi_{i+2}(\hat{t}-0) = 0$, this implies $\chi_{i+2}^c(\hat{t}-0) = -1$. If $\chi_{i+3}^c(\hat{t}-0) = -1$, then $\chi_{i+3}(\hat{t}-0) = -1$ which contradicts Proposition 2 (i). So we may assume $\chi_{i+3}^c(\hat{t}-0) = 1$. Since $u(\hat{t},\cdot) \leq \phi$ on $\mathbf{R}$, the $i+4$-th facet disapears at $t = \hat{t}$ so that $(u_x)_{i+5} = p_k$ and $\chi_{i+4}^c(\hat{t}-0) = -1$. Since both the $i+2$ and $i+4$-th facets disappear, $\chi_{i+5}(\hat{t}-0) = -1$ holds, which again contradicts Proposition 2 (i). We have thus proved $\chi^+(\phi,\hat{x}) = 1$. ∎

We now complete the proof of Theorem in Case C. We shall only present the proof for cases (I)-$(I\!I\!I)$ in Fig.3 since the proof for case (I')-$(I\!I\!I')$ is obtained by changing x by $-x$.

Case (I). By Proposition 2 (i), $R_{i-1}(\hat{t}-0)^o$ does not intersect $R(\phi,\hat{x})^o$. It follows that

$$x_i(\hat{t}-0) \leq \hat{x} < x_{i+1}(\hat{t}-0).$$

By Proposition 2 (ii), we have $\chi(\phi,\hat{x}) \neq -1$, so that $\chi(\phi,\hat{x}) \geq 0$. Since $g'(\hat{t}) \leq (u_t)_{i+1}(\hat{t}-0) = 0$, we observe that $g'(\hat{t}) - a(p_k)\Lambda(\phi,\hat{x}) \leq g'(\hat{t}) \leq 0$, which implies $(*)$.

Case $(I\!I)$. As in Case (I) we may assume $\chi(\phi,\hat{x}) \geq 0$ and $R(\phi,\hat{x})^o$ does not intersect $R_{i-1}(\hat{t}-0)^o$.

Suppose that $\chi(\phi,\hat{x}) = 0$. Since $R(\phi,\hat{x})^o$ does not intersect $R_{i-1}(\hat{t}-0)^o$, we have $\chi^-(\phi,\hat{x}) = 1$ and $\chi^+(\phi,\hat{x}) = -1$. However, this contradicts Proposition 3.

We may now assume $\chi(\phi,\hat{x}) = 1$.
(a) If $R_{i+1}(\hat{t}-0)$ includes $R(\phi,\hat{x})$, then $L(\phi,\hat{x}) \leq L_{i+1}(\hat{t}-0)$, so that $\Lambda(\phi,\hat{x}) \geq \Lambda_{i+1}(\hat{t}-0)$(since $\chi_{i+1}(\hat{t}-0) = 1$). We thus observe that

$$g'(\hat{t}) - a(p_k)\Lambda(\phi,\hat{x}) \leq (u_t)_{i+1}(\hat{t}-0) - a(p_k)\Lambda_{i+1}(\hat{t}-0) = 0,$$

which implies $(*)$.
(b) If $R_{i+1}(\hat{t}-0)$ does not include $R(\phi,\hat{x})$ (so that $R(\phi,\hat{x}) \setminus R_{i+1}(\hat{t}-0) \neq \phi$) then the $i+2$-th facet must disappear at $t = \hat{t}$, so that $(u_x)_{i+3} = p_k$ and $\chi_{i+3}(\hat{t}-0) \leq 0$. We thus obtain

$$g'(\hat{t}) - a(p_k)\Lambda(\phi,\hat{x}) \leq g'(\hat{t})$$
$$\leq (u_t)_{i+3}(\hat{t}-0) = a(p_k)\Lambda_{i+3}(\hat{t}-0) \leq 0,$$

which implies $(*)$.

Case (III). Suppose that $\chi(\phi, \hat{x}) \leq 0$. Then $\chi^+(\phi, \hat{x}) = -1$ and/or $\chi^-(\phi, \hat{x}) = -1$. This contradicts Proposition 3, so we may assume $\chi(\phi, \hat{x}) = 1$. If $R(\phi, \hat{x})^o$ intersects $R_{i-1}(\hat{t} - 0)^o$, then

$$g'(\hat{t}) - a(p_k)\Lambda(\phi, \hat{x}) \leq g'(\hat{t})$$
$$\leq (u_t)_{i-1}(\hat{t} - 0) = a(p_k)\Lambda_{i-1}(\hat{t} - 0) = 0.$$

Otherwise, $R(\phi, \hat{x})^o$ does not intersect $R_{i-1}(\hat{t} - 0)^o$, then we have $(*)$ as in the same way of the proof of Case (II) with $\chi(\phi, \hat{x}) = 1$. This completes the proof for Case C.

Case D. It turns out that there are only two possible pictures (cases (i) and (ii) in Fig.4 and Fig.5, respectively) of the graph of u just before $\hat{t}$. We observe that $\chi_i(\hat{t}-0) = \chi_{i+1}(\hat{t}-0) = 0$ since the transition number of disappearing facets is zero.

Case (i). First, suppose that $x_{i-2}(\hat{t} - 0) < \hat{x} < x_{i-1}(\hat{t} - 0)$. If $\chi(\phi, \hat{x}) \geq 0$ then

$$g'(\hat{t}) - a(p_k)\Lambda(\phi, \hat{x}) \leq g'(\hat{t}) \leq (u_t)_{i-1}(\hat{t} - 0) = a(p_k)\Lambda_{i-1}(\hat{t} - 0) < 0,$$

where $p_k = (u_x)_{i-1}(\hat{t} - 0)$. If $\chi(\phi, \hat{x}) = -1$ then $R(\phi, \hat{x}) \supset R_{i-1}(\hat{t} - 0)$, so that $L(\phi, \hat{t}) \geq L_{i-1}(\hat{t} - 0)$. We now obtain $\Lambda(\phi, \hat{x}) \geq \Lambda_{i-1}(\hat{t} - 0)$ to get

$$g'(\hat{t}) - a(p_k)\Lambda(\phi, \hat{x}) \leq (u_t)_{i-1}(\hat{t} - 0) - a(p_k)\Lambda_{i-1}(\hat{t} - 0) = 0.$$

Next, we note that the proof when $x_{i+1}(\hat{t} - 0) < \hat{x} < x_{i+2}(\hat{t} - 0)$ is the same as the preceeding case. It remains to consider the case when $\hat{x} = x_i(\hat{t}-0)$. Since $u(\hat{t}, \cdot) - \phi$ takes its maximum at $\hat{x}$, we have $p_{k-1} \leq \phi'(\hat{x}) \leq p_k$. Since ϕ is faceted at $\hat{x}$ by the assumption of Step 2, either $\phi'(\hat{x}) = p_{k-1}$ or $\phi'(\hat{x}) = p_k$. If $\phi'(\hat{x}) = p_{k-1}$, then by Lemma 2 for $y \in (x_{i+1}(\hat{t}-0), x_{i+2}(\hat{t}-0))$ we have $\Lambda(\phi, \hat{x}) \geq \Lambda(u(\hat{t}, \cdot), y) \geq \Lambda_{i+2}(\hat{t}-0)$. We now observe that

$$g'(\hat{t}) - a(p_{k-1})\Lambda(\phi, \hat{x}) \leq (u_t)_{i+2}(\hat{t} - 0) - a(p_{k-1})\Lambda_{i+2}(\hat{t} - 0) = 0,$$

which is the same as $(*)$. If $\phi'(\hat{x}) = p_k$, then the argument similar to the preceeding case leads $g'(\hat{t}) \leq (u_t)_{i-1}(\hat{t} - 0)$ and $\Lambda(\phi, \hat{x}) \geq \Lambda_{i-1}(\hat{t} - 0)$ by Lemma 2. We obtain

$$g'(\hat{t}) - a(p_k)\Lambda(\phi, \hat{x}) \leq (u_t)_{i-1}(\hat{t} - 0) - a(p_k)\Lambda_{i-1}(\hat{t} - 0) = 0,$$

which is the same as $(*)$.

Case (ii). First, suppose that $x_{i-2}(\hat{t} - 0) < \hat{x} < x_{i-1}(\hat{t} - 0)$. Then we have $\chi(\phi, \hat{x}) \geq 0$ and $R(\phi, \hat{x})^o \cap R_{i+2}(\hat{t} - 0)^o = \phi$. As in the proof of Case C (II) (a), if $R(\phi, \hat{x}) \subset R_{i-1}(\hat{t} - 0)$ then $\chi(\phi, \hat{x}) = 1$, $L(\phi, \hat{x}) \leq L_{i-1}(\hat{t} - 0)$, so that $\Lambda(\phi, \hat{x}) \geq \Lambda_{i-1}(\hat{t} - 0)$. It then follows that

$$g'(\hat{t}) - a(p_k)\Lambda(\phi, \hat{x}) \leq (u_t)_{i-1}(\hat{t} - 0) - a(p_k)\Lambda_{i-1}(\hat{t} - 0) = 0,$$

where $p_k = (u_x)_{i-1}(\hat{t}-0)$. If $R(\phi, \hat{x}) \setminus R_{i-1}(\hat{t}-0) \neq \phi$, then $i-2$-th facet disappears at $t = \hat{t}$, so that $(u_x)_{i-3} = p_k$ and $\chi_{i-3}(\hat{t}-0) \leq 0$. We thus obtain

$$g'(\hat{t}) - a(p_k)\Lambda(\phi, \hat{x}) \leq g'(\hat{t}) \leq (u_t)_{i-3}(\hat{t}-0) = a(p_k)\Lambda_{i-3}(\hat{t}-0) \leq 0.$$

Next, if $x_{i+1}(\hat{t}-0) < \hat{x} < x_{i+2}(\hat{t}-0)$, then the proof parallels the preceeding case. Finally, we note that the point $\hat{x}$ does not agree $x_i(\hat{t}-0)$, since $u - \psi$ attains its maximum at $(\hat{t}, \hat{x})$. The proof of our main Theorem is now complete. ∎

Remark. The proof of our Theorem is simplified if we use our Existence Theorem when u is periodic in x. Indeed, we only have to prove that an admissible evolving crystal solving (4) is a generalized solution. Let u be a weakly admissible evolving crystal solving (4) and let t_0 be the first time that some facets disappear. Let $\bar{u}$ be a generalized solution with $\bar{u}(0, x) = u(0, x)$. By the unique existence theorem $\bar{u}$ has the semigroup property, i.e. $\bar{u}(t, x)$ is a generalized solution with initial data $\bar{u}(t_0, x)$ at $t = t_0$. Since an admissible evolving crystal solving (4) is a generalized solution, $u(t, x) = \bar{u}(t, x)$ for $t < t_0$. By continuity of u and $\bar{u}$, $u(t_0, x) = \bar{u}(t_0, x)$. Since u is a generalized solution for $t > t_0$ where t is close to t_0, by the semigroup property we have $u = \bar{u}$ for such t. Thus u is a generalized solution across the time $t = t_0$. Repeating this argument we conclude that u is a generalized solution in Q.

This argument is of course simpler than our proof for cases C and D. However, the latter has two advantages.

(i) It is a direct proof without appealing any nontrivial results like Existence Theorem.

(ii) The argument is local so it does not require that u is periodic in x. Moreover it is possible to prove a local version of our main Theorem although we do not present its form here.

Remark. Even if C exists, the proof of Step 2 is similar. The proof of Step 1 needs the extra condition for a as we have pointed out at the end of the proof of Step1. This condition guarantees that the corner point stays corner in the weakly sense [G]; if this condition is missing then at the corner point with $\chi_j^c = -1$ (for $C < 0$) u is not a generalized solution. Such a condition is also appeared in [GSS].

Acknowledgement. The second author was partially supported by THE SUHARA MEMORIAL FOUNDATION and a Grant-in-Aid for Scientific Research (07044094), International Science Research Program, Joint Research, the Ministry of Education, Science and Culture.

References

[AG] S. B. Angenent and M. E. Gurtin, *Multiphase thermomechanics with interfacial structure 2.*, Evolution of an isothermal interface, Arch. Rational Mech. Anal. **108** (1989), 323-391.

[EGS] C. M. Elliott, A. R. Gardiner and R. Schätzle, *Crystalline curvature flow of a graph in a variational setting*, Univ. of Sussex at Brighton, Centre for mathematical analysis and its applications, Research report No.**95/06**.

[FG]   T. Fukui and Y. Giga, *Motion of a graph by nonsmooth weighted curvature*, Proc. First World Congress of Nonlinear Analysis, (ed. V. Lakshmikantham) Walter de Gruyter, Berlin, vol.I(1995), 47-56 **I** (1995), 47-56.

[GMHG] M. -H. Giga and Y. Giga, *Geometric evolution by nonsmooth interfacial energy*, Banach Center Pub., to appear.

[G]   Y. Giga, *Motion of a graph by convexified energy*, Hokkaido Math. J. **23** (1994), 185-212.

[GG]   Y. Giga and M. E. Gurtin, *A comparison theorem for crystalline evolution in the plane*, Quart J. Appl. Math., to appear.

[GirK1] P. S. Girão and R. V. Kohn, *Convergence of a crystalline algorithm for the heat equation in one dimension and for the motion of a graph by weighted curvature*, Numer. Math. **67** (1994), 41-70.

[GirK2] P. S. Girão and R. V.Kohn, *The crystalline algorithm for computing motion by curvature, Variational Methods for Discontinuous Structure*, (eds. R. Serapioni and F. Tomarelli) Birkhauser, to appear.

[GSS]   M. E. Gurtin, H. M. Soner and P. E. Souganidis, *Anisotropic motion of an interface relaxed by formulation of infinitesimal wrinkles*, J. Differential Equations **119** (1995), 54-108.

[T1]   J. Taylor, *Constructions and conjectures in crystalline nondifferential geometry*, Proceedings of the Conference on Differential Geometry, Rio de Janeiro, pp.321-336, Pitman, London, 1988.

[T2]   J. Taylor, *Mean curvature and weighted mean curvature*, Act. Metall **40** (1992), 1475-1485.

Department of Mathematics
Hokkaido University
Sapporo 060, Japan

R GOGLIONE AND M PAOLINI

Numerical simulations of crystalline motion by mean curvature with Allen–Cahn relaxation

Abstract. In this paper we present some numerical simulations of motion by mean curvature associated to an underlying anisotropy of crystalline type. Such anisotropies are characterized by a Frank diagram (and consequently a Wulff shape) of polygonal type. Our simulations are based on an Allen-Cahn type regularization, performed in the context of a Finsler geometry. The choice of a nonregular double well potential leads to a double obstacle formulation that allows us to exploit the dynamic mesh algorithm, thus reducing considerably the computational cost. A number of simulations show the robustness of our discretization process, which is capable to deal with situations (presence of a generic forcing term) that are critical for other known numerical techniques.

1. Introduction. The interest in studying the evolution of interfaces in anisotropic materials arises from a number of phase transition phenomena in which the underlying anisotropic structure is of primary importance and dictates the resulting macroscopic behaviour.

To give an example, the existence of preferred motion directions in the human cardiac tissue is a direct consequence of the orientation of fibers, and such anisotropic microstructure must be taken into account if we want to understand the crucially important contraction dynamics [7] and motivates the theoretical investigation of motion by mean curvature in a so-called *relative geometry*. The key idea is to endow the underlying space with a so-called *Finsler metric*, which naturally leads to the concepts of *anisotropic mean curvature* and *anisotropic mean curvature flow*, see [1], [3], [24], [26], [27], and, for the 2-D case, [11], [12].

The setting of [3] allows for smooth and strictly convex metrics only; in the non-convex case, anisotropic mean curvature evolution is difficult even to be defined because the parabolic equation which models the problem becomes forward-backward, hence ill-posed. Nevertheless, it is interesting to consider more general metrics, for which smoothness and/or strict convexity conditions do not hold. We shall consider here the very important case of the so-called *crystalline anisotropy*, which corresponds to a convex but not strictly convex, nor smooth, Frank diagram. This choice allows e.g. to model crystal growth and is thus of clear practical relevance. Here the principal curvatures in the sense of differential geometry cannot in general be defined pointwise everywhere [23], and, consequently, mean curvature has no longer a local meaning; indeed, for faceted shapes with corners it is tipically either zero or infinite.

Crystalline anisotropy has been considered from a computational point of view in [14] and [22] by means of an efficient front-tracking method which, however, has the disadvantage of not beeing able to cope properly with the presence of a forcing term. Instead we shall use an indirect approach, based on an Allen-Cahn type regularization in which sharp interfaces are replaced by a thin ϵ-wide transition region. The choice of a nonregular potential in place of the usual quartic double well potential (see also [5], [21]) allows us to recover a computational complexity typical of front-tracking (by using the *dynamic mesh method* [19]), still keeping the advantage of insensitivity to singularity formation and topological changes of full-domain approaches.

In particular we shall present here some numerical simulations obtained with crystalline surface energy density, possibly with a nonzero forcing term g in the evolution law [3]. Some numerical simulations of motion with more general metrics (only locally convex or nonconvex) will be presented in a forthcoming paper [10]. The mathematical setting consists in defining a smooth and strictly convex anisotropic surface energy density $\phi^o : \Omega \times \mathbf{R}^N \to [0, +\infty)$ which defines a norm on the cotangent bundle of Ω and allows us to work in the context of a Finsler (relative) geometry. In particular, we will consider only the spatially homogeneous case (no position dependence). Under suitable smoothness conditions for ϕ^o, we consider the natural law of motion of a smooth hypersurface Σ which consists in prescribing a velocity given by

$$(1.1) \qquad V = \kappa_\phi + g$$

in the direction of some *anisotropic normal* n_ϕ, where κ_ϕ is the local anisotropic mean curvature of Σ and g is a given bounded driving force [3].

The layout of the paper is as follows. In Section 2 we give some preliminar definitions related to Finsler geometries; in Section 3 we define the notion of anisotropic mean curvature flow; we regularize the law of motion by means of a reaction-diffusion equation (Section 4) wich is then discretized by means of linear finite elements in Section 5, giving rise to a nonlinear matrix formulation. Finally, a number of numerical simulations are presented in Section 6 which exploit the robustness of our approach.

2. Preliminaries. Sticking with the notation of [3] we introduce a metric which allows us to describe the inhomogeneity and anisotropy of the space.

Given an open smooth bounded domain $\Omega \subseteq \mathbf{R}^N$, let us introduce a real function $\phi^o : \Omega \times \mathbf{R}^N \to \mathbf{R}$, which satisfies, for any fixed $x \in \Omega$, the usual properties of a norm,

$$(2.1) \qquad \phi^o(x, \xi) \geq 0 \quad \xi \in \mathbf{R}^N, \qquad \phi^o(x, \xi) = 0 \Leftrightarrow \xi = 0,$$

$$(2.2) \qquad \phi^o(x, t\xi) = |t| \phi^o(x, \xi), \quad \xi \in \mathbf{R}^N, \ t \in \mathbf{R},$$

$$(2.3) \qquad \phi^o(x, \xi_1 + \xi_2) \leq \phi^o(x, \xi_1) + \phi^o(x, \xi_2) \quad \xi_1, \xi_2 \in \mathbf{R}^N.$$

We also require the additional assumption

$$\lambda|\xi| \leq \phi^o(x, \xi) \leq \Lambda|\xi|, \quad x \in \Omega, \ \xi \in \mathbf{R}^N,$$

The unit ball

$$B_{\phi^o}(x) := \{\xi \in \mathbf{R}^N : \phi^o(x,\xi) \le 1\}$$

is often referred as the *Frank diagram* of the anisotropy. If B_{ϕ^o} is strictly convex and with smooth boundary, we shall speak of strictly convex smooth anisotropy.

By means of ϕ^o we can introduce a Finsler metric $\phi : \Omega \times \mathbf{R}^N \to [0, +\infty)$ as

$$(2.4) \qquad \phi(x,\xi) := \sup \{\xi^* \cdot \xi \ : \ \xi^* \in B_{\phi^o}(x)\}$$

where we have denoted by $\cdot$ the euclidean scalar product. The unit ball $B_{\phi}(x)$ is defined as

$$B_{\phi}(x) := \{\xi \in \mathbf{R}^N : \phi(x,\xi) \le 1\}$$

and is known as *Wulff shape*.

By means of ϕ a global distance can be introduced for any $x, y \in \Omega$ as

$$\mathrm{dist}_{\phi}(x, y) = \inf \int_0^1 \phi(\gamma(t), \gamma'(t)) \, dt$$

where the infimum is taken over all regular arcs $\gamma : [0, 1] \to \Omega$ such that $\gamma(0) = x$, $\gamma(1) = y$.

For more details about the metrics ϕ and ϕ^o see [3].

Associated to ϕ we also need a notion of volume measure $\mathcal{H}_{\phi}^N$ and a notion of surface measure $\mathcal{P}_{\phi}^{N-1}$. The volume measure $d\mathcal{H}_{\phi}^N(x)$ is given, for any $x \in \Omega$, by

$$d\mathcal{H}_{\phi}^N(x) = \omega_N \mathrm{det}_{\phi}(x) \, dx$$

where ω_N is the Lebesgue measure of the euclidean unit ball $\{\xi \in \mathbf{R}^N : |\xi| \le 1\}$ and $\mathrm{det}_{\phi} : \Omega \to (0, +\infty)$, defined as

$$(2.5) \qquad \mathrm{det}_{\phi}(x) = (\mathcal{L}(B_{\phi}(x)))^{-1}, \qquad x \in \Omega,$$

is the density function. Here $\mathcal{L}$ denotes the Lebesgue measure. Using the metric ϕ^o, we can define an anisotropic counterpart of the $(N-1)$-dimensional Hausdorff measure $d\mathcal{H}^{N-1}$ [4]. For a regular surface Σ and for any $x \in \Sigma$, this is given by

$$(2.6) \qquad d\mathcal{P}_{\phi}^{N-1}(x) = \omega_N \phi^o(x, \nu) \mathrm{det}_{\phi}(x) \, d\mathcal{H}^{N-1}(x)$$

where ν is the inner euclidean unit normal vector to the surface at x.

Finally, if ϕ^o is smooth, we define a vector valued function $T^o : \Omega \times \mathbf{R}^N \to \mathbf{R}^N$ as

$$(2.7) \qquad T^o(x,\xi) = \phi^o(x,\xi) \nabla_{\xi} \phi^o(x,\xi)$$

which is positively homogeneous of degree one. If ϕ^o is also strictly convex, $T^o(x, \cdot)$ becomes a one to one mapping of B_{ϕ^o} onto B_ϕ and is linear if we choose a riemannian metric ϕ, i.e.

$$\phi(x, \xi) = \left[\sum g_{ij}(x)\xi^i\xi^j\right]^{1/2},$$

where $g_{ij}(x)$, $i, j = 1, ..., N$ are the components of a riemannian metric tensor g.

The duality mapping $T^o(x, \cdot)$ can be defined even if ϕ^o is not smooth, in which case it becomes a multivalued maximal monotone vector valued function mapping straight sides (resp. vertices) of the boundary of the Frank diagram into vertices (resp. straight sides) of the boundary of the Wulff shape.

If the Frank diagram is a polygon, then the Wulff shape is also a polygon and we shall speak of *crystalline anisotropy*. We say that ϕ^o describes a nonconvex anisotropy if property (2.3) is not satisfied (i.e. the Frank diagram is not convex), and finally the nonconvex crystalline case corresponds to a nonconvex anisotropy for which the convex hull of the Frank diagram is a polygon.

In the nonconvex case, definition (2.4) still applies but the resulting metric ϕ is always convex. In any case the dual of $\phi(x, \cdot)$, defined by (2.4) with B_ϕ in place of B_{ϕ^o}, recovers the convex envelope of $\phi^o(x, \cdot)$. The mapping T^o can still be defined by (2.7), for smooth choices of ϕ^o, but is no longer monotone.

3. Anisotropic motion by mean curvature.

Given a surface $\Sigma \subset\subset \Omega$ defined as zero level set of some volume function u, positive inside Σ, with non-vanishing gradient on Σ, the ϕ-*co-normal* vector field to Σ is defined in a neighbourhood of Σ as

$$\nu_\phi(x) = \frac{\nabla u}{\phi^o(x, \nabla u)}.$$

Clearly $\nu_\phi(x)$ is an element of $B_{\phi^o}(x)$, normal to Σ. The ϕ-*normal* vector field is given by

$$n_\phi(x) = T^o(x, \nu_\phi(x))$$

and on Σ gives the direction of geodesics connecting the surface Σ to some nearby point $y \notin \Sigma$ such that x is the nearest point to y in Σ with respect to the ϕ-distance dist_ϕ.

Finally, we need to introduce an appropriate *anisotropic* divergence operator: given a vector field $\eta : \Omega \to \mathbf{R}^N$, we define

$$\text{div}_\phi\eta = \text{div}\eta + \eta \cdot \nabla(\log(\det_\phi))$$

which, in the homogeneous case (no position dependence), becomes the usual divergence operator. The anisotropic mean curvature is then defined as

$$\kappa_\phi(x) = -\text{div}_\phi n_\phi(x), \qquad x \in \Sigma.$$

206

It can be shown [3] that these definitions of ν_ϕ, n_ϕ and κ_ϕ are intrinsic properties of the surface Σ since they do not depend on the function u, but just on its zero level set Σ.

The particular choice $\phi^o(x,\xi) = |\xi|$ and, consequently, $T^o(x,\xi) = \xi$ gives the homogeneous and isotropic case.

Our definition of mean curvature can be justified by computing the first variation of the anisotropic surface area along a given smooth vector field $h : \Sigma \to \mathbf{R}^N$. For any λ sufficiently small we can define the surface $\Sigma_{\lambda h}$ by moving any $x \in \Sigma$ to the new position $x + \lambda h(x)$. Tedious computations [3] then show that

$$\left[\frac{d}{d\lambda} \mathcal{P}_\phi(\Sigma_{\lambda h})\right]_{\lambda=0} = -\int_\Sigma \kappa_\phi(x) h(x) \cdot \nu_\phi(x) \, d\mathcal{P}_\phi(x),$$

which essentially says that the most convenient law of motion in order to minimize the ϕ-surface area is given by (1.1) with $g = 0$.

4. Regularization. To introduce an approximating PDE for (1.1), we will follow the same path as in [20] in the more general case of a non zero forcing term g. Such PDE is similar to the well known Allen-Cahn model, which approximates the classical mean curvature motion in the sense that the zero level set of the solution approximates with an error of order $\mathcal{O}(\epsilon^2)$, where ϵ is the relaxation parameter, a surface evolving by mean curvature. This fact was rigorously stated and proved for the homogeneous and isotropic case [8], [9], [16], [17], [18], and is still formally true for a general smooth and strictly convex anisotropy ϕ^o [3].

We consider the *nonregular* double well potential $\Psi : \mathbf{R} \to \mathbf{R}^+$ defined as

$$\Psi(t) = \begin{cases} 1 - t^2 & \text{if } t \in [-1,1] \\ +\infty & \text{elsewhere} \end{cases}$$

and also introduce the derivative

$$\psi(t) = \frac{1}{2}\Psi'(t)$$

which has to be understood in the sense of the subdifferential, obtaining the multi-valued function

$$\psi(t) = \begin{cases} -t & \text{if } t \in (-1,1), \\ (-\infty, 1] & \text{if } t = -1, \\ [-1,\infty) & \text{if } t = 1. \end{cases}$$

Other choices are possible, but the nonregular potential allows us to exploit the *dynamic mesh algorithm* of [19], thus reducing considerably the computational cost.

For any $\epsilon > 0$ (relaxation parameter), the approximating reaction-diffusion equation reads formally as follows:

Find $u \in H^1(\Omega)$ such that

(4.1)
$$\epsilon a(x) \frac{du}{dt} - \epsilon \operatorname{div}_\phi a(x) T^o(x, \nabla u) + \frac{\psi(u)}{\epsilon a(x)} \ni \frac{c_0}{2} g,$$

where $a : \Omega \to \mathbf{R}^+$ is a space dependent *density* function satisfying $0 < a_{min} \le a(x) \le 1$ for any $x \in \Omega$, (it will be chosen to be small where we need better accuracy), $c_0 = \int_{-1}^{1} \sqrt{\Psi(s)} ds$ and $g \in L^\infty(\Omega)$. Problem (4.1) can be stated rigorously with the following double obstacle variational inequality.

Find $u \in H^1(\Omega; [-1, 1])$ such that $\forall w \in H^1(\Omega; [-1, 1])$

$$
\text{(4.2)} \quad \begin{aligned}
&\epsilon \Big(a(x)\frac{du}{dt}, w - u \Big)_\phi + \epsilon \Big(a(x)T^o(x, \nabla u), \nabla w - \nabla u \Big)_\phi \\
&- \frac{1}{\epsilon} \Big(\frac{u}{a(x)}, w - u \Big)_\phi - \frac{\pi}{4} \Big(g, w - u \Big)_\phi \ge 0,
\end{aligned}
$$

where $(\cdot, \cdot)_\phi$ denotes the $L^2(\Omega)$ scalar product weighted with $\omega_N \det_\phi(x)$:

$$
(f, g)_\phi = \int_\Omega fg \; d\mathcal{H}^N_\phi(x) = \omega_N \int_\Omega fg \det_\phi(x) \; dx
$$

Eqs. (4.1) and (4.2) should be completed with appropriate initial and boundary conditions. Wellposedness of problem (4.1) is guaranteed if T^o is a monotone operator, which corresponds to convexity of ϕ^o.

5. Discretization.
Following [20] we discretize (4.2) in space using piecewise linear globally continuous finite elements over a subdivision of Ω into triangles of size $\le h$, and in time by an explicit Euler scheme (forward differences) with time step τ. Given $h > 0$ (spatial discretization parameter) denote by $\mathcal{S}_h = \{S_i\}_{i=1}^I$ a partition of Ω into triangles of diameter $\le h$ satisfying the usual compatibility conditions [6], by V_h the corresponding finite element space of piecewise linear, globally continuous functions over $\mathcal{S}_h$, generated by the *hat* functions $\{\chi_j\}_{j=1}^J$, associated to the nodes $\{x_j\}_{j=1}^J$, and by V_h^0 the space of piecewise constant functions over $\mathcal{S}_h$. Also, denote by $\Pi_h : C^0(\bar{\Omega}) \to V_h$ the usual Lagrange interpolation operator and by $\Pi_h^0 : C^0(\bar{\Omega}) \to V_h^0$ the piecewise constant interpolant defined for any $f \in C^0(\bar{\Omega})$ as $\Pi_h^0 f(x) = f(\mathbf{b}_S)$, where $S \in \mathcal{S}_h$ is such that $x \in S$ and $\mathbf{b}_S$ is the baricenter of S.

We intend to approximate the variational inequality (4.2) by the usual Galerkin procedure. The integrals involved in (4.2) are then approximated by first taking the piecewise constant interpolant $a_h = \Pi_h^0(a)$, and then using the trapezoidal quadrature rule (mass lumping). We also interpolate on V_h^0 the metric $\phi(x, \xi)$ with respect to x: $\phi_h(\cdot, \xi) = \Pi_h^0 \phi(\cdot, \xi)$. The corresponding duality operator $T_h^o(x, \xi^\star)$ will also become piecewise constant in x. After time discretization we end up with the following variational inequality, $n = 1, ..., N; N = T/\tau$.

Find $U^n \in V_h \cap H^1(\Omega; [-1, 1])$ such that $\forall W \in V_h \cap H^1(\Omega; [-1, 1])$

$$
\text{(5.1)} \quad \begin{aligned}
&\epsilon \Big(a_h(x)(U^n - U^{n-1}), W - U^n \Big)_{\phi_h}^h + \tau\epsilon \Big(a_h(x)T_h^o(x, \nabla U^{n-1}), \nabla W - \nabla U^n \Big)_{\phi_h} \\
&- \frac{\tau}{\epsilon} \Big(\frac{U^*}{a_h(x)}, W - U^n \Big)_{\phi_h}^h - \tau\frac{\pi}{4} \Big(g^n, W - U^n \Big)_{\phi_h}^h \ge 0.
\end{aligned}
$$

The notation $(\cdot, \cdot)^h_{\phi h}$ stands for the use of the trapezoidal quadrature rule: $(f, g)^h_{\phi h} = \int_\Omega \Pi_h(fg) \, d\mathcal{H}^N_{\phi h}(x)$. Explicit treatment of the last term corresponds to the choice

$$U^* = U^{n-1},$$

but a slightly different choice for U^* improving stability of the scheme is possible (see [20], Section 7). The initial datum U^0 is a suitable approximation of $u(\cdot, 0)$, e.g. $U^0 = \Pi_h(u(\cdot, 0))$.

The solution U^n of (5.1) can be obtained by first solving the variational problem for $U^{n-1/2} \in V_h$

(5.2)
$$\epsilon\left(a_h(x)(U^{n-1/2} - U^{n-1}), V\right)^h_{\phi h} + \tau\epsilon\left(a_h(x)T^o_h(x, \nabla U^{n-1}), \nabla V\right)_{\phi h}$$
$$- \frac{\tau}{\epsilon}\left(\frac{U^*}{a_h(x)}, V\right)^h_{\phi h} - \tau\frac{\pi}{4}\left(g^n, W - U^n\right)^h_{\phi h} = 0 \qquad \forall V \in V_h,$$

and then projecting $U^{n-1/2}$ on $V_h \cap H^1(\Omega; [-1, 1])$ node by node:

(5.3)
$$U^n = \Pi_h \sigma(U^{n-1/2}),$$

where $\sigma(t) = \max(-1, \min(1, t))$.

Now the dynamic mesh strategy described in [19] can be applied: the computation can be avoided where U^{n-1} is locally constant ± 1 and the solution can be computed only on the transition region $\{|U^{n-1}| < 1\}$ surrounded by a *security* strip one triangle wide. This mesh is then updated at each time step by adding/removing triangles where needed. The time increment τ itself is adapted at each time step in order to meet a stability constraint; this allows the use of large time steps for as long as the surface remains smooth. The time step selection procedure is described in [20].

For notational convenience we shall describe the algorithm for a fixed global mesh (as we already did for (5.1) and (5.2)), with the understanding that the simulations of Section 6 were actually obtained with the use of the dynamic mesh strategy.

By identifying any function $U \in V_h$ with the vector $\{U_j\}^J_{j=1}$ of its nodal values, so that $U = \sum^J_{j=1} U_j\chi_j$, we can define the nonlinear function $K = (K_1, ..., K_J) : \mathbf{R}^J \to \mathbf{R}^J$ by

$$K_j(U) = \epsilon \int_\Omega a_h(x)T^o_h(x, \nabla U) \cdot \nabla\chi_j \, d\mathcal{H}^N_{\phi h}(x), \qquad \forall U \in V_h, \quad j = 1, ..., J.$$

If we define the three (diagonalized) mass matrices $M_i = \{m^i_{jk}\}^J_{j,k=1}$, $i = -1, 0, 1$, as

$$m^i_{jk} = \int_\Omega [\epsilon a_h(x)]^i \, \Pi_h(\chi_j\chi_k) \, d\mathcal{H}^N_{\phi h}(x),$$

the variational formulation (5.2) is then equivalent to the following matrix formulation

(5.4) $$M_1 U^{n-1/2} - M_1 U^{n-1} + \tau K(U^{n-1}) - \tau M_{-1} U^{n-1} - \tau\frac{\pi}{4} M_0 g^n = 0.$$

An efficient way for computing $K(U^{n-1})$ is described in [20].

6. Numerical simulations. Since we shall present only numerical simulations for homogeneous anisotropies (no dependence on the position), we shall simply write $\phi^o(\xi)$ and $T^o(\xi)$ in place of $\phi^o(x,\xi)$ and $T^o(x,\xi)$ and we shall denote by B_{ϕ^o} and B_ϕ the Frank diagram and the Wulff shape of a given metric ϕ^o. All numerical simulations are done in dimension $N = 2$, on a Spark Station 10, using the C programming language.

Concerning the graphic presentation of the Wulff shape a preliminar remark is required. If we write the metric ϕ^o as $\phi^o(\xi) = \rho\psi(\theta)$, where $\rho = |\xi|$ and θ is such that $\xi = \rho(cos\theta, sin\theta)$, it is well known that the strict convexity property (2.3) is equivalent, analitically, to $\psi + \psi'' > 0$ and, geometrically, to strict convexity of the Frank diagram. In such case, as already observed, T^o is a bijection from B_{ϕ^o} to B_ϕ, since $\phi(T^o(\xi)) = \phi^o(\xi)$, see [3]. The Wulff shape B_ϕ can then be easily obtained by plotting the set $\{T^o(\xi) : \phi^o(\xi) = 1\}$.

When ϕ^o is not convex, the situation is different: T^o is no longer one to one and, if ν_1 and ν_2 are the contact points of a bitangent with the Frank diagram, we have $T^o(\nu_1) = T^o(\nu_2) = n_\phi$, where n_ϕ is a corner point of the Wulff shape. Moreover,

$$\partial B_\phi \subset \{T^o(\xi) : \phi^o(\xi) = 1\}$$

and the inclusion is strict. Selfintersections of $\{T^o(\xi) : \phi^o(\xi) = 1\}$ are corners of ∂B_ϕ and points beyond such selfintersections are not part of ∂B_ϕ.

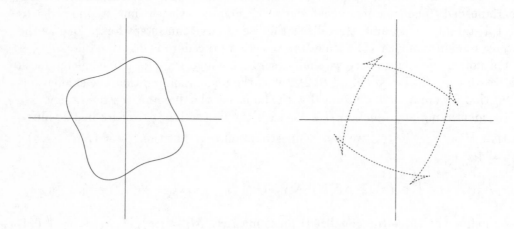

FIGURE 6.1. A nonconvex metric.
a: The Frank diagram.
b: The set $\{T^o(\xi) : \phi^o(\xi) = 1\}$.

In Figure 6.1 the Frank diagram (Figure 6.1a) and the set $\{T^o(\xi) : \phi^o(\xi) = 1\}$ (Figure 6.1b) of a nonconvex metric are depicted. It is clear that each nonconvex region in the Frank diagram corresponds to an "ear", i.e. to a triangular-shaped region located between two cusps and a selfintersection point in the dashed line.

The two cusps in the dashed line correspond to the inflection points of ∂B_{ϕ^o}. The piece of curve between two cusps corresponds to the locally concave part of ∂B_{ϕ^o} between two inflection points. The arc of curve between a selfintersection point and a cusp corresponds to the arc between a contact point of the bitangent and an inflection point in the Frank diagram.

Consequently, the Wulff shape of a nonconvex metric ϕ^o (or of its convexified) is the region obtained from the set $\{T^o(\xi) : \phi^o(\xi) = 1\}$ by removing the "ears" extending beyond its cross point. For convenience, in the following figures, we will call Wulff shape the plotted set $\{T^o(\xi) : \phi^o(\xi) = 1\}$ with the understanding that, when ϕ^o is nonconvex, the actual Wulff shape is obtained by the procedure described above.

We will show some simulations of fronts evolutions: we start from a locally concave everywhere metric (nonconvex crystalline case, Example 1) already proposed in [20]; then, we will consider a number of examples with a six-fold crystalline anisotropy both in lack of driving forcing term g (Example 2 and Example 3) and with a nonconstant $g \neq 0$ (Example 4).

Example 1: Nonconvex crystalline. We begin considering a star-shaped metric ϕ^o, locally concave everywhere. It represents an example of crystalline energy because its Frank diagram touches the boundary of its convex hull only at discrete points. If we use polar coordinates in the ξ−plane, an example of such anisotropy is obtained choosing

$$\psi(\theta) = B + A\cos(k\hat{\theta}),$$

where $\hat{\theta}$ is θ reduced modulo $\pi/3$ to the interval $[-\pi/6, \pi/6]$, $k = 2$, $B = 3/4$ and $A = 9/16$. The values of k, B, A have been chosen so that the convexity condition $\psi + \psi'' \geq 0$ is never satisfied. In Figure 6.2a the Frank diagram B_{ϕ^o} (solid line) and the Wulff shape B_φ (dashed line) of the metric are depicted.

Now let us consider a circular initial datum $\Sigma = \{|x| = 1\}$ with no driving force ($g \equiv 0$). The discretization parameters are $h = 0.05$ and $\epsilon = 0.22468$. We used the density function $a(x) = max(0.1, |x|)$ knowing that a singularity will appear at the origin.

Since B_{ϕ^o} is not convex, the evolution istantaneously develops wrinkled regions (see Figure 6.2b), i.e. a faceting with fixed orientations θ_1, θ_2 depending on the concave region width in the Frank diagram. Indeed, the "expensive" normal directions corresponding to points of the unit ball of ϕ^o strictly inside its convex hull are avoided and, on the contrary, the more stable θ_i directions are preferred [15]. "The lenghts of the initial and terminal facets decrease with time, while the lenghts of the internal facets remain constant" [2]. After a while the evolving curve assumes a hexagonal shape (the initial condition is completely forgotten), moves homotetically to the Wulff shape and shrinks to a point.

Example 2: A crystalline metric. Here, the metric ϕ^o is obtained from the previous one by convexification; as expected from Figure 6.2a, the unit ball of ϕ is a rotated regular hexagon (Figure 6.3a). Again we consider a circular initial datum

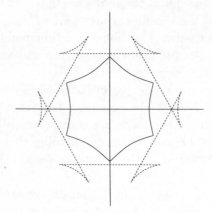
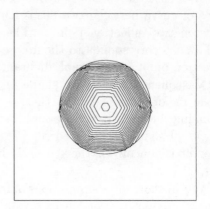

FIGURE 6.2 (Ex. 1). A nonconvex crystalline energy.
a: Frank diagram (solid) and Wulff shape (dashed).
b: Evolution with initial circular datum and $g \equiv 0$ at $t = 0.02i$, $i \geq 0$.

and the driving force $g \equiv 0$. The discretization parameters are the same as in Example 1. In a short time the evolving curve assumes a hexagonal shape, but there is no wrinkling now.because the metric is convex. Figure 6.3b shows that each side moves parallel to itself as expected.

By graphically comparing Figures 6.2b and 6.3b it could be conjectured that the two evolutions tend to coincide as ϵ, h, τ tend to zero. This is a very important issue and deserves further investigation.

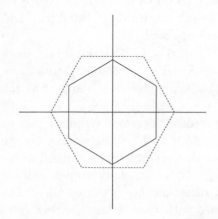
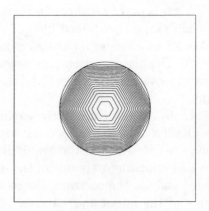

FIGURE 6.3 (Ex. 2). Crystalline metric.
a: Frank diagram (solid) and Wulff shape (dashed).
b: Evolution with circular datum at $t = 0.02i$, $i \geq 0$.

Example 3: Polygonal initial datum ($g \equiv 0$). With the same choice of anisotropy of Example 2 and no forcing term we choose two different polygonal initial interfaces with sides parallel to the sides of the Wulff shape.

In such situation it has been shown that the evolution remains polygonal with sides moving parallel to themselves with speed given by the so called *weighted mean curvature* wmc [25]. In particular, if S_i is the i-th facet of the evolving polygonal Σ with normal n_i, we have

(6.1)
$$\text{wmc}(S_i) = -\sigma_i \Lambda(n_i)/\text{Lenght}(S_i);$$

here $\Lambda(n_i)$ is the lenght of the edge in the Wulff shape B_ϕ with normal n_i, σ_i is 1 (resp. -1) if at both ends of the segment S_i the polygonal interface is convex (resp. concave), and σ_i is 0 if the polygonal interface is convex at one end of the segment and concave at the other.

The total number of segments cannot increase, thus not allowing creation of new edges. In this framework the algorithm proposed by Roosen and Taylor [22], which takes advantage of the polygonal structure of the front, can be successfully applied. Convergence of such algorithm has been proved by Girão [13] and Girão and Kohn [14] in the case of a convex initial polygon or when the initial surface is the graph of a piecewise linear function respectively. It should be stressed that the assumption $g \equiv 0$ is essential for such convergence results in that a general forcing can lead to creation of new faces. To deal with the possibility of creation of new faces in their crystalline algorithm, Roosen and Taylor added the so-called *shattering* of the interfacial edges, which is a delicate part of the algorithm; our approach based on relaxation, instead, works well in such general situations also (see Example 4 below).

For both simulations that we are going to present, the discretization parameters are the same as in Example 1 and the mesh is uniform.

In the first case we choose the polygonal initial interface of Figure 6.4a in the rectangular domain

$$\Omega := \{x = (x_1, x_2) \in \mathbf{R} \times \mathbf{R} : 0 \le x_1 \le 2, 0 \le x_2 \le 3\sqrt{3}\},$$

with a reflection condition on the sides $x_1 = 0$ and $x_1 = 2$.

In accordance with law (6.1), just the horizontal edges move, since the middle one has wmc $= 0$ (Figure 6.4b). Moreover, the evolution slows down since the lenght of the segments increases. The interface evolves as long as it becomes completely horizontal.

In the second case we start with the initial condition of Figure 6.5a in

$$\Omega := \{x = (x_1, x_2) \in \mathbf{R} \times \mathbf{R} : 0 \le x_1 \le 4, 0 \le x_2 \le 3\sqrt{3}\};$$

it is again a polygonal interface with sides parallel to the sides of the Wulff shape, but the horizontal edges stand at different heights. The evolution is depicted in Figure 6.5b for $0 \le t \le 0.6$ and in Figure 6.5c for $t > 0.6$. It is clear that, for $t > 0.6$, the motion is similar to the one depicted in Figure 6.4b.

213

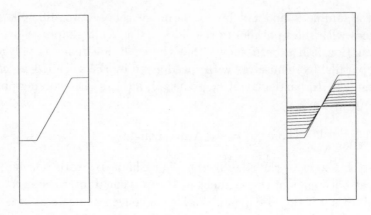

FIGURE 6.4 (Ex. 3). Crystalline metric.
a: Polygonal initial datum.
b: Evolution with no forcing term at $t = 0.1i$, $i \geq 0$.

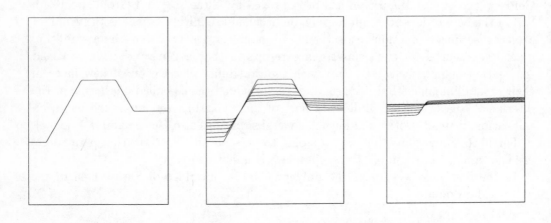

FIGURE 6.5 (Ex. 3). Crystalline metric.
a: Initial datum.
Evolution with no forcing term for $0 \leq t \leq 0.6$ (5b) and for $t > 0.6$ (5c).

Example 4: Polygonal initial datum with noncostant $g \neq 0$. We shall now present an example where the presence of a general forcing term can lead to quite critical situations. Again we choose $h = 0.05$, $\epsilon = 0.22468$ and a uniform mesh.

Starting with the same initial datum given in the first case of Example 3 in the domain

$$\Omega := \{x = (x_1, x_2) \in \mathbf{R} \times \mathbf{R} : 0 \leq x_1 \leq 2, 0 \leq x_2 \leq 3\sqrt{3}\},$$

as forcing term we choose

$$g(x) = 3(1 - x)$$

214

so that we have a positive pressure in the left half of the domain and a negative pressure in the remaining region.

For $t \in [0, 0.2]$ the resulting evolution is depicted in Figure 6.6a. Now, the middle segment rotates because it is undergoing the action of the forcing term. It is then clear that the interface becomes a polygonal having the middle segment not parallel to any side of the Wulff shape. More general choices of g could well lead to evolutions which are no longer polygonal in shape.

FIGURE 6.6 (Ex. 4). Crystalline metric.
$(g(x) = 3(1 - x))$. Evolution for $0 \le t \le 0.2$ (a), for $0.2 < t \le 1.5$ (b), for $1.5 < t \le 2.2$ (c).

At $t \approx 0.2$ the interface is completely flat, but afterwards a new edge is created which again rotates around its middle point untill, at $t \approx 1.5$, it splits in two parts (each moving parallel to a side of the Wulff shape) separated by a vertical face (not in the Wulff shape!) of increasing length. This simulation clearly shows that evolutions with crystalline anisotropy can become quite complex when we allow for the presence of a space dependent forcing term. This is almost always the case for (e.g.) phase transition problems, where typically the driving force is itself solution of a parabolic equation coupled with the evolving surface.

References

1. F. Almgren, J.E. Taylor, and L. Wang, *Curvature-driven flows: a variational approach*, SIAM J. Control Optim. **31** (1993), 387–437.
2. S.B. Angenent and M.E. Gurtin, *Multiphase thermomechanics with an interfacial structure 2. Evolution of an isothermal interface*, Arch. Rational Mech. Anal. **108** (1989), 323–391.
3. G. Bellettini and M. Paolini, *Anisotropic motion by mean curvature in the context of Finsler geometry*, Hokkaido Math. J. (to appear).
4. G. Bellettini, M. Paolini, and S. Venturini, *Some results on surface measures in Calculus of Variations*, Ann. Mat. Pura Appl. (to appear).
5. X. Chen and C.M. Elliott, *Asymptotics for a parabolic double obstacle problem*, Proc. Roy. Soc. London Ser. A **444** (1994), 429–445.

6. P.G. Ciarlet, *The Finite Element Method for Elliptic Problems*, North-Holland, Amsterdam, 1978.

7. P. Colli Franzone, L. Guerri, and S. Rovida, *Wavefront propagation in an activation model of the anisotropic cardiac tissue: asymptotic analysis and numerical simulations*, J. Math. Biol. **28** (1990), 121–176.

8. P. De Mottoni and M. Schatzman, *Geometrical evolution of developed interfaces*, Trans. Amer. Math. Soc. **347** (1995), 1533–1589.

9. L.C. Evans, H.-M. Soner, and P.E. Souganidis, *Phase transitions and generalized motion by mean curvature*, Comm. Pure Appl. Math. **45** (1992), 1097–1123.

10. F. Fierro, R. Goglione, and M. Paolini, *Numerical simulations of mean curvature flow in presence of a nonconvex anisotropy*, (in preparation).

11. M.E. Gage, *Evolving plane curves by curvature in relative geometries*, Duke Math. J. **72** (1993).

12. M.E. Gage and Yi Li, *Evolving plane curves by curvature in relative geometries II*, Duke Math. J. **75** (1994), 79–98.

13. P.M. Girao, *Convergence of a crystalline algorithm for the motion of a simple closed convex curve by weighted curvature*, SIAM J. Numer. Analysis **32** (1995), 886–899.

14. P.M. Girao and R.V. Kohn, *Convergence of a crystalline algorithm for the heat equation in one dimension and for the motion of a graph by weighted curvature*, Numer. Math. **67** (1994), 41–70.

15. M.E. Gurtin, H.-M. Soner, and P.E. Souganidis, *Anisotropic motion of an interface relaxed by the formation of infinitesimal wrinkles*, J. Differential Equations **119** (1995), 54–108.

16. R.H. Nochetto, M. Paolini, and C. Verdi, *Sharp error analysis for curvature dependent evolving fronts*, Math. Models Methods Appl. Sci. **3** (1993), 711–723.

17. _____, *Optimal interface error estimates for the mean curvature flow*, Ann. Scuola Norm. Sup. Pisa Cl. Sci. (4) **21** (1994), 193–212.

18. _____, *Double obstacle formulation with variable relaxation parameter for smooth geometric front evolutions: asymptotic interface error estimates*, Asymptotic Anal. **10** (1995), 173–198.

19. _____, *A dynamic mesh method for curvature dependent evolving interfaces*, J. Comput. Phys. (to appear).

20. M. Paolini, *An efficient algorithm for computing anisotropic evolution by mean curvature*, Curvature flows and related topics, Mathematical Sciences and Applications, Vol. 5 (A. Damlamian, J. Spruck, and A. Visintin, eds.), Gakkotosho, Tokyo, 1995.

21. M. Paolini and C. Verdi, *Asymptotic and numerical analyses of the mean curvature flow with a space-dependent relaxation parameter*, Asymptotic Anal. **5** (1992), 553–574.

22. A.R. Roosen and J.E. Taylor, *Modeling crystal growth in a diffusion field using fully-facetted interfaces*, J. Comput. Phys. (1993), 54–108.

23. R. Schneider, *Convex Bodies: the Brunn-Minkowsky Theory*, Encyclopedia of Mathematics and its Applications, Cambridge Univ. Press, 1993.

24. J.E. Taylor, *Crystalline variational problems*, Bull. Amer. Math. Soc. (N.S.) **84** (1978), 568–588.

25. _____, *Motion by crystalline curvature*, Computing Optimal Geometries, Selected Lectures in Math., Amer. Math. Soc., Providence, RI, 1991, pp. 63–65.

26. _____, *Mean curvature and weighted mean curvature II*, Acta Metall. **40** (1992), 1475–1485.

27. _____, *Motion of curves by crystalline curvature, including triple junctions and boundary points*, Proc. of Symposia in Pure Math. **54** (1993), 417–438.

DIPARTIMENTO DI MATEMATICA, UNIVERSITÀ DI MILANO, 20133 MILANO, ITALY

E-mail: goglione@isa.mat.unimi.it

DIPARTIMENTO DI MATEMATICA E INFORMATICA, UNIVERSITÀ DI UDINE, 33100 UDINE, ITALY

E-mail: paolini@dimi.uniud.it

J KAČUR

Solution of degenerate parabolic problems by relaxation schemes

Abstract. A numerical method is proposed to solve a large number of FBP modelled by degenerate doubly nonlinear parabolic, parabolic-elliptic equations and systems. The method is related to that one proposed in [13],[14] and is based on a nonstandard time discretization including two relaxation functions by means of which the diffusion degeneracies (slow and fast) are controlled.

1. Introduction

We consider the following mathematical model

$$\partial_t b(x, u) - \nabla a(t, x, u, \nabla \beta(x, u)) = f(t, x, u)$$
$$\text{in} \quad I \times \Omega, \quad I \equiv (0, T), \quad T < \infty$$

(1.1)

with the mixed boundary conditions

$$\beta(x, u) = 0 \quad \text{on} \quad \Gamma_1 \times I,$$
$$a(t, x, u, \nabla \beta(x, u)).\nu = g(t, x, \beta(x, u)) \quad \text{on} \quad \Gamma_2 \times I$$

(1.2)

and with the initial condition

$$b(x, u(x, 0)) = b(x, u_0)$$

(1.3)

Here $\Omega \subset R^N$ is a bounded domain with Lipschitz continuous boundary $\partial \Omega$ and $\Gamma_1, \Gamma_2 \subset \partial \Omega$ are opened with $mes_{N-1}\Gamma_1 + mes_{N-1}\Gamma_2 = mes_{N-1}\partial\Omega$, $\Gamma_1 \cap \Gamma_2 = 0$, $mes_{N-1}\Gamma_1 > 0$. The main goal of this paper is to extend the convergence results for the approximation schemes studied in [13],[14] to the case when $b(x, s)$ and $\beta(x, s)$ are absolutely continuous in s for all $x \in \Omega$. In [13], [14] we have assumed b, β to be Lipschitz continuous in s.

One of the following structural assumptions will be considered:

I $b(x, s), \beta(x, s)$ are strictly monotone in s, $\quad \forall x \in \Omega$;

II b is nondecreasing in s, $\beta(x, s) \equiv s$. In that case the terms a, f in (1.1) are considered in the form $a \equiv a(t, x, b(x, u), \nabla u), f \equiv f(t, x, b(x, u)), g \equiv g(t, x)$;

III $\beta(x, s)$ is nondecreasing in $s, \beta'_s(x, s) \leq L_\beta < \infty, b(x, s) \equiv q(x)s, q(x) \geq q_0 > 0$.
In that case the terms a, f, g are considered in the form
$a \equiv a(t, x, \beta(x, u), \nabla\beta(x, u)), f \equiv f(t, x, \beta(x, u)), g \equiv g(t, x, \beta(x, u))$.

217

The elliptic part generated by $a(t, x, \eta, \xi)$ is monotone and coercive in ξ and a polynomial growth of a in ξ is assumed, $|a(., \xi)| \approx |\xi|^{p-1}$ with $p > 1$. The growth of f, g in u (resp. in b, β) is dominated by elliptic and parabolic terms in (1.1) - see Section 2.

Thus, (1.1) represents double nonlinear, degenerate parabolic, parabolic-elliptic (in Case II) problems. The cases I,II,III include a large number of free bounda-ry problems as: flow in porous media, unsaturated-saturated flows, Stefan prob-lem (phase-change problem), flow through porous media in turbulent regime — see [1], [4], [6], [7] etc. Problem (1.1)-(1.3) includes locally both fast and slow dif-fusion. Neither $b(x, \beta^{-1}(x, s))$, nor $\beta(x, b^{-1}(x, s))$ needs to be Lipschitz continu-ous in s. By $\beta^{-1}(x, s), b^{-1}(x, s)$ we understand the inverse of $\beta(x, s), b(x, s)$ (x is fixed), respectively. If $\beta(x, R) = (r, s)$ (x is fixed) and r or s is finite then we put $\beta^{-1}(x, y) = +\infty(-\infty)$ for $y \geq s$ ($y \leq r$).

The existence of a variational solution to (1.1)-(1.3) has been studied in [2], [3], [4], [5], [7], [11], [13], [18] etc.

We use a nonstandard time discretization with a relaxation functions by means of which we control the degeneracy of $b'_s(x, s) = 0$ and $\beta'_s(x, s) = 0$.

We shall regularize b and β by b_n and β_n where $b_n(x, s) = b(x, s) + \tau^d s$ and $\beta_n(x, s)$ is absolutely continuous in s for all $x \in \Omega$ satisfying: $\beta_n(x, s) \to \beta(x, s)$ locally uniformly in s from a bounded set in R; moreover $\tau^d \leq \beta'_{n,s}(x, s) \leq \tau^{-d}$ and $min\{\beta'_s(x, s), 1\} \leq C\beta'_{n,s}(x, s)$ where C is some positive constant.

Then we reduce (1.1) to a sequence of regular elliptic problems coupled with algebraic conditions for relaxation functions. Our approximation scheme reads as follows: Let $\tau = \frac{T}{n}, n \in N$ be the time step, $u_i \approx u(t_i, x)$ for $t_i = i\tau, i = 1, ..., n$, $\Theta_i \approx \beta(x, u_i)$. On the time level $t = t_i$ we determine Θ_i from the regular elliptic problem

$$\lambda_i(\Theta_i - \beta_n(x, u_{i-1})) - \tau\nabla a(t_i, x, u_{i-1}, \nabla\Theta_i) = \tau f(t_i, x, u_{i-1}) +$$

(1.4)
$$\tau\tilde{f}_i \equiv \tau f_i + \tau\tilde{f}_i$$

$$\Theta_i = 0 \quad \text{on} \quad \Gamma_1, \quad a(t_i, x, u_{i-1}, \nabla\Theta_i).\nu = g(t_i, x, \Theta_{i-1}) \equiv g_i \quad \text{on} \quad \Gamma_2$$

where $\lambda_i \in L_\infty(\Omega)$ has to satisfy the "convergence condition"

(1.5) $\quad \frac{1}{2}\tau^d \leq \lambda_i \leq \min\{\tau^{-d}, \dfrac{b(x, u_{i-1} + \mu_i(\Theta_i - \beta_n(x, u_{i-1}))) - b(x, u_{i-1})}{\Theta_i - \beta_n(x, u_{i-1})}\} + \tau^d$

with a relaxation function $\mu_i \in L_\infty(\Omega)$ satisfying

(1.6) $\quad \dfrac{1}{2}\tau^d \leq \mu_i \leq \dfrac{\beta_n^{-1}(x, \beta_n(x, u_{i-1}) + \alpha(\Theta_i - \beta_n(x, u_{i-1}))) - u_{i-1}}{\Theta_i - \beta_n(x, u_{i-1})}$

where $d \in (0, 1)$, $\alpha \in (0, 1)$ (α close to 1), $0 \leq K$ (big), are parameters of the method, $i = 1, ..., n$. We determine u_i from

$$(1.7) \qquad b_n(x, u_i) := b_n(x, u_{i-1}) + \lambda_i(\Theta_i - \beta_n(x, u_{i-1}))$$

In the points $x \in \Omega$ where $\Theta_i = \beta_n(x, u_{i-1})$ we define the difference quotients in (1.5), (1.6) by $b'_s(x, u_{i-1}) \cdot \frac{\alpha}{\beta'_{n,s}(x, u_{i-1})}$, $\frac{\alpha}{\beta'_{n,s}(x, u_{i-1})}$, respectively. The term $\tilde{f}_i$ in (1.4) represents a small term (error) satisfying

$$(1.8) \qquad |\tilde{f}_i|_2 = o(1), \quad |\tilde{f}_i|_{q'} \leq c, \quad \forall n, i = 1, ..., n, \quad q' > q, \quad (p^{-1} + q^{-1} = 1)$$

where $o(1)$ is Landau's symbol denoting any term converging to 0 for $\tau \to 0$.

If we put $\mu_i = \lambda_i = \frac{1}{2}\tau^d$ then (1.5), (1.6) are satisfied However, a good approximation requires to reach equalities on the R.H.S where special cases have been treated.

To determine λ_i, μ_i in an "optimal way" we can use the iterations in (1.4)-(1.6) with parameter $k = 1, ...,$ where we replace λ_i, Θ_i in (1.4) by $\lambda_{i,k-1}, \Theta_{i,k}$ and then we define $\lambda_{i,k}, \mu_{i,k}$ by equalities with the corresponding difference quotients in (1.5), (1.6), respectively, where Θ_i is replaced by $\Theta_{i,k}$. The determination of λ_i, Θ_i in (1.4)-(1.6) will be discussed in Section 2.

By means of the relaxation functions λ_i, μ_i we control the degeneracy of $b'_s(x, s) = 0$, $\beta'_s(x, s) = 0$. When $b(x, s) \equiv s$ then we can take $\lambda_i \equiv \mu_i$ and we obtain the approximation scheme used in [11], [15]. If $\beta(x, s) \equiv s$ then we can take $\mu_i \equiv \alpha$ and we obtain the approximation scheme used in [12]. The Stefan type problems III (in a more special case where the elliptic operator is linear) have been approximated in [20], [22], [23] etc. Analogously as in [14] we can approximate (1.1)-(1.3) in the form (1.4) where the "convergence condition" concerning λ_i is

$$(1.9) \qquad \left| \lambda_i - \frac{b_n(x, u_{i-1} + \mu_i(\Theta_i - \beta_n(u_{i-1}))) - b_n(x, u_{i-1})}{\Theta_i - \beta_n(u_{i-1})} \right| < \tau$$

and μ_i satisfies (1.6). Then we define

$$u_i = u_{i-1} + \mu_i(\Theta_i - \beta_n(x, u_{i-1})), \quad i = 1, ..., n.$$

This scheme is more economical as (1.4)-(1.6). However, (1.9) represents a very strong restriction for λ_i. In [14] we have assumed $b(x, s), \beta(x, s)$ to be Lipschitz continuous in s and in (1.9), (1.6), (1.4) we have used β in the place of β_n. In the case of approximation (1.4), (1.6), (1.9) the model (1.1) has to be restricted to the cases where the polynomial growth of a in ξ is of degree $p \geq 2$. Then, models of flow with turbulent regime (-see [7]) are not encluded. Moreover, we were able to guarantee (1.9) only in more special cases of (1.1) -see [14].

In Section 2 we formulate the convergence of $\lambda_{i,k} \to \lambda_i$, $\Theta_{i,k} \to \Theta_i$, $\mu_{i,k} \to \mu_i$ where $\lambda_i, \mu_i, \Theta_i$ satisfy (1.4)-(1.6). In Section 3 we discuss the convergence of (1.4)-(1.6) in the Case I. The obtained convergence results can be extended to systems of the form

$(1.1)_S$

$$\partial_t b^j(x, u^j) - \nabla a^j(t, x, u, \nabla \beta(x, u)) = f^j(t, x, u)$$

$$\beta^j(x, u^j) = 0 \quad \text{on} \quad \Gamma_1^j \times I, a^j(t, x, u, \nabla \beta(x, u)).\nu = g^j(t, x, \beta(x, u)) \quad \text{on} \quad \Gamma_2^j \times I$$

for $j = 1, ..., m$, where $\beta(x, u) = (\beta^j(x, u^j), j = 1, ..., m)$ and $u \equiv (u^j, j = 1, ..., m)$ are vector functions. The convergence of the approximation scheme (1.4), (1.6), (1.9) can be proved following [14] and the present arguments.

2. Assumptions and convergence of iterations

The following assumption will be used in the Case I. To describe the growth of a, f in u we shall make use of the following function (see [2], [14])

(2.1)

$$B^*(x, s) := b(x, \beta^{-1}(x, s))s - \int_0^s b(x, \beta^{-1}(x, z))dz$$

$$\text{for} \quad s \in \{y \in R : y = \beta(x, z)\}$$

Analogously we define $B^*(x, s)$ replacing b, β by b_n, β_n, respectively. We assume that

$H_1)$ $b(x, s)$ is strictly monotone in s for a.e. $x \in \Omega, b(x, 0) = 0$,

$$0 \le b_s'(x, s), \quad |b_s'(x, s) - b_s'(\bar{x}, s)| \le \omega(|x - \bar{x}|)b_s'(x, s)$$

where $\omega : R_+ \to R_+$ is continuous, $\omega(0) = 0$ and

(2.2) $$|b_n(x, s)|^2 \le C_1 B_n^*(x, \beta_n(x, s)) + C_2 \quad \forall s \in R; \forall n.$$

$H_2)$ $\beta(x, s)$ is strictly monotone in s for a.e. $x \in \Omega, \beta(x, 0) = 0$
$\quad 0 \le \beta_s'(x, s) < \infty \quad a.e.s \in R, |\beta_s'(x, s) - \beta_s'(\bar{x}, s)| \le \omega(|x - \bar{x}|)\beta_s'(x, s).$
$H_3)$ $a(t, x, \eta, \xi) : I \times \Omega \times R \times R^N \to R^N$ is continuous,

$$(a(t, x, \eta, \xi_1) - a(t, x, \eta, \xi_2)).(\xi_1 - \xi_2) \ge 0,$$

$$|a(t, x, \eta, \xi)| \le C(1 + |b(x, \eta)|^{2\gamma} + |\xi|^{p-1}), \quad p > 1, \quad 0 < \gamma < \frac{p-1}{p};$$

$$a(t, x, \eta, \xi).\xi \ge C_1|\xi|^p - C_2$$

220

uniformly for $(t, x) \in Q_T \equiv I \times \Omega$, $\quad \eta \in R, \xi \in R^N$;

$H_4)$ $f(t, x, \eta) : I \times \Omega \times R \to R$ is continuous and $|f(t, x, \eta)| \leq C(1 + |b(x, \eta)|^{2\gamma})$
$0 < \gamma < \frac{p-1}{p}$;

$H_5)$ $g(t, x, s) : I \times \Gamma_2 \times R \to R$ is continuous and $|g(t, x, s)| \leq C(1 + |s|^\gamma)$,
$0 < \gamma < p - 1$;

$H_6)$ $u_0 \in L_2(\Omega)$.

The stronger convergence results will be obtained under the strong monotonicity

$H_7)$ $(a(t, x, \eta, \xi_1) - a(t, x, \eta, \xi_2)) \cdot (\xi_1 - \xi_2) \geq C|\xi_1 - \xi_2|^p$ which requires $p \geq 2$.

2.3 Remark. *The assumption (2.2) is the only structural restriction due to our method. If b is bounded, then (2.2) is satisfied. If $|b(x, s)| \approx c_1|s|^\alpha$, $|\beta(x, s)| \approx c_2|s|^\beta$ (asymptotically for $|s| \to \infty$) then (2.2) requires $\alpha \leq \beta$. If $\int_0^s b'_s(x, z)[\beta_n(x, z) - \beta(x, z)]dz \geq 0$ for all $s \in R, x \in \Omega$ then the term $|b(x, \eta)|^{2\gamma}$ can be replaced by $(B^*(x, \beta(x, \eta)))^\eta$.*

We shall use the standard functional spaces $L_\infty \equiv L_\infty(\Omega), L_p \equiv L_p(\Omega), W_p^1(\Omega)$ (Sobolev space), $V = \{u \in W_p^1; v|_{\Gamma_1} = 0\}, C^{0,\beta}(\bar{\Omega})$ (Hölder space), $L_p(I, V)$ - see, e.g. [19]. We denote $x.y = \sum_{i=1}^N x_i y_i$, $(u, v) = \int_\Omega u.v$, $(u, v)_{\Gamma_2} = \int_{\Gamma_2} u.v$ and V^* is the dual space to V with the duality $< u, v >$ for $u \in V^*, v \in V$.

Let $|\cdot|_\infty, |\cdot|_p, \|\cdot\|, \|\cdot\|_*, |\cdot|_{\Gamma,p}, \|\cdot\|_{L_p(I,V)}, \|\cdot\|_{0,\alpha}$ denote the norms in L_∞, L_p, V, V^*, $L_p(\Gamma), L_p(I, V), C^{0,\alpha}$, respectively. By C we denote a generic nonnegative constant.

In the next we drop the variable x in terms $a, f, g, \lambda_i, \mu_i$. Our concept of solving (1.1)-(1.3) is based on the notion of variational solution and an energy type a priori estimates — see [2], [13].

2.4 Definition. A measurable function $u : Q_T \to R$ is a variational solution of (1.1)-(1.3) iff

(i) $b(x, u) \in L_2(Q_T), \partial_t b(x, u) \in L_q(I, V^*)$ $(p^{-1} + q^{-1} = 1)$,
$\beta(u) \in L_p(I, V)$;

(ii) $\int_I < \partial_t b(x, u), v >= - \int_{Q_T}(b(x, u) - b(x, u_0)).\partial_t v$, $\quad \forall v \in V \cap L_\infty(Q_T)$
with $\partial_t v \in L_\infty(Q_T)$, $v(x, T) = 0$;

(iii) $\int_I < \partial_t b(x, u), v > + \int_I(a(t, u, \nabla \beta(x, u)), \nabla v) + \int_I (g(t, \beta(x, u), v)_{\Gamma_2} = \int_I(f(t, u), v)$, $\quad \forall v \in L_p(I, V)$.

Also the solution $\Theta_i \in V$ of (1.4) is understood in the variational sense

(2.5)
$$(\lambda_i(\Theta_i - \beta_n(x, u_{i-1})), v) + \tau(a(t_i, u_{i-1}, \nabla \Theta_i), \nabla v) + \tau(g_i, v)_{\Gamma_2} = \tau(f_i + \tilde{f}_i, v),$$

for all $v \in V$.

The existence of a variational solution $\Theta_i \in V \cap L_2$ of (2.5) is guaranteed by the theory of monotone operators (-see, e.g.[17]) provided $\lambda_i \in L_\infty$, $u_{i-1} \in L_2$, $\Theta_{i-1} \in V \cap L_2$. The assumptions $H_1) - H_6)$ and (1.5)-(1.7) then guarantee the existence of $\{\Theta_i\}_{i=1}^n \in V \cap L_2$.

Now we shall be concerned with the determination of $\lambda_i, \mu_i, \Theta_i$ satisfying (1.4)–(1.6). We propose the following iteration scheme

$$(2.6) \qquad \begin{aligned} (\lambda_{i,k-1}(\Theta_{i,k} - \beta_n(u_{i-1})), v) + \tau(a(t_i, u_{i-1}, \nabla\Theta_{i,k}), \nabla v) + \\ \tau(g_i, v)_{\Gamma_2} = \tau(f_i, v) \qquad \forall v \in V \end{aligned}$$

where

$$(2.7) \quad \bar{\lambda}_{i,k} := \min\{\tau^{-d}, \frac{b(x, u_{i-1} + \mu_{i,k}(\Theta_{i,k} - \beta_n(u_{i-1}))) - b(x, u_{i-1})}{\Theta_{i,k} - \beta_n(u_{i-1})}\} + \frac{2}{3}\tau^d$$

$$\lambda_{i,k} := min\{\bar{\lambda}_{i,k}, \lambda_{i,k-1}\} \quad \text{for} \quad k = L, L+1, \dots \quad \text{and}$$

$$\lambda_{i,k} := \bar{\lambda}_{i,k} \quad \text{for} \quad k = 1, \dots, L$$

with

$$(2.8) \qquad \bar{\mu}_{i,k} := \frac{\beta_n^{-1}(x, \beta_n(x, u_{i-1}) + \alpha(\Theta_{i,k} - \beta_n(x, u_{i-1}))) - u_{i-1}}{\Theta_{i,k} - \beta_n(x, u_{i-1})}$$

$$\mu_{i,k} := min\{\bar{\mu}_{i,k}, \mu_{i,k-1}\} \quad \text{for} \quad k = L, L+1, \dots$$

$$\mu_{i,k} := \bar{\mu}_{i,k} \quad \text{for} \quad k = 1, \dots, L$$

and $\mu_{i,0} := \frac{\alpha}{\beta'_{n,s}(x, u_{i-1})}$, $\lambda_{i,0} := min\{\tau^{-d}, b'_{n,s}(x, u_{i-1}) \, \mu_{i,0}\}, \infty > L \geq 1$ being an integer.

2.9 Remark. Generally, the choice $\lambda_{i,k} \equiv \bar{\lambda}_{i,k}$, $\mu_{i,k} \equiv \bar{\mu}_{i,k}$ is not convergent for $k \to \infty$, especially in the neighbourhood of interfaces ($\beta'_s(x, z) = 0, b'_s(x, z) = 0$). By means of the construction (2.7), (2.8) the sequences $\{\mu_{i,k}\}, \{\lambda_{i,k}\}$ are forced to be monotone and hence convergent. The parameter L can be choosen separately (with respect to the structural properties of b, β) in practical implementations. Another strategy of definition of $\lambda_{i,k}, \mu_{i,k}$ is possible, e.g.,

$$\lambda_{i,k} := \sum_{j=1}^{k-1} \alpha_j^{(k)} \lambda_{i,j} + \alpha_k^{(k)} \bar{\lambda}_{i,k}, \quad \mu_{i,k} := \sum_{j=1}^{k-1} \alpha_i^{(k)} \mu_{i,j} + \alpha_k^{(k)} \bar{\mu}_{i,k}$$

$$\sum_{1}^{k} \alpha_j^{(k)} = 1, \quad \alpha_j^{(k)} \geq 0$$

but, generally, we are not able to guarantee its convergence. The construction (2.7), (2.8) (for small L) leads to fast convergence of $\lambda_{i,k} \to \lambda_i, \mu_{i,k} \to \mu_i$ (λ_i, μ_i satisfy (1.4)-(1.6)). However, as a penalty we eventually loose the equalities on R.H.S. in (1.5), (1.6).

We formulate the convergence result concerning the iterations (2.6)-(2.8) for $k \to \infty$ in the form:

2.10 Theorem. *Let the assumptions* $H_1) - H_6), u_0 \in L_\infty(\Omega)$ *and* $p > \frac{2N}{N+2}$ *be satisfied. Let* $\{\lambda_{i,k}\}, \{\mu_{i,k}\}, \{\Theta_{i,k}\}$ *be from (2.6)-(2.8). Then* $\lambda_{i,k} \to \lambda_i$ *and* $\mu_{i,k} \to \mu_i$ *in* $L_r(\Omega)$, $\forall r > 1$ *and* $\Theta_{i,k} \to \Theta_i$ *in* $L_2(\Omega)$ *for* $k \to \infty$ *where* λ_i, μ_i, *and* Θ_i *satisfy (1.5), (1.6) and (1.4) with* $\tilde{f}_i \equiv 0$. *If* $|\Theta_{i,k}|_{q'} \leq c_i$ *with* $q' > max\{2, q\}$ *then there exist* $\tilde{f}_i$, $k_0 \equiv k_0(i, \tau) < \infty$ *such that* $\lambda_i := \lambda_{i,k_0}, \Theta_{i,k_0}$ *satisfy (1.4)-(1.6) where* $\tilde{f}_i$ *satisfies (1.8).*

The proof is the same as that one in [13].

2.12 Consequence. *If the assumptions of Theorem 2.10 are satisfied and if* $\Theta_{i,k} \in C^{0,\delta}(\bar{\Omega})$ *with* $\|\Theta_{i,k}\|_{0,\delta} \leq C(i, n)$ *uniformly in* k, *then* $\Theta_{i,k} \to \Theta_i$, $\lambda_{i,k} \to \lambda_i, \mu_{i,k} \to \mu_i$ *in* $C(\bar{\Omega})$ *where* $\Theta_i, \lambda_i, \mu_i$ *satisfy (1.4)-(1.6).*

3. Convergence of the method

In this section we prove the convergence of the approximation scheme (1.4)-(1.7) in Case I. We use some arguments from [13], [14] and thus some steps in the proof will be sketched only. By means of Θ_i, u_i from (2.5), (1.5)-(1.7) we construct approximate solutions (Rothe's functions)

$$\Theta^n(t) := \Theta_{i-1} + (t - t_{i-1})\tau^{-1}(\Theta_i - \Theta_{i-1}), \quad t \in [t_{i-1}, t_i], i = 1, ..., n$$
(3.1)
$$\bar{\Theta}^n := \Theta_i \quad \text{for} \quad t \in]t_{i-1}, t_i], i = 1, ..., n$$
$$\bar{\Theta}^n(0) \equiv \Theta_0 \equiv \beta(u_0)$$

Analogously we define u^n and $\bar{u}^n$ by means of $u_i(i = 1, ..., n)$. By $\{\bar{n}\}$ we denote a subsequence of $\{n\}$. Our main result in this section is

3.2 Theorem. *Let the assumptions* $H_1) - H_6)$ *and (1.8) be fulfilled. Then there exists a variational solution u of (1.1)-(1.3). Moreover,* $b_{\bar{n}} (x, \bar{u}^{\bar{n}}) \to b(x, u)$, $b(x, \bar{u}^{\bar{n}}) \to b(x, u)$ *in* $L_s(Q_T), s < 2$, $\bar{\Theta}^{\bar{n}} \to \Theta \equiv \beta(x, u)$ *in* $L_{p_0}(Q_T), p_0 < min\{p, 2\}$, *where* $\{\bar{\Theta}^n\}, \{\bar{u}^n\}$ *are from (1.4)-(1.7), (3.1). If the variational solution u is unique, then the original sequences* $\{b_n(x, \bar{u}^n)\}, \{\bar{\Theta}^n\}$ *are convergent.*

In the proof we shall use the "integration by parts formula" — see [2], [14] (Lemma 3.25)

(3.3)
$$\int_0^t < \partial_t b(x,u), \beta(x,u) >= \int_\Omega B^*(x, \beta(x, u(t))) - \int_\Omega B^*(x, \beta(x, u_0)), \quad \text{a.e.} t \in I$$

and some a priori estimates contained in the following lemmas.

3.4 Lemma. *The following a priori estimates hold:*

(3.5)
$$\max_{1 \le i \le n} \int_\Omega B^*(x, \beta(x, u_i)) \le C, \quad \sum_{i=1}^n \|\Theta_i\|_\tau^p \le C$$

(3.6)
$$\sum_{i=1}^n \int_\Omega \frac{1}{\lambda_i} (b_n(x, u_i) - b_n(x, u_{i-1}))^2 \le C$$

(3.7)
$$\sum_{i=1}^n |u_i - u_{i-1}|_2^2 \le C\tau^{-2d}$$

uniformly for n.

Proof. We insert (1.7) into (2.5) and put $v = \Theta_i$. We sum it up for $i = 1, ..., j$ and denote the corresponding equality $J_1 + J_2 + J_3 = J_4$.
  We rearrange the first term in the form

$$J_1 = \sum_{i=1}^j (b_n(x, u_i) - b_n(x, u_{i-1}), \Theta_i - \beta_n(x, u_{i-1}))$$

(3.8)
$$+ \sum_{i=1}^j (b_n(x, u_i) - b_n(x, u_{i-1}), \beta_n(x, u_i))$$

$$- \sum_{i=1}^j (b_n(x, u_i) - b_n(x, u_{i-1}), \beta_n(x, u_i) - \beta_n(x, u_{i-1}))$$

$$\equiv J_1^1 + J_1^2 - J_1^3.$$

We can check easily that the signs of $u_i - u_{i-1}$, $\Theta_i - \beta_n(x, u_{i-1})$, $b_n(x, u_i) - b_n(x, u_{i-1})$, and $\beta_n(x, u_i) - \beta_n(x, u_{i-1})$ are the same. Then, from (1.5)-(1.7) successively we obtain

224

$$b_n(x, u_i) - b_n(x, u_{i-1}) = \lambda_i(\Theta_i - \beta_n(x, u_{i-1})) \leq$$
$$b_n(x, u_{i-1} + \mu_i(\Theta_i - \beta_n(x, u_{i-1}))) - b_n(x, u_{i-1}) \leq$$
$$b_n(x, \beta_n^{-1}(x, \beta_n(x, u_{i-1}))) + \alpha(\Theta_i - \beta_n(x, u_{i-1}))) - b_n(x, \beta_n^{-1}(x, \beta_n(x, u_{i-1})))$$

and hence

$$|\beta_n(x, u_i) - \beta_n(x, u_{i-1})| \leq \alpha|\Theta_i - \beta_n(x, u_{i-1})|$$

Inserting it in J_1^3 we have

(3.9) $$0 \leq J_1^3 \leq \alpha J_1^1, \qquad \alpha \in (0, 1).$$

In the term J_1^2 we use Abel's summation - (see [13],[14])

(3.10)
$$J_1^2 \geq (b_n(x, u_j), \beta_n(x, u_j)) - (b_n(x, u_0), \beta_n(x, u_0))$$
$$- \sum_{i=1}^{j} \int_\Omega \left(\int_{\beta_n(x, u_{j-1})}^{\beta_n(x, u_j)} b_n(x, \beta_n^{-1}(z)) dz \right) =$$
$$\int_\Omega B_n^*(x, \beta_n(x, u_j)) - \int_\Omega B_n^*(x, \beta_n(x, u_0))$$

(H_6) and $H_2)$ guarantee that $u_0 \beta_n(x, u_0), B^*(x, \beta_n(x, u_0)) \in L_1(\Omega))$ where the monotonicity of b_n, β_n have been used. The elliptic term J_2 gives us $(mes_{N-1}\Gamma_1 > 0)$

(3.11) $$J_2 \geq C \sum_{i=1}^{j} |\nabla \Theta_i|_p^p \tau - C \geq C \sum_{i=1}^{j} \|\Theta_i\|_\tau^p - C.$$

We estimate the boundary term J_3 (-see H_5)) by

(3.12) $$|J_3| \leq \epsilon \sum_{i=1}^{j} \int_{\Gamma_2} \tau |\Theta_i|_p^p + C_\epsilon \leq \epsilon C \sum_{i=1}^{j} \|\Theta_i\|^p \tau + C_\epsilon$$

where Young's inequality and the imbedding $V \subset L_p(\partial\Omega)$ have been used. Finally, we estimate J_4 by

(3.13) $$|J_4| \leq C_\epsilon \sum_{i=1}^{j} \tau \int_\Omega B_n^*(x, \beta_n(x, u_i)) + \epsilon \sum_{i=1}^{j} \tau \|\Theta_i\|^p + C_\epsilon$$

because of $H_4)$, (2.2) and (1.8). Then (3.5)-(3.10) and Gronwall's argument imply the required a priori estimates. The last apriori estimate is a consequence of $|b_n(x, u_i) - b_n(x, u_{i-1})| \geq C\tau^d |u_i - u_{i-1}|$. Thus the proof is complete.

3.14 Consequence. *From (3.6) and (1.7) we obtain the estimate*

$$\sum_{i=1}^{n} |\Theta_i - \beta_n(x, u_{i-1})|^2 \le C\tau^{-d}, \quad \sum_{i=1}^{n} |u_i - u_{i-1}|_2^2 \le C\tau^{-d}$$

Denote by $\bar{u}_\tau^n(t) := \bar{u}^n(t - \tau)$ for $t \in I, \bar{u}^n \equiv u_0$ for $t \in (-\tau, 0)$ and

(3.15)
$$\hat{b}_n(x, \bar{u}^n) := b_n(x, u_{i-1}) + \frac{t - t_{i-1}}{\tau}(b_n(x, u_i) - b_n(x, u_{i-1}))$$
$$\text{for} \quad t \in [t_{i-1}, t_i], \quad i = 1, ..., n$$

3.16 Lemma. *The estimates*

$$\|\partial_t \hat{b}_n(x, \bar{u}^n)\|_{L_q(I, V^*)} \le C, \quad (q^{-1} + p^{-1} = 1)$$

$$\int_0^{T-z-\tau} \left(b_n(x, \bar{u}^n(t + z)) - b_n(x, \bar{u}^n(t)), \beta_n(x, \bar{u}^n(t + z)) - \beta_n(x, \bar{u}^n(t)) \right) \le$$
$$C(z + \tau^{(1-d)/2})$$

hold uniformly $\forall n, \forall 0 < z \le z_0$.

The proof goes follows the lines of [13] (Lemma 3.16).

3.17 Remark. *The estimate (3.18) holds if we replace $\bar{u}^n$ by $\bar{u}_\tau^n$. Indeed, we sum up (2.5) for $i = j, ..., j + k - 1$ and then we put $\varphi = \tau(\Theta_{j+k} - \Theta_j)$.*

Following the compactness argument in [13] we obtain

3.18 Lemma. *There exists $u : Q_T \to R$ with $\beta(x, u) \in L_p(I, V)$, $b(x, u) \in L_\infty(I, L_2(\Omega)), \partial_t b(x, u) \in L_q(I, V^*), B^*(x, \beta(x, u)) \in L_\infty(I, L_1(\Omega))$ and a subsequence $\{\bar{n}\}$ of $\{n\}$ such that*

$$\bar{u}^{\bar{n}} \to u \quad \text{a.e. in} \quad Q_T;$$
$$b_{\bar{n}}(x, \bar{u}^{\bar{n}}) \to b(x, u) \quad \text{in} \quad L_s(Q_T), \quad \forall s < 2;$$
$$\bar{\Theta}^{\bar{n}} \to \Theta \equiv \beta(x, u) \quad \text{in} \quad L_{p_0}(Q_T), \quad p_0 < min\{p, 2\};$$
$$\bar{\Theta}^{\bar{n}} \rightharpoonup \beta(x, u) \quad \text{in} \quad L_p(I, V); \quad \partial_t \hat{b}_{\bar{n}}(x, \bar{u}^{\bar{n}}) \rightharpoonup \partial_t b(x, u) \quad \text{in} \quad L_q(I, V^*).$$

Proof. We define $\rho(x, s) := min\{b_s'(x, s), \beta_s'(x, s), 1\}$ and $W(x, s) := \int_0^s \rho(x, z)dz$. The function $W(x, s)$ is strictly monotone in s and satisfies

$$|W(x, s_1) - W(x, s_2)| \le C.min\{|b_n(x, s_1) - b_n(x, s_2)|, |\beta_n(x, s_1) - \beta_n(x, s_2)|\}.$$

We prove the compactness of $\{W(x, \bar{u}_\tau^n)\}$ in $L_1(Q_T)$. As a consequence of (3.18) we have (-see Remark 3.19)

$$\int_0^{T-z-\tau} \int_\Omega |W(x, \bar{u}_\tau^n(t+z)) - W(x, \bar{u}_\tau^n(t))|^2 \le$$

(3.19)
$$\int_0^{T-z-\tau} (b_n(x, \bar{u}_\tau^n(t+z)) - b_n(x, \bar{u}_\tau^n(t)), \beta_n(x, \bar{u}_\tau^n(t+z)) - \beta_n(x, \bar{u}_\tau^n(t)) \le$$

$$C(z + \tau^{(d-1)/2}), \qquad \forall n, \quad 0 < z \le z_0$$

Now we prove

(3.22)
$$\int_I \int_\Omega |W(x+y, \bar{u}_\tau^n(t, x+y)) - W(x, \bar{u}_\tau^{\bar{n}}(t, x))|dxdt \le \bar{\omega}(|y|) + c.\tau^{(1-d)/2}$$

where $\bar{\omega} : R_+ \to R_+$ is continuous and $\bar{\omega}(0) = 0$. First, $\int_I \|\bar{\Theta}^n\|^\gamma \le C$ implies (see,e.g. [21])

$$\int_I \int_\Omega |\bar{\Theta}^n(t, x+y) - \bar{\Theta}^n(t, x)|dxdt \le \omega_1(|y|)$$

where ω_1 has the same properties as $\bar{\omega}$. Then Consequence 3.14 implies

$$\int_I \int_\Omega |\beta_n(x, \bar{u}_\tau^n(t, x+y)) - \beta_n(x, \bar{u}_\tau^n(t, x))|dxdt \le C (\omega_1(|y|) + \tau^{(1-d)/2})$$

and consequently

(3.21)
$$\int_I \int_\Omega |W(x, \bar{u}_\tau^n(t, x+y)) - W(x, \bar{u}_\tau^n(t, x))| \le C(\omega(|y|) + \tau^{(1-d)/2})$$

The continuity properties of $b_s'(x, s), \beta_s'(x, s)$ in x (-see $H_1), H_2)$) then imply

$$\int_I \int_\Omega |W(x+y, \bar{u}_\tau^n) - W(x, \bar{u}_\tau^n)|dxdt \le C\omega(|y|)$$

for both $\bar{u}_\tau^n = \bar{u}_\tau^n(t, x+y)$ and $\bar{u}_\tau^n = \bar{u}_\tau^n(t, x)$. Hence and from (3.21) we obtain (3.20). From (3.19) and from (3.20) the compactness of $\{W(x, \bar{u}_\tau^n)\}$ in $L_1(Q_T)$ follows. Strict monotonicity of $W(x, s)$ in s then implies

$$\bar{u}_\tau^{\bar{n}} \to u \quad \text{a.e. in} \quad Q_T.$$

The second a priori estimate in Consequence 3.14 can be rewritten in the form

(3.22)
$$\int_I \int_\Omega |\bar{u}^n - \bar{u}_\tau^n|^2 \le C\tau^{1-d} \to 0 \qquad \text{for} \quad n \to \infty.$$

Then we can assume $\bar{u}^{\bar{n}} \to u$ a.e. in Q_T ($\{\bar{n}\}$ is a suitable subsequence of $\{n\}$). Thus $\beta(x, \bar{u}^{\bar{n}}) \to \beta(x, u), \bar{\Theta}^{\bar{n}} \to \Theta \equiv \beta(x, u)$ in $L_{p_0}(Q_T)$ because of Consequence 3.14 and Lemma 3.4. Since $b_n(x, s) \to b(x, s)$ for $n \to \infty$ locally uniformly in s and $\bar{u}^{\bar{n}} \to u$ a.e. in Q_T we have $b_{\bar{n}}(x, \bar{u}^{\bar{n}}) \to b(x, u)$ in $L_s(Q_T)$ $\forall s < 2$ (see (2.2), (3.5)). Due to the estimate (3.17) we have $\partial_t \hat{b}_n(x, \bar{u}^{\bar{n}}) \rightharpoonup \chi$ in $L_q(I, V^*)$. From (3.5) it follows

$$\int_I |\hat{b}_n(x, \bar{u}^n) - b_n(x, \bar{u}^n)|_2^2 \le 2\tau \sum_{i=1}^n |b_n(x, u_i) - b_n(x, u_{i-1})|_2^2 \le C\tau^{1-d}$$

and hence $\chi \equiv \partial_t b(x, u)$. To prove $B^*(x, \beta(x, u)) \in L_\infty(I, L_1(\Omega))$ we use the estimate

(3.23)
$$B_n^*(x, \beta_n(x, s)) = B^*(x, \beta(x, s) + \tau^d \int_0^s \beta_n(x, z)dz + b(x, s)[\beta_n(x, s) - \beta(x, s)] -$$

$$- \int_0^s b(x, y)[\beta_{n,s}'(x, y) - \beta_s'(x, y)]dy$$

which guarantees locally uniform (with respect to s) convergence $B^*(x, \beta_n(x, s)) \to B^*(x, \beta(x, s))$. Then, pointwise convergence of $\beta(x, \bar{u}^{\bar{n}}) \to \beta(x, u)$, the estimate (3.5) and Fatou's argument imply $B^*(x, \beta(x, u)) \in L_\infty(I, L_1(\Omega))$. The rest of the proof is a consequence of Lemma 3.4.

3.24 Remark. *Taking into account that $\beta(x, u) \in L_p(I, V), \partial_t b(x, u) \in L_q(I, V^*)$ (u is from Lemma 3.18) we can prove an "integration by parts formula" (3.3) along the same lines as in [14] (-see Lemma 3.25) where $p \ge 2$ has been considered. The formula (3.26) in [14] holds also in the case $1 < p$. The only difference in the argumentation is that we have to multiply (3.26) in [14] by $\lambda_\epsilon(t) := min\{1, \frac{1}{\epsilon|\beta(x,u)|}\}$ and proceed as in [2] (Lemma 1.5) taking $\epsilon \to 0$.*

The formula (3.3) is substantially used in the proof of Theorem 3.2. The proof of Theorem 3.2 is based on Lemma 3.18 and goes in the same lines as that one in [13]. Here we have to use $\underline{\lim} \int_\Omega B_n^*(x, \beta_n(x, \bar{u}^n)) \ge \int_\Omega B^*(x, \beta(x, u))$ which is a consequence of (3.23) and Fatou's argument.

We obtain under the strong monotonicity assumption H_7) the stronger convergence results.

3.25 Theorem. *Let the assumptions of Theorem 3.2 and H_7) be satisfied. Then*

$$\bar{\Theta}^n \to \beta(x, u) \quad \text{in} \quad L_p(I, V).$$

For the proof see [13] (Theorem 3.31).

The correspondig convergence results in cases II and III can be obtained in the same way as in [13]. The convergence results correspondig to the scheme (1.4), (1.6), (1.9) can be obtained in the same way as in [14].

3.26 Remark. *The obtained convergence results in Theorem 3.2 and 3.25 can be extended to the nonhomogeneous Dirichlet boundary condition on Γ_1 and the continuity of a, f, g in their variables can be replaced by Caratheodory conditions. Moreover, the convergence results hold for systems of the form*

$$\partial_t b^j(x, u^j) - \nabla a^j(t, x, u, \nabla\beta(x, u)) = f^j(t, x, u),$$

(1.3') $$u^j = 0 \quad \text{on} \quad \Gamma_1^j \times I,$$

$$a^j(t, x, u, \nabla\beta(x, u)).\nu = g(t, x, \beta(x, u)) \quad \text{on} \quad \Gamma_2^j \times I, \quad \text{for} \quad j = 1, ..., m$$

where $u \equiv (u^1, \ldots, u^m)$, $\beta(x, u) \equiv (\beta^1(x, u^1), \ldots, \beta^m(x, u^m))$ and b^j, β^j satisfy the hypotheses $H_1), H_2)$. The assumptions $H_3) - H_6)$ have to be rewritten also for the system. The relaxation functions λ_i, μ_i are also vector functions and each component λ_i^j, μ_i^j satisfies (1.5), (1.6) with $b^j, \beta^j, (\beta^{-1})^j$ in the place of b, β, β^{-1}.

REFERENCES

[1] N.Ahmed, D.K.Sunada: *Nonlinear flow in porous media.* J.Hydraulics Div. Proc. Amer. Soc. Civil Engeg 95, (1969), 1847-1857.

[2] H.W.Alt, S.Luckhaus: *Quasilinear elliptic-parabolic differential equations.* Math. Z. 183, (1983), 311-341.

[3] H.W.Alt, S.Luckhaus, A.Visintin: *On nonstationary flow through porous media.* Annali di Matematica, 19, ..., 303-316.

[4] Y.Amirat: *Ecoulements en milieu poreux n'obeissant pas a la loi de Darcy.* M^{2}AN vol 25, u-3, 273-306, (1991).

[5] J.Bear: *Dynamics of fluids in porous media.* Elsevier, New York, 1972.

[6] D.Blanchard, G.Francfort: *Study of a double nonlinear heat equation with no growth assumptions on the parabolic term.* SIAM, J.Math.Anal. 19, (1988), 1032-1056.

[7] L.M.Chounet, D.Hilhorst, C.Jouron, Y.Kelanemer, P.Nicolas: *Couplet heat and mass transfer in porous media: Numerical simulation for soil-building transfers* Preprint to appear in "Proceedings" 10th International Heat transfer Conference (IHTC) BRIGTON (U.K.), 14-18, (1994).

[8] J.J.Diaz: *Nonlinear pde's and free boundaries Vol 1.* Elliptic Equations, Research Notes in Math. n-106. Pitman, London 1985 *Vol.2. Parabolic and Hyperbolic Equations.* (to appear).

[9] J.J.Diaz: *On a nonlinear parabolic problem arising in some models related to turbulent flows.* (to appear in SIAM, J.Math.Anal.).

[10] B.H.Gilding, M.Vallentgoed: *A nonlinear degenerating parabolic problem from the theory of type II superconductors,* Memorandum N1112, January 1993, Faculty of Applied Mathematics, University of Twente.

[11] U.Hornung: *A parabolic-elliptic variational inequality.* Manuscripta math. 39, 155-172 (1982). Acta Math. Univ. Comenianae, Vol. LXI, 1 (1992), 27-39.

[12] W.Jäger, J.Kačur: *Solution of porous medium systems by linear approximation scheme.* Num.Math. 60, 407-427 (1991).

[13] J. Kačur: *Solution of some free boundary problems by relaxation schmes,* Preprint M3-94, Comenius Univ. Faculty of Math. and Physics (1994).

[14] W.Jäger, J.Kačur: *Solution of doubly nonlinear and degenerate parabolic problems by relaxation schemes* Preprint M2-94, Comenius Univ. Faculty of Math. and Physics, (1994)

[15] J.Kačur: *On a solution of degenerate elliptic-parabolic systems in Orlicz-Sobolev spaces I,II.* I. Math.Z. 203, (1990), 153-171; II. Math.Z. 203, (1990), 569-579.

[16] J.Kačur, A.Handlovičova, M.Kačurova: *Solution of nonlinear diffusion problems by linear approximation schemes.* SIAM Num.Anal. 30, 1993, 1703-1722.

[17] J.Kačur, S.Luckhaus: *Approximation of degenerate parabolic systems by nondegenerate elliptic and parabolic systems.* Preprint M2-91, Faculty of Mathematics and Physics, Comenius University, 1991, 1-33.

[18] A.Kufner, S.Fučik: *Nonlinear differential equations.* SNTL, Praha, 1978, Elsevier, 1980.

[19] A.Kufner, O.John, S.Fučik: *Funcion spaces.* Academia CSAV, Prague, 1967.

[20] E.Magenes, R.H.Nochetto, C.Verdi: *Energy error estimates for a linear scheme to approximate nonlinear parabolic problems.* Math.Mod.Num.Anal.21, 1987, 655-678.

[21] J.Nečas: *Les methodes directes en theorie des equationes elliptiques.* Academie, Prague, 1967.

[22] R.H.Nochetto, M.Paolini, C.Verdi: *Selfadaptive mesh modification for parabolic FBPs: Theory and computation in Free Boundary Problems.* K.-H.Hoffman and J.Sprekels, eds., ISNM 95 BIckhäuser, Basel, 1990, 181-206

[23] R.H.Nochetto, C.Verdi: *Approximation of degenerate parabolic problems using numerical integration.* SIAM J.Numer.Anal. 25 (1988), 784-814.

[24] A.A.Samarskij, J.S.Nikolajev: *Methods of solution of difference equations.(in Russian)* Nauka, Moskow, 1978.

[25] M.Slodička: *Solution of nonlinear parabolic problems by linearization.* Preprint M3-92, Comenius Univ., Faculty of Math. and Physics, 1992.

[26] M.Slodička: *On a numerical approach to nonlinear degenerate parabolic problems.* Preprint M6-92, Comenius Univ. Faculty of Math. and Physics, 1992.

Department of Numerical Analysis, Faculty of Mathematics and Physics, Comenius University, Mlynska dolina, 842 15 Bratislava, Slovakia

M K KORTEN

Uniqueness for the Cauchy problem with measures as data for $u_t + \Delta(u-1)_+$ under optimal conditions of regularity and growth

1. Introduction

In this contribution we want to report a uniqueness result for the equation

$$(1.1) \qquad u_t = \Delta(u-1)_+,$$

where $0 \le u \in L^1_{loc}(\mathbb{R}^n \times (0,T))$ is a solution in the distribution sense, and the initial data are taken in the sense of measures (only!).

In [AK] a Harnack type inequality for non-negative, locally integrable distributional solutions to (1.1) was found, giving as a consequence the existence of an initial trace for such solutions, which is a non negative, locally finite measure satisfying the growth condition

$$(1.2) \qquad \int_{\mathbb{R}^n} e^{-c|x|^2} d\mu(x) \le M$$

where $c = c(T)$ and $M = M(u,n,T)$. It was also shown that any non negative measure satisfying (1.2) gives rise to a solution to (1.1). Uniqueness was shown for locally integrable initial data, attained in L^1 norm of each compact subset of $\mathbb{R}^n$. Therefore, we seek uniqueness within the growth class given by (1.2). To this aim we first develop some regularity theory for solutions to (1.1) (this is the content of sections 2 and 3), mainly a representation formula by means of the traces of u on the base and lateral boundary of any cylinder $B_r \times (t_0, t_1), r > 0$ and $0 < t_0 < t_1 < T$. Then we find $u \in L^2_{loc}(\mathbb{R}^n \times (0,T))$.

In section 4 we state a local comparison result, as a consequence of which $(u-1)_+$ is a subsolution (in the sense of $\mathcal{D}'$) of the heat equation. We use this result to obtain the local boundedness of u in $\mathbb{R}^n \times (0,T)$. The results in sections 2 to 4 hold for local solutions, defined in a cylinder.

Section 5 is concerned with the uniqueness theorem for the Cauchy problem. The technique of sequences of potentials we use goes back to Pierre [P] and Dahlberg and Kenig [DK2]. However, two additional difficulties arise: we need to construct the solution to the "adjoint" problem as a difference of solutions to two problems, increasing in the time variable. Moreover, due to the fact that $u \in L^2_{loc}(\mathbb{R}^n \times (0,T))$ only, we have to construct this solution "stripwise".

We want to point out that our uniqueness result holds under optimal conditions of regularity (see sections 2 and 3) and growth at infinity (see [AK]).

A few last words about motivations. In one space dimension, and for a given function $u_I \in C_0^\infty$ (say such that $u_I(x) > 1$ in some interval) equation (1.1) can be thought of as describing the energy per unit volume in a Stefan-type problem where the latent heat of the phase change is given by $1 - u_I(x)$. Note that discontinuous solutions should be expected for (1.1) (see [BKM]).

Complete proofs of the stated results can be found in [K].

2. Local representation.

We will in this section assume that $u \geq 0$ and

$$u \in L^1_{loc}(\Omega)$$

where Ω is a domain in $\mathbb{R}^n \times (0, \infty)$. We will also assume that u solves the equation

$$u_t = \Delta(u - 1)_+$$

in the distribution sense.

The following lemma establishes the existence of a trace at each time level.

Lemma 2.1. Let $I = (a, b), 0 < a < b < T$, and let $D \subset \mathbb{R}^n$ be a bounded domain with smooth boundary. Set $K = \overline{D}$ and assume $K \times [a, b]$ is compact and contained in Ω. Then for each $\tau \in I$ there is a unique bounded and nonnegative measure ν_τ on D such that

$$\sup_{\tau \in I} \int_D d\nu_\tau < \infty$$

and

(2.1) $$\int_{\omega \cap \{t > \tau\}} [(u - 1)_+ \Delta \eta + u \frac{\partial \eta}{\partial t}] dx dt + \int_D \eta(x, \tau) d\nu_\tau = 0$$

for each $\eta \in C_0^\infty(\omega) = \mathcal{A}$. Here $\omega = D \times (a, b)$.

We will study next the existence of a trace on the lateral sides of a cylinder in Ω.

Lemma 2.2. Let $I = (a, b), 0 < a < b < T$, and let $D \subset \mathbb{R}^n$ be a bounded domain with smooth boundary. Set $\omega = D \times (a, b)$, $K = \overline{D}$ and assume that $K \times [a, b]$ is compact and contained in Ω. Then there is a unique bounded and non-negative measure μ on $S = \partial D \times (a, b)$ such that

$$\int_\omega [(u - 1)_+ \Delta \eta + u \frac{\partial \eta}{\partial t}] dx dt + \int_S \frac{\partial \eta}{\partial n} d\mu = 0$$

for each $\eta \in \mathcal{B}$. Here $\frac{\partial}{\partial n}$ denotes differentiation with respect to the inward unit normal to ∂D and $\mathcal{B}$ denotes the class of all $\eta \in C_0^\infty (\Omega \cap \{(x, t) : a < t < b\})$ with the property that $\eta(x, t) = 0$ whenever $x \in \partial D$.

232

We want to study next the regularity of the trace ν_τ.

Lemma 2.3. *Let $I = (a,b), 0 < a < b < T$, and let $D \subset I\!\!R^n$ be a bounded domain with smooth boundary. Set $\omega = D \times (a,b)$ and assume that $\overline{\omega}$ is contained in Ω. For $a < \tau < b$ let ν_τ be the trace defined by (2.1). Suppose ψ is a continuous function with support in ω. Then*

$$\int_D \psi(x,t)d\nu_t$$

is a continuous function of t and

$$\int_a^b \Big(\int_D \psi(x,\tau)d\nu_\tau\Big)d\tau = \int u\psi dx d\tau.$$

We can now state our most general trace result.

Lemma 2.4. (Local representation) *Let $I = (a,b), 0 < a < b < T$, and let $D \subset I\!\!R^n$ be a bounded domain with smooth boundary. Set $K = \overline{D}$ and assume that $K \times [a,b]$ is compact and contained in Ω. For $a < \tau < b$ let ν_τ be the trace defined by (2.1) and set $\omega = D \times (a,b)$. Suppose $\psi \in C_0^\infty(\Omega)$ equals zero on $\partial D \times [a,b]$. Then*

$$\int_D \psi(x,b)d\nu_b = \int_D \psi(x,a)d\nu_a + \int_\omega [(u-1)_+\Delta\psi + u\frac{\partial\psi}{\partial t}]dx dt +$$

(2.2)
$$+ \int_{\partial D \times (a,b)} \frac{\partial\psi}{\partial n}d\mu,$$

where μ is defined in lemma 2.2.

3. Potential Theoretical Methods

In this section we will show the steps we follow to prove that $u \in L^2_{loc}(I\!\!R^n \times (0,\infty))$.

We will continue to assume $u \geq 0$ and

$$u \in L^1_{loc}(\Omega),$$

where Ω is a domain in $I\!\!R^{n+1}$. Let $I = (a,b), 0 < a < b < \infty$, and let $D \subset I\!\!R^n$ be a bounded domain with smooth boundary. Set $\omega = D \times (a,b)$ and assume $\overline{\omega}$ is contained in Ω.

We will also assume that u solves the equation

$$u_t = \Delta(u-1)_+$$

in the distribution sense. Let now G be the Green function of D and define w in ω by $w(x,t) = \int_D G(x,y)d\nu_t$ where ν_t is the trace define by (2.1). We will establish that w

is locally bounded. This will be achieved by first noticing that w satisfies a parabolic equation (3.1) and then using a variant of the Moser iteration technique introduced in [DK2].

Lemma 3.1. *Let $F \geq 0$ be smooth in ω and K compact with $K \subset D$. Then there is a $k > 1$ such that if $H \geq 0$ and $\Delta F \geq -H$ then*

$$\int_{K \times [a,b]} F^k dx dt \leq C \int_\omega H dx dt \quad \sup_{t \in [a,b]} \left(\int_D F(x,t) dx \right)^{k-1} +$$

$$+ \mathrm{dist}(K, \partial D)^{-n} \quad \sup_{t \in [a,b]} \left(\int_D F(x,t) dx \right)^k$$

where C only depends on ω.

For $\eta \in C_0^\infty(\omega)$ we have now by lemma 2.2 and lemma 2.3

$$\int_\omega \frac{\partial \eta}{\partial t} w dx dt = \int_\omega \frac{\partial G \eta}{\partial t} u dx dt =$$

$$= \int_\omega \eta(u-1)_+ dx dt - \int_{\partial D \times [a,b]} \frac{\partial G \eta}{\partial n} d\mu.$$

We can rewrite this as

$$(3.1) \qquad \int_\omega \frac{\partial w}{\partial t} \eta dx dt + \int_\omega \eta(u-1)_+ dx dt = \int_{\partial D \times [a,b]} \frac{\partial G \eta}{\partial n} d\mu.$$

Define now for $(x,t) \in \omega$ the function h by

$$h(x,t) = \int_{\partial D \times [a,t]} \frac{\partial G(\xi, x)}{\partial n_\xi} d\mu.$$

Then $\Delta h(x,t) = 0$ and $\int_D h(x,t) dx \leq C$ for all $t \in (a,b)$ and if we set $q = w - h - G1(x)$ then q satisfies the following relations in the distribution sense

$$(3.2.a) \qquad\qquad \frac{\partial q}{\partial t} + (u-1)_+ = 0$$

$$(3.2.b) \qquad\qquad \Delta q = -(u-1).$$

It follows from the Harnack inequality for harmonic functions and the uniform integrability of h that for every compact set $K \subset D$

$$\sup_{K \times [a,b]} h < \infty.$$

Fix now $x_0 \in D$ and set $B_p = \{x \in \mathbb{R}^n : |x - x - 0| < \rho\}$. Assume that $\overline{B_\rho} \subset D$, $a < T < \tau < b$ and set $S = B_\rho \times (\tau, b]$, $\quad R = B_\rho \times (T, b]$. Fix now M so large that

234

(3.3)
$$Q = q + M \geq 0 \quad \text{in} B_\rho \times [a, b].$$

Lemma 3.2. *Suppose P is a smooth and convex function with a bounded derivative p on $[0, \infty)$. Also suppose that $P(0) = p(0) = 0$. Then*

$$\sup_{s \in [\tau, b]} \int_{B_\rho} P(Q(x, s))dx + \int_S \max(-\Delta P(Q), 0)dxdt \leq$$

$$\leq C(\tau - T)^{-1} \int_R P(Q)dxdt,$$

where the constant C can be taken to depend only on the dimension n.

We need next lemma 4.3 of [DK1] which for our purpose reeds as follows:

Lemma 3.3. *(B.E.J. Dahlberg-C.E. Kenig) Let $\alpha \geq 0$. Then we have the following estimate*

$$\sup_{\tau < s < b} \int_{B_\rho} Q^{\alpha+1}(x, s)dx + \int_S \max(-\Delta Q^{\alpha+1}, 0)dxdt \leq$$

$$\leq C(\tau - T)^{-1} \int_R Q^{\alpha+1}dxdt,$$

where C can be taken to depend only on n.

We next prove the boundedness of w, by means of a Moser iteration applied to Q defined by (3.3).

Lemma 3.4. *Let $\Omega \subset \mathbb{R}^{n+1}$ be a domain and assume $u \geq 0$ belongs to $L^1_{loc}(\Omega)$. Suppose $D \subset \mathbb{R}^n$ is a domain with $\overline{D} \times [a, b] \subset \Omega$. Suppose furthermore that $a < T < b$ and $B_{3\rho} = \{x \in \mathbb{R}^n : |x - x_0| < 3\rho\} \subset D$. If $u_t = \Delta(u - 1)_+$ in Ω then $w = Gu$ is bounded in $R = B_\rho \times [T, b]$. Here G is the Green function of D.*

Lemma 3.5. *Let $I = (a, b), 0 < a < b < T$, and let $D \subset \mathbb{R}^n$ be a bounded domain with smooth boundary. Set $\omega = D \times I$ and assume that $\overline{\omega}$ is compact and contained in Ω. Then $w(x, t)$ and $\int_a^t (u - 1)_+(x, s)ds$ are bounded in ω . Furthermore, if $(x, t) \in \omega$ then*

$$w(x, t) = G\nu_a(x) - \int_a^t (u - 1)_+(x, s)ds + h(x, t).$$

Here h is bounded in ω and satisfies $\Delta h = 0$.

We want now to show that $u \in L^2_{loc}(\Omega)$. The proof relies on the obtained regularity results and on lemma 4.10 of [DK1].

Theorem 3.6. *Let $\Omega \subset \mathbb{R}^{n+1}$ be a domain. If $u \in L^1_{loc}(\Omega)$ is a weak solution of the equation*

$$\frac{\partial u}{\partial t} = \Delta(u-1)_+$$

in Ω, then $u \in L^2_{loc}(\Omega)$.

4. Local comparison.

Theorem 4.1. *Let $u, v \in L^1_{loc}(\mathbb{R}^n \times (0,T))$ be nonnegative solutions in the sense of distributions to equation (1.1), and let $D \subset \mathbb{R}^n$ be any bounded domain with sufficiently smooth boundary.*

Suppose that for $0 < t_0 < T$, $u(x,t_0) \geq v(x,t_0)$ holds for a.e. $x \in D$ and $\mu_u \geq \mu_v$ on $\partial D \times (t_0, T)$, where μ_u, μ_v are the traces of $(u-1)_+$ and $(v-1)_+$ respectively definded by lemma 2.2.

Then $u(x,t) \geq v(x,t)$ a.e. in $D \times (t_0, T)$.

Lemma 4.2. *Let $\bar{t} > 0$. If $u(x, \bar{t}) < 1$ for $x \in E$, where $E \subset \mathbb{R}^n$ is measurable and $|E| > 0$, then $u(x,t) \leq u(x, \bar{t})$ for a.e. $x \in E$ and $0 < t < \bar{t}$.*

The following corollary shows that if $0 \leq u \in L^1_{loc}(\mathbb{R}^n \times (0,T))$ is a solution in the sense of distributions to (1.1) then $(u-1)_+$ is a (weak) subsolution to the heat equation.

Corollary 4.3. *$(u-1)_+$ satisfies $\Delta(u-1)_+ - \frac{\partial}{\partial t}(u-1)_+ \geq 0$ in $\mathcal{D}'(\mathbb{R}^n \times (0,T))$.*

Corollary 4.4. *$u \in L^\infty_{loc}(\mathbb{R}^n \times (0,T))$.*

5. Uniqueness to the Cauchy problem with measures as initial data.

The method of sequences of potentials we will use appeared first in [P] (Newtonian potentials) and in [DK2] (Green and newtonian potentials). The main novelties in what follows are

i) the construction of the solution to the "adjoint" problem (see (5.1) and (5.2)) as difference of solutions to two problems, increasing in the time variable. This construction is imposed by our use of Green potentials, since by the growth condition at infinity (1.2) we cannot use Newton potentials, and

ii) The "stripwise" construction of this solutions in order to handle the fact that solutions can be expected to belong only to $L^2_{loc}(\mathbb{R}^n \times (0,T))$.

Theorem 5.1. (Uniqueness) *Let $0 \leq u, v \in L^1_{loc}(\mathbb{R}^n \, times(0,T))$ be solutions in the sense of $\mathcal{D}'(\mathbb{R}^n \times (0,T))$ to (1.1) and let*

$$\lim_{t \downarrow 0} \int_{\mathbb{R}^n} (u(x,t) - v(x,t))\varphi(x)dx = 0, \qquad \forall \varphi \in C_0^\infty(\mathbb{R}^n).$$

Then $u = v$ a.e. in $\mathbb{R}^n \times (0, T)$.

Outline of proof. Let u, v be two solutions to (1.1) in $\mathbb{R}^n \times (0, T), T > 0$. Let $\{\tau_i\} \subset (0, T), \tau_i \downarrow 0$ as $i \to \infty$ be such that the traces of u and v on $\mathbb{R}^n \times \{\tau_i\}$ are respectively given by $u(x, \tau_i)$ and $v(x, \tau_i)$ (see lemma 2.1.). Let $\varepsilon > 0$ and $R > R_1 > 1$ which will be fixed later, and let $c(x, t) = \frac{(u-1)_+ - (v-1)_+}{u-v}$ if $u \neq v$, $c(x, t) = 1$ if $u = v$. Let $c_i(x, t) = (c(x, t) \vee \frac{1}{2^i})$, $c_{i,m}(x, t) = (c(x, t) \vee \frac{1}{2^i}) * \rho_m \geq \frac{1}{2^i}$, and choose $m(i)$ such that

$$(5.1) \qquad \int_{B_R} \int_{\tau_{i+1}}^{\tau_i} |u - v|^2 |c_i(x, t) - c_{i,m(i)}(x, t)|^2 dx dt \leq \frac{\varepsilon}{2^{2i}}.$$

Take $\Theta(x) \in C_0^\infty(B_R)$ such that $\mathrm{supp}\Theta \subset B_{R_1}$, $0 \leq \Theta(x) \leq 1$, and let $(\Delta\Theta)_+ := \chi_{\{\Delta\Theta > 0\}} \Delta\Theta$ and $(\Delta\Theta)_- := -\chi_{\{\Delta\Theta \leq 0\}} \Delta\Theta$. Note that $(\Delta\Theta)_+$ and $(\Delta\Theta)_-$ are non-negative Lipschitz continuous functions. Define $c_\varepsilon(x, t) = c_{i,m(i)}(x, t)$ for $t \in (\tau_{i+1}, \tau_i)$, $i = 0, 1, 2, \ldots$, where we take $\tau_0 = t_1$, $0 < t_1 < T$, and $Gf = \int_{B_R} G(x, y) f(y) \, dy$ where G is the Green function of B_R. Let us consider the solutions $\psi^+(x, t)$ and $\psi^-(x, t)$, to the problems

$$(5.2) \qquad \begin{aligned} \psi_t + c_\varepsilon(x, t)\Delta\psi &= 0 && \text{in} && B_R \times (0, t_1) \\ \psi &= 0 && \text{in} && \partial B_R \times (0, t_1) \\ \psi^+(x, t_1) &= G(\Delta\Theta)_+(x) && \text{in} && B_R \times \{t_1\} \\ (\text{respectively}, \psi^-(x, t_1) &= G(\Delta\Theta)_-(x)) && \text{in} && B_R \times \{t_1\}) \end{aligned}$$

where $0 < t_1 < T$.

In each cylinder $B_R \times (\tau_{i+1}, \tau_i)$, ψ^+ and ψ_- are smooth functions. Therefore $\Delta\psi^+$ (respectively, $\Delta\psi^-$) is a weak solution to

$$(5.3) \qquad \begin{aligned} h_t + \Delta(c_\varepsilon h) &= 0 && \text{in} && B_R \times (0, t_1) \\ h &= 0 && \text{in} && \partial B_R \times (0, t_1) \\ h^+(x, t_1) &= -(\Delta\Theta)_+ \leq 0 && \text{in} && B_R \times \{t_1\} \\ (\text{respectively}, h^-(x, t_1) &= -(\Delta\Theta)_- \leq 0) && \text{in} && B_R \times \{t_1\}), \end{aligned}$$

the relations $-c_\varepsilon \Delta\psi^+ = 0 = \psi_t^+$ and $-c_\varepsilon \Delta\psi^- = 0 = \psi_t^-$ holding on $\partial B_R \times \{\tau_i\}$. Whence $0 \leq \psi^+, \psi^- \leq \|\Theta\|_{L^\infty(B_R)}$, and the bound
(5.4)

$$\|(\nabla\psi \cdot \vec{\nu})\|_{L^\infty(\partial B_R)} \leq \exp\left(\frac{-(R - R_1 - 1)^2}{8(t_1 - t)}\right) \int_{(R - R_1 - 1)/\sqrt{t_1 - t}}^\infty s^2 \exp(s^2/8) ds$$

holds for both $\psi = \psi^+$ and $\psi = \psi^-$ (see [B]). Application of theorem 4.1. to (5.3) yields $h^+, h^- \leq 0$. We obtain

$$(5.5.a) \qquad \int_{B_R} |h^+(x, t)| \, dx \leq \int_{B_R} (\Delta\Theta)_+ \, dx \leq \int_{B_R} |\Delta\Theta| \, dx, \qquad 0 < t < t_1,$$

and

$$(5.5.b) \quad \int_{B_R} |h^-(x,t)| \, dx \leq \int_{B_R} (\Delta\Theta)_- \, dx \leq \int_{B_R} |\Delta\Theta| \, dx, \qquad 0 < t < t_1,$$

ψ^+ and ψ^- are increasing functions of t, satisfying

$$0 \leq \psi^+(x,t) \leq G(\Delta\Theta)_+, \qquad 0 < t < t_1,$$

and

$$0 \leq \psi^-(x,t) \leq G(\Delta\Theta)_-, \qquad 0 < t < t_1.$$

Consider now the solutions ψ_ε^+, ψ_ε^- of (5.2) and h_ε^+ and h_ε^- of (5.3). By (5.5.a) and (5.5.b), there exist measures

$$\lambda_\varepsilon^+ = \lim_{t\downarrow 0}(-h_\varepsilon^+(x,t)),$$

$$\lambda_\varepsilon^- = \lim_{t\downarrow 0}(-h_\varepsilon^-(x,t)),$$

where the limits are taken in the sense of measures. Therefore,

$$G(-h_\varepsilon^+(x,t)) = \psi_\varepsilon^+ \to G(\lambda_\varepsilon^+(x)),$$

and

$$G(-h_\varepsilon^-(x,t)) = \psi_\varepsilon^- \to G(\lambda_\varepsilon^-(x)),$$

a.e. as $t \downarrow 0$. Let $w(x,t) = Gu(x,t)$. By the first equation of (3.3) we have that

$$w_t - g_t = -(u-1)_+ \qquad \text{in} D'(B_R),$$

and observing that $g(x,t) = \int_{\partial B_R \times [0,t]}(u-1)_+ \frac{\partial G(\xi,x)}{\partial n_\xi} ds$ is a non decreasing function of t we get $g_t \geq 0$. Therefore, if $\tau > t$,

$$\int_{B_R} u(x,t)\psi_\varepsilon^+(x,t)dx \geq \int_{B_R} u(x,\tau)\psi_\varepsilon^+(x,t)dx + \int_{B_R} g(x,\tau)h_\varepsilon^+(x,t)dx,$$

an analogous inequality holding for ψ_ε^-. Taking $\liminf_{t\downarrow 0}$ in the above formulas and then letting $\tau \downarrow 0$ we obtain

$$\liminf_{t\downarrow 0} \int_{B_R} u(x,t)\psi_\varepsilon^+(x,t)dx \geq \int_{B_R} G\mu(x)d\lambda_\varepsilon^+(x),$$

and

$$\liminf_{t\downarrow 0} \int_{B_R} u(x,t)\psi_\varepsilon^-(x,t)dx \geq \int_{B_R} G\mu(x)d\lambda_\varepsilon^-(x).$$

where μ, ν, are the initial traces of u, v, respesctively (see [AK]). For $\gamma > s > 0$, let $v_s(x,t) := v(x,t+s)$ and $b_s := u - v_s$. Let $W(x,t) = Gv(x,t)$ and observe that $0 \leq W_s \leq C_s$. Let $f(x,t) = \int_{\partial B_R \times [0,t]}(v-1)_+ \frac{\partial G(x,\xi)}{\partial n_\xi} dS$.

238

$$\limsup_{t\downarrow 0} \int_{B_R} v_s(x,0)\psi_\varepsilon^+(x,t)dx = \int_{B_R} W_s(x,0)d\lambda_\varepsilon^+(x) + \int_{B_R} f_s(x,0)d\lambda_\varepsilon^+(x).$$

Therefore

$$(5.6.a) \qquad \liminf_{t\downarrow 0} \int_{B_R} b_s(x,t)\psi_\varepsilon^+(x,t) \geq -\int f_s(x,0)\,d\lambda_\varepsilon^+(x).$$

and in turn

$$(5.6.b) \qquad \liminf_{t\downarrow 0} \int_{B_R} b_s(x,t)\psi_\varepsilon^-(x,t)dx \geq -\int f_s(x,0)\,d\lambda_\varepsilon^-(x).$$

From lemma 2.4, for $0 < t_0 < t_1 < T$, we have

$$\int_{B_R} b_s(x,t_1)\psi_\varepsilon^+(x,t_1)dx = \int_{B_R} b_s(x,t_0)\psi_\varepsilon^+(x,t_0)dx +$$

$$+ \int_{B_R}\int_{t_0}^{t_1}\{[(u-1)_+ - (v_s-1)_+]\Delta\psi_\varepsilon^+(x,t) + b_s\frac{\partial\psi_\varepsilon^+}{\partial t}\}dxdt +$$

$$+ \int_{\partial B_R\times(t_0,t_1)} \frac{\partial\psi_\varepsilon^+}{\partial n}(u-1)_+ds - \int_{\partial B_R\times(t_0,t_1)} \frac{\partial\psi_\varepsilon^+}{\partial n}(v_s-1)_+ds.$$

Due to (5.4) we may choose R such that the absolute values of the last two terms of the right hand side are smaller than ε. In order to estimate the second term of the right hand side we add and subtract

$$\int_{B_R}\int_{t_0}^{t_1} b_s(x,t)c_\varepsilon(x,t)\Delta\psi_\varepsilon^+ dxdt.$$

We find

$$|\int_{B_R}\int_{t_0}^{t_1}[c(x,t) - c_\varepsilon(x,t)]b_s(x,t)\Delta\psi_\varepsilon^+ dxdt| \leq$$

$$\leq \{\int_{B_R}\int_0^{t_1} |c_\varepsilon(x,t)||\Delta\psi_\varepsilon^+|^2 dxdt\}^{\frac{1}{2}} \cdot \{\int_{B_R}\int_0^{t_1} \frac{|c(x,t) - c_\varepsilon(x,t)|^2}{c_\varepsilon(x,t)}|b(x,t)|^2 dxdt\}^{\frac{1}{2}} \leq A^+\varepsilon,$$

where we have used

$$\{\int_{B_R}\int_0^{t_1} |c_\varepsilon(x,t)||\Delta\psi_\varepsilon^+|^2 dxdt\}^{1/2} \leq \frac{1}{2}\int_{B_R} |\nabla G(\Delta\Theta)_+(x)|^2 dx = A^+,$$

and (5.1). Then, by (5.6.a) and (5.6.b)

$$-\int_{B_R} f_s(x,0)d\lambda_\varepsilon^+(x) \leq \int_{B_R} b_s(x,t_1)G(\Delta\Theta)_+(x)dx + (2+A^+)\varepsilon.$$

Letting $s\downarrow 0$, we have

$$0 \leq \int_{B_R} b(x,t_1)G(\Delta\Theta)_+(x)dx + (2+A^+)\varepsilon.$$

A similar reasoning yields

$$0 \leq \int_{B_R} b(x, t_1) G(\Delta\Theta)_-(x) dx + (2 + A^-)\varepsilon.$$

Interchanging the roles of u and v we obtain

$$0 \leq \int_{B_R} (v - u)(x, t_1) G(\Delta\Theta)_+(x) dx + (2 + A^+)\varepsilon,$$

and

$$0 \leq \int_{B_R} (v - u)(x, t_1) G(\Delta\Theta)_-(x) dx + (2 + A^-)\varepsilon,$$

whence

$$0 \leq \int_{B_R} (u - v)(x, t_1) \Theta(x) dx + (4 + A^+ + A^-)\varepsilon, \qquad \forall \varepsilon > 0,$$

and therefore

$$0 \leq \int_{B_R} (u - v)(x, t_1) \Theta(x) dx,$$

and

$$0 \leq \int_{B_R} (v - u)(x, t_1) \Theta(x) dx.$$

Remarks. Due to formula (2.2) applied in each cylinder $B_R \times (\tau_{i+1}, \tau_i)$, ψ^+ and ψ^- are admissible test functions. Moreover, the bound (5.4) holds on the lateral surface of each cylinder. Since $(u - 1)_+$ and $(v - 1)_+$ are continuous functions in $\mathbb{R}^n \times (0, T)$ (see [AK]) and belong to $L^1_{loc}(\mathbb{R}^n \times [0, T))$ (due to the results in [AK]), in order to have $\lim_{t \downarrow 0} f(x, t) = 0 = \lim_{t \downarrow 0} g(x, t)$, by Fubini's theorem, R must be taken in the complement of a null set.

REFERENCES

[AK]. D. Andreucci, M.K. Korten, *Initial traces of solutions to a one-phase Stefan problem in an infinite strip*, Revista Matemática Iberoamericana, Vol.9, No.2 (1993), 315-332.

[B]. J.E. Bouillet, *Signed solutions to diffusion-heat conduction equations*, Trab. de Matemática, IAM – CONICET, preprint **101**, Dec. 1986, see also Free Boundary Problems: Theory and Applications, Proc. Int. Colloq., Irsee-Ger. 1987, Vol. II, Pitman Res. Notes Math. Ser. **186** (1990), 480-485.

[BKM]. J.E. Bouillet, M.K. Korten, V. Márquez, *Singular limits and the "mesa" problem*, Trab. de Matemática, IAM – CONICET, preprint **133**, June 1988.

[DK1]. B.E.J. Dahlberg and C.E. Kenig, *Weak solutions of the porous medium equation*, Trans. Am. Math. Soc. **336**, No.2 (1993), 711-725.

[DK2]. B.E.J. Dahlberg and C.E. Kenig, *Non-Negative solutions of generalized porous medium equation*, Rev. Mat. Iberoamericana, Vol. **2** , No **3** (1986), 267-305.

[K]. **M.K. Korten**, *Nonnegative solutions of $u_t = \Delta(u-1)_+$: Regularity and uniqueness to the Cauchy problem*, to appear in Nonlinear Analysis, Theory, Methods & Applications.

[F]. **A. Friedman**, *Partial Differential Equations of Parabolic Type*, Prentice Hall, Inc. (1964).

[P]. **M. Pierre**, *Uniqueness of the solution of $u_t - \Delta\phi(u) = 0$ with initial datum a measure*, J. Nonlin. Anal. Th. Meth. Applic. **6** (1982), 175-187.

Marianne K. Korten
IAM (CONICET)
Viamonte 1636 - 1er cuerpo 1er piso
1055 - Buenos Aires - ARGENTINA

and

Departamento de Matemática, FCEyN, UBA
Ciudad Universitaria, Pab. No 1, Nuñez
1428 - Buenos Aires - ARGENTINA

A LAPIN

Weak solutions for non-linear fluid flow through porous medium

A free boundary problem for the stationary fluid flow through porous medium under the gravity forces (dam problem) is studied. Non-linear dependence of the fluid velocity on the piezometric head is supposed, degenerated and discontinious cases including.

Variational formulation in the fixed domain and the existence of weak solutions for the dam problem were studied earlier for linear filtration law - Darcy's law (cf. [1,2,3] and consequent papers). The methods of these articles do not seem to be applicable in the non-linear case.

We give several weak formulations of the problems under consideration and prove the existence of weak solutions using finite-dimensional approximation and regularisation technique. For simplicity we restrict ourselves to the study of two-dimensional problems.

1. Let $D \in \mathbf{R}^2$ be the domain occupied by the dam. We suppose the boundary ∂D being Lipschitz, piecewise-smooth curve and consisted of the parts $S_i, i = 1, 2, 3,$. Here S_1 corresponds to impermeable bottom while S_2, S_3 being the parts in contact with fluid and atmosphere correspondingly. Let (x_1, x_2) be Cartesian coordinates and the gravity forces acts in the direction $-\mathbf{e}, \mathbf{e} = (0, 1)$. Let $S_2 \cup S_3$ lie over S_1 and consist of the finite number of parts, every of them being either the graph of x_1 or vertical segment. We denote by Ω unknown domain - wet part of the dam and by S_0 free boundary, separating wet and dry parts (cf. [4] for more detailed formulation of the problem in the case of linear filtration).

Problem (P_0). We look for the pair (u, Ω), where Ω is subdomain of D with piecewise Lipschitz continuous boundary and $u(x)$ is non-negative in Ω and satisfies the following relations:

$$div\mathbf{v}(x) = 0, \mathbf{v}(x) = -k(x, |\nabla u + \mathbf{e}|)(\nabla u + \mathbf{e}), x \in \Omega$$

$$\mathbf{v}(x)\mathbf{n}(x) = 0, x \in S_1, u(x) = u_0(x), x \in S_2,$$

$$\mathbf{v}(x)\mathbf{n}(x) \geq 0, u(x) = 0, x \in S_3 \cup \bar{\Omega},$$

$$\mathbf{v}(x)\mathbf{n}(x) = 0, u(x) = 0, x \in S_0$$

Here $\mathbf{n}(x)$ is unit outward normal vector, $u(x), \mathbf{v}(x)$ have the physical sense of pressure and velocity of the fluid, atmospheric pressure supposed to be zero, function

242

$u_0(x)$ is non-negative and linear by x_2 and equal to zero for the points in $\bar{S}_2 \cup \bar{S}_3$. Let the function k satisfy the following assumptions:

(1) $\qquad k(x, |\xi|)\xi \quad$ is measurable in $\qquad x \in D$ and continious in $\xi \in \mathbf{R}^2$

(2) $\qquad k(x, |\xi|)|\xi| \quad$ is nondecreasing in $|\xi|$ a.e. in D

(3) $\alpha|\xi|^{p-1} \le k(x, |\xi|)|\xi| \le \beta|\xi|^{p-1} \qquad$ a.e. in $D \qquad$ and for sufficiently large $|\xi|$,

where $1 < p < \infty; \qquad \alpha, \beta = const > 0$

We define the following convex closed subsets in W_p^1 :

$$M = \{u \in W_p^1(D) : u(x) = u_0(x), x \in S_2; u(x) \le 0, x \in S_3\},$$

$$K = \{u \in W_p^1(D) : u(x) = u_0(x), x \in S_2; u(x) = 0, x \in S_3; u(x) \ge 0, x \in D\},$$

and maximal monotone Heaviside graph $H(t) = \{1, t > 0; [0, 1], t = 0; 0, t < 0\}$.

Definition 1 *We call a pair $(u, \chi) \in K \times L_\infty(D)$ by a weak solution of the problem* (**P$_0$**) *if*

(4) $\qquad \chi(x) \in H(u(x))$ *a.e.* $x \in D$

(5) $\qquad \int_D k(x, |\nabla u + \mathbf{e}|)(\nabla u + \chi \mathbf{e})\nabla(\eta - u)dx \ge 0 \quad \forall \eta \in M$

Proposition 1 *Let (u, Ω) be the solution of* (**P$_0$**) *("classical solution"). Then (u, χ_Ω), where χ_Ω is the characteristic function of Ω, is a weak solution.*

Along with the variational inequality (5) we consider two following ones:

(6) $\qquad \int_D \chi(x)k(x, |\nabla u + \mathbf{e}|)(\nabla u + \mathbf{e})\nabla(\eta - u)dx \ge 0 \quad \forall \eta \in M, u \in M$

(7) $\qquad \int_D \chi(x)k(x, |\nabla \eta + \mathbf{e}|)(\nabla \eta + \mathbf{e})\nabla(\eta - u)dx \ge 0 \quad \forall \eta \in M, u \in M$

Proposition 2 *a) Any solution of the problem (4), (5) is also a solution of (4),(6).*
b) If (u, χ) is a solution of the problem (4),(6) , then (u^+, χ) ,$u^+(x) = \sup\{0, u(x)\}$ is a solution of (4), (5).
c) The problems (4),(6) and (4),(7) are equivalent.

Two first statements follow from equality $\chi(x)\nabla u(x) = \nabla u(x)$ a.e. in D fulfilled for $u(x) \geq 0$ and $\chi(x) \in H(u(x))$ and the fact that $\chi(x) \in H(u^+(x))$ if $\chi(x) \in H(u(x))$. To prove the last statement we use the properties (1),(2) of the function k .

2. Let us construct the finite-dimensional approximation $M_n = \{u_n \in W_p^1(D) : u_n(x) = u_0(x), x \in S_2; u_n(x) \leq 0, x \in S_3\}$ of the set M such that $M_n \subset M_{n+1}$ and $\cup_n M_n$ is dense in M and let $B_R = \{u \in W_p^1(D) : \|u\| \leq R\}$ be the ball in $W_p^1(D)$.

Proposition 3 *For any $n, R > 0$ there exists a solution of the following problem (we omit the index R in the solution) :*

$$\chi_n(x) \in H(u_n(x)) \quad a.e. x \in D \tag{8}$$

$$\int_D \chi_n(x)k(x, |\nabla u_n + \mathbf{e}|)(\nabla u_n + \mathbf{e})\nabla(\eta - u_n)dx \geq 0 \forall \eta \in M_n \cap B_R, u_n \in M_n \cap B_R \tag{9}$$

For proving we consider the regularisation of variational inequality (9) with $\chi(x)$ changed by $H_\varepsilon(u_{n\varepsilon}(x))$, where $H_\varepsilon(t) = \{0, t < 0; t/\varepsilon, 0 \leq t < \varepsilon; 1, t \geq \varepsilon\}$. The existence for the solution of the regularised problem follows from the continuity of the corresponding operator in finite-dimensional space and from the boundedness of the set $M_n \cap B_R$. Further we choose the subsequence $\{u_{n\varepsilon}\}$ strongly convergent in $W_p^1(D)$ to u_n as $\varepsilon \longrightarrow 0$ while $H_\varepsilon(u_{n\varepsilon})$ converges *-weak in $L_\infty(D)$ to χ_n . Passing to the limit in the regularised variational inequality we derive (9). It is easy to check also that $\chi_n(x) \in H(u_n(x))$.

Theorem 1 *Let the conditions (1) - (3) be satisfied. Then the problem (4), (5) has a solution.*

Sketch of the proof. We get that the pair $(u_n^+, \chi_n) \in K \times L_\infty(D)$ simultaneously with (u_n, χ_n) satisfies the following equations:

$$\chi_n(x) \in H(u_n^+(x)) \quad \text{a.e. in } D \tag{10}$$

$$\int_D \chi_n(x)k(x, |\nabla\eta + \mathbf{e}|)(\nabla\eta + \mathbf{e})\nabla(\eta - u_n^+)dx \geq 0 \quad \forall \eta \in M_n \cap B_R \tag{11}$$

Extracting the subsequences such that:

$$u_n^+ \longrightarrow u = u_R \text{ weak in } W_p^1(D), \quad \chi_n \longrightarrow \chi = \chi_R \text{ * - weak in } L_\infty(D).$$

we conclude that $u_R(x) \geq 0, \chi_R(x) \in H(u_R(x))$ a.e. in D. Further for arbitrary function $\eta \in M \cap B_R$ we take the sequence $\eta_n \in M_n \cap B_R$ strongly to η converging in $W_p^1(D)$. Passing to the limit as $n \longrightarrow \infty$ in (11) we obtain that the pair (u_R, χ_R)

satisfies the variational inequality (7) for all $\eta \in M \cap B_R$. It rests to use the assumption (3) to get the estimate

$$\|u_R\| \leq C \quad \forall R$$

with constant C independent on R . Thus (7) and as consequence (5) is true for all $\eta \in M$.

3. We consider now the problem with discontinuous dependence of the velocity on the piezometric head (cf. [5],[6]). Let the function k be defined by the equality

$$(12) \qquad k(x,\xi) = k_0(x,\xi) + aH_0(\xi - \xi_0)/\xi, \quad x \in D, \quad \xi \geq 0$$

where k_0 satisfies the assumptions (1)-(3) , $H_0(t) = \{0, t \leq 0; 1, t > 0\}$ - Heaviside function, a, ξ_0 are positive constants. As above we call by the classical solution of the problem under consideration the pair (u, Ω) satisfying all statements in $(\mathbf{P_0})$ excepting the equation $div\mathbf{v}(x) = 0$ being satisfied on the set $\{x \in \Omega : |\nabla u + \mathbf{e}| > \xi_0\}$. This problem we refer to as the problem $(\mathbf{P_1})$.

Definition 2 *We call by the weak solution of the problem* $(\mathbf{P_1})$ *a pair* $(u, \chi) \in K \times L_\infty(D)$ *satisfying the inclusion (4) and variational inequality:*

$$(13) \quad \int_D k_0(x, |\nabla u + \mathbf{e}|)(\nabla u + \chi\mathbf{e})\nabla(\eta - u)dx + F_1(\eta) - F_1(u) \geq 0 \quad \forall \eta \in M$$

$$\text{where } F_1(u) = a \int_D \chi(x)(|\nabla u + \mathbf{e}| - \xi_0)^+ dx$$

Theorem 2 *Let the assumptions (1)–(3), (12) be satisfied, then a solution for the problem (4), (13) exists.*

To prove this result we consider the regularized problem: for arbitrary $\varepsilon > 0$ find $(u_\varepsilon, \chi_\varepsilon) \in K \times L_\infty(D)$ such that inclusion (4) and following variational inequality

$$(14) \int_D \chi(x)(k_0(x, |\nabla\eta + \mathbf{e}|) + a\frac{H_\varepsilon(|\nabla\eta + \mathbf{e}| - \xi_0)}{|\nabla\eta + \mathbf{e}|})(\nabla\eta + \mathbf{e})\nabla(\eta - u_\varepsilon)dx \geq 0 \quad \forall \eta \in M$$

are valid. The existence of a solution for this problem follows from Theorem 1. We extract the subsequence of these solutions such that

$$u_\varepsilon \longrightarrow u \text{ weakly in } W_p^1(D), \quad \chi_\varepsilon \longrightarrow \chi^* \text{ - weakly in } L_\infty(D).$$

Then $u \in K, \chi(x) \in H(u(x))$ a.e. in D. We pass to the limit as $\varepsilon \longrightarrow 0$ in (14) using the equality $\chi_\varepsilon(x)\nabla u_\varepsilon(x) = \nabla u_\varepsilon(x)$ a.e. in D and the fact that

$$\frac{H_\varepsilon(|\nabla\eta + \mathbf{e}| - \xi_0)}{|\nabla\eta + \mathbf{e}|}(\nabla\eta + \mathbf{e}) \longrightarrow \frac{H_0(|\nabla\eta + \mathbf{e}| - \xi_0)}{|\nabla\eta + \mathbf{e}|}(\nabla\eta + \mathbf{e})$$

245

strongly in $L_{\bar{p}}^2(D), \bar{p} = p/(p-1)$. As result we derive

$$(15)\int_D \chi(x)(k_0(x, |\nabla\eta + \mathbf{e}|) + a\frac{H_0(|\nabla\eta + \mathbf{e}| - \xi_0)}{|\nabla\eta + \mathbf{e}|})(\nabla\eta + \mathbf{e})\nabla(\eta - u_\varepsilon)dx \geq 0 \quad \forall \eta \in M$$

Similar to [6] we establish that the function u being a solution of variational inequality (15) (for fixed χ) at the same time provides the minimum in M to the following functional:

$$F(u) = F_0(u) + F_1(u), \quad F_0(u) = \int_D \chi(x) \int_0^{|\nabla u + \mathbf{e}|} tk_0(x, t)dt.$$

As $F_0(u)$ is differentiable the function u satisfies the inequality

$$F_0'(u)(\eta - u) + F_1(\eta) - F_1(u) \geq 0$$

coinciding with (13).

Acknowledgement: This work was supported by Russian Foundation for Basic Researches, Grant no. 95-01-00448.

References

1. Baiocchi C. Su una problema a frontiera libera conesso a questione di idraulica// Ann. Mat. Pura Appl. 1972, 92, 107–127.

2. Alt H.W. Strömungen durch inhomegene poröse Medien mit freiem Rand// J.Reine Angew. Math. 1979. V.305, 89–115.

3. Brezis H., Kinderlehrer D., Stampacchia G. Sur une nouvelle formulation du problèm de l'écoulement à travers une digue // C.R. Acad. Sci. Paris 1978, 287, 711–714.

4. Friedman A. Variational principles and free boundary problems N.Y.: Wiley, 1982.

5. Lapin A.V. Study of some non-linear filtration problems // J. Comput. Math. and Math. Phys. 1979, 19, 689–700 (in Russian).

6. Ljaŝko A.D., Badriev I.B., Kartchevsky M.M. On the variational method for equations with monotone discontinouos operators// Izv. VUZ, Math. 1978, 198, 63–69. (in Russian).

Alexander Lapin
Kazan State University
Russia

M ÔTANI AND Y SUGIYAMA
Gradient estimates for solutions of some non-Newtonian filtration problems

1   Introduction

In this note, we are concerned with the following two types of nonlinear parabolic equations $^\alpha$(E) and (E)$^\ell$:

$$^\alpha(\text{E}) \begin{cases} \frac{\partial u}{\partial t} - u^\alpha \text{div} \left(|\nabla u|^{p-2} \nabla u(x,t) \right) = 0, & (x,t) \in \Omega \times [0,T], \\ u(x,t) = 0, & (x,t) \in \partial\Omega \times [0,T], \\ u(x,0) = u_0(x), & x \in \Omega, \end{cases}$$

$$(\text{E})^\ell \begin{cases} \frac{\partial u}{\partial t} - \text{div} \left(u^\ell |\nabla u|^{p-2} \nabla u(x,t) \right) = 0, & (x,t) \in \Omega \times [0,T], \\ u(x,t) = 0, & (x,t) \in \partial\Omega \times [0,T], \\ u(x,0) = u_0(x), & x \in \Omega, \end{cases}$$

where Ω is a bounded domain in $\mathbb{R}^N$ with smooth boundary $\partial\Omega$, and α, ℓ, p are non-negative parameters. We also assume that the initial data $u_0(\cdot)$ is non-negative in Ω. These equations arise from the nonlinear filtration problems for non-Newtonian fluids. As a matter of course, $^\alpha$(E) with $\alpha = 0$ or (E)$^\ell$ with $\ell = 0$ gives the evolution equation generated by the so-called p-Laplacian $\Delta_p u(\cdot) = \text{div}(|\nabla u(\cdot)|^{p-2}\nabla u(\cdot))$; and (E)$^\ell$ with $p = 2$ gives the porous medium equations. These equations have been studied by so many peoples so far.

It is easy to see that (E)$^\ell$ are transformed to the so-called doubly nonlinear equations: $(v^\delta)_t - \Delta_p v = 0$, via the change of variables $v = u^{1/\delta}$ with $\delta = (p-1)/(\ell+p-1)$.

As for the study in this direction, we refer to the paper of Prof. R. Showalter in this Proceedings and its references (see also [4, Tsutsumi]).

On the other hand, because of the degeneracy or singularity caused by the terms u^ℓ, u^α or $|\nabla u|^{p-2}$, it seems hardly possible to construct global (in time) classical solutions for $^\alpha$(E) or (E)$^\ell$ except for special cases. However [1, Ivanov] and [5, Vespri] established the $C_{loc}^\alpha(\Omega)$-regularity of weak (global) solutions for (E)$^\ell$. The main purpose of this note is to investigate the existence of much more regular (not necessarily global) solutions. More precisely, our final goal is to get (local) solutions in $W^{1,\infty}(\Omega)$.

In order to see the relationship between two equations $^\alpha$(E) and (E)$^\ell$, it is useful to note the identity : $\text{div}(u^\ell|\nabla u|^{p-2}\nabla u) - u^\ell \text{div}(|\nabla u|^{p-2}\nabla u) = \ell u^{\ell-1}|\nabla u|^p$. Therefore, if the term $\ell u^{\ell-1}|\nabla u|^p$ could be treated as a perturbation for the leading term $\text{div}(u^\ell|\nabla u|^{p-2}\nabla u)$ or $u^\ell \text{div}(|\nabla u|^{p-2}\nabla u)$, then $^\alpha$(E) and (E)$^\ell$ could be solved in a unified way and had the same nature as far as the existence of solution is concerned. However, the singularity of the term $u^\ell|\nabla u|^p$ is almost the same as that of $\mathrm{div}(u^\ell|\nabla u|^{p-2}\nabla u)$ or $u^\ell\mathrm{div}(|\nabla u|^{p-2}\nabla u)$, so the term $u^\ell|\nabla u|^p$ can not be regarded as a perturbation. Hence $^\alpha(\mathrm{E})$ and $(\mathrm{E})^\ell$ are slightly different to each others in nature as will be shown later in §2.

Our main results are given in the next section, and the sketch of their proofs is given in §3.

2   Main Results

Throughout of this note, we always assume the following conditions.

(A.1) $u_0 \in W^{1,\infty}(\Omega)$ and $u_0 \geq 0$,

(A.Ω) The mean curvature $K(x)$ (with respect to the outward normal) of $\partial\Omega$ at x is non-positive for all $x \in \partial\Omega$.

Then main results are stated as follows.

Theorem 1 (Global existence for $^\alpha(\mathrm{E})$) *Let $N = 1$ and $1 \leq \alpha < \infty$, then $^\alpha(E)$ has a solution u belonging to $L^\infty(0, T; W^{1,\infty}(\Omega)) \cap W^{1,2}(0, T; L^2(\Omega))$.*

Theorem 2 (Local existence for $^\alpha(\mathrm{E})$) *Let $N \geq 2$ and $2 \leq \alpha < \infty$, then there exists a number $T_0 \in (0, T]$ depending only on $|u_0|_{W^{1,\infty}}$ such that $^\alpha(E)$ has a solution u belonging to $L^\infty(0, T_0; W^{1,\infty}(\Omega)) \cap W^{1,2}(0, T_0; L^2(\Omega))$.*

Theorem 3 (Local existence for $(\mathrm{E})^\ell$) *Let $N \geq 1$ and $2 \leq \alpha < \infty$, then there exists a number $T_0 \in (0, T]$ depending only on $|u_0|_{W^{1,\infty}}$ such that $(E)^\ell$ has a solution u belonging to $L^\infty(0, T_0; W^{1,\infty}(\Omega)) \cap W^{1,2}(0, T_0; L^2(\Omega))$.*

Remark.   If Ω is convex, then (A.Ω) is always satisfied.

Here we are concerned with the solution of $^\alpha(\mathrm{E})$ or $(\mathrm{E})^\ell$ in the following sense.

Definition   A function $u \in L^\infty(0, S; W_0^{1,\infty}(\Omega)) \cap W^{1,2}(0, S; L^2(\Omega))$ is said to be a solution of $^\alpha(\mathrm{E})$ (resp. $(\mathrm{E})^\ell$) in [0,S] if

$$(1) \qquad \int_\Omega u_t\varphi dx + \int_\Omega u^\beta|\nabla u|^{p-2}\nabla u \cdot \nabla\varphi dx + \beta^* \int_\Omega u^{\beta-1}|\nabla u|^p\varphi dx = 0$$

holds for a.e. $t \in [0, S]$ and all $\varphi \in C_0^\infty(\Omega)$ with $\beta = \beta^* = \alpha$ (resp. $\beta = \ell$ and $\beta^* = 0$).

248

3   Proofs of theorems

We here give the sketch of proofs of our main theorems.

For a proof of Theorems 1 and 2, we first prepare the following approximate equation $^\alpha(E)_\varepsilon$ for $^\alpha(E)$:

$$(2)^\alpha(E)_\varepsilon \begin{cases} \frac{\partial u^\varepsilon}{\partial t} - ((u^\varepsilon)^2 + \varepsilon^2)^{\frac{\alpha}{2}} \operatorname{div}\left((|\nabla u^\varepsilon| + \varepsilon^2)^{\frac{p-2}{2}} \nabla u^\varepsilon\right) = 0, & (x,t) \in \Omega \times [0,T], \\ u^\varepsilon(x,t) = 0, & (x,t) \in \partial\Omega \times [0,T], \\ u^\varepsilon(x,0) = u_0^\varepsilon(x), & x \in \Omega, \end{cases}$$

where ε is a positive parameter and u_0^ε is an approximation for the initial data u_0 such that $0 \le u_0^\varepsilon \in C_0^\infty(\Omega)$, $u_0^\varepsilon \to u_0$ strongly in $W_0^{1,r}(\Omega)$ for all $r \in [1,\infty)$ as $\varepsilon \to 0$ and $|u_0^\varepsilon|_{L^\infty} \le |u_0|_{L^\infty}$, $|\nabla u_0^\varepsilon|_{L^\infty} \le |\nabla u_0|_{L^\infty}$ for all $\varepsilon \in (0,1]$. It is easy to see that the operator $A^\varepsilon : u \mapsto (u^2 + \varepsilon^2)^{\frac{\alpha}{2}} \operatorname{div}((|\nabla u|^2 + \varepsilon^2)^{\frac{p-2}{2}} \nabla u)$ satisfies all the structure conditions required in Theorem 4.1 and Theorem 5.4 of [3, Ladyzhenskaya et al.], hence $^\alpha(E)_\varepsilon$ has a unique classical solution u^ε.

A priori estimates :   In what follows, we denote u^ε by u for the sake of simplicity if no confusion arises. Multiplying (2) by $u^- = \min(u,0)$, we get

$$\begin{aligned} \frac{1}{2}\frac{d}{dt}|u^-|_{L^2}^2 &= -\int_\Omega ((u^-)^2 + \varepsilon^2)^{\frac{\alpha}{2}}(|\nabla u^-|^2 + \varepsilon^2)^{\frac{p-2}{2}}|\nabla u^-|^2 dx \\ &\quad - \int_\Omega \alpha((u^-)^2 + \varepsilon^2)^{\frac{\alpha-2}{2}}|u^-|^2(|\nabla u^-|^2 + \varepsilon^2)^{\frac{p-2}{2}}|\nabla u^-|^2 dx \\ &\le 0, \end{aligned}$$

whence follows $|u^-(\cdot,t)|_{L^2} \le |u^-(\cdot,0)|_{L^2} = 0$, that is to say, $u(x,t) \ge 0$ for all $(x,t) \in \Omega \times [0,T]$.

Furthermore, multiplying (2) by $[u - M]^+ = \max(u - M, 0)$ with $M = |u_0|_{L^\infty}$ and repeating the same procedure as above, we finally obtain

$$(3) \qquad\qquad 0 \le u(x,t) \le |u_0|_{L^\infty} \quad \text{for all } (x,t) \in [0,T].$$

We next multiply (2) by $-\operatorname{div}((|\nabla u|^2 + \varepsilon^2)^{\frac{r-2}{2}}\nabla u)$ with r sufficiently large to have

$$\frac{1}{r}\frac{d}{dt}\int_\Omega (|\nabla u|^2 + \varepsilon)^{\frac{r}{2}} dx + \int_\Omega (u^2 + \varepsilon^2)^{\frac{\alpha}{2}}\left\{(|\nabla u|^2 + \varepsilon^2)^{\frac{p-2}{2}}u_i\right\}_i \left\{(|\nabla u|^2 + \varepsilon^2)^{\frac{r-2}{2}}u_j\right\}_j dx = 0.$$

Here and henceforth, we use the summation convention and the simplified notations such as $v_i = \partial v/\partial x_i$ and $v_{ij} = \partial^2 v/\partial x_i \partial x_j$.

Then by using the integration by parts twice, we derive

$$\frac{1}{r}\frac{d}{dt}\int_\Omega (|\nabla u|^2 + \varepsilon^2)^{\frac{r}{2}} dx + J = I_\Omega + I_{\partial\Omega},$$

249

with

$$J = J_1 + J_2 + J_3,$$

$$J_1 = (p-2)(r-2)\int_\Omega (u^2+\varepsilon^2)^{\frac{\alpha}{2}}(|\nabla u|^2+\varepsilon^2)^{\frac{r+p-8}{2}}(u_i u_j u_{ij})^2 dx,$$

$$J_2 = (r+p-4)\int_\Omega (u^2+\varepsilon^2)^{\frac{\alpha}{2}}(|\nabla u|^2+\varepsilon^2)^{\frac{r+p-6}{2}}(u_j u_{ij})^2 dx,$$

$$J_3 = \int_\Omega (u^2+\varepsilon^2)^{\frac{\alpha}{2}}(|\nabla u|^2+\varepsilon^2)^{\frac{r+p-4}{2}}(u_{ij})^2 dx,$$

$$I_\Omega = \alpha \int_\Omega (u^2+\varepsilon^2)^{\frac{\alpha-2}{2}}u(|\nabla u|^2+\varepsilon^2)^{\frac{r+p-4}{2}}(u_{ii}u_j u_j - u_{ij}u_i u_j)dx,$$

$$I_{\partial\Omega} = \varepsilon^\alpha \int_{\partial\Omega}(|\nabla u|^2+\varepsilon^2)^{\frac{r+p-4}{2}}(u_{ij}u_j n_i - u_{ii}u_j n_j)dS,$$

where $n(x) = (n_1(x), \ldots, n_N(x))$ is the unit outward normal vector at $x \in \partial\Omega$. If $N = 1$, then it is clear that $I_\Omega = I_{\partial\Omega} = 0$. Hence, since $J \geq 0$, it follows that

$$\frac{d}{dt}\int_\Omega (|\nabla u|^2+\varepsilon^2)^{\frac{r}{2}}dx \leq 0 \quad \text{for all } t \in [0,T] \text{ and all } r.$$

Thus letting $r \to \infty$, we deduce

(4) $$|\nabla u(\cdot,t)|_{L^\infty} \leq |\nabla u_0|_{L^\infty} \quad \text{for a.e. } t \in [0,T].$$

As for the case where $N \geq 2$, assumption (A.Ω) assures that $I_{\partial\Omega} \leq 0$. To show this, for any $x_0 \in \partial\Omega$, choose an orthogonal matrix A so that the unit outward normal vector n can be represented as $n = (0, \ldots, 1)$ with respect to the new coordinates $y = A(x - x_0)$. Furthermore, in the neighborhood U of $y = 0$ (i.e. $x = x_0$), the points in $\partial\Omega$ satisfies $y_N = \omega(y_1, \ldots, y_{N-1})$ with smooth function ω. Then the boundary condition $u|_{\partial\Omega} = 0$ together with the fact that $n = (0, \ldots, 0, 1)$ implies

(5) $$u(y_1, \ldots, y_{N-1}, \omega(y_1, \ldots, y_{N-1})) = 0 \quad \text{in } U,$$

(6) $$\frac{\partial\omega}{\partial y_j}(0) = 0 \quad \text{for all } j = 1, 2, \ldots, N-1.$$

Making use of these relations and the fact that A is an orthogonal matrix, we can finally conclude

$$R = u_{ij}u_j n_i - u_{ii}u_j n_j = \left(\frac{\partial u}{\partial n}\right)^2 \sum_{p=1}^{N-1}\frac{\partial^2 u}{\partial y_p^2} = \left(\frac{\partial u}{\partial n}\right)^2 (N-1)K(x_0) \leq 0,$$

whence follows $I_{\partial\Omega} \leq 0$. (For details, see §5 of chapter II of [2].)

As for I_Ω, we observe that

$$|I_\Omega| \leq \alpha \int_\Omega (u^2+\varepsilon^2)^{\frac{\alpha}{4}}(|\nabla u|^2+\varepsilon^2)^{\frac{r+p-4}{4}}|u_{ii}|(u^2+\varepsilon^2)^{\frac{\alpha-2}{4}}(|\nabla u|^2+\varepsilon^2)^{\frac{r+p-4}{4}}|\nabla u|^2 dx$$

250

$$+\alpha \int_\Omega (u^2 + \varepsilon^2)^{\frac{\alpha}{4}}(|\nabla u|^2 + \varepsilon^2)^{\frac{r+p-6}{4}}|u_j u_{ij}|(u^2 + \varepsilon^2)^{\frac{\alpha-2}{4}}(|\nabla u|^2 + \varepsilon^2)^{\frac{r+p-2}{4}}|\nabla u|dx$$

$$\leq \delta \int_\Omega (u^2 + \varepsilon^2)^{\frac{\alpha}{2}}(|\nabla u|^2 + \varepsilon^2)^{\frac{r+p-4}{2}}|u_{ii}|^2 dx$$

$$+\delta \int_\Omega (u^2 + \varepsilon^2)^{\frac{\alpha}{2}}(|\nabla u|^2 + \varepsilon^2)^{\frac{r+p-6}{2}}|u_j u_{ij}|^2 dx$$

(7)
$$+C_\delta \int_\Omega (u^2 + \varepsilon^2)^{\frac{\alpha-2}{2}}(|\nabla u|^2 + \varepsilon^2)^{\frac{r+p}{2}} dx,$$

where δ is an arbitrary positive number and C_δ is a number depending only on δ. Since the first and second terms of the right hand side of (7) can be canceled out by J_3 and J_2 respectively, we finally get

$$\frac{1}{r}\frac{d}{dt}\int_\Omega (|\nabla u|^2 + \varepsilon^2)^{\frac{r}{2}} dx \leq C_\delta (M+1)^{\alpha-2}\int_\Omega (|\nabla u|^2 + \varepsilon^2)^{\frac{r+p}{2}} dx.$$

Hence we have

$$\frac{d}{dt}|(|\nabla u|^2 + \varepsilon^2)^{\frac{1}{2}}|_{L^r} \leq C_\delta (M+1)^{\alpha-2}(|\nabla u|^2_{L^\infty} + \varepsilon^2)^{\frac{p}{2}}|(|\nabla u|^2 + \varepsilon^2)^{\frac{1}{2}}|_{L^r}.$$

Consequently, by Gronwall's inequality and by letting $r \to \infty$, we see that there exists a constant C independent of ε such that

(8)
$$|\nabla u(t)|_{L^\infty} \leq |\nabla u_0|_{L^\infty} + C\int_0^t |\nabla u(s)|_{L^\infty}^{p+1} ds + C$$

Thus, by the standard argument, we can easily find that there exists a positive number T_0 depending only on $|\nabla u_0|_{L^\infty}$ such that

(9)
$$|\nabla u(\cdot, t)|_{L^\infty} \leq C \quad \text{for all } t \in [0, T_0] \text{ and } \varepsilon \in (0,1]$$

To prove Theorem 3, we introduce another approximate equation $(E)_\varepsilon^\ell$ for $(E)^\ell$:

(10) $(E)_\varepsilon^\ell$
$$\begin{cases} \dfrac{\partial u^\varepsilon}{\partial t} - ((u^\varepsilon)^2 + \varepsilon^2)^{\frac{\ell}{2}}\text{div}\left((|\nabla u^\varepsilon| + \varepsilon^2)^{\frac{p-2}{2}}\nabla u^\varepsilon\right) \\ \qquad\qquad -\ell|u^\varepsilon|^{\ell-1}|\nabla u^\varepsilon|^p = 0, & (x,t) \in \Omega \times [0,T], \\ u^\varepsilon(x,t) = 0, & (x,t) \in \partial\Omega \times [0,T], \\ u^\varepsilon(x,0) = u_0^\varepsilon(x), & x \in \Omega, \end{cases}$$

where u_0^ε is the same approximation for u_0 given as before. Since $\ell \geq 2$, the existence and uniqueness of classical solutions u^ε is assured again by Theorems 4.1 and 5.4 of [3]. Furthermore, by the same arguments as for $^\alpha(E)_\varepsilon$, we can derive the same estimate as (3). Therefore the term $|u^\varepsilon|^{\ell-1}$ in (10) can be replaced by $(u^\varepsilon)^{\ell-1}$.

To establish the estimate for ∇u (u^ε is again denoted by u), we multiply (10) again by $-\text{div}((|\nabla u|^2 + \varepsilon^2)^{\frac{r-2}{2}}\nabla u)$ to get

$$\frac{1}{r}\frac{d}{dt}\int_\Omega (|\nabla u|^2 + \varepsilon^2)^{\frac{r}{2}} dx + J = I_\Omega + I_{\partial\Omega} + II_\Omega,$$

251

where J, I_Ω and $I_{\partial\Omega}$ are the same as before (with α replaced by ℓ) and

$$II_\Omega = \ell \int_\Omega u^{\ell-1}|\nabla u|^p((|\nabla u|^2 + \varepsilon^2)^{\frac{r-2}{2}} u_i)_i dx.$$

By the integration by parts, we find

$$II_\Omega = \ell(\ell-1) \int_\Omega u^{\ell-2}|\nabla u|^p(|\nabla u|^2 + \varepsilon^2)^{\frac{r-2}{2}}|\nabla u|dx + III_\Omega,$$

$$III_\Omega = \ell\, p \int_\Omega u^{\ell-1}(|\nabla u|^2 + \varepsilon^2)^{\frac{r+p-4}{2}} u_i u_j u_{ij} dx.$$

The same arguments as for (7) yield

$$|III_\Omega| \le \delta J_2 + C_\delta \int_\Omega u^{\ell-2}(|\nabla u|^2 + \varepsilon^2)^{\frac{r+p}{2}} dx.$$

Thus we can again deduce

$$\frac{1}{r}\frac{d}{dt} \int_\Omega (|\nabla u|^2 + \varepsilon^2)^{\frac{r}{2}} dx \le C \int_\Omega (|\nabla u|^2 + \varepsilon^2)^{\frac{r+p}{2}} dx,$$

whence we obtain the same local estimate for $|\nabla u(t)|_{L^\infty}$ as (9).

In order to establish the estimate for u_t, we multiply $(E)^\ell_\varepsilon$ by $A_\varepsilon u = -\text{div}((|\nabla u|^2 + \varepsilon^2)^{\frac{p-2}{2}}\nabla u)$, then we have

$$\frac{1}{p}\frac{d}{dt} \int_\Omega (|\nabla u|^2 + \varepsilon^2)^{\frac{p}{2}} dx + \int_\Omega (u^2 + \varepsilon^2)^{\frac{p}{2}}|A_\varepsilon u|^2 dx$$

$$= \int_\Omega \ell u^{\ell-1}|\nabla u|^p A_\varepsilon u dx$$

$$\le \frac{1}{2} \int_\Omega u^\ell |A_\varepsilon u|^2 dx + \frac{\ell^2}{2} \int_\Omega u^{\ell-2}|\nabla u|^{2p} dx.$$

Hence, using (3) and (9), we can derive a priori bound for $|(u^2 + \varepsilon^2)^{\frac{\ell}{4}} A_\varepsilon u|_{L^2(0,T_0;L^2(\Omega))}$, which gives

(11) $$\int_0^{T_0} \int_\Omega |u^\varepsilon_t|^2 dx dt \le C \quad \text{for all } \varepsilon \in (0,1].$$

We can repeat the same verification for (11) also for $^\alpha(E)_\varepsilon$ much easier and find that the solution u^ε of $^\alpha(E)_\varepsilon$ satisfies (11). (For the case $N = 1$, T_0 can be replaced by T.)

Convergence: If we put $w = u^{\frac{\ell}{p-1}}u$ (or $w = u^{\frac{\alpha}{p-1}}u$), then w_t and $\text{div}(|\nabla w|^{p-2}\nabla w)$ are bounded in $L^2(0,T_0;L^2(\Omega))$. Therefore, by Ascoli's theorem and a standard argument from convex analysis, we can show that there exists a sequence u^{ε_k} denoted by u_k such that

$$(u_k)_t \rightharpoonup u_t \text{ weakly in } L^2(0,T_0;L^2(\Omega)),$$
$$u_k \rightarrow u \text{ strongly in } C([0,T_0];L^2(\Omega)),$$
$$u_k \rightharpoonup u \text{ weakly star in } L^\infty(0,T_0;W^{1,\infty}(\Omega)),$$
$$\text{div}(|\nabla w_k|^{p-2}\nabla w_k) \rightharpoonup \text{div}(|\nabla w|^{p-2}\nabla w) \text{ weakly in } L^2(0,T_0;L^2(\Omega)),$$
$$\nabla w_k \rightarrow \nabla w \text{ strongly in } L^p(0,T_0;L^p(\Omega)),$$

where $w_k = u_k^{\frac{\ell}{p-1}} u_k$ (or $u_k^{\frac{\alpha}{p-1}} u_k$) and $w = u^{\frac{\ell}{p-1}} u$ (or $u^{\frac{\alpha}{p-1}} u$).

Hence it is easy to verify that u becomes a solution of $(E)^\ell$ (or $^\alpha(E)$). $\square$

Concluding remarks

1. In Theorems 2 and 3, we assumed $(A.\Omega)$, which might be a technical assumption. However, if the equation has "the finite propagation property" (such as the porous medium equations) and the initial data have compact supports, then $(A.\Omega)$ does not give any restriction.

2. We can also derive a result on the uniqueness of solutions in the class $L^\infty(0, T; W_0^{1,\infty}(\Omega)) \cap W^{1,2}(0, T; L^2(\Omega))$ as follows:

    (a) The solution of $^\alpha(E)$ is unique, provided that $1 < p \le 2$ and $2 \le \alpha$.

    (b) The solution of $(E)^\ell$ is unique, provided that $1 < p \le 2$ and $2 \le \ell$ or that $2 \le p$ and $1 \le \ell$.

3. By using the same idea as above, we can give some estimates of higher derivatives of solutions for some special cases. Especially, for the porous medium equations $u_t - (u^\ell u_x)_x = 0$ with ℓ be even integers, we can construct a (time local) C^∞-solution in $[0, T_0]$. Here T_0 depends only on the $W^{2,\infty}$-norm of the initial data.

References

[1] A. V. IVANOV, The classes B_{ml} and Hölder estimates for quasilinear doubly degenerate parabolic equations, *Zap. Nauchn. Sem. St. Petersburg Otdel. Math. Inst. Steklov (LOMI)*, **197** (1992), 42–70.

[2] O. A. LADYZHENSKAYA, "The Boundary Value Problems of Mathematical Physics", Applies Mathematical Science vol. 49, Springer-Verlag, 1988.

[3] O. A. LADYZHENSKAYA, V. A. SOLONNIKOV AND N. N. URAL'CEVA, "Linear and Quasi-linear Equations of Parabolic Type", Transl. Math. AMS. Providence, R.I. , 1968.

[4] M. TSUTSUMI, On solutions of some doubly nonlinear degenerate parabolic equations with absorption, *J. Math. Anal. Appl.*, **132** (1988), 187–212.

[5] V. VESPRI, On the local behaviour of a certain class of doubly non-linear parabolic equations, *Manuscripta Math*, **75** (1992), 65–80.

Mitsuharu Ôtani and Yoshie Sugiyama
Department of Applied Physics
School of Science and Engineering, Waseda University
3-4-1 Okubo Tokyo, JAPAN

J F RODRIGUES AND L SANTOS
On the glacier kinematics with the shallow-ice approximation

1. Introduction

The free surface flow of a glacier or ice sheet is a typical example of a free boundary problem in theoretical glaciology, which is considered here from a mathematical point of view.

We consider a variational approach for the kinematics description, in a three dimensional model, of the surface of the glacier.

In the formulation of the mathematical model, the unknown is the height of the glacier. Its equation contains a diffusive term, arising in the accumulation/ablation function (regarded as an equivalent normal ice flux), from the linearization of the mean curvature of the boundary surface, which takes into account the surface erosion of the glacier.

Within the shallow-ice approximation, the transport in the direction of the displacement is dominant with respect to the diffusion spread. In the limit case, this corresponds to an ultraparabolic equation in an unknown domain, whose boundary is the boundary of the glacier, regarded as being not *a priori* known.

In the second section we describe the model. Its first subsection describes the kinematics of the free surface, the second one derives the *shallow-ice approximation* and the last one introduces the variational inequalities.

In Section 3, the mathematical study of the variational inequality problems is presented. The first subsection studies the steady-state case while in the second one the evolutive case is described. In particular, we discuss the asymptotic convergences with respect to the small ratio of thickness/lenght. The last subsection is dedicated to the asymptotic behaviour in time of the solution.

2. The mathematical model

Our aim in this section is to obtain the kinematic description of the free ice-air surface $z = H(x, y, t)$, supposing given the fixed bed $z = h(x, y)$, neglecting the crustal deformation at the base, as well as other mechanical and thermal effects (see, for instance, [6] and [2] for a physical introduction).

It will be considered the *shallow-ice approximation*, since the physical process involves large times and large length scales in the longitudinal direction, compared to those in the transverse directions (see [4], Chap. 5). The *shallow-ice approximation* consists in the introduction of a streching transformation in terms of a small parameter μ, $0 < \mu \ll 1$, for instance $\mu = \ell/L$ where ℓ is a mean thickness of the ice sheet and L a representative length of the glacier.

2.1. Kinematic description of the free surface

Let x, y, z denote the spatial coordinates and $\bar{t}$ denote the time. Suppose that the ice-air free surface $\mathcal{S}$ is given by the equation

$$S(x, y, z, \bar{t}) = z - H(x, y, \bar{t}) = 0. \tag{1}$$

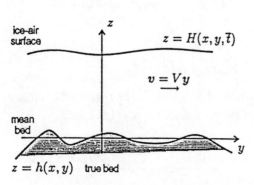

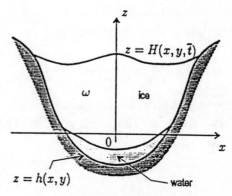

i) Vertical plan $(0; y, z)$ representing a long nearly parallel ice slab flow:

ii) Cross section, with arbitrary shape, orthogonal to flow velocity, for a glacier with melt zones.

If $\vec{n}$ denotes the outward unitary normal to $\mathcal{S}$,

$$\vec{n} = \frac{\nabla S}{\|\nabla S\|} = \frac{(-\partial_x H, -\partial_y H, 1)}{\sqrt{(\partial_x H)^2 + (\partial_y H)^2 + 1}}.$$

If $\vec{v}$ denotes the velocity of the ice and $\vec{w}$ the velocity of the free surface, then the normal influx of ice is given by

$$A = -(\vec{v} - \vec{w}).\vec{n}, \tag{2}$$

being $A > 0$ if there exists accumulation and $A < 0$ if there exists ablation.

In general, S is a non-material surface and hence dS/dt is not zero, but the derivative of S "following the surface" vanishes, which means that

$$0 = \partial_{\bar{t}} S + \vec{w}.\nabla S,$$

and, if $\vec{v} = (v_1, v_2, v_3)$, we have

$$\partial_{\bar{t}} H + v_1 \partial_x H + v_2 \partial_y H - v_3 = A\sqrt{1 + (\partial_x H)^2 + (\partial_y H)^2}. \tag{3}$$

The surface mass balance function $A = A(x, y, \bar{t}, H(x, y, \bar{t}))$ has been subjected to many theoretical discussions, particularly in the surface-wave planar approach, where a nonlinear diffusion equation for H has been considered (see [2] or [1]). On the other hand, if there is practically no surface melting over Antartica, in Greenland there is a significant ablation at the ice sheet margins and in mountain glaciers there are often strong rates of accumulation and ablation.

We will assume here that A is proportional to the radius of curvature of the free surface plus a given function, i.e.,

$$A = A(x, y, \bar{t}) = \alpha \left(\frac{1}{R_1(H)} + \frac{1}{R_2(H)} \right) + a(x, y, \bar{t}), \tag{4}$$

where a represents the external ablation/accumulation rate, α is the constant of erosion and R_1 and R_2 are the principal radii of curvature of S at $(x, y, H(x, y, \bar{t}))$.

It is well known that

$$\frac{1}{R_1} + \frac{1}{R_2} = \frac{\partial_x^2 H \left(1 + (\partial_x H)^2\right) + \partial_y^2 H \left(1 + (\partial_y H)^2\right) - 2\partial_{xy}^2 H \, \partial_x H \, \partial_y H}{(1 + (\partial_x H)^2 + (\partial_y H)^2)^{3/2}}$$

and, linearizing, we obtain

$$\|\nabla S\| = \left(1 + (\partial_x H)^2 + (\partial_y H)^2\right)^{\frac{1}{2}} \approx 1,$$

and

$$\frac{1}{R_1} + \frac{1}{R_2} \approx \partial_x^2 H + \partial_y^2 H.$$

So,

$$\partial_{\bar{t}} H + v_1 \partial_x H + v_2 \partial_y H - v_3 = \alpha(\partial_x^2 H + \partial_y^2 H) + a. \tag{5}$$

2.2. The shallow-ice approximation

Let l represent the mean thickness of the ice and L be the representative lenght of the glacier. Let us call

$$\mu = \frac{l}{L}; \tag{6}$$

since the longitudinal direction is large with respect to the transverse direction, we have $0 < \mu \ll 1$.

Following [4], introduce now a streching transformation of the variables and velocities (notice it is expected that transverse and vertical velocities are much smaller than longitudinal velocities).

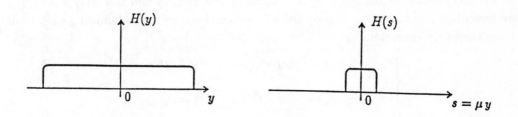

$$\begin{cases} x = x \\ s = \mu y \\ z = z \\ t = \mu \bar{t} \end{cases} \qquad \begin{cases} V_1 = \frac{1}{\mu} v_1 \\ V_2 = v_2 \\ V_3 = \frac{1}{\mu} v_3 \\ t = \mu \bar{t} \end{cases} \tag{7}$$

and, in the new variables, equation (5) is written as follows:

$$\partial_t H + V_1 \partial_x H + V_2 \partial_s H - V_3 = \frac{\alpha}{\mu} (\partial_x^2 H + \mu^2 \partial_y^2 H) + \frac{a}{\mu}. \tag{8}$$

Suppose now

- the velocity field is constant and parallel to the longitudinal direction. So,

$$(V_1, V_2, V_3) \sim (0, 1, 0);$$

- $\dfrac{\alpha}{\mu} \sim 1$ and $g = \dfrac{a}{\mu}$ is finite.

2.3. The variational inequality approach

Let

$$u(x, s, t) = H(x, s, t) - h(x, s) \geq 0, \tag{9}$$

denote the *height of the glacier*. Writing,

$$\nu = \alpha \mu \sim \mu^2,$$

257

$$f_\nu = g + \partial_x^2 h + \nu \partial_s^2 h - \partial_s h,$$

then, in the region $\{u > 0\}$, u satisfies the equation

$$\partial_t u + \partial_s u = \partial_x^2 u + \nu \partial_s^2 u + f_\nu. \tag{10}$$

The natural boundary condition $u = 0$ at the end or at the begining of the glacier, presents the difficulty that the boundary is itself part of the unknown. Where there is no ice it is natural to assume $f^\nu \leq 0$. Following Van Harten and van Hassel (see [3]), we formulate the kinematic description of the ice sheet with the following unilateral complementary conditions

$$\begin{cases} u \geq 0, \quad \partial_t u + \partial_s u - \partial_x^2 u - \nu \partial_s^2 u \geq f^\nu, \\[2mm] u(\partial_t u + \partial_s u - \partial_x^2 u - \nu \partial_s^2 u - f^\nu) = 0, \end{cases} \tag{11}$$

in a larger fixed domain $Q = \Omega \times]0, T[$, where $\Omega =]a, b[\times]0, S[$.

Adding the boundary and initial conditions, for given nonnegative functions u_0 and u_1,

$$u(a, s, t) = u(b, s, t) = 0 \qquad \forall (s, t) \in]0, S[\times]0, T[,$$

$$u(x, 0, t) = u_1(x, t) \qquad \forall (x, t) \in]a, b[\times]0, T[,$$

$$u(x, s, 0) = u_0(x, s) \qquad \forall (x, t) \in \Omega, \tag{12}$$

$$\nu \partial_s u(x, S, t) = 0 \qquad \forall (x, t) \in]a, b[\times]0, T[,$$

multiplying (11) by a nonnegative test function $v = v(x, s)$ in Ω and integrating by parts, we verify that, for each $t > 0$,

$$\int_\Omega (\partial_t u + \partial_s u - f^\nu)(v - u) + \int_\Omega \partial_x u \, \partial_x (v - u) + \nu \int_\Omega \partial_s u \, \partial_s (v - u) \geq 0, \tag{13}$$

for all $v \geq 0$ such that $v = 0$ at $x = a, b$ and $v = u_1$ at $s = 0$.

We have here four different variational inequality problems and their limit relations:

Evolutionary	$u_\nu(t)$	$\longrightarrow$	u_ν^∞	Elliptic
parabolic case	$\chi_{\{u^\nu(t) > 0\}}$	$t \to +\infty$	$\chi_{\{u^{\nu\infty} > 0\}}$	steady-state case
$(\nu > 0, T < +\infty)$				$(\nu > 0, T = +\infty)$
	$\downarrow \nu \to 0$		$\downarrow \nu \to 0$	
Ultraparabolic	$u(t)$	$\longrightarrow$	u^∞	Parabolic
case	$\chi_{\{u(t) > 0\}}$	$t \to +\infty$	$\chi_{\{u^\infty(t) > 0\}}$	limit problem
$(\nu = 0, T < +\infty)$				$(\nu = 0, T = +\infty)$

These convergences are verified, not only for the solutions, but also for the free boundaries, i.e., for the characteristic functions of the glacier regions $\{u > 0\}$, in appropriate spaces, as we shall see. Here we denote $\chi_A(P) = 1$ if $P \in A$ and $\chi_A(P) = 0$ if $P \notin A$.

3. The mathematics of the model

In the first subsection we treat the steady-state case as an elliptic obstacle problem. We prove the existence of a regular solution u_∞^ν of the steady-state problem and we study the behavior of u_∞^ν and of $\chi_{\{u_\infty^\nu > 0\}}$, when $\nu \to 0$, obtaining, in the limit, a parabolic variational inequality problem.

In subsection 2 we consider the evolutionary problem, studying also the behavior of its solution u^ν and of $\chi_{\{u^\nu > 0\}}$, when $\nu \to 0$, remarking that the limit problem is ultraparabolic.

In the third subsection, we study the asymptotic behaviour, when $t \to +\infty$ of the solution of the ultraparabolic problem.

Here we extend the approach of [8], based in a continuous casting model, which was slightly simpler due to the homogeneous Cauchy data. We refer to [7] and [9] for the detailed proofs, some using results that can be found in [5].

3.1. The steady-state case

Let

$$\Omega =]a, b[\times]0, S[, \qquad \Gamma_0 = \{a\} \times]0, S[\cup \{b\} \times]0, S[, \qquad \Gamma_1 =]a, b[\times \{0\},$$

and set

$$V = \left\{ v \in H^1(\Omega) : v|_{\Gamma_0} = 0 \right\}, \tag{14}$$

$$\mathbb{K}_\infty = \left\{ v \in V : v \geq 0, v|_{\Gamma_1} = u_\infty^1 \right\}, \tag{15}$$

where u_∞^1 is a given function such that $u_\infty^1 \geq 0$, $u_\infty^1(a) = u_\infty^1(b) = 0$,

$$a_\nu(u, v) = \int_\Omega \partial_s u v + \nu \int_\Omega \partial_s u \partial_s v + \int_\Omega \partial_x u \partial_x v, \qquad \forall u, v \in H^1(\Omega), \tag{16}$$

where $f_\nu^\infty = g^\infty + \partial_x^2 h + \nu \partial_s^2 h - \partial_s h$, $g^\infty = g^\infty(x, s)$.

The *steady-state problem* ($\nu > 0$) is an elliptic variational problem, defined as follows:

$$u_\nu^\infty \in \mathbb{K}_\infty : \qquad a_\nu(u_\nu^\infty, v - u_\nu^\infty) \geq f_\nu^\infty(v - u_\nu^\infty), \qquad \forall v \in H^1(\Omega). \tag{17}$$

259

Let

$$IK = \{v \in H_0^1(a, b) : v \geq 0\}. \tag{18}$$

The *limit problem* ($\nu = 0$) is the parabolic variational problem defined below:

$$
\begin{cases}
u^\infty(0) = u_1^\infty, \quad u^\infty(s) \in IK \text{ for a.e. } s \in]0, S[, \\
\int_a^b \partial_s u^\infty(s)(v - u^\infty(s)) + \int_a^b \partial_x u^\infty(s)\partial_x(v - u^\infty(s)) \\
\qquad \geq \int_a^b f^\infty(s)(v - u^\infty(s)), \qquad \forall v \in IK, \text{ for a.e. } s \in]0, S[.
\end{cases} \tag{19}
$$

Letting

$$\chi_\nu^\infty = \chi_{\{u_\nu^\infty > 0\}}, \qquad \chi^\infty = \chi_{\{u^\infty > 0\}}, \tag{20}$$

we have the following

Theorem 3.1 *If*

$$g^\infty \in L^2(\Omega), \qquad h \in H^2(\Omega), \qquad u_1^\infty \in IK,$$

then

$$\|u_\nu^\infty - u^\infty\|_{L^2(0,S;H_0^1(a,b))} \leq C\sqrt{\nu}. \tag{21}$$

If $f^\infty = g^\infty + \partial_x^2 h - \partial_s h \neq 0$ a.e., then

$$\chi_\nu^\infty \longrightarrow \chi^\infty, \qquad \text{when } \nu \to 0, \qquad \text{in } L^p(\Omega), \quad 1 \leq p < +\infty. \tag{22}$$

∎

3.2. The evolutionary case

Let

$$IK(t) = \{v \in V : v \geq 0, \ v|_{\Gamma_1} = u_1(t)\}. \tag{23}$$

The evolution case ($\nu > 0, T < +\infty$) corresponds to the following parabolic variational inequality

$$
\begin{cases}
u^\nu(0) = u_0, \quad u^\nu(t) \in IK(t) \text{ for a.e. } t \in]0, T[, \\
\int_\Omega \partial_t u^\nu(t)(v - u^\nu(t)) + \int_\Omega \partial_s u^\nu(t)(v - u^\nu(t)) + \int_\Omega \partial_x u^\nu(t)\partial_x(v - u^\nu(t)) \\
\quad + \nu \int_\Omega \partial_s u^\nu(t)\partial_s(v - u^\nu(t)) \geq \int_\Omega f^\nu(v - u^\nu(t)), \\
\text{for all } v \in IK(t) \quad \text{and for a.e. } t \in]0, T[,
\end{cases} \tag{24}
$$

where $f^\nu = g + \partial_x^2 h + \nu \partial_s^2 h - \partial_s h$, $g = g(x, s, t)$.

Define now

$$R =]0, S[\times]0, T[, \qquad Q =]a, b[\times R, \tag{25}$$

$$\mathcal{K} = \{v \in L^2(R; H_0^1(a, b)) : v \geq 0 \text{ a.e. in } Q\}, \tag{26}$$

$$\Lambda = \partial_s v + \partial_t v, \qquad \mathcal{W} = \{v \in L^2(Q) : \Lambda v \in L^2(Q)\}. \tag{27}$$

The *limit ultraparabolic case* $(\nu = 0, T < +\infty)$ is defined as follows:

$$\begin{cases} u \in \mathcal{K} \cap \mathcal{W}, \quad u|_{t=0} = u_0, \; u|_{s=0} = u_1, \\ \displaystyle\int_Q \Lambda u(v - u) + \int_Q \partial_x u \partial_x (v - u) \geq \int_Q f(v - u), \quad \forall v \in \mathcal{K}. \end{cases} \tag{28}$$

Proposition 3.2 *Suppose that*

$$g \in L^2(Q), \qquad h \in H^2(\Omega),$$

$$u_0 \in H^1(]a, b[\times]0, S[), \; u_1 \in H^1(]a, b[\times]0, T[), \; u_0 \geq 0, \; u_1 \geq 0.$$

Then the variational inequality (28) has a unique solution u satisfying

$$\Lambda u \in L^2(Q), \qquad u \in L^2(R; H^2(a, b)). \tag{29}$$

∎

Theorem 3.3 *Let u^ν be the unique solution of problem (24) under the assumptions of Proposition 3.2.*

Then

$$u^\nu \longrightarrow u \text{ when } \nu \to 0 \text{ in } L^2(R; H_0^1(a, b)) - weak, \tag{30}$$

$$\text{and in } L^\infty(0, T; L^2(\Omega)) - weak^*.$$

If $\chi_\nu = \chi_{\{u^\nu > 0\}}$, $\chi = \chi_{\{u > 0\}}$ and $f = g + \partial_x^2 h - \partial_s h \neq 0$ a.e., then

$$\chi_\nu \longrightarrow \chi \text{ when } \nu \to 0 \text{ in } L^p(Q), \; 1 \leq p < +\infty, \tag{31}$$

∎

3.3. The asymptotic behaviour

Finally, we consider the large time behaviour in the limit case $\nu = 0$. Under natural assumptions on the stabilization of the external ablation/accumulation rates $g(t) \longrightarrow g^\infty$ and the boundary data $u_1(t) \longrightarrow u_1^\infty$. We define

$$\delta(t) = \int_t^{t+1} \int_\Omega |g(\tau) - g^\infty|^2 + \int_t^{t+1} \int_a^b |u_1(\tau) - u_1^\infty|^2. \tag{32}$$

Theorem 3.4 *Let u be the solution of the ultraparabolic problem* (28) *and u^∞ the solution of problem* (19). *Suppose that*

$$u \in L^\infty_{loc}(0, \infty; L^2(0, S; H^1_0(a, b))),$$

$$h \in H^2(\Omega), \qquad u_0 \in H^1(]a, b[\times]0, S[), \quad u_1 \in H^1(]a, b[\times]0, T[), \quad u_0 \geq 0, \ u_1 \geq 0,$$

$$g \in L^\infty(0, \infty; L^2(\Omega)), \qquad g^\infty \in L^2(\Omega), \qquad \delta(t) \longrightarrow 0 \ when \ t \to +\infty.$$

Then

$$u(t) \longrightarrow u^\infty \ when \ t \to +\infty \ in \ L^2(\Omega). \tag{33}$$

If $f^\infty \neq 0$ a.e., $\chi(t) = \chi_{\{u(t)>0\}}$, $\chi^\infty = \chi_{\{u^\infty>0\}}$, then

$$\int_t^{t+1} \|\chi(\tau) - \chi_\infty\|^p_{L^p(\Omega)} d\tau \longrightarrow 0 \ when \ t \to +\infty. \tag{34}$$

■

References

[1] Diaz, J.I., Fowler, A.C.& Schiavi, E., *Mathematical Analysis of a Shallow Ice-Sheet Flow Model*, (to appear).

[2] Fowler, A.C., *Glacier and Ice Sheets*, in "The Mathematics of Models for Climatology and Environment", Ed. J. I. Díaz and J. L. Lions, NATO ASI Series on "Global Change", Springer-Verlag.

[3] van Harten, A. & van Hassel, R.R., *On a singularly perturbed time-dependent free boundary problem*, J. Math. Anal. and Appl., **137** (1989), 70–98.

[4] Hutter, K., *Theoretical Glaciology*, D. Reidel, Dordrecht, 1983.

[5] Lions, J.L., *Quelques Méthodes de Résolution des Problèmes aux Limites Non Linéaires*, Dunod, Paris, 1969.

[6] Morland, L.W., *The flow of ice sheets and ice shelves*, IUTAM Internat. Summer School on Continuum Mech. in Environmental Sci. and GeoPhys., Udine, June 1992.

[7] Rodrigues, J.F., *Obstacle Problems in Mathematical Physics*, North Holland, Amsterdam, 1987.

[8] Rodrigues, J.F. & Santos, L., *Asymptotic convergences in a one-phase continuous casting Stefan problem with high extraction velocity*, IMA Journal of Appl. Math., **42** (1989), 251–267.

[9] Rodrigues, J. F. & Santos, L., *Some free boundary problems in Theoretical Glaciology*, in "The Mathematics of Models for Climatology and Environment", Ed. J. I. Díaz and J. L. Lions, NATO ASI Series on "Global Change", Springer-Verlag.

José-Francisco Rodrigues
CMAF, Universidade de Lisboa
Av. Prof. Gama Pinto, 2
1699 Lisboa Codex, PORTUGAL

Lisa Santos
CMAT & University of Minho
Campus de Gualtar
4710 Braga, PORTUGAL

T ROUBÍČEK

Modelling of microstructure governed by non-quasiconvex variational problems

Abstract: This contribution surveys various numerical approximation techniques applicable to relaxed vectorial variation problems describing, e.g., a steady-state configuration of cristalline martensitic materials.

Keywords: Non-quasiconvexconvex vectorial variational problems, fast oscillations, relaxation, Young measures, numerical approximation.

1. The original problem

A steady-state configuration of elastic both geometrically and meterially nonlinear solid bodies occupying a bounded domain $\Omega \subset \mathbb{R}^n$ with a Lipschitz boundary Γ is governed by a vectorial variational problem

$$\text{(VP)} \qquad \int_\Omega \varphi(x, y(x), \nabla y(x)) \mathrm{d}x + \int_\Gamma \varphi_1(x, y(x)) \mathrm{d}S \to \inf, \qquad y \in W^{1,p}(\Omega; \mathbb{R}^m),$$

where $y : \Omega \to \mathbb{R}^m$ is a displacement, $\varphi : \Omega \times \mathbb{R}^m \times \mathbb{R}^{m \times n} \to \mathbb{R}$ is a potential-energy density and $\varphi_1 : \Gamma \times \mathbb{R}^n \to \mathbb{R}$ is a surface energy density, $1 < p < +\infty$. We admit also $n \neq m$ though in elasticity $n = m$ except some symmetrical situations like, e.g., the anti-plane deformation where $m = 1$. We are especially interested in crystalline materials composed from several phases (typically several lower-symmetry martensitic phase and possibly also higher-symmetry austenite) which may exhibit a microstructure. In this situation, the potential $\varphi(x, r, \cdot)$ has several rotationally invariant wells, each of them corresponds to one phase. Therefore, we must admit a certain nonconvexity of $\varphi(x, r, \cdot) : \mathbb{R}^{m \times n} \to \mathbb{R}$ (more precisely, $\varphi(x, r, \cdot)$ need not be quasiconvex) and then (VP) need not possess any solution so that its extension (=relaxation) must be done. Recall that a function $v : \mathbb{R}^{m \times n} \to \mathbb{R}$ is called quasiconvex if $v(A) \leq \text{meas}(\Omega)^{-1} \int_\Omega v(A + \nabla y(x)) \mathrm{d}x$ for any $A \in \mathbb{R}^{m \times n}$ and any $y \in W_0^{1,p}(\Omega; \mathbb{R}^m)$, cf. [7].

2. The relaxed problem

We will treat a continuous extension of (VP), which preserves a detailed "limit" information about the possible fine oscillations of the gradient of minimizing sequences for (VP), i.e. the so-called microstructure. Neglecting some technicalities, the continuously extended relaxed problem involves Young measures in place of ∇y (cf. e.g.

[1, 2, 4]):

$$
\text{(RP)} \quad \begin{cases} \text{Minimize} & \int_\Omega \int_{\mathbb{R}^{m \times n}} \varphi(x, y(x), A)\nu_x(\mathrm{d}A)\mathrm{d}x + \int_\Gamma \varphi_1(x, y(x))\mathrm{d}S , \\ \text{subject to} & \int_{\mathbb{R}^{m \times n}} A\nu_x(\mathrm{d}A) = \nabla y(x) \text{ for a.a. } x \in \Omega, \\ & y \in W^{1,p}(\Omega; \mathbb{R}^m), \quad \nu \in \mathcal{G}^p(\Omega; \mathbb{R}^{m \times n}), \end{cases}
$$

where $\mathcal{G}^p(\Omega; \mathbb{R}^{m \times n}) = \{\nu = \{\nu_x\}_{x \in \Omega}; \exists \{y_k\}_{k \in \mathbb{N}} \subset W^{1,p}(\Omega; \mathbb{R}^m)$ bounded & $\forall h \in L^1(\Omega; C_0(\mathbb{R}^{m \times n})) : \lim_{k \to \infty} \int_\Omega h(x, \nabla y_k) = \int_\Omega \int_{\mathbb{R}^{m \times n}} h(x, A)\nu_x(\mathrm{d}A)\mathrm{d}x\}$ denotes the set of all so-called gradient L^p-Young measures. The continuous relaxation yields a detailed information about a microstructure described by a Young measure (= a weakly measurable collection of probability measures parametrized by $x \in \Omega$) ν and also avoids a necessity to evaluate the quasiconvex envelope of $\varphi(x, r, \cdot)$, which arises within lower-semicontinuous relaxation, but creates inevitably another difficulty because the set $\mathcal{G}^p(\Omega; \mathbb{R}^{m \times n})$ is not effectively defined.

We assume the following data qualification: φ and φ_1 are Carathéodory functions with a suitable growth and coercivity, namely

$$
c_0|A|^p \leq \varphi(x, r, A) \leq a_0(x) + c_1(|r|^p + |A|^p) , \quad b(x)|r|^\beta \leq \varphi_1(x, r) \leq a_1(x) + c_1|r|^p \quad (1)
$$

with $a_0 \in L^1(\Omega)$, $a_1 \in L^1(\Gamma)$, $c_0, c_1, \beta > 0$, $b \geq 0$ nonvanishing on Γ, $1 < p < +\infty$, and such that $\varphi(x, \cdot, A)$ is Lipschitz continuous in the sense

$$
|\varphi(x, r_1, A) - \varphi(x, r_2, A)| \leq (a(x) + c|r_1|^{p-1} + c|r_2|^{p-1} + c|A|^{p-1})|r_1 - r_2| \quad (2)
$$

with some $a \in L^{p/(p-1)}(\Omega)$ and $c > 0$. Then it is possible to show that (RP) is a correct relaxation of the original problem (VP) in the sense that (RP) always posseses a solution, the set of all solutions to (RP) is stable (more precisely, upper semicontinuous) with respect to a suitable data perturbations, every minimizing sequence for (VP) has a weak* cluster point which solves (RP) and, conversely, every solution to (RP) is attainable by a minimizing net for (VP), for details see [18].

3. Approximation of the relaxed problem

The main aim of this contribution is to present a state of art in approximation theory of the relaxed vectorial variational problem (RP); due to the restricted scope, the results will be presented without proofs, referring to the references, especially to [17, 18].

We will not consider a direct finite-element approximation of (VP) (see e.g. Chipot, Collins, Gremaud, Luskin, Kinderlehrer Nicolaides, Riordan, and Wang [3, 5, 6, 8, 10, 13, 14]) which always converges to (RP) but expectedly very slowly.

Rather we can make a direct finite-element approximation of (RP) by making a triangulation $\mathcal{T}_d$ of Ω such that all elements (=simplexes) from $\mathcal{T}_d$ have diameter less than $d > 0$, and then by restriction of (RP) to y element-wise affine and ν element-wise constant; let us denote the resulted problem by (RP$_d$). It is known that, for $d \to 0$, the solution of (RP$_d$) converges to a solution of (RP) in the sense that

$$\lim_{d \to 0} \min(\text{RP}_d) = \min(\text{RP}) \qquad (3)$$

and every cluster point of every sequence of solutions to (RP$_d$) solves (RP), which can be written shortly in terms of the Kuratowski upper limit "Limsup" as

$$\text{Limsup}_{d \to 0} \text{Argmin}(\text{RP}_d) \subset \text{Argmin}(\text{RP}), \qquad (4)$$

where "Argmin" denotes the set of all solutions to the indicated problem; we refer to [17] or also to [18, Proposition 6.3.7] for details. For $\varphi(x, r, A)$ independent of x and r the scheme (RP$_d$) has been also proposed by Pedregal [16].

Anyhow, the problem how to describe effectively the set $\mathcal{G}^p(\Omega; \mathbb{R}^{m \times n})$ still remains. Therefore, further approximation is needed. The general philosophy is to replace $\mathcal{G}^p(\Omega; \mathbb{R}^{m \times n})$ by another set (either smaller or larger) which can be defined effectively.

As to the former case, one can take all 2^k-atomic pair-wise rank-one connected Young measures, the resulted set being denoted by $\mathcal{G}_k^p(\Omega; \mathbb{R}^{m \times n})$ and the resulted problem by (RP$_d^k$); i.e. this problem consists in minimization of the same functional as in (RP) but for $y \in W^{1,p}(\Omega; \mathbb{R}^n)$ element-wise affine on the triangulation $\mathcal{T}_d$ and ν element-wise constant on $\mathcal{T}_{\lceil}$ and of the form

$$\nu_x = \sum_{l=1}^{2^k} a_l(x) \delta_{A_l(x)} \qquad (5)$$

with $a_l = a_l(x)$ and $A_l = A_l(x)$ satisfying the following recursive conditions invented by Dacorogna [7] and called (H_N)-condition:

$$\left.\begin{array}{l} a_l = \prod_{j=1}^k c_{[(l-1)2^{j-k}]+1,j} \,, \qquad A_l = A_{l,k} \,, \qquad l = 1, ..., 2^k \\[2mm] c_{2i,j} A_{2i,j} + c_{2i-1,j} A_{2i-1,j} = A_{i,j-1}, \\[2mm] c_{2i,j} + c_{2i-1,j} = 1, \quad c_{2i,j}, c_{2i-1,j} \geq 0, \quad \text{Rank}(A_{2i,j} - A_{2i-1,j}) \leq 1, \\[2mm] i = 1, ..., 2^{j-1}, \quad j = 1, ..., k, \qquad A_{1,0} = \nabla y \in \mathbb{R}^{m \times n}. \end{array}\right\} \qquad (6)$$

where [·] denotes the integer part. The scheme (RP$_d^k$) has been proposed by Nicolaides and Walkington [9], see also [19], using basically the same ideas as Dacorogna [7, Section 5.1.1.2] and Kohn and Strang [11, Section 5C].

As to the latter case, one can take all L^p-Young measures which satisfy the Jensen inequality for all quasiconvex functions from a prescibed finite set X, the resulted set being denoted by $\mathcal{G}_X^p(\Omega; \mathbb{R}^{m \times n})$ and the resulted problem by (RP$_{d,X}$); i.e. this problem consists in minimization of the same functional as in (RP) but for $y \in W^{1,p}(\Omega; \mathbb{R}^n)$ element-wise affine on the triangulation $\mathcal{T}_d$ and ν element-wise constant on $\mathcal{T}_{\lceil}$ and satisfying

$$\forall v \in X : \quad \int_{\mathbb{R}^{m \times n}} v(A) \nu_x(\text{d}A) \geq v\left(\int_{\mathbb{R}^{m \times n}} A \nu_x(\text{d}A)\right). \qquad (7)$$

As $\mathcal{G}_k^p(\Omega; \mathbb{R}^{m \times n}) \subset \mathcal{G}^p(\Omega; \mathbb{R}^{m \times n}) \subset \mathcal{G}_X^p(\Omega; \mathbb{R}^{m \times n})$, we have always the two-side estimate

$$\min(\mathsf{RP}_d^k) \geq \min(\mathsf{RP}_d) \geq \min(\mathsf{RP}_{d,X}) \tag{8}$$

provided X contains all linear functions, or (which is basically equally effective) all functions $A \mapsto \pm[A]_{ij}$.

The convergence of the scheme (RP_d^k) is based on the results by Dacorogna [7] and Kohn and Strang [11]: if the rank-one convex envelope of $\varphi_d(x, r, \cdot)$, where φ_d denotes the potential φ averaged over the particular elements of $\mathcal{T}_d$, coincide with the quasiconvex one (cf. [7] for definitions of these envelopes), then

$$\lim_{k \to \infty} \min(\mathsf{RP}_d^k) = \min(\mathsf{RP}_d), \quad \text{and} \tag{9}$$

$$\limsup_{k \to \infty} \mathrm{Argmin}(\mathsf{RP}_d^k) \subset \mathrm{Argmin}(\mathsf{RP}_d). \tag{10}$$

If the rank-one and the quasi-convex envelopes differ from each other less than $\varepsilon/|\Omega|$ or if

$$\min(\mathsf{RP}_d^k) - \min(\mathsf{RP}_d) \leq \varepsilon, \tag{11}$$

then we can say at least that any solution to (RP_d^k) with k large enough is an ε-approximate solution to (RP_d). The difference (11) is actually often rather small and can be justified experimentally by using the two-side estimate (8).

The character of convergence of $(\mathsf{RP}_{d,X})$ is a bit different. We have always the convergence

$$\lim_{X \to X_\infty} \min(\mathsf{RP}_{d,X}) = \min(\mathsf{RP}_d) \tag{12}$$

where $X \to X_\infty$ indicates that X ranges the collection of all finite subsets of the set X_∞ of the quasiconvex functions with a growth less than p; of course, this collection is considered as directed by the inclusion. Then we have also

$$\limsup_{X \to X_\infty} \mathrm{Argmin}(\mathsf{RP}_{d,X}) \subset \mathrm{Argmin}(\mathsf{RP}_d). \tag{13}$$

However, X_∞ is not effectively defined so that this convergence is purely theoretical only. Taking X all $\pm$subdeterminants (and $p > \min(n, m)$), then $G_X^p(\Omega; \mathbb{R}^{m \times n})$ is composed from the so-called polyconvex Young measures (cf. Pedregal [15]) and immediately $\min(\mathsf{RP}_{d,X}) = \min(\mathsf{RP}_d)$ if the quasiconvex envelope of $\varphi_d(x, r, \cdot)$ coincides with the polyconvex one; for the definition of the polyconvex envelope see, e.g., [7]. If they differ from each other by no more than $\varepsilon/|\Omega|$ or if

$$\min(\mathsf{RP}_d) - \min(\mathsf{RP}_{d,X}) \leq \varepsilon, \tag{14}$$

then solutions to $(\mathsf{RP}_{d,X})$ are in a suitable sense also ε-approximate solutions to (RP_d). More precisely, if (y, ν) solves $\min(\mathsf{RP}_{d,X})$, then there is a modified Young measure $\tilde{\nu}$

such that the pair $(y, \tilde{\nu})$ is an ε-approximate solution to (RP_d), i.e.

$$\left.\begin{aligned}
&(y, \tilde{\nu}) \in W^{1,p}(\Omega; \mathbb{R}^m) \times G_H^p(\Omega; \mathbb{R}^{m \times n}), \\
&\int_{\mathbb{R}^{m \times n}} A\tilde{\nu}(\mathrm{d}A) = \nabla y(x), \quad \tilde{\nu} \text{ element-wise constant on } \mathcal{T}_d, \\
&\int_{\tilde{\Omega}} \int_{\mathbb{R}^{m \times n}} \varphi(x, y(x), A)\tilde{\nu}(\mathrm{d}A) \; \mathrm{d}x + \int_\Gamma \varphi_1(x, y(x))\mathrm{d}S \leq \min(\mathrm{RP}_d) + \varepsilon,
\end{aligned}\right\} \quad (15)$$

and certain momenta of ν and $\tilde{\nu}$ coincide with each other, namely

$$\forall v \text{ subdeterminant}: \quad \int_{\mathbb{R}^{m \times n}} v(A)\nu_x(\mathrm{d}A) = \int_{\mathbb{R}^{m \times n}} v(A)\tilde{\nu}_x(\mathrm{d}A). \quad (16)$$

This fact can be proved by taking $\tilde{\nu} \in G^p(\Omega; \mathbb{R}^{m \times n})$ element-wise constant such that $\int_{\mathbb{R}^{m \times n}} A\tilde{\nu}(\mathrm{d}A) = \nabla y(x)$ and $\int_\Omega \int_{\mathbb{R}^{m \times n}} \varphi_d(x, y, A)\tilde{\nu}(\mathrm{d}A)\mathrm{d}x = \varphi_d(x, y, \nabla y(x))^{\mathrm{qc}}\mathrm{d}x$, where $(\cdot)^{\mathrm{qc}}$ denotes the quasiconvex hull; such $\tilde{\nu}$ always exists due to the assumed coercivity of φ, cf. (1). The pair $(y, \tilde{\nu})$ apparently satisfies (13). Moreover, since X contains $\pm\mathrm{adj}_k$ (here adj_k denotes some subdeterminant of the order k, we have $\int_{\mathbb{R}^{m \times n}} \mathrm{adj}_k(A)\nu(\mathrm{d}A) = \mathrm{adj}_k(\nabla y(x))$. Also we have $\int_{\mathbb{R}^{m \times n}} A\nu(\mathrm{d}A) = \nabla y(x)$ at our disposal. Then every $\tilde{\nu} \in G^p(\Omega; \mathbb{R}^{m \times n})$ such that $\int_{\mathbb{R}^{m \times n}} A\tilde{\nu}(\mathrm{d}A) = \nabla y(x)$ satisfies also $\int_{\mathbb{R}^{m \times n}} \mathrm{adj}_k(A)\tilde{\nu}(\mathrm{d}A) = \mathrm{adj}_k(\nabla y(x))$ from which (16) already follows.

To implement the scheme $\min(\mathrm{RP}_{d,X})$ with X consisting from $\pm$subdeterminants, we can always consider ν as a convex combination of a finite number (namely $1 + \sum_{k=1}^{\min(m,n)} \binom{m}{k}\binom{n}{k}$ Dirac measures. The tolerance ε from (14) can be again justified experimentally by means of the two-side estimate (8).

Let us still remark that, by introducing suitable envelopes, these results can be generalized for larger X which contains, beside all $\pm$subdeterminants, also a finite number of some quasiconvex (but not polyconvex) functions.

Let us also note that the scheme (RP_d^k) results (after a suitable transformation) to a nonconvex mathematical-programming problem with several box-constraints only, while the scheme $(\mathrm{RP}_{d,X})$ with X containing all $\pm$determinants yields a nonconvex mathematical-programming problem with $mn + 1$ linear but also $\sum_{k=2}^{\min(m,n)} \binom{m}{k}\binom{n}{k}$ nonlinear equality constraints on each element. This makes the latter scheme a bit more delicate for calculations but we cannot rely only on the former scheme because no other estimate of the energy error than (8) does not exist in general situations. Numerical examples for model two-dimensional problems with two rotationally invariant wells describing materials having two phases (tetragonal, monoclinic, or cubic) which are or are not rank-one connected have been calculated by Kružík [12], where a detailed numerical experience can be found.

Acknowledgement. This research as well as its presentation on the FBP Conference has been partly supported by a grant of the Academy of Sciences of the Czech Republic No.175101.

References

[1] BALL, J.M., JAMES, R.D.: Fine phase mixtures as minimizers of energy. *Archive Rat. Mech. Anal.* **100** (1988), 13–52.

[2] BALL, J.M., JAMES, R.D.: Proposed experimental tests of a theory of fine microstructure and the two-well problem. *Phil. Trans. Royal Soc. London* A **338** (1992), 389–450.

[3] CHIPOT, M., COLLINS, C., KINDERLEHRER, D.: Numerical analysis of oscillations in multiple well problems. *Numer. Math.* **70** (1995), 259–282.

[4] CHIPOT, M., KINDERLEHRER, D.: Equilibrium configurations of crystals. *Arch. Rational Mech. Anal.* **103** (1988), 237–277.

[5] COLLINS, C., LUSKIN, M.: Numerical modelling of the microstructure of crystals with symmetry-related variants. In: *US-Japan Workshop on Smart/Inteligent Materials and Systems* (Eds. I.Ahmad et al.), Technomic Publ. Comp., Lancaster, 1990, pp.309–318.

[6] COLLINS, C., LUSKIN, M., RIORDAN J.: Computational results for a two-dimensional model of crystalline microstructure. In: IMA Vol. in Math. and Applications **54** *Microstructure and Phase Transitions* (Eds. D.Kinderlehrer, R.James, M.Luskin, J.L.Ericksen), Springer, New York, 1993, pp.51–56.

[7] DACOROGNA, B.: *Direct Methods in the Calculus of Variations*, Springer, Berlin, 1989.

[8] GREMAUD, P.: Numerical analysis of a nonconvex variational problem related to solid-solid phase transition. *SIAM J. Numer. Anal.* **31** (1994), 111–127.

[9] NICOLAIDES, R.A., WALKINGTON, N.J.: Computation of microstructure utilizing Young measure representations. In: *Recent Advances in Adaptive and Sensory Materials and their Applications.* (C.A.Rogers, R.A.Rogers, eds.) Technomic Publ., Lancaster, 1992, pp.131–141.

[10] KINDERLEHRER, D., NICOLAIDES, R.A., WANG, H.: Spurious oscillations in computing microstructures. (preprint)

[11] KOHN, R.V., STRANG, G.: Optimal design and relaxation of variational problems. *Comm. Pure Appl. Math.* **39** (1986), 113–137, 139–182, 353–377.

[12] KRUŽÍK, M.: Numerical approach to double well problems. (submitted)

[13] LUSKIN, M.: Numerical analysis of microstructure for crystals with a nonconvex energy density. In: *The Metz Days Surveys* 1989-90 (M.Chipot, J. Saint Jean Paulin, eds.), Pitman Res. Notes in Math., Longman, 1991, pp.156–165.

[14] LUSKIN, M.: Approximation of a laminated microstructure for a rotationally invariant, double well energy density. IMA Preprint No. 1325, Minneapolis, 1995.

[15] PEDREGAL, P.: Laminates and microstructure. *Euro. J. Appl. Math.* **4** (1993), 121–149.

[16] PEDREGAL, P.: On the numerical analysis of non-convex variational problems. *Numer. Anal.* (submitted)

[17] ROUBÍČEK, T.: A note about relaxation of vectorial variational problems. In: *Calculus of Variations, Applications and Computations.* (Eds.: C.Bandle et al.) Pitman Res. Notes in Math. Sci. **326**, Longmann, 1995, pp. 208–214.

[18] ROUBÍČEK, T.: *Relaxation in Optimization Theory and Variational Calculus.* W. de Gruyter, Berlin, 1996 (in preparation)

[19] WALKINGTON, N.J.: Numerical approximation of non-convex variational problems. (preprint)

Mathematical Institute, Charles University

Sokolovská 83

CZ-186 00 Praha 8, Czech Republic

and

Institute of Information Theory and Automation
Academy of Sciences
Pod vodárenskou věží 4
CZ-182 08 Praha 8, Czech Republic
e-mail: `roubicek@karlin.mff.cuni.cz`

A STANCU

Self-similarity in the deformation of planar convex curves

Abstract: This paper outlines the resemblance between the self-similar evolutions of a crystalline curvature flow for closed, convex polygonal curves in the plane and the self-similarity of the planar anisotropic curvature flow for closed, convex, smooth curves.

We announce here our result on the uniqueness of self-similar solutions of J. E. Taylor's deformation of curves by crystalline curvature, which proves, under a certain symmetry assumption, one of her earlier conjectures on the subject.

1991 Mathematics Subject Classification: Primary 53-02, 58F25, 53A04.

1. Introduction.

Motion of planar, convex curves that evolve piecewise smoothly in time appears in the dynamics of two-phase systems in the modern continuum thermodynamics, assuming the curve to be an evolving interface separating two bulk phases ([2], [10].) In the following, we restrict ourselves to convex embedded curves only.

The mechanical laws in conjunction with the thermodynamical conditions lead to an evolution equation of the interface. If $\Gamma = \Gamma(t)$ is an evolving interface, then its normal velocity V satisfies the following partial differential equation:

$$(1) \qquad \beta(\vec{n}) \cdot V = \sum_{i=1}^{2} \frac{\partial}{\partial x_i} \left(\frac{\partial}{\partial p_i} f(\vec{n}) \right) + C,$$

where

$\vec{n}$ is the inward unit normal vector to $\Gamma(t)$,

$\beta(\vec{n})$ is a positive kinetic coefficient,

$f(\vec{n})$ is the interfacial energy density, a non-negative function, which can be extended to $\mathbf{R}^2$ by $f(\lambda q) = \lambda f(q)$, for any q unit vector and $\lambda > 0$, and $f(0) = 0$,

C is a constant.

Definition 1.1: *We say that an interface $\Gamma(t)$ evolving by (1) is geometrically self-similar if $\Gamma(t) = \mu(t)\bar{\Gamma}$ for some fixed compact shape $\bar{\Gamma}$.*

271

Consider a smooth interface and let $C = 0$, $\beta = 1$ and f constant on all the unitary directions. We then obtain the curvature flow $V = k$, where k is the curvature function of $\Gamma(t)$. This is known as the curve-shortening equation corresponding to a smooth isotropic interface ([5], [7], [9]). The curvature flow is the flow on the space of smooth, closed curves along the negative gradient of the length functional called the curve shortening flow because, in a certain sense, it decreases the length of the curve in the most efficient way. Embedded planar curves moving by this equation shrink to a point and become asymptotic to a shrinking circle. If the initial curve is a circle, one can notice that the evolving curves will be smaller and smaller circles. In fact, this is the only self-similar solution of the curve shortening equation ([5]).

From many interesting points of view it is natural to consider anisotropic interface energy and kinetic coefficient. A particular case of this is the anisotropic curvature flow $V = \gamma k$, where γ is some given function of direction which is smooth and strictly positive. This case has been studied by many authors in a variety of contexts ([4], [6], [8]). One of them is M. Gage's appproach for a symmetric weight function γ, $\gamma(\vec{n}) = \gamma(-\vec{n})$ on all the unitary directions ([6], [8]). Considering convex smooth interfaces, he shows that, if γ can be decomposed as $\gamma = \frac{\tilde{h}}{\tilde{k}}$, where $\tilde{h}$ and $\tilde{k}$ are the support function and the curvature respectively of some smooth, strictly convex body $\tilde{K}$, then the associated weighted curvature flow is the curve-shortening of the unique Minkowski geometry whose isoperimetrix is $\tilde{K}$. Actually, M.Gage and Y. Li proved that every symmetric, smooth, strictly positive γ can be so decomposed ([8]). The evolution equation of a convex body of support function h and curvature k takes then the form:

$$(2) \qquad\qquad \left(\frac{h}{\tilde{h}}\right)_t = -\frac{k}{\tilde{k}},$$

where $\frac{k}{\tilde{k}}$ is the Minkowski curvature of the curve at the point with Euclidean curvature k. (Note that for $\gamma \equiv 1$ this is the motion of the interface by curvature.) Under (2) each convex shape shrinks to a point with its shape approaching the Minkowski isoperimetrix. Moreover, the assumption on the periodicity of γ, equivalent to the symmetry of $\tilde{K}$, implies that its associated curvature flow has a unique self-similar solution which is an attractor for any other solution of the flow. ([6], [8].)

Definition 1.2: *We call the Wulff shape of f, ([9]), the set*

$$\Lambda = \Lambda(f) = \{q \in \mathbf{R}^2 \mid q \cdot \vec{n} \leq f(\vec{n})\}.$$

With this terminology, the evolution of curves by weighted curvature corresponds to smooth phase boundaries in (1) interpreting $\mathcal{L} = \int \tilde{h} ds$, $ds = $ the Euclidean arclength, as the total energy of the interface and $\tilde{K}$ its Wulff shape. Mathematically, the weighted curvature flow is the negative gradient flow of the functional $\mathcal{L} = \int \tilde{h} ds$

modulo a tangential component of the flow which only reparametrizes the curve of evolution, but does not change its shape. In this sense, the flow by curvature of an anisotropic interface is a Minkowski curve-shortening flow. One other proof of the uniqueness of self-similar solutions of the anisotropic curvature flow relaxing the assumption on the smoothness of the energy density function to a boundness from both above and below by strictly positive constants can be found in [3].

2. The Crystalline Curvature Flow.

It is of a natural interest in material sciences to consider interfacial energy densities that are continous, but have derivatives with jump discontinuities. A particular class of energies with this property, called crystalline, have Wulff shapes that are polygonal. Crystalline energies are compatible with polygonal interfaces. In this case (1) becomes:

$$(3) \qquad\qquad b_i V_i = a_i L_i^{-1} - F,$$

where i refers to the i-th side of the interface ([10]). Independently, J.E. Taylor proposed (3) for the special case $F = 0$. [12]

Taylor defines the motion of polygonal curves by crystalline curvature so that the normal velocity of each segment is inversely proportional to the length of that segment [12].

Definition 2.1: *Let K and $\tilde{K}$ be planar, convex polygons with the same number of sides, $n > 4$, respectively parallel. If l_i and $\tilde{l}_i$ are the lengths of the corresponding parallel sides of K and $\tilde{K}$, and $h_i = \sup_{x \in K} < x, n_i >$, $\tilde{h}_i = \sup_{x \in \tilde{K}} < x, \tilde{n}_i >$, where n_i, $\tilde{n}_i$ are the outward normal unit vectors to l_i, respectively $\tilde{l}_i$ then:*

$$(4) \qquad\qquad (h_i)_t = -\frac{\tilde{h}_i \tilde{l}_i}{l_i} \qquad \text{for all } 1 \leq i \leq n.$$

Notice that if the origin O lies inside K and $\tilde{K}$, then $h_i = \text{dist}(O, l_i)$, $\tilde{h}_i = \text{dist}(O, \tilde{l}_i)$.

As in the smooth case, we have that for embedded curves the flow is well defined up until the time that the area vanishes ([7], [1])

Theorem 2.1:([12]) *The motion by crystalline curvature is defined and continous for any polygonal curve, up until the time that all line segments become of zero length. Polygonal curves remain polygonal under this motion.*

J. E. Taylor conjectured ([11]) that the only self-similar solutions K of (4) are homotheties of $\tilde{K}$, provided that $\tilde{K}$ is not a parallelogram. This was independently conjectured by Angenent and Gurtin. ([2]) It is easy to show that, in the case $n = 4$, there are many self-similar solutions.

273

The formula (4) in the form:

$$(5) \qquad \left(\frac{h_i}{\tilde{h}_i}\right)_t = -\frac{\tilde{l}_i}{l_i}$$

resembles strongly the symmetric form of the curvature flow of curves in a Minkowski geometry, as defined by M. Gage (2). Notice that the substitution of $\frac{k}{k}$ by $\frac{\tilde{l}_i}{l_i}$, since $k = \tilde{k} = 0$ except for a finite set of points, is motivated by the definition of the Euclidean curvature which implies $k = \frac{d\theta}{ds} \approx \frac{\Delta\theta}{\Delta s}$ for small Δs's. The similarity of the two equations led to the possibility of an approach for the uniqueness of self-similar solutions in the crystalline case close to the one for the uniqueness of self-similar solutions of the anisotropic curvature flow.

3. Results and Conclusions.

Consider now Taylor's crystalline flow as defined above (4), with a centrally symmetric reference body $\tilde{K}$. Notice that a homothety of $\tilde{K}$ will flow self-similarly under these evolution equations, so there is, at least, one self-similar solution to the flow.

Following the approach of M. Gage ([6]) for the smooth case of the anisotropic flow, we define a unique Minkowski geometry for which $\tilde{K}$ is its isoperimetrix. The existence of a self-similar solution is equivalent to a constant in time isoperimetric ratio $\frac{\mathbf{L}^2}{\mathbf{A}}$, where $\mathbf{L}$ is the Minkowski length of K, while $\mathbf{A}$ is its area.

This will imply an identically zero Bonessen functional of K relative to $\tilde{K}$, $\mathbf{B}_K(\rho) = \rho\mathbf{L} - \mathbf{A} - \rho^2 \text{Area}(\tilde{K})$, for ρ between the inner, ρ_{in}, respectively, the outer radius, ρ_{out}, of the minimal annulus of K relative to $\tilde{K}$.

The main argument is using an integral geometry result to show that, assuming K and $\tilde{K}$ are non-homothetic to each other, the Bonnesen functional can be zero on the interval $[\rho_{in}, \rho_{out}]$ only in the polygonal case $n = 4$, where was already known that more than one self-similar solution exists. It turns out that $n = 4$ is an exceptional case because the set of normal directions, non-parallel to the ones in which either ρ_{in} or ρ_{out} is reached, has measure zero.

The symmetry of $\tilde{K}$ is essential in the last step and this is the only place where it is used.

Therefore we have proved:

Theorem 3.1: *If $n > 4$ and $\tilde{K}$ is symmetric with respect to the origin, then the only self-similar solution of* (1) *is a homothety of $\tilde{K}$.*

Moreover we have:

Theorem 3.2: *Every closed polygonal embedded curve in the plane evolving by crystalline curvature becomes asymptotic to the self-similar flow.*

274

The outline of the proof is the following. We first show that as any polygonal convex curve evolves, the above isoperimetric ratio decreases to its minimum, as the area enclosed by the curve goes to zero. Since for the Minkowski geometry the isoperimetric ratio is minimized only for the isoperimetrix, the conclusion follows.

The question of uniqueness and attraction of self-similar solutions for the motion of convex curves by crystalline curvature for a nonsymmetric reference body is still open, as well as, for the curve-shortening flow in a non-symmetric Minkowski geometry. Solving one of these problems will provide a good insight into the other question.

One may also hope to use the behaviour of smooth convex curves deforming under the weighted curvature flow, to get some understanding of the evolution of polygonal curves under a more general crystalline flow (3). Even in the simple case of a rectangle while taking $F = 0$, there exist $\{a_i\}_{i=1,2,3,4}$ and $\{b_i\}_{i=1,2,3,4}$ such that there are no self-similar solutions to the associated evolution equation. It is possible that one of the dimensions of the rectangle approches zero faster than the other, so that the crystal shrinks to a point, but its asymptotic shape is that of a 'needle' ([10]). However, following the smooth case solved by M. Gage and Y. Li ([8]), we believe that for $n > 4$, any even energy function can be uniquely decomposed in $\tilde{h}_i \tilde{l}_i$ for a centrally symmetric polygonal shape, $\tilde{K}$, (the isoperimetrix of a unique Minkowski geometry) and therefore there is a unique self-similar solution associated to its flow.

References.

[1] Angenent S., 1990, *Parabolic Equations for Curves on Surfaces. Part I. Curves with p-Integrable Curvature*, Annals of Math 132, p.451-483 [2] Angenent S., Gurtin

M., 1989, *Multiphase Thermomechanics with Interfacial Structure 2. Evolution of an Isothermal Interface*, Archive for Rat. Mech. and Anal. , 108, p.323-391.

[3] Dohmen C., Giga Y., Mizoguchi N., 1993 *Existence of Selfsimilar Shrinking Curves for Anisotropic Curvature Flow Equations*, preprint, to appear in Calc. Var.

[4] Fukui T., Giga Y., 1992 *Motion of a Graph by Nonsmooth Weighted Curvature*, preprint, to appear in Proc. of the First World Congress of Nonlinear Analysts.

[5] Gage M., 1984 *Curve Shortening Makes Convex Curves Circular*, Invent. Math. 76, p.357-364.

[6] Gage M., 1993 *Evolving Plane Curves by Curvature in Relative Geometries*, Duke Math. Journal, 72, p.441-466.

[7] Gage M., Hamilton R. 1986 *The Heat equation Shrinking Convex Plane Curves*, J. Diff. Geometry 23, p.69-96.

[8] Gage M., Li Y. 1994 *Evolving Plane Curves by Curvature in Relative Geometries II* , Duke Math. Journal, 75, p.79-98.

[9] Grayson M., 1987 *The Heat Equation Shrinks Embedded Plane Curves to Points* , J. Diff. Geometry 26, p.285-314.

[10] Gurtin M., 1993 *Thermomechanics of Evolving Phase Boundaries in the Plane* , Clarendon Press, Oxford.

[11] Taylor J. E., 1991 *Constructions and conjectures in crystalline nondifferential geometry,* Proceedings of the Conference in Differential Geometry, Rio de Janeiro, 1988, Differential Geometry p.321-336, Pitman Monographs Surveys Pure Appl. Math.52, Longman, Sci. Tech., Harlow.

[12] Taylor J. E., 1993 *Motion of curves by crystalline curvature, including triple junctions and boundary points,* Diff. Geom.: Partial Diff. Eqs. on Manifolds (Los Angeles, CA, 1990) p.417-438, Proc. Sympos. Pure Math., 54, Part 1, AMS, Providence, RI.

Alina Stancu
Department of Mathematics
University of Rochester
Rochester, NY, 14627
USA
e-mail: stancu@gauss.math.rochester.edu

J L VAZQUEZ

The free boundary problem for the heat equation with fixed gradient condition

1. Introduction

The Stefan problem is a very well-known example of free boundary problem (FB problem) for the heat equation. In its one-phase formulation it consists of the determination of a *domain* in space-time, Ω, a subset of $Q_T = \mathbf{R}^N \times (0, T)$, and a *function* $u(x, t)$ defined and positive in such domain, which represents the temperature of the phase under consideration, and satisfies a parabolic PDE, typically the heat equation. We have thus

(1) $$u_t = \Delta u, \quad u > 0 \quad \text{in } \Omega.$$

In FB problems with second-order equations two conditions are given on the moving interface, Γ, which is the a priori unknown lateral boundary of Ω, $\Gamma = \partial\Omega \cap Q_T$. In the standard Stefan problem (SP) these conditions are (i) the condition of temperature continuity

(2) $$u = 0,$$

and (ii) the kinetic condition

(3) $$L\mathbf{v} = -\frac{\partial u}{\partial n}\mathbf{n},$$

where $\mathbf{v}(x, t)$ denotes the normal velocity of motion of the free boundary at a point (x, t) and $\partial u/\partial n$ denotes the gradient of u in the direction of the outward spatial normal $\mathbf{n}$ to Γ, taken as a limit value as $y \to x$, $u(y, t) > 0$ (i.e., from inside Ω). The constant L is the latent heat. In order to complete the conditions we add initial data

(4) $$u(x, 0) = u_0(x) > 0 \qquad \text{for } x \in \Omega_0,$$

where Ω_0 is the initial domain. We have posed the problem in the infinite ambient space $\mathbf{R}^N$. In case a bounded space is taken, appropriate boundary conditions should be given on the fixed boundary. Though clearly formulated in the XIX century, a rigorous theory for this problem was obtained only in the last 4 decades. There exists now an extensive literature of the problem, its numerous variants and applications, cf. e.g. [Ru], [M1].

In this article I will deal with a different free boundary problem for the heat equation that has recently attracted the interest of researchers. The problem consists of finding a function $u(x, t) > 0$ which solves the heat equation (1) in an a priori unknown domain $\Omega \subset Q_T = \mathbf{R}^N \times (0, T)$ with lateral boundary Γ. On Γ we impose the conditions (2) and

$$(5) \qquad\qquad\qquad \frac{\partial u}{\partial n} = -1,$$

This condition replaces (3) and makes the problem quite different from the Stefan problem and its numerous variants studied in the literature. The constant in the second member of (5) is put to 1 as a normalization and can be replaced by any positive number by a simple rescaling. In more general versions of the problem it is replaced by a fixed positive function of x and t. Initial conditions like (4) are also needed. Here we will refer to this problem as Problem (FGP).

Assuming that we have smooth initial data in Ω_0 which are continuous up to the smooth boundary and take the value 0 on $\Gamma_0 = \partial\Omega_0$, we understand by *classical solution* of Problem (FGP) a smooth surface Γ which starts from $\Gamma_0 = \partial\Omega_0$ and a smooth function u defined in a domain Ω with lateral boundary Γ and satisfying (1), (2), (4), (5). For brevity when the context is clear we will simply say that u and not (Ω, u) is the solution. Classical solutions to problem (FGP) in one dimension are relatively easy to construct until the possible occurrence of certain singularities which can be described. The problem is much more difficult in several space dimensions so that a concept of weak solution is needed and has been introduced in [CV].

The article presents a report of progress obtained to this date and known to the author. It also discusses the main applications which use this model under suitable physical assumptions. These applied problems were formulated decades ago, but the time was not ripe then for a complete and rigorous analysis. I will present some highlights of the mathematical development, discussing in particular the existence and properties of classical, weak and limit solutions and the methods of constructing solutions. The text touches also the questions of uniqueness, asymptotic behaviour, relation with other equations and alternative mathematical formulations.

2. The combustion model

The above problem arises in a quite natural way in combustion theory to describe the propagation of curved premixed equi-diffusional deflagration flames in the limit of high activation energy. This is an asymptotic method which simplifies the complicated system of nonlinear equations describing the process of combustion on the basis of physically sound approximations, making it thus amenable to further qualitative analysis. As in all simplified models it is of great importance to keep in mind the approximations involved when analyzing the meaning of the mathematical results. Therefore, it will not be useless to review the main lines of the well-known derivation of the thermo-diffusive model for flame propagation together with the high activation energy asymptotics. The main assumption in the present modelization is that of taking to the limit the activation energy of the chemical reaction.

As explained in the classical textbooks, the problem of flame propagation can be written in its generality as a system of PDE's (conservation laws) for the variables ρ, density of the mixture, $\mathbf{v}$, velocity, p pressure, T temperature, and Y_α, the mass fractions of the different combustible components and products, connected also by suitable constitutive relations. Obviously, the chemistry is a crucial part of the problem and usually many, even hundreds of reactions are involved, which complicates enormously the analysis. But in some basic respects such a complication is not necessary and as a first approximation we will consider the simple case of one fuel and one oxidizer giving one product according to the overall irreversible chemical reaction

$$(6) \qquad F + \nu_O O_2 \to \nu_P P + (q),$$

where a mass ν_O of oxygen is consumed per unit mass of fuel to yield a mass ν_P of products plus a thermal energy q. The system is written as follows. First, we have the hydrodynamic equations

$$(7) \qquad \frac{\partial \rho}{\partial t} + \nabla \cdot (\rho \mathbf{v}) = 0,$$

$$(8) \qquad \frac{\partial}{\partial t}(\rho \mathbf{v}) + \nabla \cdot (\rho \mathbf{v} \mathbf{v}) = -\nabla p + \nabla \cdot \tau',$$

where we have neglected the effect of gravity and τ', the viscous stress tensor is given by the Navier-Stokes law. We then have the energy balance equation for the common temperature of the premixed flame

$$(9) \qquad \frac{\partial}{\partial t}(\rho c_p T) + \nabla \cdot (\rho c_p \mathbf{v} T) = \nabla \cdot (k \nabla T) + q W_F.$$

According to the *isobaric approximation*, usually accepted in the description of slow deflagrations, pressure variations can be neglected but for the momentum equation (8). In particular, the corresponding work terms have been neglected in the second member of (9). Besides, one usually writes $k = \rho c_p D_T$, where D_T is the thermal diffusivity. Moreover, we have the component equations

$$(10) \qquad \frac{\partial}{\partial t}(\rho Y_\alpha) + \nabla \cdot (\rho \mathbf{v} Y_\alpha) = \nabla \cdot (\rho D_\alpha \nabla Y_\alpha) - W_\alpha.$$

To this we have to add the law of state

$$(11) \qquad p = \rho R T / M$$

and the chemistry is governed by the Arrhenius law

$$(12) \qquad W_F = \rho B Y_F^m Y_0^n e^{-E/RT}, \qquad W_O = \nu_O W_F,$$

where m and n are the orders of reaction. We will see later that the very precise form of the Arrhenius reaction term will not be important for our study, only its asymptotic properties will matter. In formulas (9)-(12) c_p, q, D_T, R, M, B, m, n and E will be considered as functions of T, Y and ρ, but such dependence is not crucial

279

and is neglected in this modelization. The last constant, called the *activation energy* of the reaction, will play a fundamental role in what follows. Actually, the difference in activation energies selects the reactions that really matter in the flame description in a many-step reaction chain, by eliminating those with too small or too large relative energies because their processes are too fast or too slow. Another important constant is the relation between the thermal diffusivity and the diffusivity of the species α

$$(13) \qquad Le_\alpha = \frac{D_T}{D_\alpha}.$$

It is the called the *Lewis number* and is also a crucial parameter in the combustion process. For more details on the formulation cf. [BL], [W], [Z4]. In this generality the problem is mathematically too difficult, but cf. [La].

A way to perform a further mathematical investigation and derive the basic qualitative properties of the solutions consists in passing to a limit situation. In 1938 Zeldovich and Frank-Kamenetski, [ZF], proposed to make a limit analysis for flame propagation for very large E, taking into account the high sensitivity of the Arrhenius factor of the chemical reaction with respect to the temperature when the activation energy is very large. These methods have been very popular in engineering in the study of both *premixed* and *diffusion flames*, and are explained in [Ba], [Z4], [BL], [W] or [Li]. Only recently a mathematically rigorous investigation has been addressed on those issues.

We want to describe a premixed deflagration flame in which a homogeneous mixture of fuel and oxygen is attained. We are interested in the situation of a one-species reaction and we will now assume moreover that the amount of oxygen is considered so large, $Y_O \approx 1$, that only the amount of fuel matters, hence only the fraction $Y_F = Y$ is of importance (deficient species). We can then write a system for the temperature and the fuel mass fraction in the form

$$(14) \qquad \rho(T_t + \mathbf{v} \cdot \nabla T) - \nabla \cdot (\rho D_T \nabla T) = A\rho Y^m e^{-E/RT}$$

$$(15) \qquad \rho(Y_t + \mathbf{v} \cdot \nabla Y) - \nabla \cdot (\rho D_F \nabla Y) = -B\rho Y^m e^{-E/RT}.$$

Here $A = qB/c_p$. The analysis of system (14)-(15) under appropriate initial and boundary conditions is one of the main subjects of current investigation in the mathematical theory of combustion, both for Lewis number $Le = D_T/D_F$ unity or different from one. It is apparent that the hypothesis of equi-diffusion, i.e. $Le = 1$, will simplify considerably the analysis, in fact much more progress has been achieved under this assumption. In the present generality the hydrodynamics appears through the variables $\mathbf{v}$ and ρ. Under the assumption of almost constant pressure the law of state (11) allows to explicit ρ as a function of T alone and the hydrodynamics equations can be solved independently giving $\mathbf{v}$ as a function of x and t.

Let us proceed now under the *assumption of equidiffusion*. We introduce the enthalpy function

$$(16) \qquad H = T + \frac{q}{c_p}Y,$$

which for $Le = 1$ satisfies the equation

$$(17) \qquad \rho(H_t + \mathbf{v} \cdot \nabla H) - \nabla \cdot (\rho D \nabla H) = 0,$$

where $D = D_T = D_F$. Assuming that we work in the whole space, $x \in \mathbf{R}^N$, and assuming that the initial enthalpy is constant, $H(x, 0) = H_0$, we obtain the *Lewis-Elbe* law

$$(18) \qquad H(x, t) = H_0$$

for all x and all t. This is a further simplification from the real problem. It can be justified for non-homogenoeus enthalpy when its relaxation time is much smaller than the characteristic times of the main process. In any case, it allows us to explicit Y in terms of T,

$$(19) \qquad Y = \frac{c_p}{q}(H_0 - T).$$

and substitute in (14) to get an equation for T:

$$(20) \quad \rho(T_t + \mathbf{v} \cdot \nabla T) - \nabla \cdot (\rho D_T \nabla T) = \rho f(T), \quad f(T) = d(H_0 - T)^m e^{-E/RT}.$$

We have accomplished the first stage of our modelization, obtaining a single equation for T, though it still contains the variables ρ and $\mathbf{v}$. Let us recall that we are assuming that $0 \le T \le H_0$. The reaction function in the second member of (20) is defined and positive for $0 < T < H_0$ with $f(T) = 0$ for $T = H_0$ and $T = 0$. Moreover, $f(T)$ has a single maximum for some value $T_* < H_0$.

Our next step will be to investigate the limit situation when the activation energy $E \to \infty$, i.e. when a certain distinguished limit is taken in the reaction function $f(T)$ of (20). In doing this it is important to investigate first the limit for travelling wave solutions which leads to problem (FGP).

3. Travelling wave analysis

A plane travelling-wave solution is most easily realized in a tube setup, where we assume an infinitely long tube filled with fuel and oxygen under the above assumptions, with initial hot temperature, H_0, on one end, say at $x = \infty$, and a fresh cold mixture at $T = T_0 < H_0$ on the other end, $x = \infty$. We then look for plane travelling-wave (TW) solutions, of the form

$$(21) \qquad T(x, t) = T(\xi), \quad \xi = x_1 + ct,$$

where c is the speed of the wave. The flame will travel from $x = \infty$ to $x = -\infty$ with an increasing temperature profile from the cold end to the hot end. Accordingly, the fuel concentration, given by (19), has a decreasing profile along the ξ-axis, going from $Y = (c_p/q)(H_0 - T_0)$ at $x = -\infty$ to $Y = 0$ at infinity. This represents the fact that at the hot end the flame has depleted the fuel (*burnt zone*) and reaction stops. It is also easy to see from the form of the reaction function that $f(T)$ will have a maximum at an intermediate zone, where most of the reaction takes place (*reaction zone*). The zone to the left is called the *fresh zone*.

Following [ZF], see also [Ba], equations (20), (7) are written for a TW as

$$(22) \qquad \rho(v+c)T' = (\rho DT')' + \rho f(T),$$

$$(23) \qquad (\rho(v+c))' = 0,$$

where primes denote differentiation with respect to ξ (or x_1 if you like). The last equation implies $\rho(v+c) =$const$= \rho_0 c$, with ρ_0 the density at $x = \infty$. Then

$$(24) \qquad c\rho_0 T' = D(\rho(T)T')' + \rho(T)f(T).$$

When we try to calculate the form of the TW it is convenient to introduce the functions

$$(25) \qquad U(\xi) = \frac{D}{\rho_0} \int^{\xi} \rho(T(s))T'(s)ds, \quad \Phi(\xi) = U'(\xi).$$

so that equation (24) is analysed as the system

$$(26) \qquad D\frac{\rho}{\rho_0}T' = \Phi,$$

$$(27) \qquad \Phi' = cT' - \frac{\rho}{\rho_0}f(T),$$

whose orbits in the phase plane (T, Φ) are given by the equation

$$(28) \qquad \frac{d\Phi}{dT} = c - \frac{D\rho}{\rho_0^2}\frac{f(T)}{\Phi}.$$

We want to find a connection from $(T = T_0, V = 0)$ to $(T = H_0, V = 0)$ which lies in the $\Phi \geq 0$ half-plane. Given the form of the second-member no such connection exists if $T_0 > 0$! This is the so-called *cold boundary difficulty*, and is due to the form of the reaction term and the simplifications of the problem, that does not admit a complete TW as solution (though such solutions are actually experimentally observed). Since the TW is important as asymptotics in the reaction zone, a simple correction has been found to allow for the existence of a TW. It consists in *cutting the tail* of function f, which becomes a function with compact support in an interval $T_1 = H_0 - h \leq T \leq H_0$. T_1 is the ignition temperature and h measures the temperature-range in which we take into account the reaction effect, which is switched off for values of T less than T_1. The reader will easily show that, after this correction, equation (28) admits a TW connecting $(H_0, 0)$ with a point of the horizontal axis. This point depends continuously from c. Inversely, we can calculate for given reaction function f and temperature jump $H_0 - T_0$ the speed of the corresponding TW, c.

Zeldovich and Frank-Kamenetskii [ZF] understood that a simple and the same time representative analysis of the TW behaviour can be performed in the limit where we

282

assume that the reaction function f concentrates in a narrow zone near $T = H_0$, i.e., when h is very small. In that case, putting $u = H_0 - T$, we can separate the analysis into 2 zones, the left-hand side where $u > h$ and reaction does not occur (fresh zone) and the right-hand side where $\beta(u) = f(T)$ acts but u is very small, $0 < u < h$. A most important quantity in this analysis will be the total reaction energy

$$(29) \qquad\qquad M = \int \beta(u)du.$$

Morever, we disregard in first approximation the density variations, $\rho \approx \rho_0$ (*constant density approximation*). The reader will observe that this assumption is not an essential qualitative restriction in the analysis that follows.

In the first zone which we assume to be the interval $\{-\infty < \xi < 0\}$ we have to solve the problem

$$(30) \qquad\qquad Du'' = cu', \qquad u(0) = h, u(-\infty) = A,$$

(where $A = H_0 - T_0$), which gives as solution

$$(31) \qquad\qquad u(\xi) = A(1 - e^{\lambda\xi}) + he^{\lambda\xi}.$$

with $\lambda = c/D$. On the other hand, for $\xi > 0$ we have to solve $Du'' - cu' = \beta(u)$. Multiplying by u' and integrating from ξ to ∞ we have

$$(32) \qquad\qquad \frac{D}{2}(u'(\xi))^2 + c\int_\xi^\infty (u')^2 d\xi = \int_0^h \beta(u)du.$$

Now, the last quantity is just the total β-integral, M. Hence, we obtain the estimate of the slope at the zone transition, $\xi = 0$:

$$(33) \qquad\qquad Du'(0)^2 = 2M - 2c\int_\xi^\infty (u')^2 d\xi.$$

Now, for a very concentrated function β with constant energy M the last integral can be disregarded to arrive at the asymptotic formula for the slope.

$$(34) \qquad\qquad |u'(0)| \approx \frac{\sqrt{2M}}{D}.$$

Together with the C^1 agreement of both zones and formula (31) we arrive at the limit value for the wave speed

$$(35) \qquad\qquad c = \frac{\sqrt{2M}}{AD}.$$

In fact, this limit process can be taken by considering a sequence of reaction functions β_ε with same total energy M but such that

$$(36) \qquad\qquad \beta_\varepsilon(s) \to M\delta(s), \quad \text{as } \varepsilon \to \infty.$$

In the limit $\varepsilon \to 0$ in the previous TW analysis we observe that the zone $\xi > 0$ attains a constant value $u = 0$ (i.e., $T = H_0$, it is a burned zone), and the effect of the reaction concentrates on the point $\xi = 0$ in the form of a *fixed gradient jump* given by formula (34) and which is *not* related to the wave speed. In (x, t) variables the reaction zone is the line $x = -ct$ and (34) is just the gradient jump condition (5) we were looking for.

Additional Comments. 1) It is to be noted that the division into 3 zones thus obtained and in particular the analysis of the thin reaction zone does not depend on very specific properties of the reaction term, only the concentrated character and the value of the integral M really matter. Besides, it is not a particular feature of plane travelling waves, but it is a valid paradigm for more general curved configurations. This is explained in [Ba], [Z4], by means of an asymptotic analysis with slow and fast times. For the analysis of the internal layer cf. [Fi].

2) This problem is closely related to the famous KPP problem, [KPP], [Fs], studied at the same time (1937), and originated in population dynamics. There is however an essential difference in the assumptions, since the reaction term in the KPP problem is not concentrated and morevoer $\beta'(u)$ has a maximum for $u = 0$. The conclusion is that the *minimal* wave speed is given by a different formula (with $D = A = 1$)

$$(37) \qquad c = \sqrt{2\beta'(0)}.$$

On the other hand, and this important for our presentation, there is no room for a jump condition like (34).

3) The above one-dimensional limit process was rigorously studied in [BNS] (1985), where also the limit for the system of equations for T and Y without the equidiffusion assumption is analysed. Work on many-dimensional TW's has been continued by Berestycki, Larrouturou and their collaborators, cf. [BrL], see also [Ve] and [Vo] and their references. Typically, a curved flame $T(x, y)$ is considered in $S = \{(x, y) \in \mathbf{R}^2, 0 < y < L\}$, solution of

$$(38) \qquad \Delta T - \alpha(y)T_x + g(T) = 0.$$

For more information on the cold boundary difficulty we refer to [BLR]. The analysis of the (T, Y) system with variable densities and diffusivities is done in [Ro], where its connection is shown with the equations of nonlinear diffusion. The lack of stability of the plane travelling waves for the (T, Y) system with Lewis number less than one is an important research problem studied by Sivashinski that falls out of the scope of this lecture, cf. [Si].

4. The existence of solutions with general data

We have seen the asymptotic limit derived for TWs in the combustion model by letting $\varepsilon \to 0$, i.e., $E \to \infty$. This is the origin of the name *high activation energy asymptotics*. The transition line is called the flame front, which in the limit becomes

284

an inifintely thin zone, i.e. a surface, which for a plane TW solution takes the form x_1+ct =constant. Such TWs are particular cases of the free boundary problem (FGP), i.e., (1), (2), (4), (5), where u is rescaled so that the slope $\sqrt{2M}$ becomes 1 and D is also put to 1.

Our next objective is to show that the FB problem (FGP) is attained in the limit of high activation energy in one or several space dimensions for solutions with general initial data, which not necessarily have the TW structure. This project was undertaken by Caffarelli and Vazquez, [CV], and solutions of Problem (FGP) were obtained in the limit of simplified problems (P_ε) of the form

$$(39) \qquad\qquad u_t = \Delta u_\varepsilon - \beta_\varepsilon(u_\varepsilon) \quad \text{in } Q_T,$$

with initial conditions
$$(40) \qquad\qquad u_\varepsilon(x,0) = u_{0\varepsilon}(x),$$

where the $u_{0\varepsilon}$ are C^∞-smooth and nonnegative approximations of u_0, and then taking $\varepsilon \to 0$. This generality is assumed for mathematical convenience; actually, we may take $u_{0\varepsilon} = u_0$ fixed with ε. Equation (39) is just equation (20) with $u = k(H_0 - T)$ as in (19), after assuming the constant density approximation and neglecting convection terms in comparison with the main process of reaction and diffusion (thermo-diffusive model).

Regarding the reaction term we assume that the functions $\beta_\varepsilon : \mathbf{R} \to \mathbf{R}$ are C^∞-smooth, nonnegative and bounded, with $\beta_\varepsilon(s) = 0$ for $s \le 0$ and support in a small neighbourhood of $s = 0$ of size $O(\varepsilon)$. We will keep a constant heat production M, normalized to $1/2$. A simple way of obtaining that is the following. We define the family β_ε in terms of a single function β by

$$(41) \qquad\qquad \beta_\varepsilon(s) = \frac{1}{\varepsilon}\beta(\frac{s}{\varepsilon}).$$

This is very convenient in order to use scaling arguments. We will assume that the function $\beta : \mathbf{R} \to \mathbf{R}$ satisfies the following assumptions:

(i) β is positive in the interval $I = \{0 < s < 1\}$ and 0 otherwise,
(ii) it is a C^∞ function in $[0, \infty)$,
(iii) it is increasing for $0 \le s < 1/2$, decreasing for $1/2 < s \le 1$,
(iv) the integral of β, $\int \beta(s)\, ds$, equals $1/2$.

With this choice, and if the TW analysis has general validity, formula (34) will lead precisely to the jump condition (5). Observe that the term $\beta_\varepsilon(u)$ acts as an absorption term in equation (39). Since $T = H_0 - (u/k)$, it is in fact a reaction term for the temperature, representing the effect of the exothermic chemical reaction.

Let us recall that we want problem (P_ε) to approximate the FB problem (FGP). In doing that a certain liberty exists in choosing the initial data $u_{0\varepsilon}$ and the absorption functions β_ε. But there are definite constraints. One of the difficulties we face consists in finding a solution with a free boundary Γ which starts from the initial boundary Γ_0.

Now, if for example we take compactly supported initial data u_0, functions β_ε with support in the interval $[0, \varepsilon]$ and approximations $u_{0\varepsilon}$ to the data such that $u_{0\varepsilon} \geq \varepsilon$, it clearly follows that the absorption term has no effect and we obtain just positive solutions of the heat equation in Q with no free boundary. Such a difficulty is already studied in the stationary case by Berestycki, Caffarelli and Nirenberg [BCN]. In order to represent in the limit the FB problem $\beta_\varepsilon(s)$ has to be roughly speaking concentrated in a right neighbourhood of $s = 0$ and its mass $M = \int \beta_\varepsilon(s)\,ds$ has to be directly related to the value $u_\nu = -1$ that we seek to obtain on the free boundary in the limit.

5. Limit solutions and weak solutions

Problem (P_ε) admits a unique classical solution $u_\varepsilon \in C^\infty(Q)$, $Q = \mathbf{R}^N \times (0, \infty)$, which is positive everywhere in Q. The Maximum Principle holds. We want (P_ε) to approximate (FGP) as $\varepsilon \to 0$. In the limit $\varepsilon \to 0$ we want to obtain a

$$(42) \qquad u(x, t) = \lim_{\varepsilon \to 0} u_\varepsilon(x, t).$$

This is called a *limit solution*. The question is now *whether u solves the free boundary problem (FGP) and if it does, in which sense?* Uniqueness is the next basic concern.

We want to obtain a classical solution of (FGP) if possible, but the existence of classical solutions will not be guaranteed for general initial data. A natural *weak formulation* for problem (FGP), as introduced by [CV], asks for a domain $\Omega \in \mathbf{R}^N \times (0, T)$ with Lipschitz continuous lateral boundary Γ, and a function $u \in C(\Omega \cup \Gamma)$ such that:

(i) for every test function $\phi \in C_0^\infty(\mathbf{R}^N \times [0, T))$

$$(43) \qquad \int\int_\Omega u\,(\phi_t + \Delta\phi)\,dx\,dt + \int_{\Omega_0} u_0\,\phi\,dx = \int_\Gamma \phi\,d\Sigma\,\cos\alpha,$$

(ii) u vanishes on Γ, and

(iii) the free boundary Γ starts from $\Gamma_0 = \partial\Omega_0$, i.e the section Γ_t at time t converges to Γ_0 as $t \to 0$ in some sense.

In (43) $d\Sigma$ is the area element on Γ and α is the angle formed by the exterior normal $\nu(x, t)$ at a point $(x, t) \in \Gamma$ and the hyperplane $t = $ constant, so that $dS = d\Sigma\,\cos\alpha$ is the space projection of the element $d\Sigma$. The reader will have no difficulty in checking that a classical solution is a weak solution in the sense just defined.

Corresponding definitions apply to the problem posed in a proper subspace D of $\mathbf{R}^N$, bounded or unbounded. Then, Ω is sought as a subdomain of $D \times (0, T)$ and boundary conditions have to be given on the fixed lateral boundary, either of Dirichlet or Neumann type. A typical configuration is a cylindrical domain of the form $D = \omega \times \mathbf{R}$ with homogeneous Neumann (no flux) conditions on $S = \partial\omega \times \mathbf{R}$. The set of test functions in formula (43) has to be changed accordingly.

The main result of [CV] consists in proving that limit solutions exist and under certain conditions they are weak solutions of (FGP).

THEOREM 5.1. *(i) Let u_0 be a Lipschitz continuous, bounded and nonnegative function in $\mathbf{R}^N$. Then the appoximation of previous section produces in the limit (42) a continuous function $u(x,t)$, solution of the heat equation in its positivity set, Ω. The function u is Lipschitz continuous in x and $C^{1/2}$-Hölder continuous in t. It is C^∞ smooth in x and t in Ω.*

(ii) Under the assumption that Δu_0 be strictly negative in the closed set $\overline{\Omega_0}$, and $|\nabla u_0|$ be less than 1 at the boundary of Ω_0, then the limit solution is a weak solution with free boundary Γ given by a Lipschitz continuous function $t = \tau(x)$. At all regular points of the free boundary (i.e., almost everywhere in x for almost every t) we have the boundary condition $\partial_n u = -1$.

The proof of the first part is based on a priori estimates of three kinds: a) integral estimates, b) a Bernstein estimate for ∇u, and c) a $C^{1/2}$-estimate for u_t. The passage to the limit produces a smooth solution of the heat equation in the positivity set Ω and the second member converges to a singular Borel measure supported on the lateral boundary of Ω. Identifying this measure as equivalent to the gradient jump condition (5) is the most delicate part and the proof is only performed in [CV] under the above-listed additional conditions, which are selected to imply $u_t > 0$ in Ω (contracting flame). Let us remark that the main idea of that restriction is monotonicity, and the proof is equally valid if we ask that Δu be strictly positive in Ω_0 and $\partial_n u_0$ be larger that 1 (so that $u_t > 0$ in Ω).

Subsequently, Caffarelli [C] has shown that the Lipschitz bound on spatial derivatives ∇u can be derived for general initial data as a consequence of a powerful *monotonicity formula*.

6. Typical examples of classical solutions

• The plane TWs studied in Section 3 are examples of classical solutions of Problem (FGP). Let us recall their form. Assume without loss of generality that $N = 1$ and they are monotone decreasing. Then they are given by

$$(44) \qquad u(x,t) = \frac{1}{c}(1 - e^{c(x+ct)})_+ ,$$

when they travel towards the negative x-axis (the free boundary recedes towards the support; in combustion terms, the flame advances). On the contrary, for progressing free boundary

$$(45) \qquad u(x,t) = \frac{1}{c}(e^{c(ct-x)} - 1)_+ .$$

For $c = 0$ we get the stationay profile $u(x,t) = (-x)_+$. All these solutions have a flat free boundary. As usual the symbol $(\cdot)_+$ means positive part.

• An example with curved free boundary is constructed in [CV] in the form of a *self-similar solutions* of the form

$$(46) \qquad u(x,t) = (t_1 - t)^{1/2} f(|x|/(t_1 - t)^{1/2}).$$

287

For every $t_1 > 0$ precisely one such solution with compact support is constructed, it solves (FGP) in the classical sense and vanishes identically at time T. Such a solution is the unique limit of the solutions of the approximate problems (P_ε). Moreover, a stability result is established. The free boundary has the form

$$(47) \qquad \Gamma = \{(x, t) : 0 \le t \le t_1, \ |x| = R\sqrt{t_1 - t}\}.$$

The self-similar solutions extended by zero outside of the support are globally Lipschitz functions for $0 \le t < t_1$ with maximum spatial gradient 1 taken at the free boundary. At the extinction point $(x = 0, t = t_1)$ the regularity decreases however, it is only $C^{1/2}$. Besides, it is important to remark that near extinction the free boundary has vanishing curvature radii, which are then comparable to the thickness of the transition zone even for large E, so that the model loses its validity as an asymptotic limit. It conserves however its validity as *intermediate asymptotics,* as explained by [Ba]. It is proved in [CV] that the self-similar solutions (46) are indeed the unique limit solutions of the corresponding approximate problems, and that the Maximum Principle applies to them with respect to other limit solutions. The same applies to the plane TW's. Using the self-similar solutions as comparison terms it is proved in [CV] that

THEOREM 6.1. *If the initial data is compactly supported the limit solution vanishes identically in finite time.*

• Further examples of classical solutions are the *stationary solutions with a hole.* Assume that $N > 1$. We have to find a number $R > 0$ and a function $u(x)$ defined for $x > R$ such that

$$\Delta u = 0 \quad \text{for} \quad |x| > R, \quad u(R) = 0, \quad \partial_n u(R) = -1$$

In two space dimensions, $N = 2$, such a solution takes the form

$$u(r) = R\log(r/R), \quad r = |x|.$$

Suppose now that we fix at $r = 1$ the value of the solution, say $u(1) = A$. Then, R is determined by the equation

$$A = -R\log R.$$

This equation does not admit a solution if $A > A_* = 1/e$, it has a unique solution $R = 1/e$ if $A = A_*$ and it has two solutions if $0 < A < A_*$, corresponding to a big hole and a small hole. Moreover, the radius of the small hole decreases to 0 as $A \to 0$ while the radius of the large hole goes to 1. The small holes show that we can have stationary solutions as close to 0 as we like in a certain region having holes of small radius.

A similar situation happens in $N \ge 3$. Now the solution with hole of radius $R > 0$ takes the form

$$u(r) = \frac{R^{N-1}}{N-2}\Big\{\frac{1}{R^{N-2}} - \frac{1}{r^{N-2}}\Big\}.$$

Observe that the solution is now bounded in its domain $\{|x| > R\}$. This is the famous Zeldovich flame (with $N = 3$, cf. [Z4]).

288

7. Kinetic equation on the free boundary

We now analyze the equation describing the movement of the FB. Let us consider a classical solution with smooth free boundary Γ. The kinetic equation is derived by differentiating along a normal spatial direction the equation $u(x, t) = 0$ which holds at the free boundary. If the free boundary moves along such direction with velocity $\mathbf{v}$ we get $\nabla u \cdot \mathbf{v} + u_t = 0$, from which, thanks to the boundary condition (5): $\partial_n u = -1$, it follows that

$$(48) \qquad \mathbf{v} = \mathbf{n}\, u_t = \mathbf{n}\, \Delta u\,,$$

where $\partial_t u$ and Δu are understood as limits approaching the boundary from the region $\{u > 0\}$). We now observe that for radially symmetric solutions it means (using again the condition $\partial_r u = -1$ on the free boundary)

$$(49) \qquad \mathbf{v} = \mathbf{n}\left(\partial_r^2 u\big|_\Gamma - \frac{N-1}{r(t)}\right).$$

This is in particular true for the self-similar solutions (46). With a bit more of work we prove that for general configurations the formula reads

$$(50) \qquad \mathbf{v} = \mathbf{n}\left(\partial_n^2 u\big|_\Gamma - K(x, t)\right),$$

where $K(x, t)$ is the Gauss curvature at a point of the free boundary as a surface in $\mathbf{R}^N$ for constant t. Formula (50) shows an unexpected relation of this equation with the now famous model of **motion by curvature**, cf. [CGG], [ES], [GH], [Ang].

8. Filtration in compressible porous media

The same type of free-boundary problem in one space dimension was proposed by Florin [Fl] in 1951 in the study of groundwater filtration in compressible media taking into account the effects of connate water and the *modified Darcy law with initial pressure gradient*. The first hypothesis is explained in the modeling by taking into consideration 3 constitutive elements: the free water, the solid matrix and the bound or connate water. Let x represent vertical distance measured downwards from the ground surface, let m, n and s be the respective concentrations and let u, v, w be the respective average velocities. We have $v = w$ since the connate water, located a thin layer next to the solid matrix, moves rigidly attached to it by molecular forces. The second hypothesis says that under those assumptions the standard Darcy law which relates linearly the relative velocity $u - v$ to the pressure gradient $\partial H/\partial x$ (more accurately H is the total pressure head) has to be replaced by the nonlinear Darcy law

$$(51) \qquad u - v = \frac{k}{m}\left(\frac{\partial H}{\partial x} - J_0\right),$$

valid whenever $\partial H/\partial x \geq J_0$, and $u = v$ otherwise. $J_0 > 0$ is the initial pressure gradient, also called the *threshold gradient*, below which no flow occurs. Under standard assumptions of filtration theory for which we refer to [Fl] one gets the equation for H in the filtration zone

$$(52) \qquad \frac{\partial H}{\partial t} = C\frac{\partial^2 H}{\partial x^2},$$

for some $C > 0$. This zone takes the form $\Omega = \{0 < x < r(t)\}$, where $x = r(t)$ is the a priori unknown moving interface between the filtration zone and the immobile zone. On this interface we put the conditions

$$(53) \qquad H = H_0 > 0, \quad \frac{\partial H}{\partial x} = J_0.$$

We add Dirichlet data
$$(54) \qquad H = 0 \quad \text{on } x = 0, t \geq 0.$$

We also add the condition that the interface starts at $t = 0$ at the surface $x = 0$, which eliminates the need for initial conditions on H, but these additional assumptions are not essential for the mathematical treatment. In this way we obtain problem (FGP) after setting $u = H_0 - H$.

Bear [Be, Chapter 5] discusses the above nonlinear Darcy law that he writes in the form
$$(55) \qquad \mathbf{q} = K\mathbf{J}(J - J_0)/J \quad \text{for } J = |\mathbf{J}| \geq J_0,$$

with $\mathbf{q} = 0$ for $J < J_0$. Here $\mathbf{q}$ is discharge and $\mathbf{J}$ hydraulic gradient. The need for the correction J_0 appears in different contexts, in particular in connection with fine-grained cohesive soils. Problems with nonlinear laws like (51), (55) are found in viscoplastic filtration, cf. [BER]. The corresponding movement equations in several space dimensions lead to systems of a more complex form than problem (FGP).

Existence and uniqueness of classical solutions for the one-dimensional problem proposed by Florin was proved by Ventsel' [Vn] in 1960. The free boundary and lateral data are allowed to be variable under certain precise conditions. The proof is based on discretization in time which leads to a FB problem of elliptic type, plus an integral representation for the solutions of this problem.

9. A model for the deflagration-detonation transition

In 1983 Lundford and Stewart [SL] proposed a model for the study of fast deflagration waves in order to understand the transition from deflagration to detonation (DDT), which is closely related to our investigation, see also [S]. They still deal with an ideal premixed gas undergoing a one-step Arrhenius reaction in the limit of high activation energy. But now the modelization leads to a one-dimensional two-phase problem. The normalized temperature satisfies Burger's equation

$$(56) \qquad T_t = T_{xx} + TT_x,$$

on both sides of a line $x = \zeta(t)$ where the reaction is concentrated, which is a priori unknown but for the initial location ζ_0, and is determined from the two following data: the temperature is prescribed

(57) $$T(\zeta(t), t) = T_s,$$

as well the gradient jump

(58) $$T_x(\zeta(t)+), t) - T_x(\zeta(t)-), t) = -1.$$

We add suitable intial data and contiditions as $x \to \pm\infty$. When $T(x, 0) = T_s$ for $x \geq \zeta_0$ we get the one-phase problem (FGP) after putting $u = T_s - T$ and *disregarding the convective term TT_x*. The existence and stability of travelling waves were studied by Brauner, Lunardi and Schmidt-Lainé in a series of papers, see [BLS] and its references. The question of well-posedness of the (two-phase) Cauchy Problem is solved by Bertsch, Hilhorst and Schmidt-Lainé [BHS]. Their method is based on the consideration that the problem admits an implicit formulation in the form

(59) $$T_t = T_{xx} + TT_x + (H(T - T_s))_x,$$

where H is the Heaviside function. In this way the free boundary disappears and we enter the theory of quasilinear parabolic equations. It is to be noted that such an approach does not work in several space dimensional. A multidimensional analysis of the stability of travelling waves is done in [BLS2] in the framework of fully nonlinear parabolic equations.

10. The elliptic-parabolic model for partially saturated porous media

The equation

(60) $$\partial_t c(u) = \Delta u$$

has been proposed to describe fluid flow in a partially saturated porous medium, cf. [Be]. Here u is the hydrostatic potential due to capillary suction, c is the moisture content and the dependence $c = c(u)$ takes the form of a monotone function such that $c(-\infty) = 0$, $c(\infty) = 1$ and morevoer, $c(u)$ is strictly increasing for $u < 0$ and constant for $u > 0$. Hence, the points (x, t) at which $u \geq 0$ correspond to the saturated zone where $c = 1$. In this zone $\partial_t c = 0$ and the equation becomes elliptic, $\Delta u = 0$. On the contrary, for $u < 0$ the flow is unsaturated and obeys the nonlinear parabolic equation

(61) $$\partial_t u = \frac{1}{c'(u)} \Delta u.$$

This problem has been studied in one dimension by van Duyn and Peletier [D], [DP] and by Hulshof in a series of papers, cf. [H] and its references. Suppose for definiteness that the problem is posed for $0 < x < 1$ and $0 < t < T$ with boundary conditions

(62) $$u_x(0, t) = 0, \quad u_x(1, t) = f(t) > 0,$$

and monotine initial conditions $c_0(x) \in [0, 1]$. It is then shown that there appears a continuous curve $x = \zeta(t)$ (the free boundary) separating the saturated and unsaturated zones and on the FB we obtain the conditions

$$(63) \qquad\qquad u = 0, \quad u_x = f(t).$$

They follow from observing that for $x \geq \zeta(t)$ we have $c = 1$, hence $u_{xx} = 0$, so that u_x is constant, namely the prescribed Neumann data. We have thus obtained a problem very close to (FGP) with the correct free boundary conditions. The only difference is that for (FGP) the function c has to be taken as

$$(64) \qquad\qquad c(u) = 1 - u \quad \text{for } u \leq 0.$$

This causes only minor mathematical changes. A complete theory of existence, uniqueness, comparison and regularity is developped in the framework of the elliptic-parabolic theory for equation (60), [H]. See [HH1] for convergence to travelling waves.

However, the problem in several space dimensions has in principle nothing to do with our free boundary problem (FGP), though we will show in Section 13 that it is possible to use a method of extension of the FB problem to an elliptic-parabolic form for radially symmetric solutions.

11. Relation with the Stefan problem. Undercooled solutions

In one space dimension there is a simple connection of our problem with the standard Stefan Problem (SP) described in Section 1. Namely, suppose that we have a solution of problem (FGP) with ambient space $D = \mathbf{R}_+$, positivity domain $\Omega = \{(x, t) : 0 < x < \zeta(t)\}$ and assume also we take Neumann data on the fixed lateral boundary $x = 0$:

$$(65) \qquad\qquad u_x(0, t) = g(t).$$

For instance, we can take the restriction to $x > 0$ of solutions defined in the whole line under the assumption of symmetry in the x variable and then the lateral condition is $g(t) = 0$ (think of the self-similar solutions (46)). We take as new function

$$(66) \qquad\qquad w(x, t) = -1 - u_x(x, t).$$

It safisfies equation (1) with corresponding initial data and Dirichlet data on the fixed lateral boundary:

$$(67) \qquad\qquad w(0, t) = g(t).$$

Let us check the conditions on the moving boundary $\{x = \zeta(t)\}$. Firstly, the gradient condition (5) implies a Dirichlet condition $w = 0$. The second condition is derived from the kinetic equation of Section 7. Thus,

$$(68) \qquad\qquad \mathbf{v} = u_{xx} = -w_x,$$

292

which is precisely the kinetic condition (3). We obtain in this way a solution of the Stefan problem. There is only one caveat to that construction. It is not a priori clear that $w \geq 0$, so that it can happen that we will have a nonstandard from of the problem. The condition $w \geq 0$ will be obtained if $w \geq 0$ on the fixed parabolic boundary, i.e. if $u_x(x,0) \leq -1$ and $g(t) \leq -1$. Now, if $u_x \geq -1$ everywhere in the fixed parabolic boundary we will have a case of **undercooled Stefan problem**, $w < 0$ in Ω. This is precisely what happens for the self-similar solutions (46)! In the more general case where w changes sign inside Ω we cannot interpret the problem in terms of the two variants of the (SP) mentioned.

Inversely, using formula (66) in this type of domain we can produce solutions of problem (FGP) from standard or undercooled solutions of (SP). The only crucial points are: (i) checking that u is zero (i.e, constant, we can always normalize the value). Indeed, we have

$$(69) \qquad \frac{d}{dt}u(\zeta(t),t) = 0,$$

as a consequence of the equation and the SP free-boundary conditions. (ii) Checking that u is positive in Ω. This will depend again on the data of w.

A similar analysis applies to solutions with one interface whose positivity domain extends to $x = -\infty$. Let us take a brief look at the more general one-dimensional case with two interfaces where we have a solution of (FGP) defined in a domain $\Omega = \{(x,t) : \sigma(t) < x < \zeta(t)\}$, not necessarily symmetric, formula (66) produces as before a solution of the heat equation with correct conditions for an (SP) problem on the right-hand side interface $x = \zeta(t)$. On the left-hand interface $x = \sigma(t)$ we have

$$(70) \qquad w = -2, \qquad w_x = \mathbf{v}_l,$$

the velocity of movement fo the left interface. We have a model for two phase transitions at 'temperatures' $w = 0$ and $w = -2$.

There is *no* obvious relation between problems (FGP) and (SP) in several space dimensions, even in the presence of radial symmetry.

12. Existence of classical solutions in N dimensions

In [M2] Meirmanov extended Ventsel's results to local-in-time existence for the following problem in two space dimensions. He considered the equation

$$(71) \qquad \partial_t\theta - \sum_{i,j}\partial_i(a_{ij}(x,t,\theta).\partial_j\theta) + a(x,t,\theta,\nabla\theta) = 0,$$

in a two-dimensional domain of the *special form* $G(t) = \{(x,y) : 0 < x < 1, 0 < y < R(x,t)\}$ with periodic conditions in the x variable, a condition $\theta = f(x,t)$ on the bottom boundary $y = 0$, given initial conditions $\theta = \theta_0(x,y)$ in the initial support and conditions $\theta = 0$ and

$$(72) \qquad \sum_{i,j}a_{ij}\partial_i\theta.\partial_j\theta = g(x,t) \geq a_0 > 0$$

on the free boundary $y = R(x, t)$. Under suitable assumptions on the data a classical solution of this free boundary is obtained for a small time $0 < t < T_*$. But when the equation is the heat equation, the bottom data are constant > 0 and g and θ_0 satisfy very precise assumptions the solution is *unique and global in time*. The method follows his well-known proof of classical solutions for the Stefan problem.

Andreucci and Gianni [AG] prove local in time existence and uniqueness for a two-phase problem that generalizes the above formulation. Namely, they propose to find a decomposition of the ambient domain G into two domains G_+ and G_- varying with time and separated by a smooth surface S. In each domain an equation of the above type is to be solved. Initial data are to be given as well as boundary conditions on the fixed (external) boundary. On the free boundary we have the conditions: $u = 0$ and the gradient jump condition

$$(73) \qquad [|\partial_n u|] = \partial_n^+ u - \partial_n^- u = 1.$$

Galaktionov, Hulshof, Vazquez, [GHV] solve the problem of existence and uniqueness of radially symmetric solutions supported in a ball. Classical solutions are produced that exist until they vanish identically at the origin of coordinates. The proof proceeds in two steps. In the first the *elliptic-parabolic theory* is adapted to several dimensions in the form of a non-standard boundary value problem. In order to keep the notation of the references we will work in this development with nonpositive solutions, just changing u into $-u$, which is anyway closer to the proposed physical model where temperatures in the fresh zone are below the critical temperature. In doing this we have to replace the radial free-boundary condition by

$$(74) \qquad u(\zeta(t), t) = 0, \quad u_r(\zeta(t), t) = 1,$$

where $r = \zeta(t)$ denotes the interface. The crucial observation is that radial (negative) solutions can be naturally extended as solutions to an *elliptic-parabolic* problem on a large fixed ball containing the supporting balls for all $0 \le t \le \tau$. This leads to a problem of mixed type in a fixed domain, which is easier to solve. This is done as follows. We take $N \ge 3$, the adaptations to perform in case $N = 2$ are clear. Consider the equation

$$(75) \qquad (c(u))_t = \Delta u,$$

where $c(s) = \min\{0, s\}$, and suppose that u is a solution. Then u solves the heat equation if $u < 0$, while for $u > 0$ it is harmonic in x. We can extend u to the region $r > \zeta(t)$ by setting $\Delta u = 0$ there, which by (1.1) implies that

$$(r^{N-1} u_r(r, t))_r = 0 \quad \Rightarrow \quad u_r(r, t) = \left(\frac{\zeta(t)}{r}\right)^{N-1},$$

hence

$$u(r, t) = \int_{\zeta(t)}^r \left(\frac{\zeta(t)}{\rho}\right)^{N-1} d\rho = \frac{\zeta(t)}{N-2} \left[1 - \left(\frac{\zeta(t)}{r}\right)^{N-2}\right].$$

This turns u into a radial solution of (75) which has a jump in u_t and Δu across the free boundary. Fixing a ball with radius R containing the support of the original solution for all $t \in [0, T]$, we obtain nonlocal boundary conditions on ∂B_R of the form

$$(76) \qquad\qquad u_r(R, t) = (\zeta(t)/R)^{N-1} \quad \text{(Neumann)},$$

and

$$(77) \qquad u(R, t) = \frac{\zeta(t)}{N-2}\left[1 - \left(\frac{\zeta(t)}{R}\right)^{N-2}\right] \equiv F(\zeta(t), R) \quad \text{(Dirichlet)}.$$

Here $r = \zeta(t)$ is now the a priori unknown level set of $u = 0$. We can eliminate $\zeta(t)$ from these two conditions to obtain

$$(78) \qquad u(R, t) = \frac{R}{N-2}(u_r(R, t)^{\frac{1}{N-1}} - u_r(R, t)) = \mathbf{G}(u_r(R, t)),$$

which however is not completely straightforward to work with, in view of the fact that $\mathbf{G}$ is a nonmonotone function of u_r. Both functions F and $\mathbf{G}$ depend on N. An iterative scheme based on the solution of the filtration problems (76) and (77) allows to find a solution of the mixed problem (78). We note that as a function of ζ, function $F(\zeta, R)$ is increasing for $0 \leq \zeta < R(N-1)^{-1/(N-2)}$ and decreasing for $R(N-1)^{-1/(N-2)} < \zeta \leq R$. We have to choose R appropriately to ensure that we will be in the latter situation, for then, bearing in mind that larger solutions have smaller interfaces, we can set up a monotone iteration scheme.

THEOREM 12.1. *Suppose that v_0 is continuous and radially symmetric on B_R, negative on $\{|x| < \zeta_0\}$, zero on $\{|x| \geq \zeta_0\}$, where $\zeta_0 < R$. Assume that ζ_0 lies in the interval $(R(N-1)^{-1/(N-2)}, R)$ and that $r^{N-1}v_0'(r)$ is bounded. Then there exists $T > 0$ and a unique function $u \in L^2(0, T; H^1(B_R))$ which has a continuous interface $r = \zeta(t)$ such that u is a weak solution to the Neumann problem with (76) and to the Dirichlet problem with (77). The weak formulation is equivalent to*

$$\int_0^T \int_0^{\zeta(t)} (-r^{N-1}\varphi_t c(u) + r^{N-1}u_r \varphi_r)\,dr\,dt + \int_0^{\zeta(T)} r^{N-1}\varphi(r, T)c(u(r, T))\,dr$$

$$= \int_0^{\zeta_0} r^{N-1}\varphi(r, 0)u_0(r)\,dr + \int_0^T \varphi(\zeta(t), t)\zeta(t)^{N-1}\,dt,$$

for all $\varphi \in H^1(Q_T)$. In this sense the pair (u, ζ) is the unique solution. Moreover, the comparison principle holds: if we have two solutions (u_i, ζ_i), $i = 1, 2$ and their initial data v_{0i} and ζ_{0i} are ordered, $v_{01} \leq v_{02}$ and $\zeta_{01} \geq \zeta_{02}$, then the solutions (and interfaces) are ordered in the same way.

In a second step we use von Mises variables to straighten the free boundary and prove that both the solution and its free boundary are smooth.

THEOREM 12.2. *(i) Suppose Ω_0 is a ball B, and u_0 is radially symmetric, continuously differentiable, zero on the boundary, with normal derivative equal to one. Then the combustion problem has a unique continuous solution on some interval $(0, T]$, which is real analytic for $t > 0$.*

(ii) Let $(0, T_m)$ be the maximal open time interval on which a unique analytic solution exists. Then

$$\lim_{t \to T_m} \zeta(t) = 0,$$

and the solution vanishes identically at $t = T_m$.

Similar results hold for other configurations, like data supported in an annulus or in an exterior domain, cf. [GHV]. In the latter case the hole may evolve in different ways depending on the initial data: it can stay (as in the Zeldovich flame commented above), it can expand to fill the whole space or it can contract and even disappear in finite time. This last case, called focusing, will be described in Section 14.

It is also proved there that under the conditions of Theorem 6.1 the solution coincides with the limit solutions of the approximate problems (P_ε), as constructed in [CV].

13. The problem of nonuniqueness

Consider radial symmetric initial data in the form of a hump with compact support and a bell-shaped form. As we have just seen, this gives rise to a classical solution in a certain time interval $0 < t < \tau$. If the initial gradient is very large then the support of the solution begins its evolution by expanding. This is easily shown by comparison with suitable barriers. A solution with compact supoort must eventually collapse to the origin, $r(t) \to 0$ as $t \to \tau$. Hence, there is a time of maximum expansion, say $t_1 > 0$, $t_1 < \tau$, with maximum free boundary radius r_1. Since the problem is invariant under x translations the same picture holds after shifting the solution a fixed distance in space.

Now consider the solution of problem (FGP) corresponding to initial data formed by two humps, the former one plus another similar one centered at a point $(x_1, 0, ...0)$. It is clear that for every value of x_1 larger than $2r_1$ a classical solution of the problem is constructed by just superimposing the separate evolutions of the two humps, since they have disjoint supports. Taking limits we easily conclude that for $x_1 = 2r_1$ there is a possible solution obained by superposition, which develops a point of irregular free boundary at $t = t_1$ where the two separate supports make contact. However, they separate later to undergo extinction in the form of two separate balls. This is a first weak solution for the problem with $x_1 = 2r_1$.

There is however a different continuation where the two supports merge at $t = t_1$ and do not separate in the future. This is most easily seen in 1D by considering first the case $x_1 < 2r_1$ where the two supports meet at a time t_1' before t_1 with nonzero speed, so that any physical continuation implies a nontrivial superposition immedaitgely after t_1'. The coincidence of the two flame fronts makes for a singularity that has to be resolved if we want to continue the solution. The natural way is to cancel the collapsing (inner) interfaces and to continue the solution as the solution of the (FGP) with connected support the union of the two intervals and free boundary consisting of the two outer interfaces. In this process we lose two of the four former flame fronts, precisely the

296

two inner ones which recede until they meet and the reaction freezes. This means that we will lose twice the production rate (normalized to 1) in the solution, which will evolve with different mass. There is no difficulty in checking that such a solution is a weak solution of (P) and is classical for $t \neq t_1'$. Finally, by taking the limit $x_1 \to 2r_1$ we obtain a second solution for the problem with $x_1 = 2r_1$ which coincides the former one for $t < t_1$ but is different for $t > t_1$. In fact, the mass rule gives

$$(79) \qquad \int u_2(x,t)dx - \int u_1(x,t)dx = 2,$$

so that even their extinction times are different. Clearly, one of them is the maximal and another one the minimal solution. A more general picture of nonuniqueness based on these ideas is still pending. It also unclear which solutions should be preferred in the different applications for which this problem is only an approximation.

Connected to this problem is the problem of *pulse splitting*. Suppose that in the situation of the previous example with two humps and $x_1 > 2r_1$ we add a very thin sliver of initial data uniting the initial supports along a thin strip $\{x = (x_1, x') : |x'| < \varepsilon\}$. It can be proved that the connected support splits in two after a short time which is a continuous function of the size of the connecting sliver. Detailed proofs will appear elsewhere. A complete understanding of these phenomena is still needed.

14. Asymptotic behaviour, extinction and focusing

We will describe next two asymptotic situations, known as extinction and focusing. Both are terminal situations for a flame, though for opposite reasons. Complete statements and proofs are given in [GHV].

• As we have said, when u_0 is a compactly supported function the solution vanishes in a finite time $\tau = \tau(u_0)$ called the *extinction time*, i.e.,

$$(80) \qquad u(x,\tau) \equiv 0 \quad \text{and} \quad u(x,t) \not\equiv 0 \text{ for all } t \in (0,\tau).$$

In combustion terms the burnt zone spreads to cover the whole space and the reaction stops by depletion of reactant. The paper [GHV] describes the asymptotic behaviour of the solution as $t \to \tau^-$ near an *extinction point*, x_0, i.e., a point of the set

$$(81) \quad E(u_0) = \{x \in \mathbf{R}^N : \exists \, \{x_n\} \to x \text{ and } \{t_n\} \to \tau^- \text{ such that } u(x_n, t_n) > 0\},$$

called the *extinction set*, which is nonempty under our hypotheses. Radial symmetric solutions, i.e., spheric flames, are considered. The finite-time extinction process splits into three different cases:

(i) Single-point extinction of radial symmetric solutions, $u = u(r,t)$, $r = |x|$. This happens for solutions whose initial support is a ball, but it can also be an annulus with a small inner hole. The limit profile is given by the self-similar solutions (46). In one space dimension single-point extinction can be analyzed without the assunption of radial symmetry. The symmetric one-dimensional case had been analyzed in [HH2].

(ii) Extinction on a sphere $\{|x| = r_0 > 0\}$ for initial support in the form of a thin annulus. The asymptotic profile corresponds to the one-dimensional problem, the transversal directions do not count in the limit.

(iii) The two previous asymptotic behaviours are stable (under perturbation of the data). There is a still a different type of asymptotics, in the form of a self-similar solution with the same time dependence as (46) but with a support in the form of an annulus. In other words, we have single-point extinction by means of an annular self-similar solution with converging size of the order of $O(\tau - t)$. This solution separates the basins of attraction of the two previous modes of extinction. For details and proof of these results we refer to [GHV].

- A completely different extinction mode happens when the solution loses heat and the flame is frozen in finite time. In more mathematical terms, we mean that the support (fresh zone) expands to cover the whole space. Again, working in a radial situation we consider as paradigm the case of initial data which are positive and increasing outside of a ball and leaves a hole near the origin that shrinks to zero at a time $\tau > 0$. We call this phenomenon *focusing*, because the flame front focuses at the origin.

For every $N \geq 2$ there is a self-similar solution of the form

$$(82) \qquad u(x,t) = (t_1 - t)^{1/2} f_1(|x|/(t_1 - t)^{1/2}).$$

where now f_1 is an increasing function with support in the interval $0 < \eta_1 \leq \eta < \infty$ and $f_1(\eta) \sim a_1 \eta$ as $\eta \to \infty$. In 3 dimensions we have explicit values:

$$(83) \qquad f_1(\eta) = \frac{1}{2}\left(\eta - \frac{2}{\eta}\right) \quad \text{and} \quad \eta_1 = \sqrt{2}, \ a_1 = \frac{1}{2}.$$

We can show that this self-similar solution gives the behaviour at focusing for more general (radially symmetric) data. Here is the result proved in [GHV]. Let $N \geq 2$ and let $u_0(r)$ be a smooth, bounded function satisfying

$$(84) \qquad u_0 > 0, \ u_0' > 0 \text{ on } (1,\infty); \ u_0(1) = 0; \quad |u_0'| \leq M, \ |u_0''| \leq M.$$

Then, there exists a unique solution $u(r,t)$ of the radially symmetric problem with initial data u_0 which exists for a time $0 < t < \tau$ and exhibits *finite-time focusing*: there exists $\tau = \tau(u_0) > 0$ such that the unique interface $r = \zeta(t)$ reaches the origin

$$(85) \qquad \liminf_{t \to \tau} \zeta(t) = 0 \quad \text{and } \zeta(t) > 0 \text{ for } t < \tau.$$

We can establish the following focusing description.

THEOREM 14.1. *As $t \to T^-$*

$$(86) \qquad w(\eta, t) \equiv (T - t)^{-1/2} u(\eta(T - t)^{1/2}, t) \to f_1(\eta)$$

uniformly on compact subsets in η. The interface also converges,

$$(87) \qquad \zeta(t) = \eta_1(T - t)^{1/2}(1 + o(1)) \quad \text{as } t \to T^-,$$

where $\eta_1 = \eta_1(N) > 0$ is the unique vanishing point of the function $f_1(\eta)$.

298

In terms of the original variables $\{u, r, t\}$ we have

THEOREM 14.2. *Under the above assumptions*

$$(88) \qquad u(r, \tau) = a_1 r (1 + o(1)) \quad \text{as } r \to 0,$$

where $a_1(N) > 0$ is the constant in the expansion for f_1 at infinity.

Focusing in one dimension is quite different, since then we have two independent support components located initially at $\{x \geq 1\}$ and $\{x \leq -1\}$. Generally, they arrive at the focusing time with a positive, finite speed, so that

$$(89) \qquad \zeta(t) = O(t_1 - t),$$

which is in sharp contrast with (87). The methods of proof used in [GHV] combine a priori estimates and the precise knowledge of the self-similar solutions with dynamical systems ideas and non-standard comparison arguments. These methods can have wide applicability to different free boundary problems for other semilinear and quasilinear heat equations admitting finite time extinction or blow-up. They have two main drawbacks: they have not allowed to study asymmetrical evolution (but for $N = 1$) and they are not well suited to tackle systems.

15. Concluding remarks

We have presented a new type of free boundary problem for the heat equation and advanced some of its mathematical properties. Fundamental problems are still open in several dimensions regarding e.g. uniqueness and singularity formation. Further developments presently discussed are viscosity solutions, stability considerations and a general theory for two-phase problems.

We have derived the model in the context of flame propagation at high activation energy and also in groundwater filtration and we have pointed out the main approximations involved. In combustion future and more realistic analyses must involve the (T, Y) system obtained by removing the conditions of Lewis number unity and the constant enthalpy, as well as studying the convergence as $E \to \infty$. It is also of interest to consider nonlinear diffusion instead of the linear heat equation.

Acknowledgments. These notes are an expanded version of the lecture delivered in Zakopane in June 1995. They reflect not only work of the author and colleagues who are mentioned in the text, but also comments and suggestions of various scientists. I am grateful in particular to H. Berestycki, C. M. Brauner, L. Caffarelli, V. Galaktionov, J. Hulshof, S. Luckhaus and A. Meirmanov. I am deeply indebted to G. I. Barenblatt who was kind enough to read carefully parts of this manuscript and to share with me his large experience in the physics of combustion and porous media, as well as the mathematics of free boundaries. Of course, all faults in the presentation lie with the author, who apologizes for any omissions or inaccuracies. It is to be hoped that the mathematical novelty of the subject will attract the attention of future researchers.

References

[AG] D. Andreucci and R. Gianni, *Classical solutions to a multidimensional free boundary problem arising in combustion theory*, Comm. Partial Differ. Equat., **19** (1994), pp. 803-826.

[Ang] S. Angenent, *Parabolic equations for curves on surfaces. II. Intersections, blow up and generalized solutions*, Annals Math., **131** (1991), pp. 171-215.

[Ba] G. I. Barenblatt, "Similarity, self-similarity and intermediate asymptotics", Consultants Bureau,

[BER] G.I. Barenblatt, V.M. Entov, V.M. Ryzhik, "Flow of fluids through natural rocks", Kluwer Academic Publ., 1990.

[Be] J. Bear, "Dynamics of fluids in porous media", Elsevier, New York, 1972.

[BCN] H. Berestycki, L.A. Caffarelli and L. Nirenberg, *Uniform estimates for regularization of free boundary problems*, In: "Analysis and Partial Differential Equations", Marcel Dekker, New York, 1990.

[BrL] H. Berestycki, B. Larrouturou, "Mathematical modelling of planar flame propagation" Pitman Research Notes in Mathematics, Longman, London, 1990.

[BLR] H. Berestycki, B. Larrouturou, J. M. Roquejoffre, *Mathematical investigation of the cold boundary difficulty in flame propagation*, in "Dynamical issues in combustion theory", Fife, Liñán, Williams eds., IMA volumes in Mathematics amd its Appl., **35**, Springer Verlag.

[BNS] H. Berestycki, B. Nicolaenko and B. Scheurer, *Travelling wave solutions to combustion models and their singular limits*, SIAM J. Math. Anal., **16** (1985), pp. 1207-1242.

[BHS] M. Bertsch, D. Hilhorst and C. Schmidt-Lainé, *The well-posedness of a free-boundary problem arising in combustion theory*, Nonlinear Anal., Theory, Meth. Appl., **23** (1994), pp. 1211-1224.

[BLS] C.M. Brauner, A. Lunardi and Cl. Schmidt-Lainé, *Stability of travelling waves with interface conditions*, Nonlinear Anal., Theory, Meth. Appl., **19** (1992), pp. 455-474.

[BLS2] C.M. Brauner, A. Lunardi and Cl. Schmidt-Lainé, *Multi-dimensional stability analysis of planar travelling waves*, Appl. Math. Lett., **7** (1994), pp. 1-4.

[BL] J.D. Buckmaster and G.S.S. Ludford, "Theory of Laminar Flames", Cambridge University Press, Cambridge, 1982.

[C] L.A. Caffarelli, *A monotonicity formula for heat functions in disjoint domains* , In: "Boundary Value Problems for PDE's and Applications", dedicated to E. Magenes, J.L. Lions, C. Baiocchi Eds, Masson, Paris, 1993, pp. 53-60.

[CV] L.A. Caffarelli and J.L. Vazquez, *A free boundary problem for the heat equation arising in flame propagation*, Trans. Amer. Math. Soc., **347** (1995), pp. 411-441. [UAM prepint 1993].

[CGG] Y. G. Chen, Y. Giga, S. Goto, *Uniqueness and existence of viscosity solutions of generalized mean curvature flow equations*, J. Diff. Geom., **33** (1991), pp. 749-786.

[D] C. J. van Duyn, *Nonstationary filtration in partially saturated porous media: continuity of the free boundary*, Arch.Rat. Mech. Anal., **79** (1982), 261-265.

[DP] C. J. van Duyn, L. A. Peletier, *Nonstationary filtration in partially saturated porous media*, Arch.Rat. Mech. Anal., **78** (1982), 173-198.

[ES] L. C. Evans, J. Spruck, *Motion of level sets by mean curvature*, J. Diff. Geom., **33** (1991), pp. 635-681.

300

[Fi] P. C. Fife, "Dynamics of internal layers and diffusive interfaces" CBMS-NSF Regional Conf. Series in Appied Mathematics # 53, SIAM, Philadelphia, 1988.

[Fs] R. A. Fisher, *The wave of advance of advantageous genes*, Ann. Eugenics, **7** (1937), pp. 355-369.

[Fl] V.A. Florin, *Earth compaction and seepage with variable porosity, taking into account the influence of bound water*, Izvestiya Akad. Nauk SSSR, Otdel. Tekhn. Nauk, No. bf 11 (1951) pp. 1625-1649 (in Russian).

[FK] D.A. Frank-Kamenetskii, "Diffusion and Heat Exchange in Chemical Kinetics", Princeton Univ. Press, Princeton, 1955.

[GHV] V.A. Galaktionov, J. Hulshof and J.L. Vazquez, *Extinction and focusing behaviour of spherical and annular flames described by a free boundary problem*, preprint 1995.

[GH] M. Gage, R. S. Hamilton *The heat equation shrinking convex plane curves*, J. Diff. Geometry, **23** (1986), pp. 69-96.

[HH1] D. Hilhorst and J. Hulshof, *An elliptic-parabolic problem in combustion theory: convergence to travelling waves*, Nonlinear Anal., Theory, Meth. Appl., **17** (1991), pp. 519-546.

[HH2] D. Hilhorst and J. Hulshof, *A free boundary focusing problem*, Proc. Amer. Math. Soc., **121** (1994), pp. 1193-1202.

[H] J. Hulshof, *An elliptic-parabolic free boundary problem: continuity of the interface*, Proc. Roy. Soc. Edinburgh **106A** (1987), pp. 327-339.

[KPP] A. N. Kolmogorov, I. G. Petrovski, N. S. Piskunov, *A study of the equation of diffusion with increase in the quantity of matter and its application to a biological problem*, Bul. Moskov. Gos. Univ. **17**, 1937, pp. 1-26.

[La] B. Larrouturou, *The equations of one-dimensional unsteady flame propagation: existence and uniqueness*, SIAM Jour. Math. Anal., **19** (1988), pp. 1-26.

[Li] A. Liñán, *The structure of diffusion flames*, in "Fluid dynamical aspects of combustion theory", M. Onofri and A. Tesei eds., Pitman Research Notes in Math. Series # 223, Longman Sci. Techn., 1991.

[M1] A. M. Meirmanov, "The Stefan Problem", W. De Gruyter, Berlin, 1992 (Russian edition, "Zadacha Stefana", Nauka, Novosibirsk, 1986).

[M2] A.M. Meirmanov, *On a free boundary problem for parabolic equations*, Matem. Sbornik, **115** (1981), pp. 532-543 (in Russian); English translation: Math. USSR Sbornik **43** (1982), 73-484.

[Ro] J. M. Roquejoffre, *Mathematical analysis of a planar flame model with nonlinear diffusion*, Nonlinear Analysis, TMA, **21** (1993), pp. 745-761.

[Ru] L. Rubinstein, "The Stefan Problem", Trans. Math. Monographs, AMS, vol. 27, 1971. (Russian edition, Zvaigne, Riga, 1967).

[Si] G. I. Sivashinski, *Instabilities, pattern formation and turbulence in flames*, Ann. Rev. Fluid Mech., **15** (1983), pp. 179-199.

[S] D. S. Stewart, *Transition to detonation on a model problem*, Jour. Méc. Théor. Appl., **4** (1985), pp. 103-137.

[SL] D. S. Stewart, G. S. S. Ludford, *Fast deflagration waves*, Jour. Méc. Théor. Appl., **3** (1983), pp. 463-487.

[Ve] J. M. Vega, *Travelling wavefronts of reaction-diffusion equations in cylindrical domains*, Comm. P. D. E., **18** (1993), 505-531.

[Vn] T.D. Ventsel', *A free boundary-value problem for the heat equation*, Dokl. Akad. Nauk SSSR **131** (1960), pp. 1000-1003; English transl. Soviet Math. Dokl. **1** (1960).

[Vo] A. I. Volpert, V. A. Volpert, V. A. Volpert, "Travelling wave solutions of parabolic systems", American Math. Society, Providence, RI, 1994.

[W] F. A. Williams, "Combustion Theory", 2nd. ed., Benjamin/Cummnings, Menlo Park, CA, 1985.

[ZF] Ya. B. Zeldovich, D. A. Frank-Kamenetskii *The theory of thermal propagation of flames*, Zh. Fiz. Khim., **12** (1938) pp. 100-105 (in Russian; english translation in "Collected Works of Ya. B. Zeldovich", vol. 1, Princeton Univ. Press, 1992).

[Z4] Ya. B. Zeldovich, G.I. Barenblatt, V.B. Librovich, G.M. Makhviladze, "The mathematical theory of combustion and explosions", Consultants Bureau, 1984.

Juan Luis Vazquez
Depto. Matemáticas
Univ. Autónoma de Madrid
28049 Madrid, Spain.

Part 4.

Free boundary problems
in Environment and Technology

A GLITZKY AND R HÜNLICH

Electro-reaction-diffusion systems for heterostructures

Abstract: In this paper we show the mathematical and thermodynamic correctness of a model which describes the transport of charged particles in heterostructures. An existence and uniqueness result for solutions to the corresponding equations is given. We demonstrate that the free energy decays exponentially along the trajectory of the system and that the concentrations are globally bounded from above and from below by positive constants.
MSC: 35B40, 35K45, 35K57, 78A35.

1. Introduction.

This paper is devoted to the investigation of equations modelling the migration of charged species in heterostructures via diffusion and reaction mechanisms. Such problems arise e.g. in semiconductor technology. An overview on corresponding model equations may be found in [9].

Our aim is to show that for such model equations the free energy along solutions decays monotonously and exponentially to its equilibrium value, i.e., that the models are correct from the thermodynamic point of view. The main tool in our investigations is an estimate of the free energy by the dissipation rate (see Theorem 5.1). Such estimates for reaction–diffusion equations for uncharged particles go back to [8]. For a special case with only one sort of charged dopants but using the local electroneutrality approximation analogous results have been obtained in [3,4]. Furthermore, we give a short overview on global a priori estimates for the concentrations from above and below, on existence and uniqueness results.

The notation of the occuring physical quantities is collected in Table 1. The relation between concentrations and chemical potentials is assumed to be given by the Boltzmann statistics $u_i = \bar{u}_i e^{v_i}$, $i = 1, \ldots, m$, where $\bar{u}_i$ are reference densities. Note that for heterostructures these reference densities generally depend on the space variable. The driving forces for the fluxes j_i are the gradients of the electrochemical potentials ζ_i. Thus, the fluxes contain diffusion and drift terms

$$j_i = -D_i u_i \nabla \zeta_i = -D_i u_i \nabla (v_i + q_i v_0).$$

We consider mass action type reactions of the form

$$\alpha_1 X_1 + \cdots + \alpha_m X_m \rightleftharpoons \beta_1 X_1 + \cdots + \beta_m X_m, \quad (\alpha, \beta) \in \mathcal{R},$$

$$\begin{array}{llll}
X_i, \; i = 1, \ldots, m & - & \text{mobile species} \\
q_i & - & \text{charges} \\
u_i & - & \text{concentrations} \\
v_i & - & \text{chemical potentials} \\
u_0 := \sum_{i=1}^m q_i\, u_i & - & \text{charge density of mobile species} \\
v_0 & - & \text{electrostatic potential} \\
\zeta_i := v_i + q_i v_0 & - & \text{electrochemical potentials} \\
a_i := e^{\zeta_i} & - & \text{electrochemical activities} \\
j_i & - & \text{mass fluxes}
\end{array}$$

Table 1: Notation.

where (α, β) is a pair of vectors $(\alpha_1, \ldots \alpha_m)$, $(\beta_1, \ldots, \beta_m)$ of stoichiometric coefficients which characterizes the reaction from $\sum_{i=1}^m \alpha_i X_i$ to $\sum_{i=1}^m \beta_i X_i$ and its converse reaction. Thereby $\mathcal{R}$ describes the finite set of volume reactions under consideration. We write the reaction rates in terms of products of the electrochemical activities but allow the relaxation constants to depend on the electrostatic potential,

$$(1.1) \qquad R_i(a, v_0) = \sum_{(\alpha, \beta) \in \mathcal{R}} \tilde{k}_{\alpha\beta}(\cdot, v_0) \Big(\prod_{k=1}^m a_k^{\alpha_k} - \prod_{k=1}^m a_k^{\beta_k} \Big)(\alpha_i - \beta_i).$$

Furthermore there may occur reactions on the boundary Γ of Ω. Similar to (??) we introduce $R_i^\Gamma(a, v_0)$, $\tilde{k}_{\alpha\beta}^\Gamma(\cdot, v_0)$ and $\mathcal{R}^\Gamma$. By

$$\mathcal{S} := \mathrm{span}\Big\{ \alpha - \beta : (\alpha, \beta) \in \mathcal{R} \cup \mathcal{R}^\Gamma \Big\} \subset I\!\!R^m$$

we denote the stoichiometric subspace belonging to the considered volume and boundary reactions.

The basic equations are the continuity equations for all species and the Poisson equation for the electrostatic potential,

$$\begin{aligned}
\frac{\partial u_i}{\partial t} + \nabla \cdot j_i + R_i &= 0, \quad i = 1, \ldots, m, \quad \text{on } I\!\!R_+ \times \Omega, \\
\nu \cdot j_i &= R_i^\Gamma, \quad i = 1, \ldots, m, \quad \text{on } I\!\!R_+ \times \Gamma; \\
-\nabla \cdot (\varepsilon \nabla v_0) &= f + \sum_{i=1}^m q_i u_i \quad \text{on } I\!\!R_+ \times \Omega, \\
\nu \cdot (\varepsilon \nabla v_0) + \tau v_0 &= f^\Gamma \quad \text{on } I\!\!R_+ \times \Gamma; \\
u_i(0) &= U_i, \quad i = 1, \ldots, m, \quad \text{on } \Omega
\end{aligned}$$

where ε is the dielectric permittivity, τ represents a capacity of the boundary, and the functions f and f^Γ are fixed source terms not depending on time. ¿From the above continuity equations follows the continuity equation for the charge density

$$\frac{\partial u_0}{\partial t} = \sum_{i=1}^{m} q_i \frac{\partial u_i}{\partial t} \quad \text{on } \mathbb{R}_+ \times \Omega, \quad u_0(0) = U_0 := \sum_{i=1}^{m} q_i U_i \text{ on } \Omega.$$

2. Formulation of the problem.

At first we fix basic assumptions with respect to the data of the problem.

(2.1)

$\Omega \subset \mathbb{R}^2$ bounded, Lipschitzian;

$\bar{u}_i \in L^\infty(\Omega)$, $\bar{u}_i \geq c > 0$, $U_i \in L^\infty(\Omega)$, $U_i \geq 0$,
$D_i \in L^\infty(\Omega)$, $D_i \geq c > 0$ for $i = 1, \ldots, m$;

$\mathcal{R}, \mathcal{R}^\Gamma$ finite subsets of $\mathbb{Z}_+^m \times \mathbb{Z}_+^m$,
$q \in \mathbb{Z}^m$, $\alpha \cdot q = \beta \cdot q$ for $(\alpha, \beta) \in \mathcal{R} \cup \mathcal{R}^\Gamma$;

$\tilde{k}_{\alpha\beta} \in \text{Car}(\Omega, \mathbb{R}; \mathbb{R})$, $\quad \tilde{k}_{\alpha\beta}(x, y) \leq c\, e^{c|y|}$ if $x \in \Omega$, $y \in \mathbb{R}$,
$\tilde{k}_{\alpha\beta}(x, y) \geq c_R > 0$ if $x \in \Omega$, $y \in [-R, R]$,
$\tilde{k}_{\alpha\beta}^\Gamma \in \text{Car}(\Gamma, \mathbb{R}; \mathbb{R})$, $\quad \tilde{k}_{\alpha\beta}^\Gamma(x, y) \leq c\, e^{c|y|}$ if $x \in \Gamma$, $y \in \mathbb{R}$,
$\tilde{k}_{\alpha\beta}^\Gamma(x, y) \geq c_R > 0$ if $x \in \Gamma$, $y \in [-R, R]$ for $(\alpha, \beta) \in \mathcal{R} \cup \mathcal{R}^\Gamma$;

$\varepsilon, f \in L^\infty(\Omega)$, $\tau, f^\Gamma \in L^\infty(\Gamma)$, $\varepsilon, \tau \geq c > 0$.

For the weak formulation of our problem we use the variables

$$v = (v_0, v_1, \ldots, v_m) \in \mathbb{R}^{m+1}, \quad u = (u_0, u_1, \ldots, u_m) \in \mathbb{R}^{m+1}, \quad \zeta = (\zeta_1, \ldots, \zeta_m) \in \mathbb{R}^m$$

where $\zeta_i = \zeta_{vi} := v_i + q_i v_0$. We introduce the function spaces

$$X := H^1(\Omega, \mathbb{R}^{m+1}), \quad W := \{w \in X : w_i^+ \in L^\infty(\Omega), \, i = 1, \ldots, m\}$$

and define the operators $A : W \times X \to X^*$, $E_0 : H^1 \to (H^1)^*$ and $E : X \to X^*$ by

$$\langle A(w, v), \bar{v} \rangle := \int_\Omega \Big\{ \sum_{i=1}^{m} D_i \bar{u}_i e^{w_i} \nabla \zeta_{vi} \nabla \zeta_{\bar{v}i}$$

$$+ \sum_{(\alpha, \beta) \in \mathcal{R}} \tilde{k}_{\alpha\beta}(\cdot, v_0) \big(e^{\alpha \cdot \zeta_w} - e^{\beta \cdot \zeta_w} \big) (\alpha - \beta) \cdot \zeta_{\bar{v}} \Big\} \, dx$$

$$+ \int_\Gamma \sum_{(\alpha, \beta) \in \mathcal{R}^\Gamma} \tilde{k}_{\alpha\beta}^\Gamma(\cdot, v_0) \big(e^{\alpha \cdot \zeta_w} - e^{\beta \cdot \zeta_w} \big) (\alpha - \beta) \cdot \zeta_{\bar{v}} \, d\Gamma,$$

$$\langle E_0 v_0, \bar{v}_0 \rangle := \int_\Omega \big\{ \varepsilon \nabla v_0 \nabla \bar{v}_0 - f \, \bar{v}_0 \big\} \, dx + \int_\Gamma \big\{ \tau v_0 - f^\Gamma \big\} \bar{v}_0 \, d\Gamma,$$

$$\langle E v, \bar{v} \rangle := \langle E_0 v_0, \bar{v}_0 \rangle + \int_\Omega \sum_{i=1}^{m} \bar{u}_i e^{v_i} \bar{v}_i \, dx.$$

307

The problem we shall be concerned with consists in finding a solution to
Problem (P):

$$u'(t) + A(v(t), v(t)) = 0, \quad u(t) = Ev(t) \text{ f.a.e. } t \in \mathbb{R}_+, \quad u(0) = U,$$

$$u \in H^1_{\text{loc}}(\mathbb{R}_+, X^*), \quad v \in L^2_{\text{loc}}(\mathbb{R}_+, X), \quad v_i^+ \in L^\infty_{\text{loc}}(\mathbb{R}_+, L^\infty(\Omega)), \quad i = 1, \ldots, m.$$

The 0–th components of these equations represent the continuity equation for the charge density and the Poisson equation, respectively. By 1 we denote the function with the constant value 1 on Ω. We define

$$\mathcal{U} := \Big\{ u \in X^* : u_0 = \sum_{i=1}^m q_i u_i, \quad (\langle u_1, 1 \rangle, \ldots, \langle u_m, 1 \rangle) \in \mathcal{S} \Big\}.$$

Integrating the continuity equations over $(0, t) \times \Omega$ we find the following invariance property.

Lemma 2.1. *If (u, v) is a solution to (P) then $u(t) \in \mathcal{U} + U$ for all $t \in \mathbb{R}_+$.*

Next we define the energy functionals. Let $\Phi : X \longrightarrow \overline{\mathbb{R}}$ be

$$\Phi(v) := \int_\Omega \Big\{ \frac{\varepsilon}{2} |\nabla v_0|^2 - f v_0 \Big\} \mathrm{d}x + \int_\Gamma \Big\{ \frac{\tau}{2} v_0^2 - f^\Gamma v_0 \Big\} \mathrm{d}\Gamma + \int_\Omega \sum_{i=1}^m \bar{u}_i \left(e^{v_i} - 1 \right) \mathrm{d}x.$$

Since $\Omega \subset \mathbb{R}^2$ we have dom $\Phi = X$. Moreover, the functional Φ is continuous, strictly convex and Gâteaux differentiable, hence subdifferentiable, and it holds $\partial \Phi = E$. By $F : X^* \longrightarrow \overline{\mathbb{R}}$ we denote its conjugate functional $F := \Phi^*$. Then F is proper, lower semicontinuous and convex. It holds $u = Ev = \partial \Phi(v)$ if and only if $v \in \partial F(u)$. If $u \in X^*$ and $u = Ev$ then F can be calculated as

$$F(u) = \int_\Omega \Big\{ \frac{\varepsilon}{2} |\nabla v_0|^2 + \sum_{i=1}^m \Big\{ u_i (\ln \frac{u_i}{\bar{u}_i} - 1) + \bar{u}_i \Big\} \Big\} \mathrm{d}x + \int_\Gamma \frac{\tau}{2} v_0^2 \mathrm{d}\Gamma.$$

The value $F(u)$ represents the free energy of the state u. We introduce the dissipation rate

$$D(v) := \langle A(v, v), v \rangle, \quad v \in W.$$

3. Steady states.

With regard to Lemma 2.1 it seems to be useful to discuss the steady states (u, v) for Problem (P) which satisfy $u \in \mathcal{U} + U$. These steady states of (P) are the solutions to

(3.1) $$A(v, v) = 0, \quad u = Ev, \quad u \in \mathcal{U} + U, \quad v \in W.$$

We formulate an additional assumption, some kind of a Slater condition:

(3.2) $$\int_\Omega \sum_{i=1}^m U_i \kappa_i \mathrm{d}x > 0 \quad \forall \kappa \in \mathcal{S}^\perp, \ \kappa \geq 0, \ \kappa \neq 0.$$

Theorem 3.1. *Under the assumptions (??) and (??) there exists a unique steady state* (u^*, v^*). *The element* v^* *is the unique minimizer of* $\Phi - \langle U, \cdot \rangle$ *on* $\mathcal{U}^\perp$ *while* u^* *is the unique minimizer of* F *on* $\mathcal{U} + U$. *Furthermore,*

$$u^*, v^* \in L^\infty(\Omega, I\!\!R^{m+1}), \quad u_i^* \geq c > 0 \text{ a.e. on } \Omega, \quad a_i^* := e^{v_i^* + q_i v_0^*} > 0, \ i = 1, \dots, m.$$

4. Monotonicity and boundedness of the free energy.

By means of the unique steady state (u^*, v^*) we define the functional $\Psi : X^* \longrightarrow \overline{I\!\!R}$,

$$\Psi(u) := F(u) - F(u^*) - \langle u - u^*, v^* \rangle.$$

The functional Ψ is proper, lower semicontinuous and convex, we have $\Psi(u) \geq 0$, $\Psi(u^*) = 0$. Note that for $u \in \mathcal{U} + U$ (especially for $u = u(t)$ if (u, v) is a solution to (P)) the last term vanishes and $\Psi(u) = F(u) - F(u^*)$ represents the distance of the free energy to its equilibrium value $F(u^*)$. For $u \in X^*$ with $u = Ev$ we obtain

$$\Psi(u) = \int_\Omega \left\{ \sum_{i=1}^m \left\{ u_i (\ln \frac{u_i}{u_i^*} - 1) + u_i^* \right\} + \frac{\varepsilon}{2} |\nabla(v_0 - v_0^*)|^2 \right\} dx + \int_\Gamma \frac{\tau}{2} (v_0 - v_0^*)^2 d\Gamma.$$

By the strong monotonicity of E_0 we get for such u

$$c \left\{ \sum_{i=1}^m \left\| \sqrt{u_i} - \sqrt{u_i^*} \right\|_{L^2}^2 + \|v_0 - v_0^*\|_{H^1}^2 \right\} \leq \Psi(u) \leq c \left\{ \sum_{i=1}^m \|u_i - u_i^*\|_{L^2}^2 + \|u_0 - u_0^*\|_{(H^1)^*}^2 \right\}.$$

If (u, v) is a solution to (P) then $v(t) - v^* \in \partial \Psi(u(t))$ f.a.e. $t \in I\!\!R_+$ and by the Brézis formula we obtain

$$
\begin{aligned}
e^{\lambda t_2} \Psi(u(t_2)) - e^{\lambda t_1} \Psi(u(t_1)) &= \int_{t_1}^{t_2} e^{\lambda s} \left\{ \lambda \Psi(u(s)) + \langle u'(s), v(s) - v^* \rangle \right\} ds \\
\text{(4.1)} \qquad &= \int_{t_1}^{t_2} e^{\lambda s} \left\{ \lambda \Psi(u(s)) - \langle A(v(s), v(s)), v(s) \rangle \right\} ds \\
&= \int_{t_1}^{t_2} e^{\lambda s} \left\{ \lambda \Psi(u(s)) - D(v(s)) \right\} ds.
\end{aligned}
$$

Now we set in (??) $\lambda = 0$. Since $D(v) \geq 0$ and $\Psi(u) = F(u) - F(u^*)$ for solutions to (P) we find the following energetic estimates.

Theorem 4.1. *We assume (??) and (??). Let* (u, v) *be a solution to (P). Then*

$$F(u(t_2)) \leq F(u(t_1)) \quad for \quad t_2 \geq t_1 \geq 0,$$

i.e., F *decreases monotonously along any solution to (P). In addition,*

$$F(u(t)) \leq F(U) \quad \forall t \geq 0, \quad \|D(v)\|_{L^1(I\!\!R_+)} \leq F(U) - F(u^*),$$

$$\|v_0 - v_0^*\|_{L^\infty(I\!\!R_+, H^1)}^2 + \sum_{i=1}^m \|u_i(\ln(u_i/u_i^*) - 1) + u_i^*\|_{L^\infty(I\!\!R_+, L^1)} \leq c,$$

$$\|v_0\|_{L^\infty(I\!\!R_+, L^\infty(\Omega))}, \ \|v_0\|_{L^\infty(I\!\!R_+, L^\infty(\Gamma))} \leq c$$

where c depends only on the data.

The last estimate is a consequence of regularity results for elliptic boundary value problems [6] working in this way only in the two-dimensional case.

5. Exponential decay of the free energy.

At first we formulate an additional assumption concerning the structure of the reaction system. We define the set $\mathcal{M}$

$$\mathcal{M} := \Big\{ a \in I\!R_+^m, \, v_0 \in H^1(\Omega) : \prod_{i=1}^m a_i^{\alpha_i} = \prod_{i=1}^m a_i^{\beta_i} \; \forall (\alpha, \beta) \in \mathcal{R} \cup \mathcal{R}^\Gamma,$$

$$(E_0 v_0, u_1, \ldots, u_m) \in \mathcal{U} + U \text{ where } u_i := \bar{u}_i \, a_i \, e^{-q_i v_0}, \, i = 1, \ldots, m \Big\}.$$

Supposing that

(5.1) $$\mathcal{M} \subset \text{int } I\!R_+^m \times H^1(\Omega)$$

the elements (a, v_0) of $\mathcal{M}$ correspond to steady states (u, v) in the sense of (??) and vice versa. Then by Theorem 3.1 it follows $\mathcal{M} = \{a^*, v_0^*\}$.

Theorem 5.1. *Let the assumptions (??), (??) and (??) be satisfied. Then for every $R > 0$ there exists a $c_R > 0$ such that*

$$F(Ev) - F(u^*) \leq c_R D(v)$$

for all $v \in M_R := \big\{ v \in W : F(Ev) - F(u^*) \leq R, \, Ev \in \mathcal{U} + U \big\}$.

Idea of the proof. For the complete proof see [5]. If $v \in M_R$ we have that $\|v_0\|_{L^\infty(\Omega)} \leq c(R)$ which guarantees for $u = Ev \in \mathcal{U} + U$ the two estimates

(5.2) $$c(R) \Big\{ \sum_{i=1}^m \big\| \sqrt{a_i/a_i^*} - 1 \big\|_{L^2}^2 + \|v_0 - v_0^*\|_{H^1}^2 \Big\} \leq \Psi(u) \leq c \sum_{i=1}^m \|u_i - u_i^*\|_{L^2}^2,$$

(5.3) $$D(v) \geq c(R) \, \tilde{D}(a)$$

where

$$\tilde{D}(a) = \int_\Omega \Big\{ \sum_{i=1}^m \big| \nabla \sqrt{a_i/a_i^*} \big|^2 + \sum_{(\alpha,\beta) \in \mathcal{R}} \Big[\prod_{i=1}^m \sqrt{a_i/a_i^*}^{\,\alpha_i} - \prod_{i=1}^m \sqrt{a_i/a_i^*}^{\,\beta_i} \Big]^2 \Big\} \, dx$$

$$+ \int_\Gamma \sum_{(\alpha,\beta) \in \mathcal{R}^\Gamma} \Big[\prod_{i=1}^m \sqrt{a_i/a_i^*}^{\,\alpha_i} - \prod_{i=1}^m \sqrt{a_i/a_i^*}^{\,\beta_i} \Big]^2 d\Gamma.$$

310

By an indirect proof it can be verified that for every $R > 0$ there exists a constant $\tilde{c}_R > 0$ such that

$$\Psi(u) < \tilde{c}_R \, \tilde{D}(a) \quad \forall v \in M_R, \, v \neq v^* \quad \text{(with } u, a \text{ corresponding to } v\text{)}. \quad \square$$

Theorem 5.2. *Let (??), (??) and (??) be satisfied. Then there exists a constant $\lambda > 0$ such that*

$$(5.4) \qquad\qquad F(u(t)) - F(u^*) \leq e^{-\lambda t} \left(F(U) - F(u^*) \right) \quad \forall t \geq 0,$$

for any solution (u, v) to (P), i.e., F decays exponentially to its equilibrium value along any trajectory. Moreover, for some $c > 0$ depending only on the data it holds

$$(5.5) \quad \begin{array}{c} \|v_0 - v_0^*\|_{L^2(\mathbb{R}_+, H^1)}, \; \|v_0 - v_0^*\|_{L^1(\mathbb{R}_+, L^1)}, \; \|v_0 - v_0^*\|_{L^1(\mathbb{R}_+, L^1(\Gamma))} \leq c, \\[2mm] \|u_i/u_i^* - 1\|_{L^1(\mathbb{R}_+, L^1)}, \; \|u_i/u_i^* - 1\|_{L^1(\mathbb{R}_+, L^1(\Gamma))} \leq c, \quad i = 1, \ldots, m, \end{array}$$

for any solution (u, v) to (P).

Proof. Let (u, v) be any solution to (P) and $R := \Psi(U)$. Then $v(s) \in M_R$ for a.e. s and

$$\Psi(u(s)) \leq c_R \, D(v(s)) \text{ for a.e. } s.$$

Setting now $\lambda = 1/c_R$ in (??) we obtain (??). The integrability properties in (??) can be derived from (??), (??), (??) and Theorem 4.1. $\quad \square$

6. Existence, uniqueness and global estimates.

In a very short summary we collect the existence, uniqueness and asymptotic results. We suppose additionally that

$$U_i \geq c > 0, \, i = 1, \ldots, m,$$

$$(6.1)$$

$$\sum_{i=1}^{m} \alpha_i, \, \sum_{i=1}^{m} \beta_i \leq 2 \quad \forall (\alpha, \beta) \in \mathcal{R}, \; \sum_{i=1}^{m} \alpha_i, \, \sum_{i=1}^{m} \beta_i \leq 1 \quad \forall (\alpha, \beta) \in \mathcal{R}^\Gamma.$$

The condition concerning the initial values is sharper than (??).

Theorem 6.1. *Let (??), (??) and (??) be satisfied. Then there exists a unique solution (u, v) to (P). Moreover there exists a constant c depending only on the data such that for the solution (u, v) to (P)*

$$\sum_{i=1}^{m} \left(\|u_i(t)\|_{L^\infty} + \|v_i(t)\|_{L^\infty} \right) \leq c \qquad \forall t \in \mathbb{R}_+.$$

Proof. A priori estimates for the concentrations u from above can be found by the energetic estimates of Theorem 4.1 and by Moser technique (see [1] for the van Roosbroeck system and [2] for general systems as considered here). The integrability properties (??) enable us to show that $\ln u_i$, $i = 1, \ldots, m$, may be estimated in $L^\infty(I\!R_+, L^1)$ by a constant only depending on the data. By Moser iteration lower bounds for the concentrations are obtained. By regularization and time discretization we prove global existence of solutions. For the uniqueness we use the $L^\infty(I\!R_+, W^{1,p})$–regularity of v_0 which results from [7]. $\square$

Theorem 5.2 and the L^∞–estimates lead to the following asymptotic estimates.

Theorem 6.2. *We assume (??), (??) and (??). Let (u, v) be the solution to (P) and $p \in [1, +\infty)$. Then there exist constants c, $\lambda_p > 0$ such that*

$$\sum_{i=1}^m \|u_i(t) - u_i^*\|_{L^p} + \sum_{i=0}^m \|v_i(t) - v_i^*\|_{L^p} \le c\, e^{-\lambda_p t} \quad \forall t \in I\!R_+.$$

References.

[1] Gajewski H. and Gröger K., 1990, *Initial boundary value problems modelling heterogeneous semiconductor devices*, in Schulze B.-W. and Triebel H., *Surveys on analysis, geometry and mathematical physics*, Teubner-Texte zur Mathematik, vol.117, p.4.

[2] Gajewski H. and Gröger K., 1994, *Reaction–diffusion processes of electrically charged species*, Preprint 118, Weierstraß–Institut für Angewandte Analysis und Stochastik, Berlin.

[3] Glitzky A., Gröger K. and Hünlich R., 1994, *Existence, uniqueness and asymptotic behaviour of solutions to equations modelling transport of dopants in semiconductors*. In: Frehse J. and Gajewski H., 1994, *Special topics in semiconductor analysis*, Bonner Mathematische Schriften, no.258, p.49.

[4] Glitzky A., Gröger K. and Hünlich R., *Discrete–time methods for equations modelling transport of foreign–atoms in semiconductors*, Nonlinear Anal., to appear.

[5] Glitzky A. and Hünlich R., 1995, *Energetic estimates and asymptotics for electro-reaction–diffusion systems*, Preprint 168, Weierstraß–Institut für Angewandte Analysis und Stochastik, Berlin.

[6] Gröger K., *Boundedness and continuity of solutions to second order elliptic boundary value problems*, to appear.

[7] Gröger K., 1989, *A $W^{1,p}$–estimate for solutions to mixed boundary value problems for second order elliptic differential equations*, Math. Ann., vol.283, p.679.

[8] Gröger K., 1992, *Free energy estimates and asymptotic behaviour of reaction–diffusion processes*, Preprint 20, Institut für Angewandte Analysis und Stochastik, Berlin.

[9] Höfler A. and Strecker N., 1994, *On the coupled diffusion of dopants and silicon point defects*, Technical Report 94/11, ETH Integrated Systems Laboratory, Zurich.

Annegret Glitzky
Weierstrass Institute
for Applied Analysis and Stochastics
D–10117 Berlin
Germany
e-mail: glitzky@wias-berlin.de

Rolf Hünlich
Weierstrass Institute
for Applied Analysis and Stochastics
D–10117 Berlin
Germany
e-mail: huenlich@wias-berlin.de

D HÖMBERG
An extended model for phase transitions in steel

Abstract: We present a general model for phase transitions in steel, which is capable of describing a complete heat treatment cycle of heating and cooling down. It consists of four ODEs for the phase fractions coupled with a nonlinear parabolic energy balance equation.

We discuss existence and uniqueness results. As an application, we present numerical simulations of the Jominy end-quench test in the case of eutectoid carbon steel.

1. Introduction.

The general aim is to control the microstructure of steel. In practice, this is realized by a process named heat treatment, i.e. heating up and then cooling down a workpiece to get a desired distribution of phases. Basic types are annealing, to reduce internal stresses (e.g. in welding seams) and hardening, to increase the strength of the structure (e.g. the outer part of gear wheels, saw blades, ...).

During such a process, five phases may occur: austenite, ferrite, pearlite, bainite, martensite. To understand the principle mechanisms of the occuring phase transitions, we will first consider the special case of plain carbon steel in the so–called eutectoid composition of 0.8% carbon content. Here, only three phases may occur: austenite, pearlite, martensite.

2. Special case: cooling of eutectoid carbon steel

In this section we will assume that the steel initially is in the high temperature phase, called austenite. Then, cooling down slowly, a diffusive transition to pearlite starts. Cooling down very fast, after reaching a threshold temperature M_s, martensite begins to grow caused by a non–diffusive phase transition. The growth of martensite only happens during non–isothermal stages of the cooling process, i.e. only when the temperature is really decreasing. The transformation kinetics for eutectoid carbon steel are depicted in Figure 1.

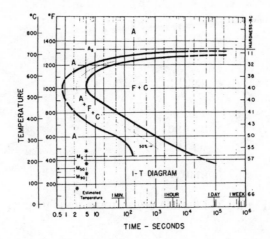

Figure 1: Isothermal–transformation diagram for the plain carbon steel C 1080 (from [2])

The diffusive austenite–pearlite phase transition may be described mathematically by the *Johnson – Mehl equation*

$$p(t) = 1 - e^{-c_1(\theta)t^{c_2(\theta)}},$$

and *Scheil's Additivity Rule*

$$\int_0^t \frac{1}{\tau(\theta(\xi), p(t))} d\xi = 1,$$

where $\tau(\theta, p)$ is the time to transform the volume fraction p to pearlite isothermally at temperature θ.

These formulas lead to initial–value problems of the form (cf. [2],[3]):

$$p(0) = p_0, \qquad \dot{p}(t) = -(1 - p)\ln(1 - p)g_1(t, \theta, p).$$

A typical initial–value problem to describe the non–diffusive austenite–martensite transformation is given by (cf. [3]):

$$m(0) = m_0, \qquad \dot{m}(t) = (1 - m)g_2(\theta)H(-\theta_t),$$

where H is the heaviside–function.

3. General case Figure 2 depicts the emerging phase transitions in the general case. Here, d is a diffusive and nd a non–diffusive phase transition, described by the initial–value problems of the previous section.

315

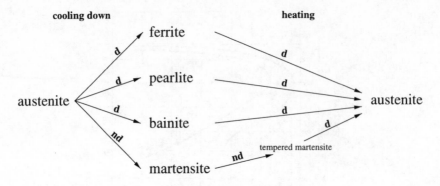

Figure 2: Possible phase transitions in steel

In the sequel, we will use the notations $a(t), f(t), p(t), b(t), m(t)$ for the volume fractions of the different phases. Hence we may eliminate one of them by claiming

$$a(t) + f(t) + p(t) + b(t) + m(t) = 1.$$

Neglecting the formation of tempered martensite and assuming instead that martensite is directly transformed back to austenite by a diffusive transition, we arrive at the following initial–value problem for phase transitions in steel:

$$(*) \begin{cases} f(0) = f_0, \ p(0) = p_0, \ b(0) = b_0, \ m(0) = m_0, \\ f_0 + p_0 + b_0 + m_0 < 1, \\ \dot{f}(t) = -a(t) \ln a(t) g_{11}(t, a, \theta) + f \ln f g_{12}(t, f, \theta), \\ \dot{p}(t) = -a(t) \ln a(t) g_{21}(t, a, \theta) + p \ln p g_{22}(t, p, \theta), \\ \dot{b}(t) = -a(t) \ln a(t) g_{31}(t, a, \theta) + b \ln b g_{32}(t, b, \theta), \\ \dot{m}(t) = a(t) g_{41}(\theta) H(-\theta_t) + m \ln m g_{42}(t, m, \theta). \end{cases}$$

Of course, there is a temperature hysteresis connected with the transitions. Hence, we may assume that depending on temperature θ, either $g_{i1} = 0$ and $G_{i2} \neq 0$ or vice versa. Then, imposing the usual Lipschitz and Carathéodory conditions on g_{ij} (cf. [3]), it is an easy exercise to prove the following

Lemma 3.1 *Let $\theta \in H^1(0, T)$, then the initial-value problem $(*)$ has a unique solution $(f, p, b, m) \in [W^{1,\infty}(0, T)]^4$, satisfying*

$$0 < f(t) + p(t) + b(t) + m(t) < 1 \quad \text{for all } t \in [0, T].$$

4. Three-dimensional model In a spatial model, the latent heats of the phase changes have to be considered. Assuming for reasons of space that the latent heat is the same for all phase changes, this leads to the following balance of energy:

$$\rho c \theta_t - \nabla \cdot \left(k \nabla \theta \right) = \rho L \left(f_t + p_t + b_t + m_t \right).$$

Next, we replace $H(-\theta_t)$ with $H_\delta(-\theta_t)$, where $H_\delta(.)$ is the Yosida approximation of the heaviside graph, put $A_\delta(\theta_t) = -H_\delta(-.)$ and consider the following regularized problem:

$$(P_\delta) \begin{cases} \rho c\theta_t + \rho Lag_{41}(\theta)A_\delta(\theta_t) - \nabla.\left(k\nabla\theta\right) = \rho L\left(f_t + p_t + b_t + m_t\right) \\ \qquad + \text{ boundary and initial conditions,} \\ \qquad \text{where } (f, p, b, m) \text{ is a solution to } (*). \end{cases}$$

Theorem 4.1 *(Cf. [3]) Under standard assumptions on the data, (P_δ) admits a unique solution $\theta \in H^{2,1}(Q)$.*

Remark 4.2 *Passing to the limit with δ yields a solution to a problem (P), which is not explicated here (cf. [3] for details).*

5. Application: The Jominy test

In this test, a cylindrical steel bar is heated up to its austenitic state, then put in a fixation and quenched by spraying water on its lower end. The resulting hardenability curve, where hardness is plotted against distance from the quenched end, serves as a quality feature for the steel.

The following figures show the cooling device according to ISO 642 (3), and a comparison between measured and simulated hardenability curves (4) for this steel (cf. [4]).

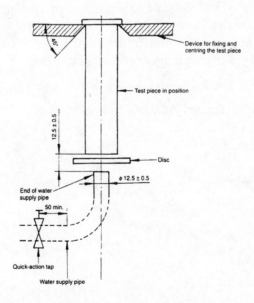

Figure 3: Diagram of the cooling device (from [5])

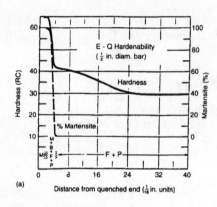

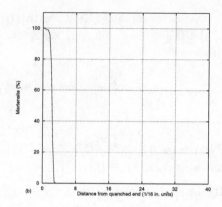

(a)

(b)

Figure 4: (a) Hardenability curve for the steel C 1080 (from [1]), (b) numerically calculated martensite fraction plotted against the distance from the quenched end.

References.

[1] American Society for Metals, *Atlas of Isothermal Transformation and Cooling Transformation Diagrams*, Ohio, 1977.

[2] Hömberg, D., *A mathematical model for the phase transitions in eutectoid carbon steel*, IMA J. Appl. Math., 54 (1995), 31–57.

[3] Hömberg, D., *Irreversible phase transitions in steel*, Math. Meth. Appl. Sc., to appear.

[4] Hömberg, D., *A numerical simulation of the Jominy end–quench test*, Acta Metallurgica et Materialia, to appear.

[5] International Organization for Standardization, *International Standard 642, Steel – Hardenability test by end quenching (Jominy test)*, 1979.

Dietmar Hömberg
Weierstrass Institute for Applied Analysis and Stochastics
Mohrenstr. 39
D – 10117 Berlin
Germany
e-mail: hoemberg@wias-berlin.de

M NAWALANY AND A TRYKOZKO
Free boundaries of environmental engineering

1. Introduction.

A number of the free boundary related questions which need to be addressed in the field of environmental engineering increases with the rate which can hardly be handled by engineers themselves. An intuition of engineers is at work all the time still capacity of answering most of these questions falls behind continuously increasing domain of technologies and engineering methods being proposed or invented. These all-increasing discipline's frontiers we call the "Free Boundary *of* Environmental Engineering" and address according to our best knowledge and experience. The list of the free boundary problems encountered in environmental engineering is traditionally subdivided into: **air protection, surface water protection and ground water protection**. Each of the three area has its own specific free boundary problems typically related to time and space scales in which the major physical phenomena occur in the given component of the environment. Intrusion of engineering activities into natural environment generates changes/deformations of the original physical processes and systems. The intrusion, which is supposedly aimed towards improving a quality of the environment (and thus our lives), frequently results in the side effects which cause deterioration in efficiency of the technical measures applied or which, in some instances, must be admitted as being detrimental. Many of these effects are due to ill-handling of the free-boundary effects at the design or operational stages of engineering activity. Some problems are generated already at the problem formulation stage when engineers encounter the inherent difficulties in posing the free boundary problem well. And later, at modelling stage some of the free boundary problems emerging in environmental engineering are caused by oversimplifications in the domain geometry. Some other problems are due to numerical approximations which are incompatible with the problem variables and/or with the domain geometry. Possibly, some solutions to the above mentioned difficulties already exist but are not known to engineers. Still, for some of them, the appropriate numerical (or analytical) methods may not exist and hence the relevant research is needed.

2. Dilemmas of the Environmental Impact Assessment (EIA).

The basic questions asked by the authorities responsible for environmental protection to the experts are normally formulated in terms of possible consequences of human intervention into environment. The consequences that need to be quantified are of different nature: physical, biological, social, cultural, legal, economical and political ones. The task, called the Environmental Impact Assessment, embraces all these aspects and all relations between them. Actually the core of the EIA is to find out, in quantifiable and verifiable way, the extent to which some human activities influence biological lives of other people. Impacts on social, cultural, legal and economic structures can be considered, in general, as derivatives of this basic impact – the impact on human biology. Even altruistic "nature protection" is actually **our** business – we know perfectly well that by destroying too much around we undermine our own existence. We fill it but still we are far away from quantifying the relations between all aspects of our lives. Now we will leave all these considerations, however exciting, unfinished and pass to quantifiable phenomena as more appropriate for this conference.

Naturally, physical and chemical phenomena and impacts can be quantified in most strict sense of the word "quantification". It simply means that the language of mathematics is structurally compatible for describing and possibly also predicting these phenomena. We will therefore spend most of this presentation on physical phenomena. It must be remarked that recently, also biological processes are becoming "partly explainable" in mathematical terms. Mathematical biology is just another example of expanding free boundary of the Environmental Engineering. Embracing biological processes into the realm of mathematics and, at the same time expanding mathematical notions so they can describe biology, is considered the greatest challenge of today's science. When biological phenomena can be approached sufficiently close by the "mathematical biology" it would become the greatest achievement of science. Needles to say that then and only then most of the questions asked within the Environmental Impact Assessment procedures could be answered in quantitative and verifiable way. This day is still far ahead of us.

Now, for the sake of decency we will confine ourself to physico-chemical phenomena referring where necessary or promising to biology. Before doing so I should mention one practical dilemma concerning the reliability of the experts' answers. This reliability is in many cases strongly related to the Free Boundary Problem.

Let us assume we, engineers, are asked to design a course of action which should ultimately lead to cleaning a part of the environment that has been polluted by some toxic substance. For instance, we can imagine that there was a leak from the storage tank into the underlying aquifer. Soil and ground water have been polluted to a degree that cannot be tolerated. With modern monitoring techniques we may assume that a concentration and a spatial extent of the pollution plume are known. It is expected from the expert to advise on the series of technical measures that could prevent the plume from further spreading and possibly also on cleaning the aquifer (soil and ground water) so the concentration of toxic substance in subsoil is lowered below some specified level. What the expert can advise? He definitely will break all the actions he proposes into three phases:

Phase I — stopping the source

Phase II — confining the plume
Phase III — remediating subsoil.

Except, possibly, Phase I the two other phases may involve the Free Boundary Problems. If hydraulic isolation or remediation is chosen one can, for instance choose a group of pumping wells (see Figure 1) to create a local system of groundwater flow which does not allow the polluted water particles to escape from some predefined and finite subregion. After the pollution source is stopped the same installation can be used for remediation by flashing the polluted subsoil with clean water. Modern design is based on numerical models which are used for estimating potential (piezometric head) or velocity field of groundwater in the vicinity of the pollution plume. Standard finite element or finite difference methods and software are used to calculate ground water flow corresponding to specified arrangements of wells and their yields. Even more sophisticated water management computer software is used at present to minimize costs of the entire operation. The concept looks clear and simple. However when the engineer designs the installation he/she must also answer the question of the side-effects. In this case one needs to estimate so called "radius of influence" *i.e.* to say how far the installation will influence the regional groundwater flow for given pumping scheme. In general case the action of the installation may effect boundary potential or flux at the systems boundary. It means that boundary conditions can be effected depending on the yield of the wells. This usually results in the iterative approach in which both geometry of the ground water system's boundaries as well as the boundary conditions need to be changed and the whole problem is to be recalculated. Clearly, this is the Free Boundary situation. New questions can be asked. For instance: Where to cut the environment for our local flow system? Which numerical methods can handle such situations? Are for instance the infinite elements a remedy? All environmental engineers are facing these and other practical questions while designing technical installations in the presence of the Free Boundaries. Good question is whether this iterative approach, which until now is mostly heuristic, can be improved and if so, what could be the "proper" way of solving the Free Boundary problems in engineering design?

3. Free Boundaries of Air Pollution Protection.

To your surprise we are not going to speak about the "puffs" emitted by chimneys of the power stations, steel plants, refineries or chemical factories. The subject of pollution transport in the atmosphere can be considered classical although still not completely solved. So the problems of solving advection-dispersion-diffusion equation with all imaginable chemical and photo-chemical reactions of air pollution and with all possible interactions between precipitation of whatever form and the pollution we leave aside. However interesting for themselves the air pollution transport processes will not be discussed. We would rather stress these technical installations and processes that normally proceed the "puffs'" emission. If properly controlled the processes may lead to more efficient air pollution control than just filters set on the industrial chimneys or the operating dedusting installations.

Coming to the source of air pollution two major processes should be considered that endanger quality of atmospheric air: dust emission and emission of gases.

For dust the steel industry is the major polluter. Melting steel in electrical furnaces generates a considerable flux of dust. Amount of dust emitted depends on three factors : on chemical contents of the ore, on the chemical contents of other melting row materials and on the conditions of the melting process (like temperature and geometry of the electrical arcs). Designing the "proper" or "optimal" parameters of the melting process is the subject of many years of research. Only recently engineers started to think about the design which can minimize the flux of dust that accompanies steel melting. Also overall geometry of the melting place *(e.g.* electrical furnace) does change in course of the melting process since solids are turned into liquid state. As the steel melting process is clearly the Free Boundary problem attacking it from the environmental protection point of view definitely has a flavor of novelty.

Similar situation holds for energy production by burning lignite, coal or petroleum products. Burning process is obviously Free Boundary Process that can be also optimized from the environmental protection point of view. Which means minimization of gaseous emission, especially those gases which are considered harmful for humans, animals or plants. As for dust the burning process needs to be mathematically described, its parameters related to gas emission and finally optimized. Clearly, the geometry of fireplace is changing during the burning process for the solid or liquid particles being incinerated are being changed into gases.

For filters which are still in use there is a clogging process that can be considered Free Boundary Problem. The longer filters operate the more dust remains in pores of the filters thus increasing the resistance to air flow. Consequently the efficiency of the industrial filters change in time depending on the growth of "dust cake" within the filters. Many experimental work has been done on designing "optimal filters" still good theory based on Free Boundary approach could be helpful.

Concluding this part we want to stress again that environmental protection of atmospheric air has its major potential in stopping or limiting the air pollution sources.

4. Free Boundaries of Surface Waters Protection.

Here the same philosophy applies. It is more efficient to influence the sources than cure the environmental effects. Many scientists tend to be fascinated with the natural processes disregarding obvious necessity of keeping the environment clean for the next generations. To them we can say "perhaps what they are solving is interesting but, at the moment, less relevant". However "down-to-earth" it sounds such approach is in the interest of environment we leave in – the environment that cannot coupe itself with the waste we produce. If we do not start thinking seriously about the environmental protection and in helping the environment in this struggle we can easily commit a crime of negligence.

Now back to water. As water is the most important medium for our life practically all the domestic, agricultural and industrial processes are based on this liquid. Flow of surface water is by definition the Free Boundary process – one boundary (upper boundary) of flowing water is not restricted. Water flowing in a river or lake or sea may occupy as much space as it needs. Calculating a position of water table is a part of the tasks undertaken by hydrodynamics. And this challenge is met by thousands of models that routinely solve the Navier-Stokes equations or simplified Saint-Venant equations using numerical methods. For the velocity field calculated one can solve additionally Advection-Dispersion-Diffusion equation for concentration of a pollutant and infer about the pollution transport in the river or lake. There is however only one good way of protecting water environment – again by minimizing amount of waste water generated. Here intervention in all human activities is a must. The intervention can be through optimizing the water-hungry processes with the criterion of minimizing amount of waste water (or waste in the water). Many of the water related problems are the Free Boundary Problems. We mention only the few related to phase changes and which, if not operated properly, cause the major damages to our surface waters

The essential examples are:

i) sugar production installations

ii) inverse osmosis installations for saline waters from mines

iii) cooling systems for power stations

iv) bio-chemical reactors in purification stations for industrial and municipal waste water

v) agricultural practices – fertilizers and pesticides use.

Of course there are myriads of other examples as well. All these human activities consist of processes that are generically Free Boundary processes. To optimize them one definitely needs to apply the Free Boundary Control procedures.

Specific group of Free Boundary problems related to protection of surface waters are the algae blooming phenomena which occasionally spoil our lakes and reservoirs. Patches of algae biomass growing in unstoppable rate and then dying out cause the quality of water deteriorating rapidly. Question of how to limit the growth of biomass in reservoirs of drinking water are strongly related to biomass dynamics. Which is clearly Free Boundary dynamics. This problem is one of the most challenging problems at the moment.

5. Free Boundaries of Ground Water Protection.

As for surface waters ground water flow can be in many instances described as the Free Boundary Problem. Typical examples are :

  i) dynamics of the phreatic ground water table in an unconfined aquifer

 ii) dynamics of fresh-water – saline-water interface

iii) flow in unsaturated zone

 iv) flow in fractured rocks

  v) seepage through dams and dikes

 vi) change in water divide caused by pumping (mentioned before)

vii) interaction between rivers and aquifers

viii) artificial recharge of groundwater from the infiltration ponds

and many others. We must admit that a free boundary flow of ground water is in most cases we know simplified considerably without thorough analysis that would allow for such simplifications. The most common approach taken is the Dupuit-Forchheimer horizontal flow approximation which in case of water quantity (water balance) works well but is useless for the ground water quality problems solving. Actually, the 3D-approach in groundwater is the only justified approach when ground water quality processes need to be analyzed as well. The three dimensional approach is clearly of Free Boundary type.

What you might expect in this moment is that we start repeating the philosophy of stopping the pollution source as the most effective way of protecting ground water. Yes, this is the absolute truth : "Do not create sources of ground water pollution first of all !". We should influence (or optimize) all the industrial, agricultural and municipal processes with the criterion of environmental protection in mind. And we should do as much for stopping the existing sources of ground water pollution as we do for surface waters and for air. However here, in case of ground water protection, things change: the pollution plume in subsoil does not decay nor dilute nor is dispersed so easily and so quickly as it does in atmospheric air and in surface waters. Dynamics of the pollution transport in groundwater is of orders of magnitude slower. In terms of the impact on human health polluted ground water remains toxic for generations. If nothing is done. Therefore confinement and remediation of polluted parts of aquifer is an essential remedy.

Recently biotechnological ground water remediation has become very popular. Local groundwater system created by wells or by some other means is used not only to confine the pollution plume but also to deliver necessary nutrients and oxygen for microorganisms that are cultivated to "eat" the pollutant. Using local bacteria has proven to be more effective in destroying toxic substances in the subsoil than imported bacteria as the former are better adapted to the local chemical environment. Growing biomass of bacteria in subsoil is again the Free Boundary Phenomenon. How to make this technology more efficient and more robust is also a theoretical challenge.

6. Hydraulic isolation of Waste Disposal Sites.

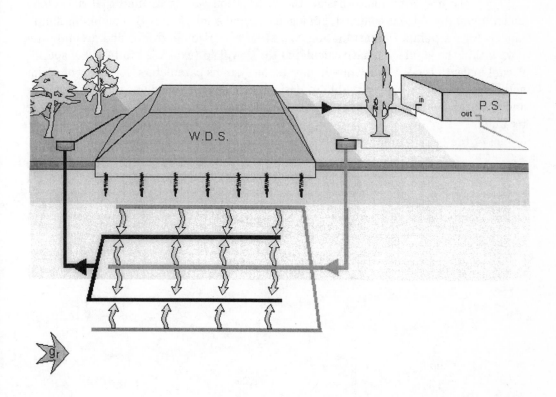

Figure 1. Concept of hydraulic isolation.

To give an example of the groundwater confinement having all the feature of the Free Boundary we are going to present part of the ongoing project of our Institute of Environmental Systems Engineering WUT and the TNO Institute of Applied Geoscience in Delft. Classical hydraulic isolation and remediation that has been mentioned already was based on injecting-pumping wells. What we consider is based on horizontal drains. Modern trenchless technologies allow us to make this kind od installation feasible. The idea is to create a local flow systems beneath the waste disposal site as shown in

Fig.1. The idea of the installation is very simple: water is pumped into the grid of parallel horizontal drains and is at the same time abstracted by the pump from the twin drains thus forming a local system of groundwater flow beneath the waste disposal site (WDS) – Fig.2. At the designing stage it was necessary to find locations of the drains, their depth and the pumping rate which could create the local flow preventing polluted water from the WDS to mix with the rest of the groundwater. Another words, the local system should form kind of capturing zone that would attract **all** particles of the (polluted) water.

6.1. Confined aquifer (simplified approach)

The very first attempt to design the installation was to assume that an aquifer underlaying the WDS is confined, homogeneous and semiinfinite. By combining simple models for : a pump, the installation consisting of horizontal drains and groundwater flow within the aquifer we have calculated the flow field (expressed in terms of specific flow q) beneath the WDS. By manipulating the systems parameters (locations of drains and their diameters) we were able to design the capturing zone for a given geometry and pollution emission rate of the WDS. Below major formulae used in the design are recalled.

<u>Model of the installation</u>

$$\Delta H = f(Q) \tag{1}$$

$$Q = \sum_{i=1,3,5,\dots} Q_i(\Delta H) \tag{2}$$

$$\sum_{i=1}^{N_d} s_i Q_i = 0, \text{with } s_i = (-1)^{i+1}. \tag{3}$$

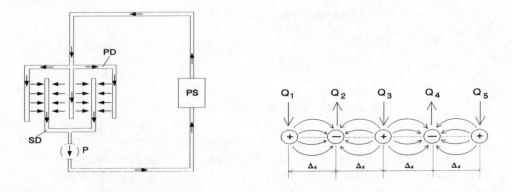

Figure 2. Horizontal and vertical views of the installation.

where: $f(.)$ — nonlinear characteristics of the pump

ΔH — lifting hight of the pump, (m)
Q — total pumping rate, (m^3/s)
Q_i — flow rate of the i-th drain
N_d — number of drains (m^3/s).

The third equation indicates that a water balance of the installation is zero (thus not causing detrimental side effects in the aquifer).

<u>Model of groundwater flow</u>
Steady state flow assumed under WDS is completely described by the piezometric head $\phi = \phi(x, z)$:

$$\phi(x, z) = \phi_r(x, z) + \phi_i(x, z) + \phi_w(x, z) + \phi_d(x, z) \tag{4}$$

where: ϕ_r — regional groundwater piezometric head, (m)
      ϕ_i — correction due to infiltration, (m)
      ϕ_w — correction due to recharge (emission) from the WDS, (m)
      ϕ_d — correction due to drains operation, (m).

Formulae for particular components are listed below:
Model for regional flow:

$$\phi_r(x, z) = \phi_o - \frac{q_r^o}{K_a} x \tag{5}$$

where q_r^o — regional specific discharge, (m/s)
Model for infiltration:

$$\phi_i(x, z) = const. - Inf \cdot \frac{z}{K_a} \tag{6}$$

where: Inf — infiltration rate, (m/s)
      K_a — hydraulic conductivity, (m/s).

Model for the recharge from the WDS:

$$\phi_w(x, z) = const. + \frac{R_w z}{2\pi K_a} \{I(-) - I(+)\} \tag{7}$$

where:

$$I(-) = I\left(\frac{x - L_w/2}{z}\right),$$

$$I(+) = I\left(\frac{x + L_w/2}{z}\right), \tag{8}$$

$$I(t) = t \ln(t^2 + 1) - 2t + 2 \arctan t. \tag{9}$$

Model for active drains:

$$\phi_d(x,z) = const. - \sum_{i=1}^{N_d} \frac{s_i Q_i}{4\pi B K_a} \left\{ \ln[(x-x_i)^2 + (z-z_i)^2] + \ln[(x-x_i)^2 + (z+z_i)^2] \right\},$$
(10)

where B — length of the drains, (m).

By combining the installation model (1) with the groundwater flow model (4) we obtain the following system of (N_d+1) nonlinear equations with (N_d+1) unknowns: Q_i, $(i=1,\ldots,N_d)$ and ΔH:

$$\Delta\phi_d(k) = \Delta H - \Delta\phi_r(k) - \Delta\phi_i(k) - \Delta\phi_w(k) \quad \text{for} \quad (k=1,3,\ldots,N_d-2)$$
$$\Delta\phi_d(k) = -\Delta H - \Delta\phi_r(k) - \Delta\phi_i(k) - \Delta\phi_w(k) \quad \text{for} \quad (k=2,4,\ldots,N_d-1) \quad (11)$$

$$\sum_{i=1}^{N_d} s_i Q_i = 0 \tag{12}$$

$$\Delta H = f(Q) = f\left(\sum_{i=1,3,\ldots} Q_i \right) \tag{13}$$

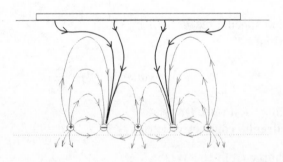

Figure 3. Local flow system capturing polluted water

Gauss-Seidel procedure has been used to solve the system of equations (11) — (13). Calculated flow rates in drains and the lifting height of the pump ΔH allowed to calculate back a piezometric head ϕ in the aquifer from formulae (4)-(10). Application of the Darcy Law allows to calculate specific discharge q in the aquifer and consequently also velocity field and particles' trajectories. Fig.3 shows water particles that start from the bottom of the WDS and are being captured by drains. Local groundwater flow created as the result of the installation operation forms typical cells which "focus" the trajectories of the polluted water particles forcing them to "hit" the sucking drains.

6.2. Unconfined aquifer (Free Boundary Approach)

Problem 6.1 has been formulated in confined aquifer, and this was a simplified approach. Most often, while solving groundwater problems, the top boundary is not known a priori and thus the problem should be considered as an unconfined flow case. Artificial recharge as a means of managing the water resources (wastewater outlets, irrigation, groundwater pumping, river flooding, etc.) has an influence on the water table height and, consequently, can be the reason of engineering problems.

Therefore it is of great interest to estimate the water table rise above the initial depth of saturation. This usually involves concerns for the water table approaching too closely to the ground surface. In agriculture, high water tables can be harmful to plant growth and drainage. For septic drain fields, high water tables can drown the unsaturated zone needed to provide waste-water treatment. For other facilities, they can adversely affect the strength of foundations.

The most often applied Dupuit approximation is inadequate when the flow has a strong vertical component, which happens near local sources or sinks. In such cases the 3D Laplace equation should be solved.

Two different numerical approaches can be considered:

— solving the full saturated – unsaturated flow system, with the saturation as an unknown. The water table shape can be traced out *a posteriori*, with respect to the computed saturation values,

— only saturated part of the domain is considered. An iterative numerical procedure is used to follow the water table shape.

In this contribution the second approach is applied.

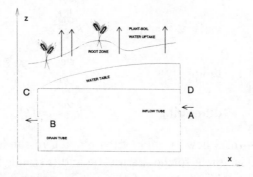

Figure 4.

As an example we shall consider the flow problem of dual-pipe subirrigation and drainage, as illustrated in Fig. 4. Cylindrical tubes are used for both subirrigation water supply (A) and subsurface drainage sinks (B). The recharge originates at the irrigation tube (the source), whereas some part of the discharge occurs by the drainage outflow (a sink) and some is due to evapotranspiration (another sink) at the top part of the boundary. The evapotranspiration $e(x)$ is assumed to be constant in the vertical direction.

The computational domain is confined by the impermeable bottom, the vertical surface of symmetry interrupted by a half-circle with radius β, another vertical surface of symmetry with a half-circle with radius b and the free groundwater table.

This problem originates from agricultural application, with the objective to provide a given quantity of water to plants. However, the system considered in this section can be regarded as a part of the system of Fig. 1, with different boundary condition imposed at the top boundary (infiltration from the WDS instead of evapotranspiration).

The governing equation for water movement in the saturated zone is a Laplace's equation:

$$\nabla \cdot (k \cdot \nabla \phi(x, z)\,) = 0 \tag{14}$$

with $\phi(x, z)$ denoting the piezometric head at point (x, z) and $\mathbf{k}$ — the conductivity coefficient, assumed to be a constant scalar.

The following boundary conditions are imposed:

$$
\begin{aligned}
\text{top boundary} \qquad & \phi = 0 \\[2mm]
& k\frac{\partial \phi}{\partial n} = e(x), \\[2mm]
\text{bottom boundary} \qquad & k\frac{\partial \phi}{\partial n} = 0, \\[2mm]
\text{vertical boundaries} \qquad & k\frac{\partial \phi}{\partial n} = 0, \\[2mm]
\text{inflow tube} \qquad & k\frac{\partial \phi}{\partial n} = -Q, \\[2mm]
\text{drain tube} \qquad & k\frac{\partial \phi}{\partial n} = fQ,
\end{aligned}
\tag{15}
$$

where Q denotes the total flow of inflow tube moving to the left into the flow region, f is the fraction of flow going to the drain tube and n stands for the outward normal to the boundary. There is a balance of the water coming in and out of the system: $es = (1 - f)Q$, where s denotes the distance between an adjacent pair of tube centers.

In order to make the problem well-defined, we assume the pressure value equal to 0 at the point C.

The steps involved in solving the problem are as follows:

1. An initial water table location is assumed;

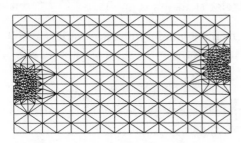

initial domain

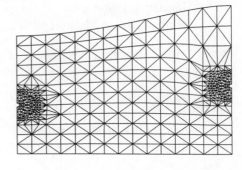

Iteration 1

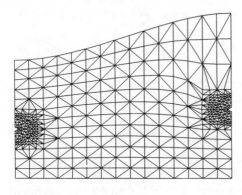

Iteration 2

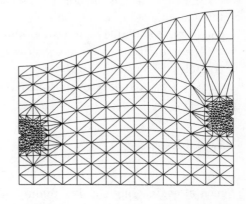

Iteration 3

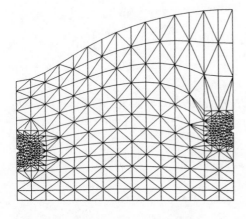

Iteration 10

Figure 5. Free boundary evolution

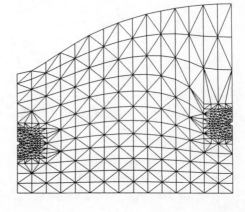

Iteration 15

2. Equation (14) together with the boundary conditions (15) is solved in the fixed domain to get values of the potential over the whole domain. The standard Finite Element Method with linear triangular elements is applied;

3. Adjustment in the free water table for all the free boundary nodes is performed by changing their z coordinates accordingly to the formula: $z_i^{new} = z_i^{old} + \phi(x_i, z_i^{old})$;

4. The modified computational domain is discretized;

5. A new flux at the top boundary is computed, such that evapotranspiration in the direction of the z-axis is constant (and equal e).

6. Steps (2)-(5) are repeated till ϕ_i becomes less than a given small value in all the free boundary nodes.

In computations the normal velocity (evapotranspiration) condition along the water table is satisfied in a natural way, so the iterations are governed by the equal pressure condition $\phi = 0$ along the water table.

We started our computations with the rectangular domain, which is divided into triangles.

The free boundary evolution is given in Figure 5.

7. Address

Not surprisingly, the presented list of the free boundary related problems in environmental engineering consists mainly of question marks. Hopefully this conference will contribute in catching-up the all-inflating Free Boundaries of Environmental Engineering and result in sharpening the theoretical as well as practical aspects of the free boundary problems in this discipline.

References

1. Nawalany M., Loch J., Sinicyn G.: *Active isolation of waste disposal sites by hydraulic means, part II: Models*, TNO-report OS 91-42-C, Delft 1992.

M. Nawalany, A. Trykozko
Institute of Environmental Engineering Systems
Warsaw University of Technology, Poland

B NEDJAR

Damage, gradient of damage and free boundaries

Abstract. In a damageable structure the boundaries between the undamaged zone and the partially damaged zone, the boundary between the partially damaged zone and the completely damaged zone are free boundaries. Within the framework of continuum damage mechanics, a model involving the gradient of the damage quantity is investigated. This model results from the principle of virtual power involving new quantities. We show its coherence from the mechanical and numerical points of view. The numerical computations show no mesh sensitivity. They describe with good agreement the main experimental properties. Concrete is choosen as an example to illustrate the theory.

1. INTRODUCTION.

Continuous damage mechanics describes, at the macroscopic level, the effect of the microfractures and microvoids on the mechanical properties of the material. Damage is defined by internal quantities.

In a damageable structure, the boundary between the undamaged zone and the partially damaged zone, the boundary between the partially damaged zone and the completely damaged zone are free boundaries. The classical continuum mechanics predictive theories of damage have a strange feature: the numerical results obtained from finite element procedures are strongly dependant on the mesh. This property can be foreseen on the set of partial differential equations of the theory. We give a predictive theory which avoids this drawback and which is coherent from the mechanical, mathematical and numerical points of view and is in agreement with experiments. The damage theory we use has been described in [3-6 and 13-14].

Damage, for instance damage of concrete, results from microscopic movements. Our basic idea is that the power of those microscopic movements must be accounted for in a predictive theory. Thus we decide to modify the expression of the power of the internal forces. We assume that this power also depends on the damage rate which is clearly related to the microscopic movements. Furthermore we assume that it depends also on the gradient of the damage rate to account for microscopic interactions.

The models issued from this formulation are free of spurious mesh sensitivity and are able, when compared to experimental results, to predict correctly the behaviour of

333

concrete structures. Accounting for the gradient of damage leads to good predictions of the structural size effect which is particulary important in civil engineering [13, 14].

The paper is organized as follows: the next section is devoted to description of the equations of the movement based on the principle of virtual powers. In the third section, a damage model is developped within the framework of the previous theory. In the fouth section, a set of numerical examples on concrete structures are given where some possibilities of the model are emphasized.

2. EQUATIONS OF MOTION AND CONSTITUTIVE LAWS.

In this section, the formulation of the damage theory developped in [3-6 and 13-14] is recalled.

Let the scalar $\beta(\mathbf{x},t)$ be a damage quantity with value 1 when the material is undamaged and value 0 when it is completely damaged. This quantity describes at the macroscopic level the decrease of the material stiffness.

Damage in a solid results from microscopic movements. We think that the power of these microscopic movements must be taken into account in the power of the internal forces. Thus, we choose this power to depend, besides on the strain rates $\mathbf{D(u)}$ ($\mathbf{u}$ is the macroscopic velocity), also on $\dfrac{d\beta}{dt}$ and $\mathbf{grad}\dfrac{d\beta}{dt}$. These latter quantities are clearly related to the microscopic movements. The gradient of damage velocity is introduced to take into account the influence of the damage at a material point on the damage of its neighbourhood. Thus, for a domain $\mathcal{D}$ interior to the solid Ω, the power of the internal forces P_i is given by:

$$P_i(\mathcal{D},\mathbf{u},\frac{d\beta}{dt}) = -\int_{\mathcal{D}} \sigma{:}\mathbf{D(u)}\, d\mathcal{D} - \int_{\mathcal{D}} (B\frac{d\beta}{dt} + \mathbf{H}.\mathbf{grad}\frac{d\beta}{dt})\, d\mathcal{D}, \qquad (1),$$

where σ is the stress tensor. The two non-classical quantities are: B, the internal work of damage and $\mathbf{H}$, the flux vector of internal work of damage.

Let the domain $\mathcal{D}$ be submitted to the volumetric external forces $\mathbf{f}$ and to the surfacic external forces $\mathbf{F}$ in its boundary $\partial\mathcal{D}$. The power of the external forces is given by:

$$P_e(\mathcal{D},\mathbf{u},\frac{d\beta}{dt}) = \int_{\mathcal{D}} \mathbf{f}.\mathbf{u}\, d\mathcal{D} + \int_{\partial\mathcal{D}} \mathbf{F}.\mathbf{u}\, d\Gamma + \int_{\mathcal{D}} A\frac{d\beta}{dt}\, d\mathcal{D} + \int_{\partial\mathcal{D}} b\frac{d\beta}{dt}\, d\Gamma, \qquad (2),$$

where A and b are respectively the volumetric and surfacic external sources of damage work. A source of damage work A or b can be produced by chemical (or in some cases electrical) actions which break the links inside a material, concrete for instance, without

334

macroscopic deformations. One can think for instance of the so-called alcali aggregate reaction which damages concrete. In all what follows, we consider only the situations where the damage is produced by mechanical actions. Thus, we take A = 0 and b = 0. The power of the external forces in (2) is then given by its classical expression.

For quasi-static evolutions, the principle of virtual power:

$$\forall \mathcal{D} \subset \Omega, \forall \mathbf{v}, \forall \gamma, \qquad P_i(\mathcal{D}, \mathbf{v}, \gamma) + P_e(\mathcal{D}, \mathbf{v}, \gamma) = 0,$$

($\mathbf{v}$ and γ are virtual velocities) gives two sets of movement equations:

$$\text{div}\sigma + \mathbf{f} = 0, \quad \text{in } \mathcal{D}, \qquad \sigma.\mathbf{n} = \mathbf{F}, \quad \text{in } \partial\mathcal{D}, \qquad (3),$$

$$\text{div}\mathbf{H} - B = 0, \text{ in } \mathcal{D}, \qquad \mathbf{H}.\mathbf{n} = 0, \quad \text{in } \partial\mathcal{D}, \qquad (4),$$

where $\mathbf{n}$ is the outward normal unit vector to $\mathcal{D}$. Equations (4) are new and non-classical. They describe the damage, *i.e.* the microscopic movements, in the domain $\mathcal{D}$.

Within the framework of continuum thermodynamics, the state of the material is characterized by its free energy Ψ. For the sake of simplicity we make the small perturbation assumption and let ε be the small strains tensor. In the context of the theory, it is natural to assume that Ψ depends also on the gradient of the state quantity β, $\Psi = \Psi(\varepsilon, \beta, \mathbf{grad}\beta)$. Also for the sake of simplicity, we assume that there is no dissipation with respect to the small strains ε, (i.e., the material is elastic) and with respect to the gradient of the damage quantity. We assume that there are only dissipative phenomenons, viscous phenomenons for instance, with respect to the damage quantity β.

A very productive way to define dissipative forces is to assume that there exist a pseudo-potential of dissipation as intoduced by Moreau (1970) [12]. A pseudo-potential of dissipation $\Phi(x)$ is a positive, convex and sub-differentiable function, with value 0 for $x = 0$.

The constitutive relations are given by [2, 3]:

$$\sigma = \frac{\partial \Psi}{\partial \varepsilon}, \qquad \mathbf{H} = \frac{\partial \Psi}{\partial(\mathbf{grad}\beta)}, \qquad B = \frac{\partial \Psi}{\partial \beta} + \frac{\partial \Phi}{\partial \dot{\beta}}, \qquad (5),$$

where $\dot{(\;)}$ denotes the time derivative of the quantity $(\;)$.

It results from the properties of the pseudo-potential of dissipation Φ, that the Clausius-Duhem inequality (we assume the temperature to be constant),

$$\frac{d\Psi}{dt} \leq \sigma : \dot{\varepsilon} + B\dot{\beta} + \mathbf{H} \cdot \mathbf{grad}\dot{\beta},$$ (6),

is satisfied. The equations describing the evolution of a piece of material are then: (3), (4) and (5) completed by adequate boundary and initial conditions.

3. A CONCRETE DAMAGE MODEL.

It is mainly observed that damage is produced by the extensions into the material when a certain threshold is acheaved. Such situations are observed in the experimental tests on concrete [9]. Another characteristic of this material is its softening behaviour when submitted to tension and compression.

We propose in this section a model developped in [13-14] which is able to describe such a behaviour. This model is a variant of the models given in [3-6].

Within the framework of the damage formulation described in the last section, the choices of the volumetric free energy Ψ and the pseudo-potential of dissipation Φ are:

$$\Psi = \frac{1}{2}\beta\{2\mu tr[\varepsilon.\varepsilon] + \lambda(tr[\varepsilon])^2\} + W(1-\beta) + \frac{k}{2}(\mathbf{grad}\beta)^2 + I(\beta),$$ (7),

$$\Phi = \frac{1}{2}c\dot{\beta}^2 - \frac{1}{2}\dot{\beta}\{2\mu tr[\varepsilon^-.\varepsilon^-] + \lambda(\langle tr[\varepsilon]\rangle^-)^2 +$$

$$(\frac{1-\beta}{1-M\beta})[2\mu tr[\varepsilon^+.\varepsilon^+] + \lambda(\langle tr[\varepsilon]\rangle^+)^2]\} + I_-(\dot{\beta}),$$ (8).

The first term of Ψ is a quadratic function with respect to the strain tensor and a linear function with respect to the damage quantity. It constitutes the simplest model where the damage is coupled with elasticity (λ and μ are the Lamé parameters).

The quantity W is the initial damage threshold expressed here in terms of volumetric energy, M describes the softening behaviour of the material, it is a quantity without dimension and its value is strictly less than 1 (M < 1) to avoid the change of the sign of 1 - Mβ, c is the viscosity parameter of damage and k measures the influence of damage at a material point on the damage of its neighbourhood. The influence of the parameters W, M and c on the behaviour under uniaxial tension is described in [13].

The expression of the pseudo-potential of dissipation (8) is choosen in such a way that damage results only from extension as it will be seen in what follows.

The functions $\langle . \rangle^+$ and $\langle . \rangle^-$ are respectively the positive part and the negative part of the scalar $\langle . \rangle$. The positive part ε^+ and the negative part ε^- of the strain tensor are obtained after diagonalisation [11]. One has the following useful properties:

$$\langle tr[\varepsilon] \rangle = \langle tr[\varepsilon] \rangle^+ - \langle tr[\varepsilon] \rangle^- \quad \text{and} \quad \langle tr[\varepsilon] \rangle^+ \langle tr[\varepsilon] \rangle^- = 0,$$

$$\varepsilon = \varepsilon^+ - \varepsilon^- \qquad \text{and} \quad tr[\varepsilon^+.\varepsilon^-] = 0,$$

$$\frac{1}{2} \frac{\partial\, tr[\varepsilon^+.\varepsilon^+]}{\partial \varepsilon} = \varepsilon^+, \qquad \frac{1}{2} \frac{\partial\, (\langle tr[\varepsilon] \rangle^+)^2}{\partial \varepsilon} = \langle tr[\varepsilon] \rangle^+ \mathbf{I_d}, \qquad (9),$$

where $\mathbf{I_d}$ is the identity second order tensor.

The functions I and I_ are the indicator functions of the intervals $[0,1]$ and $]-\infty, 0]$ respectively (the indicator function I_A of the set A is defined by, $I_A(x) = 0$ if $x \in A$ and $I_A(x) = +\infty$ if $x \notin A$). With these functions, the free energy and the pseudo-potential of dissipation have their physical values for any actual or physical value of β and of its velocity. The free energy has the value $+\infty$ for any value of β which is physically impossible ($\beta > 1$ or $\beta < 0$) and the pseudo-potential of dissipation has the value $+\infty$ for $\dot{\beta}$ positive, which is physically impossible because the damage is assumed to be irreversible.

The constitutive relations given by (5) are:

$$\sigma = \beta\{2\mu\varepsilon + \lambda(tr[\varepsilon])\mathbf{I_d}\}, \qquad \mathbf{H} = k\, \mathbf{grad}\beta, \qquad B = \frac{\partial \Psi}{\partial \beta} + \frac{\partial \Phi}{\partial \dot{\beta}}, \qquad (10),$$

where the generalized derivatives of Ψ and Φ [2, 12] are:

$$\frac{\partial \Psi}{\partial \beta} \in \frac{1}{2} \left\{ 2\mu tr[\varepsilon.\varepsilon] + \lambda(tr[\varepsilon])^2 \right\} - W + \partial I(\beta),$$

and $\quad \dfrac{\partial \Phi}{\partial \dot{\beta}} \in c\dot{\beta} - \dfrac{1}{2} \left\{ 2\mu tr[\varepsilon^-.\varepsilon^-] + \lambda(\langle tr[\varepsilon] \rangle^-)^2 + \right.$

$$\left. (\frac{1-\beta}{1-M\beta})[2\mu tr[\varepsilon^+.\varepsilon^+] + \lambda(\langle tr[\varepsilon] \rangle^+)^2] \right\} + \partial I_-(\dot{\beta}),$$

where the subdifferentials [12] ∂I and ∂I_- are defined by: $\partial I(x) = \{0\}$, if $0 < x < 1$, $\partial I(0) =]-\infty, 0]$, $\partial I(1) = [0,+\infty[$ and with $\partial I_-(x) = \{0\}$, if $x < 0$, $\partial I_-(0) = [0,+\infty[$. The elements of $\partial I(\beta)$ and $\partial I_-(\dot{\beta})$ are reactions which force β to remain between 0 and 1 and $\dot{\beta}$ to be negative.

In a domain Ω occupied by a structure, the equations of the movement are obtained by using (7), (8) and (10) in the equations (3) and (4). We get then:

$$\text{div}(\beta\{2\mu\varepsilon + \lambda(\text{tr}[\varepsilon])\mathbf{I}_d\}) + \mathbf{f} = 0, \quad \text{in } \Omega,$$

$$\sigma.\mathbf{n} = \mathbf{F}, \quad \text{in } \partial\Omega, \tag{11},$$

$$c\dot{\beta} - k\Delta\beta + \partial I(\beta) + \partial I\_(\dot{\beta}) \ni -\frac{1}{2}(1 - \frac{1-\beta}{1-M\beta})\{2\mu\text{tr}[\varepsilon^+.\varepsilon^+] + \lambda(\langle\text{tr}[\varepsilon]\rangle^+)^2\} + W, \quad \text{in } \Omega,$$

$$k\frac{\partial\beta}{\partial\mathbf{n}} = 0, \quad \text{in } \partial\Omega,$$

$$\beta(\mathbf{x},0) = \beta_0(\mathbf{x}), \quad \text{in } \Omega, \tag{12},$$

where $\Delta\beta$ is the Laplacian of β. The function β_0 is the initial value of the damage in Ω, with $\beta_0(\mathbf{x}) = 1$ when the structure is initially undamaged.

The equations (12) describe the evolution of damage in the domain Ω. In equation $(12)_1$, the source of damage in the right hand side is a strain energy depending on extensions. That agrees with the experimental observations mentioned above.

It is important to note that this model exhibits different behaviours in tension and compression. That is due to the Poisson's ratio, as already checked in [3, 5, 6, 13]. The threshold of damage is reached with a larger loading in compression than in tension (in this case the extensions are perpendicular to the load direction).

This model is then sufficient to describe the damage of solids submitted to multiaxial sollicitations (loading-unloading without changing the sign of the load).

4. EXAMPLES OF CONCRETE DAMAGE STRUCTURES.

In this section we give examples based on the one damage quantity model described in section 3. The loadings are monotone and do not change sign. Two examples of damage of structures are investigated. The specimens are analyzed as two-dimensional. Plane strain is assumed.

4.1. First example:

In concrete, most of microcracks start from an uncracked surface and grow through the depth of the specimen. Thus damage mechanics, when applied to concrete, should be able to predict the formation of damage in a specimen which is not notched or precracked. It must also predict the influence of the imposed deformation and the damage growth [7]. For this purpose, two bending tests under imposed displacements on two identical beams without notch are analyzed [6]. The first one is a three points

338

bending test and the second one is a four points bending test. The geometry and the sollicitations are shown in figure 1.

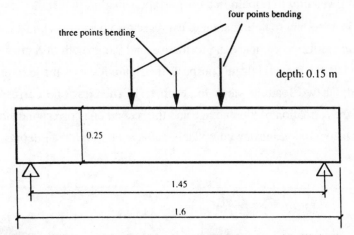

Fig. 1 - Three and four points bend. Geometry and sollicitations.

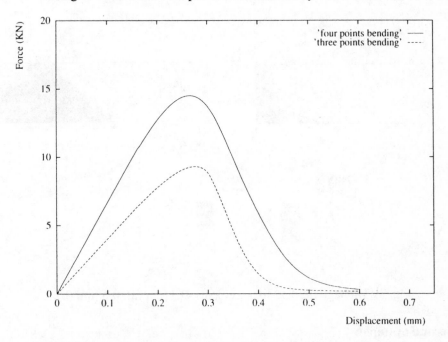

Fig. 2 - Force versus displacement curves for the three and four points bending tests.

The mechanical characteristics are: E = 27000 MPa (Young's modulus) and ν = 0.2 (Poisson's ratio). For the model we use: W = 0.5 10^{-4} MPa, M = 0.25 10^{-3} MPa, c = 0.001 MPa.s and the factor of influence of damage k = 0.2 MPa.mm^2.

The force versus displacement curves of the two tests are superposed and plotted in figure 2. The finite element results were performed with prescribed displacements u(t) at slow loading velocity to remain in a quasi-static situation ($\dot{u} = 0.001$ mm.s^{-1}, for the two tests). It is important to note that these curves show no snap-back instability.

It is interresting to compare the formation and the growth of damage for the two tests predicted by the model. For this purpose, the figure 3 shows the juxtaposition of the damage fields at two loading steps for each test. The first one corresponds to the beginning of the formation of the damage and the second one corresponds to a post-peak situation just before the beams are completely damaged through their depths.

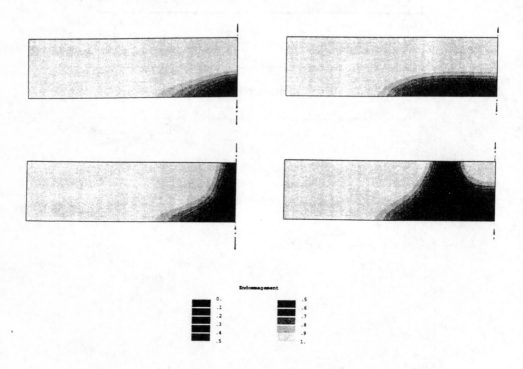

Fig. 3 - Damage field at displacements: 0.3 mm (a), 0.6 mm (b).

4.2. Second example:

This example intends to show the no mesh-sensivity of the predictive theory. The same physical problem is solved with different meshes to see whether or not the solution converges. The computations are carried out for the notched concrete plate shown in figure 4 with three meshes.

The material characteristics of the concrete used are: E = 33 GPa and ν = 0.2, for the characteristics of the model we have: W = 0.75 10^{-4} MPa, M = 0.2 10^{-3} MPa, c = 0.005 MPa.s and k = 0.2 MPa.mm^2.

The three "opening force F versus the aperture a of the notch" curves corresponding to the three meshes are plotted together with experimental results [10] in figure 5. Like in the previous example, the numerical results were performed with prescribed displacements at slow loading velocities. One can note the objectivity of these results. there is no mesh sensitivity with regard to the global behaviour of the structure. One can remark the good prediction of the actual maximum load.

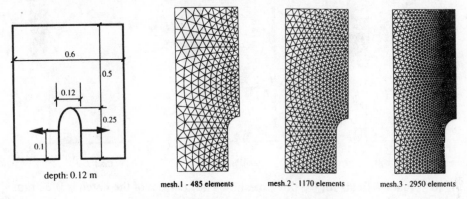

Fig. 4 - The plate and the three meshes.

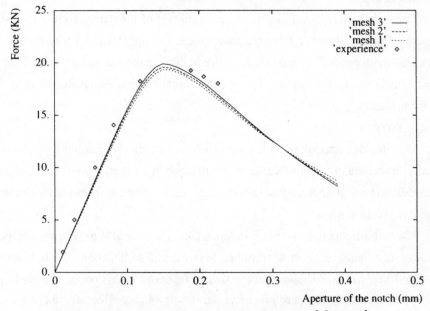

Fig. 5 - The opening force versus the aperture of the notch.
Experimental and numerical results for the three meshes.

341

To illustrate the no mesh sensitivity of the damage process, figure 6 shows the damage field predicted by the model for the three meshes at the aperture of the notch a = 0.22 mm (see figure 5).

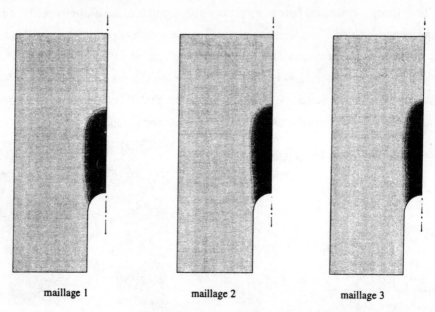

maillage 1   maillage 2   maillage 3

Fig. 6 - Damage fields for the three meshes. The aperture of the notch is 0.22 mm.

4.3. Comments.

These two examples illustrate the possibilities of the damage model issued from the theory and its coherence from the mechanical and numerical points of view. It does not exhibit mesh sensitivity and is able to predict the formation and growth of damage in concrete under multiaxial loading. Let us also emphasize that the results are in agreement with experiments.

CONCLUSION

The damage model we have described involves the gradient of the damage quantity: it accounts for the influence of the damage at a material point on the damage of its neighbourhood. It appears that this formulation is coherent from the mechanical and numerical points of view.

The resulting models are mesh independant. They are able to describe correctly the behaviour of concrete structures submitted to multiaxial sollicitations. Also, in references [6, 13 and 14], we have shown that the theory involved in this paper, allows to predict the important civil engineering property: the structural size effect in concrete structures [1, 7, 9].

REFERENCES

[1] Z. P. Bazant, O. Ozbolt, "Non local microplane model for fracture, damage and size effect in structures", J. of Engrg. Mech. ASCE, 116, pp. 2485-2505 (1990).

[2] M. Frémond, "Sur l'inégalité de Clausius-Duhem", C. R. Acad. Sci. Paris, serie II, t.311, pp. 757-764 (1990).

[3] M. Frémond, B. Nedjar,"Endommagement et principe des puissances virtuelles", C. R. Acad. Sci. Paris, serie II, t.317, n° 7, pp. 857-864 (1993).

[4] M. Frémond, B. Nedjar,"Damage and gradient of damage. The unilateral phenomenon", SMiRT12 (12th Structural Mechanics in Reactor Technology), ed. F. Kussmaul, North-holland, vol. H, pp. 375-380 (1993).

[5] M. Frémond, B. Nedjar, "Damage of concrete, the unilateral phenomenon", Nuclear Engineering and Design, **156**, pp. 323-335 (1995).

[6] M. Frémond, B. Nedjar, "Damage, gradient of damage and principle of virtual power", Int. Jrn. of Solids. Structures, (1995), in press.

[7] A. Hillerborg,"Analysis of a single crack", In Fracture Mechanics of concrete, ed. F. H. Wittman, Elsevier Pubs., 223-249 (1983).

[8] P. Ladevèze, "Sur une théorie de l'endommagement anisotrope", internal report n°34, Laboratoire de Mécanique et Technologie, Ecole Normale Supérieur de Cachan, France, (1983).

[9] J. Mazars, Z. P. Bazant, Eds. "Cracking and damage, Strain localization and size effect". Elsevier Pubs (1988).

[10] J. Mazars, D. Walter, "Endommagement mécanique du béton", Délégation Générale à la Recherche Scientifique et Technique, n° 78.7.2697 et 78.7.2698 (1980).

[11] J. J. Moreau, "Fonctionnelle convexes. Séminaire sur les équations aux dérivées partielles", Collège de France, Paris (1966).

[12] J. J. Moreau, "Sur les lois de Frottement, de Viscosité et de Plasticité", C. R. Acad. Sci. Paris, vol.271, pp. 608-611 (1970).

[13] B. Nedjar, "Mécanique de l'endommagement. Théorie du premier gradient et application au béton", Thèse de doctorat, Ecole Nationale des Ponts et Chaussées, Paris (1995).

[14] B. Nedjar, M. Frémond, "Damage, gradient of damage and structural size effect", submitted for publication to, Int. J. of Fracture (1995).

Boumediene NEDJAR, Laboratoire des matériaux et des structures du génie civil, Laboratoire mixte LCPC/CNRS, UMR 113, 2, allée Kepler, Cité Descartes, 77420, Champs sur Marne, France.

J STEINBACH

Simulation of the mould filling process by means of a temperature-dependent, viscous flow

1 Introduction

In this contribution we consider a generalized Hele-Shaw flow in compression moulding which contains non-isothermal behaviour of the molten polymers and non-symmetric effects. Beside injection moulding compression is another essential forming technique for polymers. We analyze a variational inequality approach for the flow problem. Furthermore, we present some results from the numerical analysis of the corresponding evolutionary inequality based on a finite volume element (box) method.

The classical Hele-Shaw flow arises when an incompressible viscous, isothermal fluid is moving within a thin region between slightly seperated plates. The mathematical problem in its simplest form is given by $-\triangle p = f$ in $\Omega(t)$, $p = 0$ and $\nabla p \cdot \vec{n} = -V_n$ on $\partial\Omega(t)$, where $p = p(x,t)$, $x = (x_1, x_2)$ denotes the pressure in the liquid $\Omega(t)$ and V_n is the normal velocity of $\partial\Omega(t)$ in the outward direction $\vec{n}$. The liquid movement is driven by sources or sinks introduced in the field equation by a right-hand f side or by using boundary conditions along a domain surrounded by the liquid. We recall that this problem can be regarded as a zero-specific-heat, one-phase Stefan problem such that besides the usual methods known for Stefan problems also complex variable theory provides a starting point for solving the two-dimensional problem (see [9] for a survey). It is not our aim to survey the extensive bibliography concerning the various mathematical approaches for this problem. Let us only mention, that the injection/compression case is well-posed, whereas the suction situation is ill-posed ([5]). Furthermore, different aspects of singularities of the free boundary are studied in several recently published papers, e.g. [8], [10].

2 Mathematical model for compression moulding

In the compression moulding process, schematically shown in figure 1, a premeasured initial polymer charge $\Omega(0)$ is placed in the cavity between the halves of a mould, for instance by injection. Then, these halves are brought together with the closing speed $\partial_t d(x,t) = \partial_t(d_2(x,t) - d_1(x,t))$ to squeeze and transform the polymer melt until the cavity is filled. It is beyond the scope of this contribution to give a more detailed description of the technological aspects of this forming method (see [12], [11], [20]). Here, we should only mention some different operating conditions like: filling of a closed (i.e. lateral walls appear) or a partially open mould, both mould halves are moved or one surface being stationary, cooling regime in the mould (thermoplastics) or a heated mould (thermosets). Due to the just mentioned possible asymmetric movement of the lower and upper walls or the asymmetric heating/cooling, one should consider the non-symmetric situation with respect to $z = 0$. According to figure 1, let

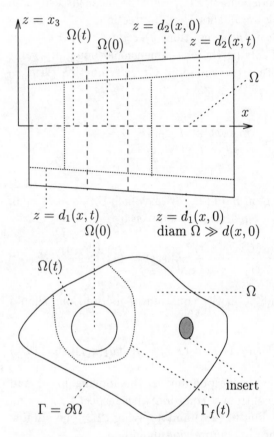

Figure 1: *Compression moulding, Side view and ground plan, initial charge $\Omega(0)$, charge shape $\Omega(t)$ after some time t*

the cavity be described by a two-dimensional domain Ω and the given gap thickness $d(x,t) = d_2(x,t) - d_1(x,t)$ decreasing with respect to time. At time $t \in [0,T]$ the plastic melt occupies the region $\Omega(t)$, which is limited by the advancing melt front $\Gamma_f(t)$ and possibly (for a closed mould) by a part $\Gamma \cap \partial\Omega(t)$ of the fixed boundary $\Gamma = \partial\Omega$.

The basic principles describing the dynamics of a viscous fluid flow are the conservation of mass and momentum together with a constitutive equation. But the geometrical assumption that the diameter of the cavity is much larger than the gap thickness allows the application of the lubrication (Hele-Shaw) theory (see also [12]). Hence, by an order-of-magnitude analysis one verifies that the equation of motion can be simplified to

$$\frac{\partial p(x,t)}{\partial x_j} = \frac{\partial}{\partial z}\left(\eta(x,z,t)\frac{\partial v_j(x,z,t)}{\partial z}\right), \quad j = 1,2, \tag{2.1}$$

where $p = p(x,t)$ denotes the pressure (independent of z), η is the shear viscosity and $\vec{v} = (v_1, v_2, v_3)$ is the velocity vector with components v_j in x_j direction, $j = 1,2,3$, $z = x_3$, $v_3 \ll v_j, j = 1,2$. Taking into account no-slip conditions $v_j(x, d_i(x,t), t) = 0$, $i,j = 1,2$ the velocity profile can be found by solving the one-dimensional boundary value problems (2.1) such that by analyzing the corresponding Green's function $G(\eta;.,.)$ one obtains

$$v_j(x,z,t) = \frac{\partial p(x,t)}{\partial x_j}\int_{d_1}^{d_2} G(\eta;z,\zeta)\,d\zeta = \frac{\partial p(x,t)}{\partial x_j}\begin{cases}\int_{d_1}^{z}\frac{\zeta-\delta}{\eta(x,\zeta,t)}d\zeta & z \in (d_1,\delta)\\[2mm]\int_{z}^{d_2}\frac{\delta-\zeta}{\eta(x,\zeta,t)}d\zeta & z \in (\delta,d_2)\end{cases} \tag{2.2}$$

where δ is the local value of z (depending on η and d_1, d_2) at which $\partial_z v_j(x,\delta,t) = 0$, $j = 1,2$ holds. Obviously, the gapwise-averaged velocity is given by

$$\bar{v}_j(x,t) = d^{-1}\int_{d_1}^{d_2} v_j(x,z,t)\,dz = \frac{-k(x,t)}{d(x,t)}\frac{\partial p(x,t)}{\partial x_j}, \quad k(x,t) = \int_{d_1}^{d_2}\frac{(z-\delta)^2}{\eta(x,z,t)}\,dz. \tag{2.3}$$

Having neglected the velocity component v_3 in the equation of motion, one should introduce a (linear) approximation

$$v_3(x,z,t) = d(x,t)^{-1}[(z - d_1(x,t))\,\partial_t d_2(x,t) + (d_2(x,t) - z)\,\partial_t d_1(x,t)], \tag{2.4}$$

for the energy equation (discussed below) corresponding to the moving lower and upper surface walls. By integrating the continuity equation with respect to z, using $v_3(x, d_i(x,t), t) = \partial_t d_i(x,t)$, $i = 1,2$ according to (2.4) and applying (2.3) one obtains the free boundary problem for the pressure in compression moulding

$$-\sum_{i=1}^{2}\frac{\partial}{\partial x_i}\left(k(x,t)\frac{\partial p(x,t)}{\partial x_i}\right) = f(x,t) = -\frac{\partial d(x,t)}{\partial t} \quad \text{in } \Omega(t),$$

$$p = 0 \quad \text{and} \quad k\nabla p \cdot \vec{n} = -(d_2(x,t) - d_1(x,t))\,\bar{v}_n \quad \text{on } \Gamma_f(t) = \Omega \cap \partial\Omega(t), \tag{2.5}$$

$$p = 0 \text{ on } \Gamma_D, \quad k\,\nabla p \cdot \vec{n} + \alpha\,p = 0,\ 0 \le \alpha \ll 1 \text{ on } \Gamma_N; \quad t \in (0,T],$$

where $\Omega(0)$ is given and the condition on the fixed boundary $\Gamma = \partial\Omega$ corresponds to an open part Γ_D and a closed (for $0 < \alpha \ll 1$ not completely impermeable, which can be justified from the practical point of view) part Γ_N of the mould ($\Gamma = \Gamma_D \cup \Gamma_N$). The conditions on $\Gamma_f(t)$ arise from the assumption that the capillarity (surface tension) is negligible and the condition across a surface of contact (separating the melt and the empty part (air)), where $\bar{v}_n = \vec{v} \cdot \vec{n}$ is the (thickness-averaged) normal interface velocity.

346

Usually, the rheological behaviour of polymer melts (cf. [20], chap. 2) is characterized by thermal effects ($\eta = \eta(\theta)$, θ as temperature) and sometimes also by non-Newtonian effects, i.e. $\eta = \eta(\theta, \dot{\gamma})$ with $\dot{\gamma}$ as the square root of one half of the second invariant of the strain tensor. Here, we restrict ourselves to the first case, for the second we refer to [18]. The relation between viscosity η and temperature θ is generally considered to follow an Arrhenius-type expression

$$\eta(\theta) = m(\theta) = c_1 \exp\left(\frac{c_2}{\theta}\right), \quad 0 < c_i = \text{const.}, \ i = 1, 2, \tag{2.6}$$

such that the flow conductivity $k = k(d, \eta(\theta))$ in (2.3) depends on the gap thickness and temperature. Based on the energy balance, especially taking into account the viscous heating term (frictional heat) given in a Hele-Shaw flow by $r = \eta \dot{\gamma}^2$, the temperature equation in the three-dimensional spatial melt region $\mathcal{L} = \Omega(t) \times (d_1(x,t), d_2(x,t)) \times (0, T)$ becomes

$$\rho\, c\, (\partial_t \theta + \vec{v} \cdot \nabla \theta) - \sum_{i=1}^{2} \frac{\partial}{\partial x_i}\left(\lambda \frac{\partial \theta}{\partial x_i}\right) - \frac{\partial}{\partial z}\left(\lambda \frac{\partial \theta}{\partial z}\right) = r = |\nabla p|^2 \frac{(z - \delta)^2}{\eta}, \tag{2.7}$$

where $\vec{v}$ is given in (2.2), (2.4). Hence, the temperature has to be found in a moving region determined by the flow problem. Other nonlinearities occur in the right-hand side r of (2.7) with $r = r(\eta, |\nabla p|^2)$ and in the convective term for which one needs the viscosity (see (2.2)). The temperature equation (2.7) is completed by an initial condition $\theta(x, z, 0) = \theta_0$ for $x \in \Omega(0)$, $z \in (d_1(x, 0), d_2(x, 0))$ and furthermore by appropriate boundary conditions at the lower and upper walls, the lateral walls and at the interface between plastic and air, which should not be discussed in detail here, cf. [18].

Again, by an order-of-magnitude analysis one verifies that in equation (2.7) the heat conduction is dominated by the z-direction, whilst the $x_1 x_2$- flow has a determining influence on the convective term. Hence, by neglecting the small diffusive terms in $x_1 x_2$-direction a two-dimensional nonlinear convection problem for the average temperature $\bar{\theta}(x, t)$ defined by $\bar{\theta}(x, t) = d^{-1} \int_{d_1}^{d_2} \theta(x, z, t)\, dz$ can be derived. To be more precisely, using the average velocity $\vec{\bar{v}}$ (cf. (2.3)) in the convective term and assuming boundary conditions of the form $\mp \lambda\, \partial_z \theta = \sigma_{PW}(\theta_W - \theta)$ for $z = d_i(x, t)$, $i = 1, 2$, where $\theta_W = \theta_W(x, t)$ is the given mould temperature and σ_{PW} denotes the heat transfer coefficient between plastic melt and mould, integration with respect to (d_1, d_2) in (2.7) leads to the nonlinear advection equation

$$\rho\, c\, (\partial_t \bar{\theta} + \vec{\bar{v}} \cdot \nabla \bar{\theta}) + \frac{2\, \sigma_{PW}}{d}\, \bar{\theta} = \frac{\sigma_{PW}}{d} \sum_{i=1}^{2} \theta_W(x, t) + \int_{d_1}^{d_2} r\, dz, \quad x \in \Omega(t), t \in (0, T), \tag{2.8}$$

completed by an initial condition $\bar{\theta}(x, 0) = \theta_0(x)$ in $\Omega(0)$.

3 Variational inequality approach for the pressure

Beside an enthalpy type formulation for the problem (2.5) another fixed domain formulation can be constructed by performing an integral transformation to the pressure in (2.5). The application of such a transformation originally introduced by Baiocchi for the dam problem leads for (one- or two-phase) Stefan problems to parabolic inequalities of first or second kind (see [15]). In terms of an Stefan-type problem the situation in the flow problem (2.5) is characterized by a 'vanishing heat capacity' and furthermore the coefficient function k depends explicitely on x and t.

Due to the technological assumption that the lower and upper surface wall move towards each other, i.e. $f \geq 0$ in $\Omega(t)$, from the maximum principle it follows $p \geq 0$ in $\bar{\Omega}(t)$ and furthermore an advancing flow front ($\Omega(t) \subseteq \Omega(t')$ for $t < t'$) is guaranteed. Consequently, let us introduce the new unknown function u by

$$u(x,t) = \int_0^t p(x,t')\, dt' = \int_{s(x)}^t p(x,t')\, dt', \quad (x,t) \in \bar{Q}, \quad Q = \Omega \times (0,T),$$

where p is extended by continuity and set equal to zero outside the melt region on $\Omega \backslash \Omega(t)$, $0 \leq t \leq T$. The melt region $\Omega(t)$ and its free boundary should be given by

$$\Omega(t) = \{x \in \Omega : t > s(x)\}, \quad \Gamma_f(t) = \{x \in \Omega : t = s(x)\}, \quad s(x) := 0 \text{ for } x \in \Omega(0)$$

for each time $t \in (0,T]$. By applying the just mentioned transformation to (2.5) and taking especially into consideration that $k = k(x,t)$ (integration by parts with respect to time is necessary), one arrives at

$$d(x,t)\, \chi_{\Omega(t)} = \int_0^t (Bu)(x,t')\, dt' + d(x,0)\, \chi_{\Omega(0)} - (Au)(x,t) \quad \text{in } \Omega,$$

$$(k\, \nabla u \cdot \vec{n})(x,t) + \alpha\, u(x,t) = \int_0^t (\partial_t k\, \nabla u \cdot \vec{n})(x,t')\, dt' \text{ on } \Gamma_N,$$

$$u(x,t) = 0 \text{ on } \Gamma_D, \ t \in (0,T]; \quad u(x,0) = 0 \quad \text{in } \Omega, \quad \Omega(0) \text{ given},$$

$$(Av)(x,t) = -\operatorname{div}(k(x,t)\, \nabla v(x,t)), \quad (Bv)(x,t') = -\operatorname{div}(\partial_t k(x,t')\, \nabla v(x,t')).$$

$$\tag{3.1}$$

The application of Green's formula in (3.1) leads to an evolutionary obstacle problem:

$$\text{Find} \quad u(t) \in K = \{w \in H^1(\Omega) : \ w \geq 0 \text{ in } \Omega, \ w = 0 \text{ on } \Gamma_D\}, \quad t \in [0,T],$$
$$a(t; u(t), v - u(t)) \geq (F(t), v - u(t)) + \int_0^t b(t'; u(t'), v - u(t))\, dt' \quad \forall\, v \in K \tag{3.2}$$

with the bilinear forms and the right-hand-side

$$a(t; v(t), w) = \int_\Omega k(x,t)\, \nabla v(x,t)\, \nabla w(x)\, dx + \alpha \int_{\Gamma_N} v(x,t)\, w(x)\, dx,$$

$$b(t'; v(t'), w) = \int_\Omega \partial_t k(x,t')\, \nabla v(x,t')\, \nabla w(x)\, dx, \quad F(t) = d(x,0)\, \chi_{\Omega(0)} - d(x,t). \tag{3.3}$$

Due to $F(0) \leq 0$ the initial condition $u(0) = 0$ in (3.1) is automatically satisfied, if one solves the obstacle problem (3.2) for $t = 0$.

Remark 3.1 It is easy to see, that for $k = k(x)$ (isothermal injection moulding) the bilinear form b as memory term vanishes in (3.2), such that elliptic inequalities containing time only as parameter are obtained (see [4], [6]). In [13], such inequalities with non-local boundary conditions describing the total flux through the boundary of the gate region together with the condition that the pressure is constant but unknown there, are investigated from the analytical point of view. On the other hand, in isothermal compression moulding, i.e. k can be separated in a product $k(x,t) = k_1(x) \, k_2(t)$, elliptic inequalities (without b) can be derived by using the transformation $u(x,t) = \int_0^t k_2(t') \, p(x,t') \, dt'$ ([3]). $\qquad \square$

Now we want to present some results concerning evolutionary obstacle problems of the form (**3**.2). At first, we state the existence of a unique solution u to (**3**.2), (**3**.3) as continuous function $u : [0,T] \to V \subset H^1(\Omega)$, where $V = \{w \in H^1(\Omega) : w = 0 \text{ on } \Gamma_D\}$. The following result is proved in [18] by exploiting a fixed point argument.

Theorem 3.2 *Let $k \in W_\infty^1((0,T); L_\infty(\Omega))$ and $F \in C([0,T]; L_2(\Omega))$ be fulfilled and let the bilinear form a be V-coercive for all $t \in [0,T]$ with a constant $m > 0$ independent of t. Then, there exists a unique solution $u(t) \in K \; \forall t \in [0,T]$ of problem (**3**.2) with $u \in C([0,T]; H^1(\Omega))$.*
Let us now demonstrate, that the solution u of (**3**.2) is actually a Lipschitz continuous function $u : [0,T] \to H^1(\Omega)$.

Theorem 3.3 *Under the assumptions of Theorem **3**.2 and $F \in W_\infty^1((0,T); L_2(\Omega))$ the unique solution u of (**3**.2) is such that*

$$u \in W_\infty^1((0,T); H^1(\Omega)) = C^{0,1}([0,T]; H^1(\Omega)).$$

Proof. To compare two solutions $u(t_i)$ for $t_i \in [0,T]$, $i = 1,2$, one takes $v = u(t_{3-i})$ in the inequality corresponding to t_i, $i = 1,2$. By subtraction one obtains

$$
\begin{aligned}
m \, \|u(t_1) - u(t_2)\|_{H^1(\Omega)}^2 \leq \; & (F(t_1) - F(t_2), u(t_1) - u(t_2)) \\
& + \int_{t_1}^{t_2} b(t'; u(t'), u(t_1) - u(t_2)) \, dt' \\
& + a(t_2; u(t_2), u(t_1) - u(t_2)) - a(t_1; u(t_2), u(t_1) - u(t_2)).
\end{aligned}
$$

From here, for $t_1 \neq t_2$ one derives

$$
\begin{aligned}
m \frac{\|u(t_1) - u(t_2)\|_{H^1(\Omega)}}{|t_1 - t_2|} \leq \; & \left\| \tfrac{F(t_1) - F(t_2)}{t_1 - t_2} \right\|_{L_2(\Omega)} + \left\| \tfrac{k(t_1) - k(t_2)}{t_1 - t_2} \right\|_{L_\infty(\Omega)} \|u(t_2)\|_{H^1(\Omega)} + \\
& + \|\partial_t k\|_{L_\infty((0,T); L_\infty(\Omega))} \, \|u\|_{C([0,T]; H^1(\Omega))}
\end{aligned}
$$

and therefore the solution u is Lipschitz continuous from $[0,T] \to V \subset H^1(\Omega)$. $\qquad \blacksquare$

In order to investigate monotony properties and the spatial regularity (see remark **3**.6), we consider a Lewy-Stampacchia type regularization (cf. [14], sect. 5.3), defined in our case by

Find $z_\varepsilon(t) \in V$: $\quad a(t; z_\varepsilon(t), v) + (\xi \, \beta_\varepsilon(z_\varepsilon(t)), v) = (F(t) + \xi, v) +$

$$+ \int_0^t b(t'; z_\varepsilon(t'), v) \, dt' \quad \forall \, v \in V, \ t \in [0, T], \quad \beta_\varepsilon(w) = \beta(\frac{w}{\varepsilon}) = \frac{(w/\varepsilon)^+}{(w/\varepsilon) + 1} = \frac{w^+}{\varepsilon + w}$$

(3.4)

with $\xi \in L_2(\Omega)$ such that $\xi \geq (-F(0))^+ \geq 0$ and $\varepsilon > 0$. Let us now summarize some properties of this auxiliary problem.

Lemma 3.4 *Let the assumptions of Theorem **3.3** be fulfilled. Then the unique solution z_ε of (**3.4**) is such that $z_\varepsilon \in W_\infty^1((0, T), H^1(\Omega))$ uniformly in ε. Under the additional assumption $\partial_t F \geq 0$ in Q, there holds $\partial_t z_\varepsilon \geq 0$ a.e. in Q. Furthermore, z_ε preserves the obstacle, i.e. $z_\varepsilon(t) \in K \ \forall t \in [0, T]$ and z_ε converges strongly in $C([0, T]; H^1(\Omega))$ as $\varepsilon \to 0$ to the solution u of the obstacle problem (**3.2**) with the error estimate*

$$\|u - z_\varepsilon\|_{C([0,T];H^1(\Omega))} \leq C\sqrt{\varepsilon} \, \|\xi\|_{L_1(\Omega)}^{1/2}.$$

Proof. The first result is obtained by using the same technique as in the proofs of Theorems **3.2**, **3.3** and, in particular, the monotony of the penalty operator β. A nondecreasing solution $z_\varepsilon(t)$ is deduced by differentiating (**3.4**) with respect to time and using $\partial_t F \geq 0$.

To show $z_\varepsilon(t) \in K$, the elliptic problem $a(0; z_\varepsilon(0), v) + (\xi \, \beta_\varepsilon(z_\varepsilon(0)), v) = (F(0) + \xi, v)$ $\forall v \in V$ for $t = 0$ is considered. Taking $V \ni v = (z_\varepsilon(0))^- = \min\{z_\varepsilon(0), 0\} \leq 0$ and using $F(0) + \xi \geq 0$, one verifies $z_\varepsilon(0) \in K$. Hence, by $z_\varepsilon(t) \geq z_\varepsilon(0)$ the inclusion $z_\varepsilon(t) \in K$ for all $t \in [0, T]$ is guaranteed. Therefore, it is allowed to take $v = z_\varepsilon(t) \in K$ in (**3.2**). Combining this with $v = (u - z_\varepsilon)(t) \in V$ in (**3.4**) one obtains by subtraction

$$m \, \|(z_\varepsilon - u)(t)\|_{H^1(\Omega)}^2 \leq (\xi \, (1 - \beta_\varepsilon(z_\varepsilon(t))), (z_\varepsilon - u)(t)) + \int_0^t b(t'; (z_\varepsilon - u)(t'), (z_\varepsilon - u)(t)) \, dt'.$$

Due to the estimate $(1 - \beta(r)) \, r \leq 1$ for $r \geq 0$ and furthermore $\xi \, (1 - \beta_\varepsilon(z_\varepsilon)) \, (-u) \geq 0$, the first term on the right-hand side can be estimated by $(\xi \, (1 - \beta_\varepsilon(z_\varepsilon(t))), (z_\varepsilon - u)(t)) \leq \varepsilon \, (\xi, 1)$. Now, using Young's inequality one derives

$$m \, \|(z_\varepsilon - u)(t)\|_{H^1(\Omega)}^2 \leq \varepsilon \, \|\xi\|_{L_1(\Omega)} + \frac{CT\gamma}{2} \, \|(z_\varepsilon - u)(t)\|_{H^1(\Omega)}^2$$

$$+ \frac{C}{2\gamma} \int_0^t \|(z_\varepsilon - u)(t')\|_{H^1(\Omega)}^2 \, dt'$$

for all $t \in [0, T]$. Finally, choosing a suitable value for γ and applying the Gronwall inequality, the proof is completed. ∎

Based on Lemma **3.4** there exists a subsequence, again denoted by z_ε with $z_\varepsilon \rightharpoonup \tilde{u}$ weakly in $H^1((0, T); H^1(\Omega))$ and moreover $z_\varepsilon \to u$ strongly in $C([0, T]; H^1(\Omega))$ such that $\tilde{u} = u$. Hence, one concludes the following statement.

350

Corollary 3.5 *Let the assumptions of Theorem* **3**.*3 together with* $\partial_t F \geq 0$ *in* Q *be fulfilled. Then, the unique solution* $u \in W_\infty^1((0,T); H^1(\Omega))$ *of the evolutionary variational inequality* **(3**.*2) is such that* $\partial_t u \geq 0$ *a.e. in* Q.

Obviously, the right-hand-side $F = d(x,0)\, \chi_{\Omega(0)} - d(x,t)$ in **(3**.3) satisfies $\partial_t F = -\partial_t d \geq 0$ such that the results of Theorems **3**.2, **3**.3 and Corollary **3**.5 can be applied to the compression moulding inequality **(3**.2), **(3**.3).

Remark 3.6 Concerning the spatial regularity, the inequalities with Dirichlet ($\Gamma_D = \partial\Omega$) and Robin/Neumann ($\Gamma_N = \partial\Omega$) boundary conditions should be considered separately. Then applying the Rothe method (semidiscretization in time) to the penalization problem **(3**.4) together with the elliptic regularity theory and passing to the limit as $\varepsilon \to 0$ under consideration of Lemma **3**.4 one concludes, that u as solution of **(3**.2), **(3**.3) belongs to $L_\infty((0,T); H^2(\Omega))$. For details we refer to [19], [16]. □

4   Numerical treatment of evolutionary inequalities

Let us now sketch the outline of the numerical solution of the obstacle problem **(3**.2), **(3**.3) by means of a finite volume element method (box scheme). For ease and brevity of presentation we restrict the discussion to problem **(3**.2), **(3**.3) with Dirichlet boundary conditions in a polygonal domain $\Omega \subset \mathbb{R}^2$. The box methed is characterized by a local modelling of balance equations on finite control regions (boxes) which causes its widespread utilization in application problems. Applied to second order boundary value problems this discretization technique is very similar to the finite element method at least for the principal term (stiffness matrix), whereas the approximation of first-order (upwinding) and zero-order terms (lumping) is different. For linear elliptic equations direct derivations of corresponding error estimates of box schemes can be found e.g. in [1], while in [7], [2] there is shown that the box method solution is of comparable accuracy to the Galerkin solution.

Let $\mathcal{T}_h$ denote an admissible and quasi-uniform triangulation of Ω consisting of triangles. Furthermore, we will use $\bar{\omega}$ and $\omega = \bar{\omega} \cap \Omega$ as the sets of the vertices and interior vertices of this triangulation, respectively. Each vertex $\xi \in \bar{\omega}$ a region Ω_ξ is associated consisting of those triangles $\mathcal{K} \in \mathcal{T}_h$ which have ξ as a vertex. Then the dual mesh $\mathcal{B}_h$ for $\mathcal{T}_h$ consisting of the control regions (boxes) $\mathcal{H}$ is constructed as follows. For each $\mathcal{K} \in \mathcal{T}_h$ one selects a distinguished point $p \in \bar{\mathcal{K}}$ and this point p is connected to the edge midpoints of $\mathcal{K}$ by straight-line segments. Hence, one obtains a partition of $\mathcal{K}$ into three subregions. Now, the box $\mathcal{H}_\xi \in \mathcal{B}_h$, $\mathcal{H}_\xi \subseteq \Omega_\xi$ associated with a vertex $\xi \in \bar{\omega}$ is defined by the union of these subregions in Ω_ξ which have ξ as a corner.

In order to illustrate how a discrete inequality and a corresponding penalization problem is derived we consider the equation $Au = -\mathrm{div}\,(k\,\nabla u) + q\,u = f$ in Ω with homogeneous Dirichlet conditions on $\Gamma = \partial\Omega$. By multiplying $Au - f$ with $v \in Q(\mathcal{B}_h) = \{w \in L_2(\Omega) : w \in H^1(\mathcal{H}) \,\forall\, \mathcal{H} \in \mathcal{B}_h,\ w = 0 \text{ on } \Gamma\}$ and integrating by parts, one obtains

$$\sum_{\mathcal{H} \in \mathcal{B}_h} \int_{\mathcal{H}} k \, \nabla u \, \nabla v \, dx - \int_{\partial \mathcal{H}} k \, \nabla u \cdot \vec{n} \, v \, dx + \int_{\mathcal{H}} (qu - f) \, v \, dx = 0 \quad \forall \, v \in Q(\mathcal{B}_h). \quad (4.1)$$

Now, the test space consists of piecewise constant functions with respect to the dual mesh $\mathcal{B}_h$, i.e. $(P^B v_h)(x) = \sum_{\xi \in \omega} v_{h,\xi} \, \Phi_\xi^B(x)$ with the grid function (vector) $(v_h)_{\xi \in \omega} \in L_2(\omega) = \mathrm{R}^{\#\omega}$, $v_{h,\xi} = v_h(\xi)$ for $x \in \omega$ and the basic functions $\Phi_\xi^B(x) = 1$ for $\xi \in \mathcal{H}_\xi$ and $\Phi_\xi^B(x) = 0$ otherwise. Furthermore, as trial space for the second term in (4.1) we use piecewise linear finite elements, i.e. $(P^G u_h)(x) = \sum_{\xi \in \omega} u_{h,\xi} \, \Phi_\xi^G(x)$ with Φ_ξ^G as the usual nodal basis, whereas for the zero-order term qu the trial function $P^B u_h$ is used to obtain a lumped diagonal mass matrix. Consequently, the discrete system to (4.1) is given by

$$\langle (L + M) \, u_h, v_h \rangle = \langle f_h, v_h \rangle \quad \forall v_h \in L_2(\omega); \quad M_{\xi,\xi} = \int_{\mathcal{H}_\xi} q \, dx, \; M_{\xi,\eta} = 0, \; \xi \neq \eta,$$

$$L_{\xi,\eta} = -\int_{\partial \mathcal{H}_\eta} k \, \partial_n \Phi_\xi^G \, dx = -\sum_{\mathcal{H}} \int_{\partial \mathcal{H}} k \, \partial_n \Phi_\xi^G \, \Phi_\eta^B \, dx, \quad f_{h,\xi} = \int_{\mathcal{H}_\xi} f \, dx$$

$$(4.2)$$

with the stiffness $L = L_{\xi,\eta}$, the mass matrix , $M = M_{\xi,\eta}$, $\xi, \eta \in \omega$ and the Euclidean scalar product $\langle u_h, v_h \rangle = \sum_{\xi \in \omega} u_{h,\xi} \, v_{h,\xi}$. Although L is obtained from line integrals, this matrix coincides with the usual Galerkin-FEM stiffness matrix (independent of the location of the points p !), provided $k \in L_\infty(\Omega)$ is a coefficient function which takes a constant value on each triangle $\mathcal{K}$.

In order to discretize problem (3.2), (3.3) the time integral on the right-hand side is splitted into subintegrals, i.e. $\int_0^{t^j} b(t'; u(t'), v) \, dt' = \sum_{i=0}^{j-1} \int_{t^i}^{t^{i+1}} b(t'; u(t'), v) dt'$ and each of these subintegrals is approximated by a trapezoidal rule weighted with a parameter $\sigma \in [0,1]$. Then, for $t = t^j$ all terms containing the unknown $u(t^j)$ are collected in the left-hand side of our inequality and we end up with

Find $y_h^j \in K_h = \{w_h \in L_2(\bar{\omega}) : w_h(\xi) \geq 0, \; \xi \in \bar{\omega}; \; w_h(\xi) = 0, \; \xi \in \gamma = \bar{\omega} \setminus \omega\}$,

$$\langle \Lambda^j y_h^j, v_h - y_h^j \rangle \geq \langle F_h^j, v_h - y_h^j \rangle + \sum_{i=0}^{j-1} \langle (\Lambda^{i+1} - \Lambda^i) \, y_h^i, v_h - y_h^j \rangle \quad \forall v_h \in K_h, \quad (4.3)$$

with the grid function $y_h^j = y_h(\xi, t^j)$, $\xi \in \bar{\omega}$, the time discretization $t^j = j\tau$, $j = 0, \ldots, N$, $N\tau = T$ and the matrices $\Lambda^i = (1 - \sigma)L^{i-1} + \sigma L^i$, $\sigma \in [0,1]$, where L^i is the stiffness matrix (cf. (4.2)) corresponding to the coefficient $k = k(x, t^i)$. We approach the discrete inequality (4.3) by penalty equations in which the condition $y_h^j \in K_h$ is replaced by penalization terms becoming progressively larger as the solution moves away from K_h.

Find $y_{\varepsilon h}^j \in L_2(\bar{\omega})$; $y_{\varepsilon h}^j(\xi) = 0, \; \xi \in \gamma$

$$\langle \Lambda^j y_{\varepsilon h}^j, v_h \rangle + \varepsilon^{-1} \langle \beta_1(y_{\varepsilon h}^j), v \rangle = \langle F_h^j, v_h \rangle + \sum_{i=0}^{j-1} \langle (\Lambda^{i+1} - \Lambda^i) \, y_{\varepsilon h}^i, v_h \rangle \quad \forall v_h \in L_2(\omega)$$

$$(4.4)$$

352

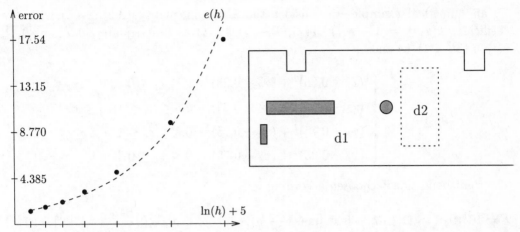

Figure 2: *Left: Numerical example, dots: error for* $\tau = h = \frac{1}{120}, \frac{1}{100}, \frac{1}{80}, \frac{1}{60}, \frac{1}{40}, \frac{1}{20}, \frac{1}{10},$ $\varepsilon = h^2$; *dashed curve:* $e(h) = 220 \cdot h^{1.06}$; *Right: real plastic part with inserts (shaded) and gap thickness inside (d2) : outside (d1) of the dotted rectangular area: d2 : d1 = 1:2*

with $\beta_1(w_h(\xi)) = (w_h(\xi))^- = \min\{w_h(\xi), 0\}, \xi \in \omega$.

An essential tool for the investigations of the problems (**4.3**), (**4.4**) is a discrete maximum principle. Hence, the corresponding triangulation should be weakly acute (obtuse triangles do not appear). Then, for practical reasons the above mentioned points p can be choosen as the circumcenters of the triangles $\mathcal{K}$ such that the control regions are $\mathcal{H}_\xi = \{x \in \Omega : \text{dist}(x, \xi) < \text{dist}(x, \eta) \; \forall \eta \in \bar{\omega}\}$. This choice, known as Voronoi polyhedrons or Dirichlet regions, is closely related to triangulation techniques (Delauney triangulation $\mathcal{T}_h$).

Let us now summarize some results from the numerical anlysis. For details and the proofs we refer to [17]. The discrete problems (**4.3**) and (**4.4**) are uniquely solvable. The solutions are stable with respect to the right-hand side F_h and furthermore y_h and $y_{\varepsilon h}$ are nondecreasing functions with respect to the time. For the solution of the discrete penalty problem (**4.4**) we construct a *finite* iteration method (for each time step), where each iteration step corresponds to the solution of a sparse linear system of equations. Let us recall from above that for a grid function y_h the piecewise linear polynomial associated with $\mathcal{T}_h$ is denoted by $P^G y_h$. Then, the obtained error estimates (cf. [17]) are of the form

$$\max_{j=0,\dots,N} \left\| P^G y_h^j - P^G y_{\varepsilon h}^j \right\|_{H^1(\Omega)} \le C \, \varepsilon \, h^{-1}, \qquad \max_{j=0,\dots,N} \left\| u^j - P^G y_h^j \right\|_{H^1(\Omega)} \le C_1 \, h + C_2 \, \tau,$$

where $u^j = u(x, t^j)$ is the solution of (**3.2**), y_h of (**4.3**) and $y_{\varepsilon h}$ of the penalty problem (**4.4**). Hence, by coupling the penalization and discretization parameters one finally derives

$$\max_{j=0,\dots,N} \left\| u^j - P^G y_{\varepsilon h}^j \right\|_{H^1(\Omega)} \le C \, h \quad \text{for } \tau = h, \varepsilon \le h^2. \tag{4.5}$$

As an numerical example we consider the above inequality (**3.2**), (**3.3**) with the coefficient $k(x,t) = 1 + \kappa(t) \cdot \rho(x)$, $\Gamma_D = \partial\Omega$, Ω as unit square, $[0,T] = [0,1]$, $\kappa(t) = (3t^2 + 1)/64$ and

$$
\rho(x) = \rho(x_1, x_2) = \begin{cases}
(x_1 - 0.25)^2 + (x_2 - 0.25)^2 & 0 \le x_1 \le 0.5,\ 0 \le x_2 \le 0.5 \\
(x_1 - 0.75)^2 + (x_2 - 0.25)^2 & 0.5 \le x_1 \le 1,\ 0 \le x_2 \le 0.5 \\
(x_1 - 0.75)^2 + (x_2 - 0.75)^2 & 0.5 \le x_1 \le 1,\ 0.5 \le x_2 \le 1 \\
(x_1 - 0.25)^2 + (x_2 - 0.75)^2 & 0 \le x_1 \le 0.5,\ 0.5 \le x_2 \le 1.
\end{cases}
$$

The right-hand side is choosen according to

$$
F = 10^4 \{ 9\, \rho^{\frac{1}{2}} \kappa^{\frac{3}{2}} (1 + d\rho) + 6\, \rho^{\frac{3}{2}} \left[\max\{\tfrac{1}{64}, \rho\}\right]^{\frac{5}{2}} - \tfrac{3}{8}\, \rho^3 \left[3\max\{0, \tfrac{64\rho - 1}{3}\} + 1\right] - 18\, \rho^2 \}
$$

such that the exact solution turns out to be

$$
u(x,t) = 0.5 \cdot 10^4 \left[(\kappa(t))^{3/2} - (\rho(x))^{3/2}\right]^2 \text{ if } \rho(x) \le \kappa(t), \quad u(x,t) = 0 \text{ if } \rho(x) \ge \kappa(t).
$$

The free boundary $\Gamma_f(t)$ is described by the circumferences of the four circles with the centres $(0.25, 0.25)$, $(0.75, 0.25)$, $(0.25, 0.75)$, $(0.75, 0.75)$ and the radius $r(t) = \sqrt{\kappa(t)}$. Some numerical result for this example are reported in the left part of figure 2, where the dots represent the obtained error $\max_{j=0,\ldots,N} \left\| I_h u^j - P^G y^j_{\varepsilon h} \right\|_{H^1(\Omega)}$ (I_h as interpolant) in comparison to the dashed curve $e(h)$. The results are in satisfactory agreement with the error behaviour $O(h) + O(\tau)$ predicted in (**4.5**) (see also [17]). Finally, in figure 3 we present simulation results for a injection-compression process of a plastic part, the geometry of which is depicted in the right part of figure 2. The initial charge for compression, consisting of the first two filling zones taken together, is obtained by numerical simulation of injection of one third of the volume (i.e. $\frac{1}{3} \text{meas}_2\Omega \times d(x,0)$) through the corresponding gates represented by dots in figure 3. Then, we switch over to compression (the next four filling zones), reducing the gap thickness continuously to one third of $d(x,0)$. To observe the remaining air the last filling zone corresponds to a filling degree of about 97%. Beside an air inclusion arising in the left and middle case, the situation with the gates located in the four corners, is characterized by joining of streams of the melt which causes a bad-looking, undesirable seam.

References

[1] L. Angermann, *Numerical solution of second order elliptic equations on plane domains*, M^{2}AN, Vol. 25, No. 2, 1991, 169-191

[2] R.E. Bank and D.J. Rose, *Some error estimates for the box method*, SIAM J. Num. Anal. 24(1987), No. 4, 777-787

[3] G. Bayada, M. Boukrouche and M. El-A. Talibi, *The transient lubrication problem as a generalized Hele-Shaw type problem*, Journ. for Anal. and its Appl., 14(1995), 59-87

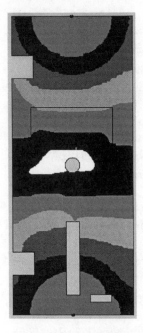  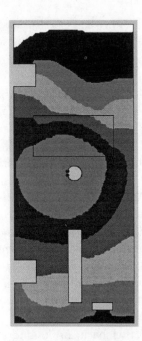

Figure 3: *Simulation results (filling zones for six time instants, degree of filling ≈ 97%, white area = air); left and middle situation ⇒ air inclusion, right: suitable flow behaviour*

[4] P. Cizek and V. Janovsky, *Hele-Shaw flow model of the injection by a point source*, Proc. of the Royal Soc. of Edinburgh, 91A, 1981, 147-159

[5] E. Di Benedetto and A. Friedman, *The ill-posed Hele-Shaw model and the Stefan problem for supercooled water*, Transact. of the ASME., Vol. 282, No. 1, 1984, 183-204

[6] C.M. Elliott and V. Janovsky, *A variational inequality approach to Hele-Shaw flow with a moving boundary*, Proc. of the Royal Soc. of Edinburgh, 88A, 1981, 93-107

[7] W. Hackbusch, *On first and second order box schemes*, Computing 41(1989), 277-296

[8] Yu.E. Hohlov, S.D. Howison, C. Huntingford, J.R. Ockendon and A.A. Lacey *A model for non-smooth free boundaries in Hele-Shaw flows*, Quart. J. Mech. Appl. Math, Vol. 47(1994), 107-128

[9] S.D. Howison, *Complex variable methods in Hele-Shaw moving boundary problems*, Euro. J. of Applied Mathematics, Vol. 3(1992), 209-224

[10] J.R. King, A.A. Lacey and J.L. Vazquez, *Persistence of corners in free boundaries in Hele-Shaw flow*, to appear in Euro. J. of Applied Mathematics

[11] C.C. Lee, F. Folgar and C.L. Tucker III, *Simulation of compression molding for fiber-reinforced thermosetting polymers*, Journ. of Engineering for Industry, Transactions of the ASME, 106(1984), 114-125

[12] C.C. Lee and C.L. Tucker III, *Flow and heat transfer in compression mold filling*, Journ. of Non-Newtonian Fluid Mechanics, 24(1987), 245-264

[13] M. Primicerio and J.F. Rodrigues, *The Hele-Shaw problem with nonlocal injection condition*, Gakuto Int. Ser., Math. Sci. Appl., Vol. 2 (1993), 375-390

[14] J.F. Rodrigues, *Obstacle problems in mathematical physics*, North Holland Mathematics studies 134, North Holland Publishing Company, Amsterdam, New York, Oxford, Tokyo, 1987

[15] J.F. Rodrigues, *Variational methods in the Stefan problem*, Lecture Notes in Mathematics 1584, Springer-Verlag, New York, Berlin, Heidelberg, London, 1994, 147-212

[16] J. Steinbach, *Evolutionary variational inequalities with a Volterra term*, Numer. Funct. Anal. and Optimiz., 17(3&4) 145-161 (1991)

[17] J. Steinbach, *Numerical solution of elliptic and evolutionary variational inequalities by means of the box method*, Report No. 413 (1994), Schwerpunktprogramm der DFG 'Anwendungsbezogene Optimierung und Steuerung', TU Munich, 1-89.

[18] J. Steinbach, *A generalized temperature-dependent, non-Newtonian Hele-Shaw flow in injection and compression moulding*, Report Nr. 96-02, TU Munich, Faculty for Mathematics, Dezember 1995; slightly shortened version to appear in Math. Meth. in the Appl. Sci.

[19] J. Steinbach, *Evolutionary variational inequalities arising in non-isothermal Hele-Shaw flow or in electrochemical machining*, in preparation

[20] C.L. Tucker III (editor), *Computer modeling for polymer processing, Fundamentals*, Hanser Publishers, Munich, Vienna, New York, 1989

Jörg Steinbach
Chair of Applied Mathematics, Technical University of Munich
Dachauer Str. 9a, 80335 Munich, Germany

Part 5.

Free boundary problems in Science

D ANDREUCCI, A FASANO, R GIANNI, M PRIMICERIO AND R RICCI
Diffusion driven crystallization in polymers

1. Introduction

Mathematical models for polymer crystallization are generally built by coupling the heat conduction with the dynamics for crystalline germs production and crystal growth. Germs production is often described by defining the nucleation rate as a phenomenological function of temperature (see e.g., [2], [3]). Instead, the model we consider here is characterized by a peculiar approach to the nucleation process, which is borrowed from works by Ziabicki (starting from [7]), which are partially built on previous classical results by Frenkel and Turnbull (see the quoted paper for the precise references, and also [4] for more information on the mathematical setting). According to such a scheme, the evolution of crystalline germs, leading to nucleation, i.e., to the appearance of new macroscopic nuclei, is governed by a Fokker-Planck type equation, set in a space of parameters (different from the physical space). If f denotes the distribution function of crystalline germs, which is defined (for any point x in the polymeric sample) in a space of state coordinates g (g can be a scalar variable like the volume of a spherulitic crystal or a vector valued variable accounting for the crystal orientation, see [7]), then, in this state space, f evolves due to statistical fluctuations and under the influence of a potential ΔF. This potential defines a zone of stable germs and a zone of unstable germs. Then the nucleation rate $\dot{N}$ is obtained as the positive part of the net flow of f through the boundary $g = g_*$ of the stable germs region.

The major interest of this approach is that it can be used to investigate memory effects in the crystallization process [8]. In fact, since the distribution of the germs is governed by a parabolic equation, the dynamics of crystallization is affected by the initial distribution of the germs themselves. In a complex process of subsequent steps of heating and cooling of the polymeric sample, by the same token, the distribution function f keeps trace of the history of the solidification process.

The mathematical setting of the problem consists in finding the temperature distribution, the crystal volume fraction and the crystal germs distribution by solving the following systems of equations (1.1), (1.2), (1.3), (1.4). In particular the temperature T solves

$$(1.1a) \qquad CT_t - k\,\Delta T = \lambda \rho w_t, \qquad x \in \Omega, 0 < t < t^*,$$

$$(1.1b) \qquad -k\frac{\partial T}{\partial \mathbf{n}} = H(T - T_a), \quad \text{on } \partial\Omega, 0 < t < t^*,$$

$$(1.1c) \qquad T(x,0) = T_0(x), \qquad \text{in } \Omega.$$

Here Ω is a sufficiently smooth bounded spatial domain and t^* is a given positive number; the specific heat C, the conductivity k, the latent heat of crystallization per unit of mass λ, the density ρ, H are positive constants. In (1.1b) $\mathbf{n}$ denotes the outer normal to $\partial\Omega$. The assumptions on the data T_a and T_0 will be specified below. The source term $\lambda\rho w_t$ arises from the release of latent heat of crystallization. Indeed w is the crystalline fraction of the polymer, and it is determined by the crystallization kinetic

$$(1.2a) \qquad w_t(x,t) = 4\pi \dot{R}(T(x,t), w(x,t)) \mathcal{I}_2(x,t), \quad 0 < t < t^*,$$

$$(1.2b) \qquad w(x,0) = w_0(x), \qquad\qquad\qquad t = 0,$$

where

$$\mathcal{I}_k(x,t) = \int_0^t \dot{N}(x,\tau)(1 - w(x,\tau)) \left(\int_\tau^t \dot{R}(T(x,s), w(x,s))\, ds \right)^k d\tau, \quad k = 1,2,3.$$

Clearly in (1.2a) we neglected the contribution of the volume of the nuclei appearing at time t. Since we consider the simple case of the scalar state variable g (volume of the spherical germs), we have

$$(1.3) \qquad \dot{N}(x,t) = \left[\frac{\partial}{\partial t} \int_{g_*(T)}^{+\infty} f(g,t,x)\, dg \right]_+ = [-cTe^{-\frac{\varepsilon}{T-T_g}} g_*^{\frac{2}{3}} f_g - g_*'(T) T_t f]_+ ,$$

where $T = T(x,t)$, f, $f_g = f$, $f_g(g_*, x, t)$, $f_0 \equiv f_0(g,x)$, and f is the solution of the evolution problem

$$(1.4a) \quad f_t - cTe^{-\frac{\varepsilon}{T-T_g}} \left[g^{\frac{2}{3}} \left(f_g + \frac{c_1}{T} \frac{\partial \Delta F}{\partial g} f \right) \right]_g = 0, \qquad g > 1, t \in (0,t^*),$$

$$(1.4b) \qquad\qquad\qquad f(g,0,x) = f_0, \qquad g > 1,$$

$$(1.4c) \qquad\qquad\qquad f(1,t,x) = 1, \qquad t \in (0,t^*),$$

$$(1.4d) \qquad\qquad\qquad f(g,t,x) \to 0, \qquad g \to \infty, t \in (0,t^*),$$

where the constant $c > 0$ is characteristic of the polymeric species, c_1 is an absolute constant, $\varepsilon > 0$ has the meaning of an activation energy. The glassification temperature T_g is defined so that for $T < T_g$ no nucleation can take place. We recall that the radial growth rate of spherulites $\dot{R}$ is a *given* function fulfilling

$$\dot{R} \equiv 0, \qquad \text{for } T \notin (T_g, T_m),$$

where T_m is the critical melting temperature.

The driving force $\frac{\partial \Delta F}{\partial g}$ is given by the derivative with respect to g of the potential

$$(1.5) \qquad \Delta F = a(T)g^{2/3} + b(T)g.$$

Here $a(T)g^{\frac{2}{3}}$ is the surface contribution, and $b(T)g$ is the volume contribution. While $a(T)$ is positive for all temperatures, $b(T)$ changes its sign at the critical melting temperature T_m. In the temperature range $T > T_m$ where $b > 0$, no nucleation (i.e., appearance of stable crystals) may occur, as $\frac{\partial \Delta F}{\partial g} > 0$ for $g > 1$. But when $T < T_m$ and $b < 0$, we have

$$\frac{\partial \Delta F}{\partial g} > 0 \quad \text{if } g < g_*, \qquad \frac{\partial \Delta F}{\partial g} < 0 \quad \text{if } g > g_*, \qquad g_* = \left(\frac{2}{3} \frac{a(T)}{|b(T)|} \right)^3.$$

Hence in the region $g > g_*$ clusters are stable and keep growing.

Comparing (1.3) with (1.4), we see that $\dot{N}$ is just the flux of f through $g = g_*$, i.e., the number of germs per unit of time, entering the stable region. The term $g'_*(T)T_t f$ in (1.3) is sometimes referred to as *athermal nucleation*: it originates from the clusters reaching $g = g_*$ because of the change in time of g_*, rather than because of their actual growth.

Condition (1.4d) is often replaced in the literature with

$$(1.6) \qquad f(G, t, x) = 0, \qquad 0 < t < t^*,$$

where $G > 1$ is a finite value larger than any possible value assumed by g_*. In this case, g in (1.4a), (1.4b) is subjected to the restriction $g < G$.

From the mathematical viewpoint, the problem presents some non standard coupling features. From one side, (1.1) and (1.4) are not a system of parabolic equations in the standard sense, since the "spatial" variable in (1.4) is g instead of x and x acts there just as a parameter. Moreover, the equations are coupled through the values of f and its derivative at $g = g_*$, which, in turn, depends on temperature.

We stipulate the following assumptions on the data of our problem; for a given $\alpha \in (0, 1)$

$$a, b \in C^1(\mathbf{R}), \quad |g_*|^{(0)} < \infty, \quad |g_*|^{(2)} < \infty,$$
$$\dot{R} \in C^1(\mathbf{R}^2), \quad |\dot{R}|^{(1)} < \infty, \quad f_0 \in H^{3+\alpha}(\mathbf{R}), \quad |f_0|_x^{(\alpha)} + |f_{0g}|_x^{(\alpha)} < \infty,$$
$$T_0 \in H^{2+\alpha}(\overline{\Omega}), \quad T_a \in H^{1+\alpha, \frac{1+\alpha}{2}}(\overline{\Omega_{t^*}}), \quad T_0, T_a \geq \vartheta > T_g, \quad w_0 \in H^\alpha(\overline{\Omega}).$$

(The notation of [5] is used here; when this is appropriate we indicate explicitly the variables involved in the norms: for instance, $|f|_x^{(\alpha)}$ denotes the Hölder norm of f as a function of x.) Moreover, the data for problem (1.1) (resp. (1.4)) are required to fulfil compatibility conditions of order 2 (resp. of order 3) at $t = 0$. For the sake of simplicity, if condition (1.4d) is prescribed, we require $f_0 \in C_0(\mathbf{R})$. In the following we denote by $\Lambda > 0$ a constant bounding all the norms of the data in the respective spaces.

The requirement that T_0, T_a are bounded away from T_g is needed to ensure that (1.4a) is uniformly parabolic. Typically, the polymer is molten at time $t = 0$, and thus $w_0 \equiv 0$.

Our main result is the following

THEOREM 1.1. *Under the assumptions given above, problem (1.1), (1.2), (1.3), (1.4), has a classical solution. If condition (1.4d) is replaced with (1.6) the solution is unique.*

2. Proof of the existence theorem

2.1 STEP 1: SMALL TIME EXISTENCE IN BOUNDED PARAMETER DOMAINS

We look first at the problem with condition (1.6). If on the contrary condition (1.4d) is prescribed, in the following G is to be understood as an arbitrarily fixed value $G > |g_*|^{(0)}$. The solution to the original problem will be recovered through an approximation procedure, on letting $G \to \infty$; of course in the following we assume that f_0 is modified near $g = G$ to fulfil the compatibility conditions there. We also define $J = [1, G] \times [0, t^*]$.

For a constant $M > 0$ to be chosen, let us define the closed subspace

$$\mathcal{K} = \{T \in H^{2+\alpha, \frac{2+\alpha}{2}}(\overline{\Omega_{t^*}}) \mid T \text{ takes the data in (1.1) and } |T|^{(2+\alpha)}_{\Omega_{t^*}} \leq M, \, T \geq \vartheta \},$$

of the space

$$\mathcal{C} = \{T \in C(\Omega_{t^*}) \mid T_t \in C(\Omega_{t^*})\},$$

endowed with the norm

$$|T|_{\mathcal{C}} = |T|^{(0)} + |T_t|^{(0)}.$$

We find a solution to our modified problem as a fixed point of a mapping $\mathcal{A}: \mathcal{K} \to \mathcal{K}$ defined as follows. Given $T_1 \in \mathcal{K}$, we consider the unique solution f of (1.4) (where T is now substituted with T_1). Then, we find w as the solution to (1.2a), where we replace T with T_1 and f with the just defined solution of (1.4). Finally, we use w to define the right hand side of (1.1a), and find the solution T_2 to (1.1) (note that the resulting problem is linear).

Then we set $\mathcal{A}(T_1) = T_2$. We have to check that $w_t \in H^{\alpha, \frac{\alpha}{2}}(\overline{\Omega_{t^*}})$, in order to make sure that T_2 has the required regularity.

Let us begin by stating the following quite obvious estimates, which can be derived by standard parabolic theory:

(2.1) $\qquad |f|^{(0)} \leq \gamma_1(t^\sigma M, \Lambda); \qquad |f|^{(3+\alpha)}_{g,t} \leq \gamma_2(t^\sigma M, \Lambda);$

362

from now on $\sigma = \sigma(\alpha)$ will denote a small positive constant (possibly changing in different occurrences), which could be made explicit; also the symbol γ_i denotes an increasing function of its arguments, which could be specified a priori. Our next task is estimating the variation of f with the space variable x. We let $\tilde{f}$ (resp. $\bar{f}$) denote the solution of (1.4) corresponding to $x = \tilde{x}$ (resp. $x = \bar{x}$). Then, setting $z = \tilde{f} - \bar{f}$ we have

$$z_t - \tilde{a}z_{gg} + \tilde{b}z_g + \tilde{c}z = (\bar{a} - \tilde{a})\bar{f}_{gg} + (\tilde{b} - \bar{b})\bar{f}_g + (\tilde{c} - \bar{c})\bar{f},$$

where a, b, c are the coefficients of the linear equation (1.4a), and $\tilde{}$ (resp. $\bar{}$) denotes evaluation at point $\tilde{x}$ (resp. $\bar{x}$). We infer from (2.1) and from the equation above that

(2.2a) $$|z|^{(0)} \leq \gamma_3(Mt^\sigma, \Lambda)\left(|\tilde{f}_0 - \bar{f}_0|^{(0)} + |\tilde{T}_1 - \bar{T}_1|^{(0)}\right).$$

Reasoning in the same way for the difference $\tilde{f}_g - \bar{f}_g$ we get

(2.2b) $$|\tilde{f}_g - \bar{f}_g|^{(0)} \leq \gamma_4(Mt^\sigma, \Lambda)\left(|\tilde{f}_{0g} - \bar{f}_{0g}|^{(0)} + |\tilde{T}_1 - \bar{T}_1|^{(0)}\right).$$

Exploiting (2.2) one derives easily that

(2.3) $$|f|_x^{(\alpha)} + |f_g|_x^{(\alpha)} \leq \gamma_5(t^\sigma M, \Lambda).$$

We also note, following [6], that the integral differential equation (1.2a) is equivalent to the ODE system

(2.4a) $$\dot{w} = 4\pi M_2 \dot{R}(T_1, w),$$

(2.4b) $$\dot{M}_2 = 2M_1 \dot{R}(T_1, w),$$

(2.4c) $$\dot{M}_1 = M_0 \dot{R}(T_1, w),$$

(2.4d) $$\dot{M}_0 = \dot{N}[f, T_1](1 - w),$$

equipped with initial data $M_0 = M_1 = M_2 = 0$ and $w = w_0$ for $t = 0$. This immediately yields, on recalling standard results of continuous dependence for ODE, that

$$|w|_x^{(\alpha)} \leq \gamma_6(t^\sigma M, \Lambda).$$

Using the estimate above in (2.4a), and then differentiating again the same equation, we find

$$|w_t|_x^{(\alpha)} \leq \gamma_7(t^\sigma M, \Lambda), \qquad |w_{tt}|^{(0)} \leq \gamma_8(t^\sigma M, \Lambda).$$

Therefore

$$|w_t|_{x,t}^{(\alpha)} \leq \gamma_9(t^\sigma M, \Lambda),$$

implying

$$|T_2|_{x,t}^{(2+\alpha)} \leq \gamma_{10}(\Lambda) + \gamma_{11}(t^\sigma M, \Lambda) \leq M,$$

where the last inequality holds provided we choose

(2.5) $\qquad M = \gamma_{10}(\Lambda) + \gamma_{11}(1, \Lambda)$, and $t^{\sigma} \leq 1/M$.

We also remark that $T_2 \geq \vartheta$ follows from the maximum principle (note that $w_t \geq 0$). Thus we have shown that $\mathcal{A} : \mathcal{K} \to \mathcal{K}$ for t small as in (2.5).

Next we show that $\mathcal{A}$ is contractive for small times in the norm $|T|_{\mathcal{C}}$. We choose $\bar{T}_1, \tilde{T}_1 \in \mathcal{K}$ and write again

$$z_t - \tilde{a} z_{gg} + \tilde{b} z_g + \tilde{c} z = (\bar{a} - \tilde{a}) \bar{f}_{gg} + (\bar{b} - \tilde{b}) \bar{f}_g + (\tilde{c} - \bar{c}) \bar{f},$$

where $z = \tilde{f} - \bar{f}$, $\bar{f}$ (resp. $\tilde{f}$) corresponds to $\bar{T}_1$ (resp. $\tilde{T}_1$), and $\bar{}$ (resp. $\tilde{}$) denotes evaluation at $T = \bar{T}_1$ (resp. $T = \tilde{T}_1$). We infer as before (provided t is small)

$$|\tilde{f} - \bar{f}|^{(0)} + |\tilde{f}_g - \bar{f}_g|^{(0)} \leq \gamma_{12}(\Lambda) |\tilde{T}_1 - \bar{T}_1|^{(0)} .$$

Thus, again looking at the ODE system (2.4), we get

$$|\tilde{w}_t - \bar{w}_t|^{(0)} \leq \gamma_{13}(\Lambda) |\tilde{T}_1 - \bar{T}_1|_{\mathcal{C}} t,$$

finally implying

$$|\tilde{T}_2 - \bar{T}_2|^{(0)} \leq \frac{1}{4} |\tilde{T}_1 - \bar{T}_1|_{\mathcal{C}} ,$$

for t small enough. The similar inequality

$$|\tilde{T}_{2t} - \bar{T}_{2t}|^{(0)} \leq \frac{1}{4} |\tilde{T}_1 - \bar{T}_1|_{\mathcal{C}} ,$$

still valid for t small, follows from a direct inspection of the problem satisfied by $v = T_t$, i.e.,

(2.6a) $\qquad C v_t - k \Delta v = \mathcal{V}$, $\qquad\qquad$ in $\Omega \times (0, t^*)$

(2.6b) $\qquad -k \dfrac{\partial v}{\partial \mathbf{n}} = H v$, $\qquad\qquad$ on $\partial \Omega \times (0, t^*)$,

(2.6c) $\qquad v(x, 0) = v_0(x) = \dfrac{k}{C} \Delta T_0(x)$, $\qquad$ in Ω,

where

$$\mathcal{V} = 4\pi \dot{R}_T(T, w) \mathcal{I}_2(x, t) \, v + (4\pi)^2 \dot{R}_w(T, w) \dot{R}(T, w) \mathcal{I}_2(x, t)^2 + 8\pi \dot{R}(T, w)^2 \mathcal{I}_1(x, t) .$$

It is immediately seen that $\mathcal{A}$ is actually contractive in the norm of $\mathcal{C}$. Therefore the problem set in $J_{t_0} = [0, G] \times [0, t_0]$ has got a unique solution for t_0 small enough. Note that the size of the time step t_0 depends solely on the norms of the initial (and

364

boundary) data. Therefore, if we can find an a priori bound for those norms at any time level, the time step will be independent of the iteration of the argument above, and the proof of global in time existence will have been accomplished. This is the program of next subsection.

2.II Step 2: A Priori Estimates and Conclusion of the Proof

In this subsection we let (T, w, f) denote a solution to the problem (1.2), (1.1), (1.4) (with (1.4d) replaced by (1.6)), defined for $t < t^*$, where $G > |g_*|^{(0)}$. Note that the a priori estimates

$$(2.7) \qquad |T|^{(0)} \le C_0, \qquad |f|^{(0)} \le C_0,$$

follow immediately from the weak maximum principle (because the source in (1.1a) vanishes for $T \notin (T_g, T_m)$). Here and below, we denote by C_0 a generic positive constant depending on the data, but not on G. Then the a priori estimates in [5] (Thm. 8.1 p. 193), together with an obvious estimate of $\|f_g\|_{L^2(J_*)}$ following from (1.4a), guarantee that

$$(2.8) \qquad |f_g|_{J_*}^{(0)} \le C_0, \qquad J_* = [1, |g_*|^{(0)}] \times [0, t^*].$$

The estimates above, when taken into account in (1.2), (1.3), allow us to conclude that

$$0 \le w_t(x, t) \le C_0 \int_0^t |T_\tau(x, \tau)| \, \mathrm{d}\tau + C_0,$$

whence, by the results of [5] chapter IV, we get for any $q > 1$,

$$\|T\|_{W_q^{2,1}(\Omega_t)} \le C_0 + C_0 \Big(\int_0^t \int_\Omega \Big(\int_0^\tau |T_s(x, s)| \, \mathrm{d}s \Big)^q \, \mathrm{d}x \, \mathrm{d}\tau \Big)^{1/q}$$

$$\le C_0 + C_0 \Big(\int_0^t \int_\Omega \tau^{q-1} \int_0^\tau |T_s(x, s)|^q \, \mathrm{d}s \, \mathrm{d}x \, \mathrm{d}\tau \Big)^{1/q}$$

$$\le C_0 + C_0 \Big(\int_0^t \Big(\int_0^\tau \int_\Omega |T_s(x, s)|^q \, \mathrm{d}x \, \mathrm{d}s \Big) \mathrm{d}\tau \Big)^{1/q}.$$

On applying Gronwall's inequality to the last estimate we infer at once,

$$(2.9) \qquad \|T\|_{W_q^{2,1}(\Omega_t)} \le C_0,$$

where C_0 depends on q too. If we take $q > 5/(1 - \alpha)$, standard embedding theorems yield

$$(2.10) \qquad |T|^{(1+\alpha)} \leq C_0 ,$$

for any fixed $\alpha \in (0, 1)$. On one hand, (2.10) gives enough regularity of the coefficients in (1.4a) so as to get

$$(2.11) \qquad |f|_{J_*}^{(3+\alpha)} \leq C_0 .$$

On the other hand, we remark that (2.9) implies that the right hand side of the differential equation satisfied by T_t belongs to $L^{q/2}(\Omega_{t^*})$, so that

$$(2.12) \qquad |T_t|^{(\alpha)} \leq C_0 ,$$

(we are also using the regularity of the data here).
Finally, estimates (2.10), (2.11), (2.12), prove the Hölder regularity of the right hand side of (1.1a); thus we may conclude

$$(2.13) \qquad |T|^{(2+\alpha)} \leq C_0 ,$$

thereby completing the a priori estimates needed to prove global solvability of the problem at hand (in domains with g bounded).

In order to complete the proof of existence in $1 < g < \infty$, we recall that all the estimates collected above do not depend on G. Let us denote by $\{(T_G , w_G , f_G)\}$ the solution to the problem introduced in last subsection. Then, a priori bounds in the form

$$|f_G|_K^{(3+\alpha)} \leq C_0(K) , \qquad K \subset [1, \infty) \times [0, t^*] \text{ compact} ,$$

can be derived as above for any fixed K and large enough G. Finally, on letting $G \to \infty$ we see that a suitable subsequence of $\{(T_G , w_G , f_G)\}$ converges to a solution of the original problem. Indeed, condition (1.4d), which is only left to be proven, follows from classical results (or see [1]).

References

[1] D. Andreucci, *Vanishing rate of solutions of non uniformly parabolic equations*, to appear,
[2] D. Andreucci, A. Fasano, M. Paolini, M. Primicerio, C. Verdi, *Numerical simulation of polymer crystallization*, Mathematical Methods and Models in Applied Sciences, **4** (1994), 135–145.
[3] D. Andreucci, A. Fasano, M. Primicerio, *On a mathematical model for the crystallization of polymers*, Proceedings of the 4th E.C.M.I. Meeting (H.J. Wacker, W. Zulehner eds), 3–16. (1991),

[4] D. Andreucci, A. Fasano, M. Primicerio, R. Ricci, C. Verdi, *Modelling nucleation in crystallization of polymers*, to apper on "Free boundary problems: theory and applications" Diaz, J.I., Herrero, M. and Vazquez, J.L. eds., Longman,

[5] O.A. Ladyzenskaja, V.A. Solonnikov, N.N. Ural'tzeva, *Linear and quasilinear equations of parabolic type*, Transl. of Math. Mon. Providence, R.I. AMS, **23** (1968).

[6] J. Berger, A. Köppl, W. Schneider, *Non isothermal crystallization. Crystallization of Polymers. System of rate equations.*, Intern. Polymer Processing, **2** (1988), 151–154.

[7] A. Ziabicki, *Generalized theory of nucleation kinetics. I. General formulations*, J. Chemical Physics, **48** (1968), 4368–4374.

[8] A. Ziabicki, G.C. Alfonso, *Memory effects in isothermal crystallization. I. Theory*, Colloid Polym. Sci., **272** (1994), 1027–1042.

D. Andreucci
Dip. Metodi e Modelli Matematici
via A. Scarpa 16
00161 Roma Italy
e-mail: andreucc@itcaspur.caspur.it

A. Fasano, R. Gianni, M. Primicerio
Dip. di Matematica U. Dini
viale Morgagni 67/a
50134 Firenze Italy
e-mail: fismat@udini.math.unifi.it

R. Ricci
Dip. di Matematica F. Enriques
via C. Saldini 50
20133 Milano Italy
e-mail: ricci@vmimat.mat.unimi.it

A BONAMI, D HILHORST, E LOGAK, M MIMURA

A free boundary problem arising in a chemotaxis model

1   Introduction

In this note, we consider a diffusion-chemotaxis system of equations which describes aggregating patterns of biological individuals which move by diffusion, chemotaxis and growth. We formally derive a free boundary problem as a singular limit of this system. The interface then describes the boundary of the aggregating individuals. We show the existence and uniqueness of a smooth solution of the free boundary problem locally in time.

Mimura and Tsujikawa propose in [8] the following system.

$$(S_\epsilon) \begin{cases} u_t = \epsilon \Delta u - \nabla.(u \nabla \chi(v)) + \dfrac{1}{\epsilon} u(1-u)(u-a) \text{ in } \Omega \times (0,T) \\ \epsilon v_t = \Delta v + u - \gamma v \text{ in } \Omega \times (0,T), \end{cases}$$

where $\Omega \subset \mathbb{R}^N$ is a smooth bounded domain and $a \in (0,1)$ is a fixed constant. The functions u and v are respectively the population density and the concentration of chemotactic substance. Here, χ is a smooth function such that $\chi(v) > 0$ and $\chi'(v) > 0$ for $v > 0$. The population is submitted to three competitive effects: diffusion, growth induced by the nonlinear term $u(1-u)(u-a)$ and a tendency of migrating towards higher gradients of the chemotactic substance induced by the advection term in the equation for u. The positive constant γ is the degradation rate of v. Moreover, the functions u and v satisfy homogeneous Neumann boundary conditions.

In the absence of growth term, System (S_ϵ) is called the Keller-Segel model and is proposed to describe slime mold aggregation ([7]). It has been investigated by many authors ([5],[6],[9], [10],[11]). A typical phenomenon is the localization of u due to the chemotaxis effect, which implies the aggregation of individuals. Especially, if the initial condition for u is close to the delta distribution at some point in $\mathbb{R}^N$ ($N \geq 2$), the solution $u(x,t)$ tends to infinity at one point in finite time ([6], [10]). This is called chemotaxis collapse in biology.

From System (S_ϵ), Mimura and Tsujikawa formally derive in [8] the following free boundary problem,

$$(S_0) \begin{cases} V_n = -\epsilon(N-1)K + \dfrac{\partial \chi(v)}{\partial n} + c_0(1/2 - a) \text{ on } \Gamma_t, t \in [0,T] \\ -\Delta v(x,t) + \gamma v(x,t) = \begin{cases} 0 & x \in \Omega_t^0, \quad t \in [0,T] \\ 1 & x \in \Omega_t^1, \quad t \in [0,T], \end{cases} \end{cases}$$

368

where, for each $t \in [0, T]$, $\Gamma_t \subset\subset \Omega$ is a closed hypersurface without boundaries such that $\Omega \setminus \Gamma_t = \Omega_t^0 \cup \Omega_t^1$. The sets Ω_t^i, $i = 0, 1$, are two open disjoint subsets of Ω and Ω_t^0 reaches the boundary of Ω. We denote by n the unit normal vector on Γ_t, which points from Ω_t^1 towards Ω_t^0. V_n is the normal velocity on Γ_t and K is the algebraic mean curvature on Γ_t (positive if Ω_t^1 is convex). The term $c_0(1/2 - a)$ is the velocity of the one-dimensional travelling front solution $w(x, t) = U(x - ct)$ of the scalar bistable equation

$$w_t = w_{xx} + w(1 - w)(w - a), \ x \in \mathbb{R}, \ t > 0$$
$$w(-\infty, t) = 1, \ w(+\infty, t) = 0$$

In this case, one can check that $c_0 = \sqrt{2}$.

In section 2, we give a formal derivation of the interface motion equation, assuming that v is a smooth given function. We present in section 3 our local existence and uniqueness result. We refer to [2] for a complete proof.

2 Formal derivation of the interface motion equation

In order to give an intuitive feeling of the way System (S_0) arises from System (S_ϵ), we consider for a given smooth function v and a fixed positive constant ϵ_0 the following equation for $u = u^\epsilon$,

(1) $$u_t = \epsilon_0 \Delta u - \nabla.(u \nabla \chi(v)) + \frac{\epsilon_0}{\epsilon^2} u(1 - u)(u - 1/2 + \frac{\epsilon}{\epsilon_0} \delta)$$

with homogeneous Neumann boundary condition and suitable initial data. We show heuristically how to derive from this equation the motion equation

(2) $$V_n = c_0 \delta + \frac{\partial \chi(v)}{\partial n} - \epsilon_0(N - 1)K \ \text{ on } \Gamma_t, t \in [0, T]$$

as ϵ tends to 0. We rewrite equation (1) as

$$L^\epsilon u = u_t - \epsilon_0 \Delta u + \nabla u.\nabla \chi(v) - \frac{\epsilon_0}{\epsilon^2}[u(1 - u)(u - 1/2 + \frac{\epsilon}{\epsilon_0} \delta) - \frac{\epsilon^2}{\epsilon_0} \Delta \chi(v) u] = 0.$$

Note that the equation $u(1 - u)(u - 1/2 + \delta) - \eta u = 0$ has for $\delta \in (0, 1)$ and $|\eta|$ small enough three solutions

$$0 = u_-(\delta, \eta) < u_0(\delta, \eta) < u_+(\delta, \eta)$$

with $u_+(\delta, 0) = 1$. We then denote by $U(z, \delta, \eta)$ the travelling wave solution associated to this cubic nonlinearity, namely the unique solution of

$$U_{zz} + W(\delta, \eta)U_z + U(1 - U)(U - 1/2 + \delta) - \eta U = 0$$
(3) $\quad U(-\infty, \delta, \eta) = u_+(\delta, \eta), \ U(0, \delta, \eta) = 1/2, \ U(+\infty, \delta, \eta) = u_-(\delta, \eta) = 0.$

The function $W(\delta, \eta)$ is the travelling wave velocity and satisfies

(4)
$$W(\delta, \eta) = c_0\delta + c_1\eta(1 + c_2\delta + c_3\eta) + O((|\delta| + |\eta|)^3)$$

for some constants $c_0 = \sqrt{2}$, c_1, c_2, c_3 and for (δ, η) close to $(0, 0)$.

Let us consider a smooth moving boundary Γ_t and let d be the signed distance function to Γ_t defined in the neighborhood of Γ_t by

$$d(x, t) = \begin{cases} dist(x, \Gamma_t) & \text{for } x \in \Omega_t^0 \\ -dist(x, \Gamma_t) & \text{for } x \in \Omega_t^1 \end{cases}$$

and smoothly extended in $\bar{\Omega}$ in order to satisfy

$$d(x, t) < 0 \text{ for } x \in \Omega_t^1; \quad d(x, t) > 0 \text{ for } x \in \Omega_t^0 \cup \partial\Omega.$$

In particular, $d = 0$ on Γ_t and $|\nabla d(x, t)| = 1$ in a neighborhood of Γ_t.

We make the assumption that for ϵ small enough, the function u^ϵ can be approximated by the function

(5)
$$\tilde{u}^\epsilon(x, t) = U(\frac{d(x, t)}{\epsilon}, \frac{\epsilon}{\epsilon_0}\delta, \frac{\epsilon^2}{\epsilon_0}\Delta\chi(v)).$$

A similar ansatz has been proposed by several authors for slightly different equations ([3], [4], [1]). An easy computation gives that

$$\tilde{u}_t^\epsilon = \bar{U}_z\frac{d_t}{\epsilon} + \frac{\epsilon^2}{\epsilon_0}\bar{U}_\eta(\Delta\chi(v))_t = \bar{U}_z\frac{d_t}{\epsilon} + O(\epsilon^2),$$

$$\nabla\tilde{u}^\epsilon = \bar{U}_z\frac{\nabla d}{\epsilon} + \frac{\epsilon^2}{\epsilon_0}\bar{U}_\eta\nabla(\Delta\chi(v)) = \bar{U}_z\frac{\nabla d}{\epsilon} + O(\epsilon^2),$$

$$\nabla\tilde{u}^\epsilon.\nabla\chi(v) = \bar{U}_z\frac{\nabla d.\nabla\chi(v)}{\epsilon} + O(\epsilon^2),$$

and

$$\Delta\tilde{u}^\epsilon = \bar{U}_{zz}\frac{|\nabla d|^2}{\epsilon^2} + \bar{U}_z\frac{\Delta d}{\epsilon} + O(\epsilon),$$

where the notation $\bar{U}$, $\bar{U}_z$, etc... means that we take the value of the functions U, U_z, etc... at the point $(z, \delta, \eta) = (\frac{d(x,t)}{\epsilon}, \frac{\epsilon}{\epsilon_0}\delta, \frac{\epsilon^2}{\epsilon_0}\Delta\chi(v))$.

Hence, using (3) and (4), we get

$$\begin{aligned} L^\epsilon\tilde{u}^\epsilon &= \epsilon_0\frac{\bar{U}_{zz}}{\epsilon^2}(1 - |\nabla d|^2) \\ &+ \frac{\bar{U}_z}{\epsilon}(d_t - \epsilon_0\Delta d + \nabla d.\nabla\chi(v) + c_0\delta) \\ &+ \bar{U}_z c_1\Delta\chi(v) \\ &+ O(\epsilon). \end{aligned}$$

370

Since u^ϵ satisfies $L^\epsilon u^\epsilon = 0$, we want to cancel the higher order terms in the expression above. We note that for x close to Γ_t, the first term is 0. Setting to 0 the coefficient of $\frac{1}{\epsilon}$ in the second term gives

$$-d_t = -\epsilon_0 \Delta d + \nabla d . \nabla \chi(v) + c_0 \delta.$$

If we consider this equation on Γ_t, it can be rewritten as

$$V_n = -\epsilon_0 (N-1) K + \frac{\partial \chi(v)}{\partial n} + c_0 \delta,$$

since $\nabla d = n$, $\Delta d = (N-1)K$ and $-d_t = V_n$ on Γ_t. This is exactly the interface motion equation (2).

We note that the third term in $L^\epsilon \tilde{u}^\epsilon$ does not vanish on Γ_t but converges pointwise to 0 away from Γ_t so that it converges weakly to 0 as a function of the x variable. Assuming that Γ_t is a classical solution to the motion equation (2), we can actually show that, for each $t \in [0,T]$, $L^\epsilon \tilde{u}^\epsilon(.,t)$ converges weakly to 0 in the sense of Radon measures. The proof relies on the fact that $|U_z|$ and $|U_{zz}|$ decrease exponentially fast to 0 in the z variable. In the case that v is fixed, rigorous convergence results to the viscosity solution of equation (2) will be shown in a forthcoming paper.

3 Existence and uniqueness of a smooth solution locally in time

We consider the initial value problem (P_0),

$$(P_0) \begin{cases} V_n = -\epsilon_0 (N-1) K + \dfrac{\partial \chi(v)}{\partial n} + c_0 \delta \ \ \text{on } \Gamma_t, t \in [0,T] \\ \Gamma_t|_{t=0} = \Gamma_0 \\ -\Delta v(x,t) + \gamma v(x,t) = \begin{cases} 0 & x \in \Omega_t^0, \ \ t \in [0,T] \\ 1 & x \in \Omega_t^1, \ \ t \in [0,T] \end{cases} \\ \dfrac{\partial v}{\partial \nu} = 0 \text{ on } \partial\Omega \times [0,T]. \end{cases}$$

The vector ν is the unit outward normal vector to $\partial\Omega$. We state below the result that we prove in [2].

Theorem 3.1 *Let $\Gamma_0 = \partial\Omega_0^0$, where Ω_0^0 is a $C^{2m+\alpha}$ subdomain of Ω, with $m \geq 1$ and $\alpha \in (0,1)$. Then there exists a time $T > 0$ such that Problem (P_0) has a unique solution (v, Γ) on $[0,T]$ with*

$$\Gamma = (\Gamma_t \times \{t\})_{t \in [0,T]} \in C^{2m+\alpha,(2m+\alpha)/2},$$

$$v|_\Gamma \in C^{2m+\alpha,(2m+\alpha)/2}.$$

The proof of this result makes use of a contraction fixed-point argument in suitable Hölder spaces, which relies on regularity estimates for the auxiliary subproblem (P_a)

$$(P_a) \begin{cases} -\Delta v(x,t) + \gamma v(x,t) = \begin{cases} 0 & x \in \Omega_t^0, \quad t \in [0,T] \\ 1 & x \in \Omega_t^1, \quad t \in [0,T] \end{cases} \\ \dfrac{\partial v}{\partial \nu} = 0 \text{ on } \partial\Omega \times [0,T], \end{cases}$$

where $\Gamma = (\Gamma_t \times \{t\})_{t \in [0,T]}$ is given. More precisely, we have the following results.

Lemma 3.1 *Let* $\Gamma = (\Gamma_t \times \{t\})_{t \in [0,T]}$ *be given as above. We assume that*

$$\Gamma \in C^{2m+\alpha, \frac{2m+\alpha}{2}},$$

with $m \in \mathbb{N}$ *and* $\alpha \in (0,1)$. *Let* v *be the solution of Problem* (P_a). *Then* v *satisfies*

$$v|_\Gamma \in C^{2m+\alpha, \frac{2m+\alpha}{2}} \text{ and } \nabla v|_\Gamma \in C^{2m+\beta, \frac{2m+\beta}{2}} \text{ for all } 0 < \beta < \alpha.$$

Moreover, we prove that the mapping $\Gamma \to (v|_\Gamma, \nabla v|_\Gamma)$ is Lipschitz continuous in Hölder spaces.

Lemma 3.2 *Let* Γ^1 *and* Γ^2 *be two given interfaces in* $C^{2+\alpha, \frac{2+\alpha}{2}}$, *for some* $\alpha \in (0,1)$. *Let* v_1 *and* v_2 *be the corresponding solutions of Problem* (P_a). *Let* $w_i = v_i|_{\Gamma^i}$ *and* $z_i = \nabla v_i|_{\Gamma^i}$ *for* $i = 1, 2$. *Then there exists* $C > 0$ *such that*

$$\|w_1 - w_2\|_{C^{1,1}} \leq C \|\Gamma^1 - \Gamma^2\|_{C^{1,1}}$$

and

$$\|z_1 - z_2\|_{C^{1,1}} \leq C \|\Gamma^1 - \Gamma^2\|_{C^{2+\alpha,(2+\alpha)/2}}.$$

References

[1] G. BARLES, H. M. SONER and P. E. SOUGANIDIS, Front propagation and phase field theory, *SIAM J. Control and Optimization* **31** (1993) 439–469.

[2] A. BONAMI, D. HILHORST, E. LOGAK and M. MIMURA, A chemotaxis-growth model and perturbed motion by mean curvature, in preparation.

[3] X. CHEN, Generation and propagation of interfaces in reaction-diffusion systems, *Trans. Amer. Math. Soc.* **334** (1992) 877–913.

[4] X. CHEN, D. HILHORST and E. LOGAK, Asymptotic behaviour of an Allen-Cahn equation with a non–local term, to appear in *Nonlinear Analysis, TMA*.

[5] S. CHILDRESS and J. K. PERCUS, Nonlinear aspects of chemotaxis, *Math. Biosci.* **56** (1986) 217–237.

[6] W. JAGER and S. LUCKHAUS, On explosions of solutions to a system of partial differential equations modelling chemotaxis, *Trans. Amer. Math. Soc.* **329** (1992) 819 – 824.

[7] E. F. KELLER and L. A. SEGEL, Initiation of slime mold aggregation viewed as an instability, *J. Theor. Biol.* **26** (1970), 399–415.

[8] M. MIMURA and T. TSUJIKAWA, Aggregating pattern dynamics in a chemotaxis model including growth, to appear in *Physica A*.

[9] M. MIMURA , T. TSUJIKAWA , R. KOBAYASHI and D. UEYAMA, Dynamics of aggregation patterns in a chemotaxis-diffusion-growth model equation, *Forma* **8** (1993) 179 –195.

[10] T. NAGAI, Blow-up of radially symmetric solutions to a chemotaxis system, preprint.

[11] R. SCHWAAF, Stationary solutions of chemotaxis systems, *Trans. Amer. Math. Soc.* **292** (1985) 531–556.

A. Bonami
MAPMO - URA 1803
Université d'Orléans BP 6759
45067 Orléans Cedex 2, France

D. Hilhorst
Laboratoire d'Analyse Numérique
CNRS et Université Paris-Sud
91405 Orsay Cedex, France

E. Logak
Département de Mathématiques et d'Informatique
Ecole Normale Supérieure
45 rue d'Ulm
75230 Paris Cedex 05, France

M. Mimura
Department of Mathematical Sciences
University of Tokyo
1-3-8 Komaba, Meguro-ku
Tokyo 153, Japan

K KUCZERA

Free energy simulations in chemistry and biology

Abstract: A brief review of the background of free energy simulations in chemistry and biology is presented. The scope of the methods is illustrated by two examples of conformational free energy simulations. The first example presents several different approaches to simulation of the *gauche–trans* equilibrium in n-butane. The second example involves exploration of the 18-dimensional free energy surfaces of two model peptides - decaalanine and deca-α-methylalanine.

1. Introduction

Free energy simulations have emerged as a powerful tool in physical, chemical and biological studies of molecular systems [1, 2, 3]. Calculating free energy changes is especially valuable, because they are basic observable thermodynamic quantities, related to equilibrium constants and rate constants of processes.

Free energy simulations may be roughly divided into two categories, involving chemical or conformational change. In the latter class processes involving changes of chemical structure are considered, also known as "computer alchemy" [3, 4]. Past applications include calculations of differences of solvation energies between different amino acids, nucleic acid bases, small organic molecules and organic and inorganic ions [5, 6, 7, 8], studies of host–guest complexes [9], chemical reactions [10, 11], design of enzymes and inhibitors [12], as well as studies of influence of point mutations on various properties of proteins such as substrate binding, thermal stability, quaternary structural changes and aggregation [4, 13, 14, 15].

Conformational changes in molecular systems occur without modifications of chemical structure, and include rotational isomerism, molecular association and a variety of other processes. Conformational free energy simulations mainly employ four approaches - direct sampling, umbrella sampling and thermodynamic perturbation theory (TP) and thermodynamic integration (TI) [1, 2, 3, 16].

In the simplest direct sampling approach, frequencies of occurrence of different values of a conformational coordinate determined from a standard molecular dynamics (MD) or Monte Carlo simulation are converted to a potential of mean force. The *trans–gauche* equilibrium in butane has been studied in this way (see [2, 3, 17] for reviews and recent applications). The direct approach fails if the different important conformational states of the studied system are not explored during the time scale of the simulation, e.g. because they are separated by significant barriers. In the umbrella sampling method an auxiliary potential is added to the Hamiltonian, biasing the system towards a selected region of conformational space which might not be accessible to direct sampling [2, 18]. Probability distributions for conformational coordinates are then calculated from molecular dynamics or Monte Carlo simulations and corrected for the bias introduced by the auxiliary potential. The thermodynamic perturbation (TP) and thermodynamic integration (TI) methods are based on the coupling parameter approach, which is described below. A number of applications of conformational free energy simulations have been reported – from the hydrophobic effect to diffusion of dioxygen through myoglobin [2, 16, 19]. The methods discussed are in principle applicable to a wider range of phenomena, including structure formation in molecular systems and phase transitions.

2. Theoretical background

Consider a molecular system of N atoms, e.g. a solute in a solvent or a macromolecule, with a separable Hamiltonian $H(p,q) = K(p) + U(q)$, where K is the kinetic and U the potential energy, q are the 3N coordinates and p the 3N conjugate momenta. The coupling parameter approach proceeds through defining a hybrid potential energy function $U(q,\lambda)$, where λ is called the coupling parameter. $U(q,\lambda)$ has the property of smoothly varying from the initial state $U_0(q)$ to the final state $U_1(q)$ as the coupling parameter λ changes from 0 to 1. When changes to the kinetic energy are neglected, the free energy differences between the states corresponding to coupling parameter values λ' and λ may be evaluated using the "thermodynamic perturbation" formula [1, 2, 20]:

$$(1) \qquad \Delta A(\lambda \to \lambda') = -kT \ln \langle e^{-\beta[U(q,\lambda')-U(q,\lambda)]} \rangle_\lambda$$

where k is the Boltzmann constant, T the temperature and $\langle ... \rangle_\lambda$ denotes an average over the ensemble corresponding to the system with potential $U(q,\lambda)$. The free energy difference between the initial and final states may then be obtained by accumulating a series of free energy changes spanning the range of λ from 0 to 1. An alternative use of the coupling parameter approach is the thermodynamic integration formula (TI), for the derivative of the free energy with respect to the coupling parameter [1, 2, 20]:

$$(2) \qquad \frac{\partial A(\lambda)}{\partial \lambda} = \langle \frac{\partial U(q, \lambda)}{\partial \lambda} \rangle_\lambda$$

In this case the free energy difference between the initial and final states may be obtained by integrating the derivative from Eq.(2) over λ from 0 to 1.

The coupling parameter method is applicable to both mutations and conformational change [1, 2]. For mutation type processes the initial and final state differ by deletion, addition or substitution of chemical groups. The coupling parameter is an extraneous coordinate and the intermediate potentials $U(q,\lambda)$ usually correspond to unphysical states. For conformational changes, λ amy be equated with the conformational coordinate of interest ξ, e.g. a distance, a planar or dihedral angle, or a more complicated construct; ξ then defines conformations of the molecular system along some path, and analogues of Eqs. (1,2) give the free energy profile, or potential of mean force, along the path. This can be seen by evaluating the probability distribution of ξ in the canonical ensemble. Starting with the basic equation of the canonical ensemble for the probability distribution of coordinates q [21]:

$$\rho(q) = \frac{1}{Z} \int dq \; e^{-U(q)/kT} \quad ; \quad Z = \int dq \; e^{-U(q)/kT}$$

and considering a general conformational coordinate ξ, defined by specifying a function $\Xi(q)$, which gives the value of ξ corresponding to configuration q of the system, we obtain [22]:

$$(3) \quad \rho(\xi) = \int dq \; \rho(q)\delta(\Xi(q) - \xi) = \frac{1}{Z} \int dq \; e^{-U(q)/kT} \; \delta(\Xi(q) - \xi) = \frac{Z(\xi)}{Z}$$

where

$$(4) \qquad Z(\xi) = \int dq \; e^{-U(q)/kT} \; \delta(\Xi(q) - \xi)$$

The conformational free energy profile $A_c(\xi)$ may then be defined as [23, 24, 25]:

$$(5) \qquad \Delta A_c(\xi) = A_c(\xi) - A_c(\xi_0) = -kT \; \ln \frac{Z(\xi)}{Z(\xi_0)} = -kT \; \ln \frac{\rho(\xi)}{\rho(\xi_0)}$$

which allows to determine the free energy of an arbitrary conformation ξ relative to some fixed reference state ξ_0. Alternately, $A_c(\xi)$ may be viewed as the potential of mean force along ξ [1, 16, 19]:

$$(6) \qquad \frac{\partial \Delta A_c(\xi)}{\partial \xi} = \frac{\partial A_c(\xi)}{\partial \xi} = \langle \frac{\partial U}{\partial \xi} \rangle_\xi$$

According to Eq. (6), in the TI scheme the derivatives of the conformational free energy with respect to the chosen conformational coordinate ξ are calculated by performing simulations with fixed values of and evaluating averages of the derivative $\partial U/\partial \xi$. Eq. (6) can be derived as the special case of the general formula Eq. (12) given in Appendix A, corresponding to m=1 and X(q)=1.

In umbrella sampling, TP, and TI approaches it is possible to decompose free energy differences into contributions from energy and entropy. The TI method additionally allows decomposition of the free energy profiles into contributions from different energy terms (internal deformation, van der Waals, electrostatic) and parts of the system (solute, solvent, individual chemical groups). Such decompositions are not rigorous, but have been employed successfully in mutation-type processes to gain insight into the microscopic mechanism of various phenomena involving macromolecules [4, 14, 15].

It is also possible to straightforwardly generalize the TI algorithm to multidimensional problems – the evaluation of free energy gradients, and higher derivatives of the

376

conformational free energy surface. We can introduce the joint probability distribution of a set of conformational coordinates ξ_i, i=1,...,m :

$$(7) \qquad \rho(\xi_1, \xi_2, ..., \xi_m) = \int dq \; \rho(q) \prod_{i=1}^{m} \delta(\Xi_i(q) - \xi_i)$$

where the functions $\Xi_i(q)$ define coordinates ξ_i in terms of the Cartesian coordinates q. Using the equation Eq. (12) from Appendix A with X(q) = 1, it is easy to show that the gradient of the free energy surface $A_c(\xi_1, \xi_2, ..., \xi_m)$ with respect to the set of coordinates is:

$$(8) \qquad \frac{\partial A_c(\xi_1, \xi_2, ..., \xi_m)}{\partial \xi_k} = \langle \frac{\partial U}{\partial \xi_k} \rangle_{\xi_i, i=1,...m} \qquad k = 1, ..., m$$

I.e. to calculate the derivative $\partial A_c / \partial \xi_k$ of the conformational free energy with respect to the k-th coordinate ξ_k of the set, we have to calculate the average of the derivative $\partial U / \partial \xi_k$ of the potential energy with respect to ξ_k over the set of conformations with all coordinates ξ_i, i=1,...,m fixed. This result is a significant step forward in simulations of molecular conformational processes. It is a starting point for exploration of multidimensional free energy landscapes of molecular systems by enabling the location and characterization of free energy minima (stable conformers), saddle points (transition states) and minimum free energy paths connecting these points. For a more in depth discussion of the methods see [26, 27].

3. Methods

The molecular model

The new CHARMM Version 22 all-hydrogen model was used to describe the potential energy of the simulated systems [31, 32]; this parameterization is identical to the one referred to as CHARMM94 in [33]. In this model butane consists of 14 atoms; in molecular dynamics simulations with SHAKE [34] constraints imposed on all C–H bonds, there were 26 conformational degrees of freedom per butane molecule. The two decapeptides studied here are acetylated at the N terminus and amidated at the C terminus; their chemical formulae are ALA10: $CH_3CO-(Ala)_{10}-CONH_2$ and AIB10: $CH_3CO-(Aib)_{10}-CONH_2$, where Aib is α-methylalanine. The ALA10 system consists of 109 atoms and has 248 degrees of freedom after application of SHAKE constraints to bonds involving hydrogen atoms and fixing all ϕ and ψ backbone dihedrals, while AIB10 consists of 139 atoms and has 318 degrees of freedom under the same conditions.

Conformational free energy simulations for n-butane

Liquid butane calculations were performed using periodic boundary conditions for a truncated cubic cell [35] based on a cube of edge 28.437 Å containing 72 butane molecules, at a density of 0.602 g/cm^3, corresponding to liquid butane at -0.5 °C and

1 atm [36]. A series of 19 simulations were performed, in which the value of the central C–C dihedral ϕ of one of the molecules was fixed consecutively at $\phi = 180, 170, ..., 10,$ $0°$ using the holonomic constraint method of Tobias and Brooks [37]. The remaining 71 butane molecules had unconstrained dihedrals.

Each simulation consisted of a 20 ps equilibration and 40 ps molecular dynamics trajectory production run. The Verlet algorithm was used with a time step of 2 fs and SHAKE [34] constraints imposed on all C–H bonds. An analogous simulation of an isolated butane molecule was performed to obtain the gas phase free energy profile.

Direct simulation of the butane conformational equilibrium

The MD free energy profiles were based on dihedral angle probability distributions. In the case of liquid butane, the distribution of the dihedral angle ϕ were obtained for the 71 unconstrained molecules over the 760 ps total production phase trajectory generated for the free energy simulations. In the gas phase the dihedral distribution was generated from a 150 ns trajectory for an isolated, unconstrained butane.

Equilibrium constants were calculated as ratios of probabilities of the *trans* state $(120°–180°)$ to the *gauche* state $(0°–120°)$.

Decapeptide free energy gradient maps

The two decapeptides ALA10 and AIB10 were simulated in vacuum, with all nine ϕ and nine ψ backbone dihedrals fixed at constant values: $\phi_i = \phi$, $\psi_i = \psi$, i=1,...,9. The conformations of the system may thus be identified by the pair of values (ϕ,ψ). For each decapeptide simulations were performed at 117 points on a square grid with $3°$ spacing, for ϕ varying from -75° to -51° and ψ from -20° to -56°. At each point a 20 ps equilibration a 40 ps trajectory generation was performed at 300 K and free energy gradients evaluated according to the multidimensional thermodynamic integration approach.

Further, the free energy gradient components corresponding to all ϕ and all ψ were added up to give

$$\frac{\partial A}{\partial \phi} = \sum_{i=1}^{9} \frac{\partial A}{\partial \phi_i} \quad ; \quad \frac{\partial A}{\partial \psi} = \sum_{i=1}^{9} \frac{\partial A}{\partial \psi_i}$$

This procedure leads to an effective two-dimensional free energy surface describing concerted conformational transitions of the peptides. Reducing the dimensionality of the free energy surface from 18 to two allows for a more facile analysis of results. By following only concerted changes in the ϕ and ψ we obtain only a partial description of the 18-dimensional conformational space. However, the complete 18-dimensional gradient is available at all explored points to enable an in-depth characterization of the visited parts of the surface.

4. Results and discussion

4.1. Butane

n-butane is a simple model system for the general problem of conformational equilibria flexible molecules. We will to compare the predictions of TI, TP and MD approaches for this well-studied system.

Butane in the gas phase

The free energy profiles $\Delta A_c(\phi)$ for rotation around the central C–C bond of n-butane in the gas phase obtained using TI, TP and direct molecular dynamics (MD) simulations are presented in Table I. The conformational free energy profiles are also compared to the adiabatic rotation profile $\Delta U_m(\phi)$, which gives the lowest possible potential energy values for a given ϕ obtained by energy minimization.

The TI gas phase free energy calculations at 275 K yield values of 3.44 and 5.60 kcal/mol for the *trans–gauche* and *syn* barriers, respectively. The free energy difference between the *gauche* minimum at $\phi=65°$ and the *trans* minimum at $\phi=180°$ was 0.96 kcal/mol. The TI, TP and direct MD results for $\Delta A_c(\phi)$ are essentially identical, agreeing within several standard deviations (see Table I).

The *gauche–trans* equilibrium constants in the gas phase are 0.31 for direct MD (at 280 K), 0.30 for TI, and 0.34 for TP (both at 275 K). Gas phase equilibrium constants found here are somewhat lower than previous calculations and experimental measurements, which mostly fall in the 0.4-0.6 range. Simulations using an older version CHARMM potential yielded an equilibrium constant of 0.59 in the gas phase [38].

Component analysis shows that the entropic contribution to the free energy profile is small. The dihedral term gives the largest contribution to the profile for *phi* $> 70°$, with the nonbonded interactions increasing in importance as the *syn* region at $\phi = 0°$ is approached.

Liquid butane

The simulation results for liquid butane are presented in Table I and Figs. 1-2. The TI, TP and direct MD free energy profiles are essentially identical, agreeing within several standard deviations. The largest differences occur in the vicinity of the *syn* barrier. The three free energy profiles are also quite similar to the gas phase results (see Table I). The general agreement between all three simulation methods used in this work indicates that the TI approach yields correct conformational free energy profiles.

The liquid phase *gauche–trans* equilibrium constants are 0.28 for TI, 0.31 for TP, and 0.37 for the direct MD method. The equilibrium constants calculated here are lower than those of previous calculations, which range from 0.47 to 0.85 [17, 39, 36, 40, 41, 42]. The main reason for these differences appears to be the higher *gauche* energy of the new CHARMM potential. Since the calculated liquid equilibrium constants are indistinguishable from the gas phase values within errors , the overall conclusion from our simulations is that there is no equilibrium change between n-butane in the gaseous

Fig. 1

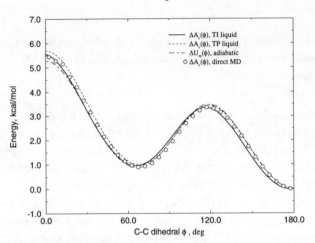

Fig. 1. Free energy profiles $\Delta A_c(\phi)$ for rotation around the central C–C bond in liquid n-butane at 272 K. (———) : TI free energy simulations; ($\cdots$): TP free energy simulations ; ($\cdot - \cdot - \cdot$): gas phase adiabatic profile $\Delta U_m(\phi)$; ($\bullet$): direct molecular dynamics simulation.

and liquid state. The absence of a conformer population shift towards the *gauche* form in the liquid has been found in a some previous studies [36, 43, 44], while significant shifts have been found in others [17, 39, 40, 41, 42]. Since a lack of equilibrium shift is present in all of our TI, TP and direct simulation results, it is clearly a property of the molecular model used, which differs in details of description of internal deformations and nonbonded interactions from those employed in other reported butane studies.

The entropic contribution to the free energy profile is small, and the TI and TP results are in good agreement (Table I). The simulations indicate an entropic destabilization of the *trans–gauche* transition state by 0.3-0.4 kcal/mol, a stabilization of the *cis* maximum by a similar amount, and approximately zero contribution to the *gauche* minimum. The statistical fluctuations of the entropy are quite large (Table I), thus these results are much less reliable than the free energies.

Fig. 2 shows the decomposition of the liquid n-butane conformational free energy profile $\Delta A_c(\phi)$ into contributions from internal deformations, and both intra- and intermolecular nonbonded interactions. Within their statistical fluctuations internal deformation and intramolecular interaction components found in the liquid are essentially identical to those found in the gas phase, while the liquid intermolecular component is indistinguishable from zero . As in the gas phase simulations, the liquid intramolecular interaction components for angles $\phi > 70^o$ are essentially identical to the adiabatic interaction term, while for $\phi < 70^o$ the liquid simulation results lie above the adiabatic contribution. This may be explained by conformational crowding in the conformations ranging between *gauche* and *cis*. The crowding may be correlated to the decreased molecular volume of these conformations relative to *trans* by about 0.5 $\overset{\circ}{A}^3$ (0.4%), apparently leading to disordering of the neighboring molecules

Table 1: Characterization of rotational profile around central C–C dihedral ϕ in n-butane. Statistical errors of the calculated quantities are given, calculated as described in text. The *trans–gauche* barrier maximum is at $\phi=120°$ in the adiabatic and TP profiles and at $\phi=115°$ in TI; the *gauche* minimum is at $\phi=65°$ in the adiabatic and TI profiles and at $\phi=67.5°$ in TP.

Feature	Adiabatic	Gas			Liquid, 272 K		
	ΔU_m	Direct MD 280 K	Free energy, 275 K		Direct MD	Free energy	
			TI	TP		TI	TP
			Free energy	ΔA_c	kcal/mol	Free energy	
		kcal/mol					
trans–gauche barrier	3.49	3.50 ± 0.01	3.44 ± 0.01	3.49 ± 0.01	3.36 ± 0.03	3.35 ± 0.02	3.44 ± 0.03
Δ(*trans–gauche*)	0.88	1.03 ± 0.02	0.96 ± 0.01	0.96 ± 0.02	0.90 ± 0.02	0.95 ± 0.02	1.00 ± 0.04
syn barrier	5.30	6.00 ± 0.19	5.60 ± 0.02	5.72 ± 0.03	5.45 ± 0.16	5.58 ± 0.03	5.70 ± 0.05
			Energy	ΔU	kcal/mol		
trans–gauche barrier			3.40 ± 0.03	3.46 ± 0.20		3.06 ± 0.80	2.99 ± 2.38
Δ(*trans–gauche*)			0.95 ± 0.04	0.97 ± 0.36		0.90 ± 1.00	0.86 ± 3.61
syn barrier			5.50 ± 0.06	5.63 ± 0.55		5.95 ± 1.26	6.00 ± 4.38
			Entropy	$-T\Delta S_c$	kcal/mol		
trans–gauche barrier			0.04 ± 0.03	0.03 ± 0.20		0.29 ± 0.80	0.45 ± 2.38
Δ(*trans–gauche*)			0.01 ± 0.04	-0.01 ± 0.36		0.05 ± 1.00	0.14 ± 3.61
syn barrier			0.10 ± 0.06	0.09 ± 0.55		-0.37 ± 1.26	-0.29 ± 4.38
trans fraction, %	71.8[a]	76.5	77.2	74.5	73.1	78.2	76.3
K_{tg}	0.39[a]	0.31	0.30	0.34	0.37	0.28	0.31

[a] Calculated at T=272 K.

Fig. 2

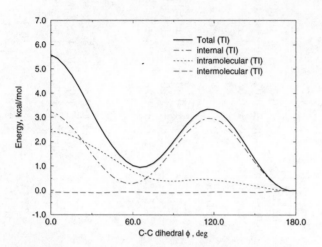

Fig. 2. Decomposition of the TI free energy profile $\Delta A_c(\phi)$ of liquid butane into components from internal deformations and nonbonded interactions, 272 K. (———) : total TI free energy profile $\Delta A_c(\phi)$; $(\cdot - \cdot - \cdot)$: TI internal deformation energy component; $(\cdots)$: TI intramolecular interaction component; $(- - -)$: TI intermolecular interaction component.

and entropy increase for conformers in the *syn* range.

4.2. Model decapeptides

ALA10 free energy gradient map

The reduced, two-dimensional free energy gradient map describing concerted transitions of ϕ/ψ dihedrals in the right-handed helical region of the ALA10 peptide $(CH_3CO\text{-}(Ala)_{10}\text{-}CONH_2)$ is presented in Fig. 3; the arrows show the direction and rate of increase of the system conformational free energy evaluated on a grid. The multidimensional TI results are used to explore several features of the ALA10 free energy surface.

A single free energy minimum was identified in the studied region, corresponding to the α-helix structure, in the vicinity of $(\phi,\text{psi}) = (-64.5^o,-42.5^o)$, in excellent agreement with average values found in α-helices in protein crystals [45]. Since all 18 gradient components change sign upon passing through this point, this is a true free energy minimum on the full 18-dimensional free energy surface. No stable 3_{10}-helix state was found.

Fig. 4 shows the profile obtained by integrating the free energy gradient along the 'α-helix trough' – the low free energy region of α-helix type conformations running diagonally through Fig. 3. In this shallow profile, anticorrelated changes of ϕ, ψ by $\pm 4.5^o$ lead to free energy changes of only about 0.7 kcal/mol, indicating that the α-helix structure can easily be deformed along the trough direction, as has been found

Fig. 3

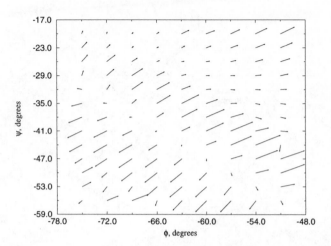

Fig. 3. The reduced two-dimensional free energy gradient map of the ALA10 peptide. Arrows indicate the direction of increasing conformational free energy $(\partial A_c/\partial\phi, \partial A_c/\partial\psi)$. For orientation, the gradient at $(\phi,\psi) = (-60°,-41°)$ is $(106.0, 87.0)$ kcal/(mol rad).

in molecular dynamics simulations [46].

Fig. 5 shows the free energy profile along the ψ=-60° line, which connects the α-helix and 3_{10}-helix regions. Integrating along the "α-helix trough" from (-64.5°,-42.5°) to (-60°,-47°) and then along $\phi = -60°$ to (-60°,-29°), we find that within our model the α-helix is more stable than the 3_{10}-helix by 13.7 ± 0.1 kcal/mol for the ALA10 peptide in vacuum. Our α-helix /3_{10}-helix free energy difference is in agreement with previous simulations, which found a preference for the α-helix conformation in alanine homopeptides [47, 29, 30].

The internal strain contribution to the gradient was small and approximately constant over the mapped area; the features of the total free energy map – minima and maxima – were determined primarily by the nonbonded terms.

AIB10 free energy gradient map

The reduced, two-dimensional free energy gradient map describing concerted transitions of ϕ/ψ dihedrals in the right-handed helical region of the AIB10 peptide (CH$_3$CO-(Aib)$_{10}$-CONH$_2$) is presented in Fig. 6. As in the ALA10 case, we present a brief analysis of the most important features of the surface.

Two free energy minima can be identified, in the vicinity of (-55.5°,-51.5°) in the α-helix region, and close to (-54°,-29°) in the 3_{10}-helix region. These structures are in excellent agreement with crystal conformations [48, 49]. Analysis of the full 18-dimensional free energy gradient shows that all gradient components change sign from negative to positive as we pass through the (-57°,-50°) and (-54°,-29°) points, indicating that minima of the complete 18-dimensional free energy hypersurface exist close to the

383

Fig. 4

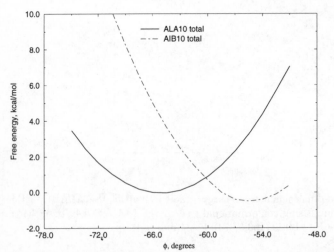

Fig. 4. Free energy profiles obtained by integration of the conformational free energy gradients along the α-helix troughs $\phi + \psi = -107^{\circ}$. (———) : ALA10 ; ($\cdot - \cdot - \cdot$) AIB10 .

minima of the reduced surface. Thus, both the α-helix and the 3_{10}-helix are stable molecular states of the AIB10 peptide in vacuum.

The free energy profile which was obtained by integrating the gradient along the 'α-helix trough' is presented in Fig. 4. For AIB10 anticorrelated changes of ϕ, ψ by $\pm$ 4.5° lead to free energy changes of about 1.2 kcal/mol, about twice as much as found for ALA10. The α-methyl substitution, which decreases the conformational flexibility of the system, thus makes the α-helix more rigid, in accord with chemical intuition.

Integrating along the "α-helix trough" from $(-55.5^{\circ},-51.5^{\circ})$ to $(-54^{\circ},-53^{\circ})$ and then along $\phi = -54^{\circ}$ to $(-54^{\circ},-29^{\circ})$, we find that within our model the α-helix is more stable than the 3_{10}-helix by 3.9 ± 0.1 kcal/mol for the AIB10 peptide in vacuum. In Fig. 5 the barrier for the concerted transition between the two helical states is about 3.1 ± 0.1 kcal/mol above the 3_{10}-helix . The smaller α-helix /3_{10}-helix free energy difference in AIB10 relative to ALA10 is in qualitative agreement with experimental observations of both α-helical and 3_{10}-helical conformations in Aib containing peptides (see [30] for a discussion).

Comparison of the ALA10 and AIB10 results allows to draw some general conclusions of the influence of the α-methyl substitution on peptide conformations. The main effect is the stabilization of the 3_{10}-helix structure. The 3_{10}-helix is unstable for ALA10 and becomes a stable state for AIB10, the α-helix /3_{10}-helix free energy difference decreases from about 14 to bout 4 kcal/mol. Further changes include a shift

Fig. 5

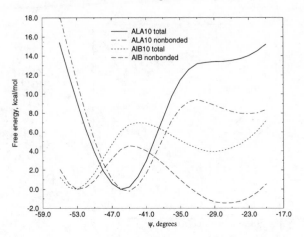

Fig. 5. Free energy profiles obtained by integration of the conformational free energy gradients along paths connecting the α-helix and 3_{10}-helix regions. (——) : ALA10 total free energy profile along line $\phi = $ -60°; ($\cdot - \cdot - \cdot$): ALA10 nonbonded interaction component along line $\phi = $ -60°; ($\cdots$): AIB10 total free energy profile along line $\phi = $ -54°; (– – –) : AIB10 nonbonded interaction component along line $\phi = $ -54°.

of the α-helix and 3_{10}-helix states toward higher ϕ in AIB10 relative to ALA10 and an increased rigidity of the α-helix with respect to anticorrelated changes of ϕ and ψ in AIB10 compared to ALA10.

5. Conclusions

A review of the theoretical background of free energy simulations and some applications in chemistry and biology have been presented. The application examples illustrate the range of problems that can be addressed using these methods.

In the first example the population ratios of the two stable conformers of n-butane were calculated in the gas and liquid phase using three different approaches – thermodynamic perturbation (TP), thermodynamic integration (TI) and direct simulation (MD). Within the molecular model used, all three simulation methods predict that there is essentially no shift in the conformational equilibrium of n-butane between the gas and liquid. Free energy decomposition analysis was used to detect and characterize conformational crowding in the *cis* transition state of n-butane.

The second example involved exploration of the 18-dimensional free energy surfaces of two model peptides - decaalanine and deca-α-methylalanine. For decaalanine the α-helix was found to be the only stable molecular state in the studied region. The α-helix structures occupied a relatively long trough, with soft deformations for anticorrelated changes of ϕ and ψ and hard deformations for correlated changes. A number of differences between ALA10 and AIB10 conformational free energy surfaces

Fig. 6

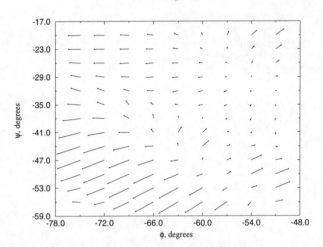

Fig. 6. The reduced two-dimensional free energy gradient map of the AIB10 peptide. Scale is identical as in Fig. 4.

was found, attributable to the effects of α-methyl substitution. The most important difference was that both the α-helix and the 3_{10}-helix were stable molecular states for AIB10, in qualitative agreement with observations of both types of helical conformations in Aib containing peptides. The α-helix trough in AIB10 was steeper and shifted towards higher ϕ values compared to ALA10, in accord with crystallographic results. Finally, the free energy difference between the α-helix and 3_{10}-helix states was only 3.9 kcal/mol in AIB10, while the barrier for the concerted 3_{10}-helix $\rightarrow \alpha$-helix transition was 3.1 kcal/mol.

In summary, based on the presented simulations we can conclude that the TI method systematically gives numerical results of comparable quality to other established conformational free energy simulation algorithms. With its algorithmic simplicity, computational efficiency, possibility of decomposition analysis of free energy differences, and generalizations to many dimensions and calculations of higher free energy derivatives, the TI approach emerges as a highly useful tool for present and future investigations of conformational equilibria of flexible molecules in condensed phases. Especially interesting is the application of the TI approach to characterize multidimensional conformational free energy surfaces. The potential usefulness of this method goes well beyond the relatively simple applications described here.

Acknowledgments: The author wishes to thank Professor Martin Karplus for making available CHARMM version 22 parameters prior to publication. This work was supported in part by the Kansas Institute for Theoretical and Computational Science.

Appendix A. Derivation of the basic TI formula

Starting with the joint probability distribution $\rho(\xi_1, \xi_2, ..., \xi_m)$ given in Eq. (7), we can

386

define the "slice" configuration integral $Z(\xi_1, \xi_2, ..., \xi_m)$ in analogy to Eq. (4):

(9) $$\rho(\xi_1, \xi_2, ..., \xi_m) = Z(\xi_1, \xi_2, ..., \xi_m)/Z$$

The multidimensional conformational free energy surface is then given by

(10) $$A_c(\xi_1, \xi_2, ..., \xi_m) = -kT \ln Z(\xi_1, \xi_2, ..., \xi_m)$$

and its gradient is

(11) $$\frac{\partial A_c(\xi_1, \xi_2, ..., \xi_m)}{\partial \xi_k} = -kT \frac{1}{Z(\xi_1, \xi_2, ..., \xi_m)} \frac{\partial Z(\xi_1, \xi_2, ..., \xi_m)}{\partial \xi_k}$$

To calculate the derivative of the "slice" configuration integral $Z(\xi_1, \xi_2, ..., \xi_m)$ with respect to ξ_k we will derive a general formula for any function $X(q)$:

(12)
$$\frac{1}{Z(\xi_1, \xi_2, ..., \xi_m)} \frac{\partial}{\partial \xi_k} [Z(\xi_1, \xi_2, ..., \xi_m) \langle X(q) \rangle_\xi] = \langle \frac{\partial}{\partial \xi_k} \left[X(q)e^{-U(q)/kT} \right] e^{U(q)/kT} \rangle_\xi$$

where $\langle ... \rangle_\xi$ denotes the canonical average over the subset of conformations with fixed values of ξ_i, $i = 1, ..., m$. To prove Eq. (12) we use the definition of the derivative on the left hand side:

(13)
$$\frac{1}{Z(\xi_1, \xi_2, ..., \xi_m)} \frac{\partial}{\partial \xi_k} [Z(\xi_1, \xi_2, ..., \xi_m)\langle X(q)\rangle_\xi] =$$

$$= \frac{1}{Z(\xi_1, \xi_2, ..., \xi_m)} \frac{\partial}{\partial \xi_k} \int dq X(q)e^{-U(q)/kT} \prod_{i=1}^{m} \delta(\Xi_i(q) - \xi_i)$$

$$= \frac{1}{Z(\xi_1, \xi_2, ..., \xi_m)} \lim_{\Delta\xi \to 0} \frac{1}{\Delta\xi} \int dq \left[X(q^*)e^{-U(q^*)/kT} - X(q)e^{-U(q)/kT} \right] \prod_{i=1}^{m} \delta(\Xi_i(q) - \xi_i)$$

$$= \frac{1}{Z(\xi_1, \xi_2, ..., \xi_m)} \int dq \frac{\partial}{\partial \xi_k} \left[X(q)e^{-U(q)/kT} \right] \prod_{i=1}^{m} \delta(\Xi_i(q) - \xi_i)$$

$$= \langle \frac{\partial}{\partial \xi_k} \left[X(q)e^{-U(q)/kT} \right] e^{U(q)/kT} \rangle_\xi$$

where in line 3 above $X(q)e^{-U(q)/kT} \delta(\Xi_k(q) - \xi_k - \Delta\xi)$ was replaced by the with the equivalent expression $X(q^*)e^{-U(q^*)/kT} \delta(\Xi_k(q) - \xi_k)$; q^* is the perturbed configuration q, corresponding to changing $\xi_k \to \xi_k + \Delta\xi$ and leaving all the other degrees of freedom unchanged. The derivative of any function of q with respect to ξ_k such as that appearing in the last two lines of Eq. (13) may be evaluated by the chain rule, e.g. for $X(q)$:

(14)
$$\frac{\partial X(q)}{\partial \xi_k} = \sum_{i=1}^{3N} \frac{\partial X(q)}{\partial q_i} \frac{\partial q_i}{\partial \xi_k}$$

where $\partial q_i/\partial \xi_k$ indicates how coordinate q_i varies with change of the conformational coordinate ξ_k.

References

[1] M. Mezei and D. L. Beveridge. Free energy simulations. *Ann. N. Y. Acad. Sci.*, 482:1–23, 1986.

[2] D. L. Beveridge and F. M. DiCapua. Free energy via molecular simulation: Applications to chemical and biomolecular systems. *Annu. Rev. Biophys. Biophys. Chem.*, 18:431–492, 1989.

[3] T. P. Straatsma and J. A. McCammon. Computational alchemy. *Annu. Rev. Phys. Chem.*, 43:407–435, 1992.

[4] J. Gao, K. Kuczera, B. Tidor, and M. Karplus. Hidden thermodynamics of mutant proteins: A molecular dynamics analysis. *Science*, 244:1069–1072, 1989.

[5] W. L. Jorgensen and C. Ravimohan. Monte Carlo simulation of differences in free energies of hydration. *J. Am. Chem. Soc.*, 83:3050–3054, 1985.

[6] P. A. Bash, U. C. Singh, R. Langridge, and P. A. Kollman. Free energy calculations by computer simulation. *Science*, 236:564–568, 1987.

[7] T. P. Lybrand, I. Ghosh, and J. A. McCammon. Hydration of chloride and bromide anions: determination of relative free energy by computer simulation. *J. Am. Chem. Soc.*, 107:7793–7794, 1985.

[8] C. L. Brooks III. Thermodynamics of ionic solvation: Monte Carlo simulations of aqueous chloride and bromide ions. *J. Phys. Chem.*, 90:6680–6684, 1986.

[9] T. P. Lybrand, J. A. McCammon, and G. Wipff. Theoretical calculation of relative binding affinity in host–guest systems. *Proc. Natl. Acad. Sci. USA*, 83:833–835, 1986.

[10] P. A. Bash, M. J. Field, and M. Karplus. Free energy perturbation method for chemical reactions in the condensed pahse: a dynamical approach based on a combined quantum and molecular mechanics potential. *J. Am. Chem. Soc.*, 109:8092–8094, 1987.

[11] M. J. Field, P. A. Bash, and M. Karplus. A combined quantum mechanical and molecular mechanical potential for molecular dynamics simulations. *J. Comp. Chem.*, 11:700–733, 1990.

[12] C. F. Wong and J. A. McCammon. Dynamics and design of enzymes and inhibitors. *J. Am. Chem. Soc.*, 108:3830–3832, 1986.

[13] V. Daggett, F. Brown, and P. A. Kollman. Free energy component analysis: Study of the glutamic acid 165 $\rightarrow$ aspartic acid 165 mutation in triosephosphate isomerase. *J. Am. Chem. Soc.*, 111:8247–8256, 1989.

[14] B. Tidor and M. Karplus. Simulation analysis of the stability mutant R96H of T4 lysozyme. *Biochemistry*, 30:3217–3228, 1991.

[15] K. Kuczera, J. Gao, B. Tidor, and M. Karplus. Free energy of sickling: A simulation analysis. *Proc. Natl. Acad. Sci. USA*, 87:8481–8485, 1990.

[16] T.P. Straatsma and J.A. McCammon. Multiconfiguration thermodynamic integration. *J. Chem. Phys.*, 95:1175–1188, 1991.

[17] H. Hayashi, H. Tanaka, and K. Nakanishi. A study of conformational equilibria in chain molecules. i. Liquid n-butane. *Molecular Simulations*, 9:401–415, 1993.

[18] J. P. Valleau and G. M. Torrie. A guide to Monte Carlo for statistical mechanics 2. Byways. In Bruce J. Berne, editor, *Statistical Mechanics. Part A. Equilibrium Techniques*. Plenum, New York, 1977.

[19] D. E. Smith and A. D. J. Haymet. Free energy, entropy and internal energy of hydrophobic interactions: computer simulations. *J. Chem. Phys.*, 98:6445–6454, 1993.

[20] W. F. van Gunsteren and H. J. C. Berendsen. Computer simulations of molecular dynamics: Methodology, applications and perspectives in chemistry. *Angew. Chem. Int. Ed. Engl.*, 29:992–1023, 1990.

[21] D. A. McQuarrie. *Statistical Mechanics*. Harper and Row, New York, 1976.

[22] L. R. Pratt, C. S. Hsu, and D. Chandler. Statistical mechanics of small chain molecules in liquids I. effects of liquid packing on conformational structures. *J. Chem. Phys.*, 68:4202–4212, 1978.

[23] W. L. Jorgensen. Theoretical studies of medium effects o conformational equilibria. *J. Phys. Chem.*, 87:5304–5314, 1983.

[24] D. Chandler. Effects of liquid structures on chemical reactions and conformational changes of non-rigid molecules in condensed phases. *Faraday Disc. Chem. Soc.*, 66:184–190, 1978.

[25] C. L. III Brooks and David A. Case. Simulations of peptide conformational dynamics and thermodynamics. *Chem, Rev.*, 93:2487–2502, 1993.

[26] K. Kuczera. Dynamics and thermodynamics of the globins. In R. Elber, editor, *New Developments in Theoretical Studies of Proteins*. World Scientific, 1995, in press.

[27] K. Kuczera. One- and multidimensional conformational free energy simulations. *J. Comp. Chem.*, in press.

[28] J. Hermans. Molecular dynamics simulations of helix and turn propensity in model peptides. *Curr. Opin. Struct. Biol.*, 3:270–276, 1993.

[29] Julian Tirado-Rives, David S. Maxwell, and William L. Jorgensen. Molecular dynamics and monte carlo simulations favor the α-helical form for alanine-based peptide in water. *J. Am. Chem. Soc.*, 115:11590–11593, 1993.

[30] S. E. Huston and G. E. Marshall. $\alpha/3_{10}$-helix transitions in α-methylalanine homopeptides: conformational transition pathway and potential of mean force. *Biopolymers*, 34:75–90, 1994.

[31] B. R. Brooks, R. Bruccoleri, B. Olafson, D. States, S. Swaminathan, and M. Karplus. CHARMM: A program for macromolecular energy, minimization and dynamics calculations. *J. Comp. Chem.*, 4:187–217, 1983.

[32] A.D. MacKerell, Jr., M. Field, S. Fischer, M. Watanabe, and M. Karplus. All-hydrogen alkane potential for use in aliphatic groups of macromolecules. *manuscript in preparation*.

[33] G. Kaminski, E. M. Duffy, T. Matsui, and W. L. Jorgensen. Free energies of hydration and pure liquid properties of hydrocarbons from the opls all–atom model. *J. Phys. Chem.*, 98:13077–13082, 1994.

[34] J. P. Ryckaert, G. Ciccotti, and H. J. C. Berendsen. Numerical integration of the cartesian equations of motion with constraints: molecular dynamics of n-alkanes. *J. Comp. Phys.*, 23:327–341, 1977.

[35] M. P. Allen and D. J. Tildesley. *Computer Simulations of Liquids*. Oxford Science Publications, London, 1987.

[36] W.L. Jorgensen. Pressure dependence of the structure and properties of liquid n-butane. *J. Am. Chem. Soc.*, 103:4721–4726, 1981.

[37] D. J. Tobias and C. L. Brooks III. Molecular dynamics with internal constraints. *J. Chem. Phys.*, 89:5115–5127, 1988.

[38] D. J. Tobias and C. L. Brooks III. Thermodynamics of solvophobic effects: A molecular-dynamics study of n-butane in carbon tetrachloride and water. *J. Chem. Phys.*, 92:2582–2592, 1990.

[39] D. Brown and J.H. Clarke. A direct method of studying reaction rates by equilibrium molecular dynamics: Application to the kinetics of isomerization in liquid n-butane. *J. Chem. Phys.*, 92:3062–3073, 1990.

[40] J.-P. Ryckaert and A. Bellemans. Molecular dynmaics of liquid alkanes. *Disscuss. Faraday Soc.*, 66:95–106, 1978.

[41] R. Edberg, D.J. Evans, and G.P. Moriss. Constrained molecular dynamics: Simulations of liquid alkanes with a new algorithm. *J. Chem. Phys.*, 84:6933–6939, 1986.

[42] P. A. Wielopolski and E. R. Smith. Dihedral angle distribution in liquid n-butane: molecular dynamics simulations. *J. Chem. Phys.*, 84:6940–6942, 1986.

[43] W. L. Jorgensen, J. D. Madura, and C. J. Swenson. Optimized intermolecular potential functions for liquid hydrocarbons. *J. Am. Chem. Soc.*, 106:6638–6646, 1984.

[44] Banon A., F. Serrano Adan, and J. Santamaria. *J. Chem. Phys.*, 83:297, 1985.

[45] C. Chotia. *Annu. Rev. Biochem.*, 53:537–572, 1984.

[46] C. L. Brooks III, M. Karplus, and B. M. Petitt. *Proteins: A Theoretical Perspective of Dynamics, Structure, and Thermodynamics.* John Wiley and Sons, New York, 1988.

[47] J. Tirado-Reves and W. L. Jorgensen. Molecular dynamics simulations of the unfolding of an α-helical analogue of ribonuclease A S-peptide in water. *Biochemistry*, 30:3864–3871, 1991.

[48] I. L. Karle, J. Flippen-Andersen, K. Uma, and P. Balaram. *Curr. Sci.*, 59:875–885, 1990.

[49] A. Bavioso, E. Benedetti, B. DiBlasio, V. Pavone, C. Pedone, C. Toniolo, and G. M. Bonora. *Proc. Natl. Acad. Sci. USA*, 83:1988–1992, 1986.

Krzysztof Kuczera
Departments of Chemistry and Biochemistry
University of Kansas
2010 Malott Hall, Lawrence, KS 66045
email: `kuczera@tedybr.chem.ukans.edu`

B LESYNG

Structure and dynamics of biomolecular systems. Basic problems for biologists – challenges for mathematicians

Abstract: A brief review is presented of simulation and computer modelling methods that play an increasing role in biomolecular sciences. Limitations of the models and methods are discussed. Problems of potential interest for mathematicians are indicated.

1 .   General remarks

Whereas for past few centuries mathematics and physico-chemical sciences have been developed based on formal theories, biological sciences have been constructed based mostly on qualitative, descriptive-type models. A half-century ago the situation started changing. Greatly this was caused by the Schroedinger's famous book, "What is Life ?" [1]. The book had a large effect on biomolecular studies. At the time the book was written the genetic code was unknown and the role of catalytic, enzymatic processes was not taken into account. In fact Schroedinger didn't discuss the enzymatic processes at all. Nevertheless, he analyzed several other key problems such as mechanisms of structural stability of biomolecules, mechanisms of biomolecular structure formation, the role of quantum effects in biological processes and asked also whether life processes involve other laws of physics hitherto unknown. This inspired a few generations of physicists, mathematicians and biologists to study basic biological problems based on well-established formal mathematical and physical theories. In the meantime considerable progress in classification and understanding of physical, elementary interactions has been achieved. Although a grand unification theory of all known interactions, strong, electromagnetic, weak and gravitational is still far from complete, the electromagnetic interactions responsible, amongst others, for development of biomolecular processes, got within the electroweak theory a well established theoretical background. On the other hand, during the past decade there has been an explosive development of experimental biomolecular and biotechnological techniques. To a large extent this is due to applications of methods that are typical domains of physics and mathematics. In particular applications of diffraction, nuclear magnetic resonance and molecular spectroscopy techniques provide information on 3D structures of biomolecules, and to a limited extent also on their internal dynamics. In this way structures of a few hundreds of proteins (mostly enzymes) and large fragments of DNA and RNA (carriers of the genetic code) were determined. Such macromolecules typically exist as a dynamic equilibrium between local energy minima which can be classified as their conformational or isomeric states. Within these molecules quantum proton or electron transfer processes (quantum hoppings) occur. Such hoppings are often coupled with non equilibrium classical conformational/isomeric transitions. This shows how complex basic biological events can be.

2 .   Specific properties of biomolecular systems

Biomolecular systems can be classified as *complex systems*, and in most cases they are more, or much more, complex than systems typically studied in the physical sciences. A few most common difficulties in formulation of the biomolecular models are listed below.

First of all, formulation of theoretical models of biomolecular systems may in some cases be difficult since <u>details of the molecular structure</u> under study, crucial for its function, like counterions, water of hydration, etc., <u>may not be known</u>. Therefore using all available experimental data and experimental measurements is strongly recommended for construction of the theoretical models.

<u>Typical molecular systems of biological importance are large</u>. They contain from hundreds to tenths of thousands of atoms. Getting analytical solutions for such systems can be difficult. Typically the computational modelling follows the mathematical formulation of the problems.

In most cases, structural properties of the most close environment of the biomolecular systems cannot be neglected. A discrete, atomic-type description of an interface region between the "core of the system" and the "remote unstructured, continuous environment", is often required. We may formulate this difficulty in the form of the following question: <u>where does the biomolecular system terminate</u> ? For example, in order to properly describe macromolecular, enzymatic systems, it is good to include a hydration shell in the molecular modelling studies. We may say that the hydration shell is an integral part of the biomolecular structure. This has strong mathematical consequences. Definition of the domain boundaries is not the unique one, and depends on precision of the developed models. This is the reason that such <u>mathematical operations like taking the limits can in some cases be difficult</u>.

In cells, <u>different biomolecular objects can be closely packed</u>. Assumption that the objects are diluted is often a too far going approximation.

Since coupled biological events may occur in different time-scales the models of such phenomena must have a <u>hierarchical structure</u> (think for example of switching the genes on/off by gene activators/repressors and think of the cell differentiation processes).

Despite the difficulties mentioned above a class of effective microscopic and mesoscopic theories and computational models have been developed. Fig.1 presents schematically a typical strategy and the most popular methods used in the modelling of biomolecular systems and high technology materials.

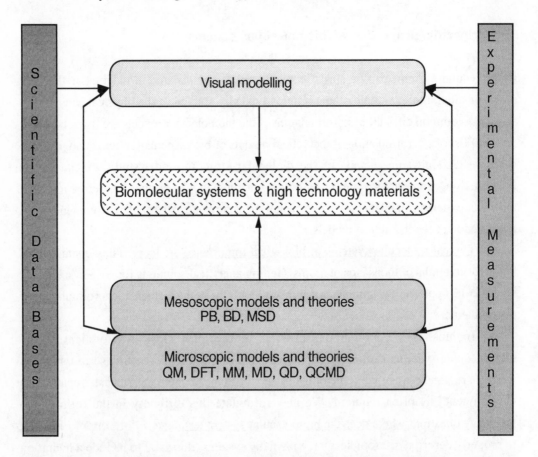

Fig.1 Theories and computational models used in studies of biomolecular systems and high-technology materials. The scheme presents a modelling strategy. The following abbreviations are applied: QM (Quantum Mechanics), DFT (Density Functional Theory), MM (Molecular Mechanics), MD (Molecular Dynamics), QD (Quantum Dynamics), QCMD (Quantum-Classical Molecular Dynamics), PB (Poisson-Boltzmann methods), BD (Brownian Dynamics), MSD (Multi-Scale Diffusion theories).

3. A brief overview of theories and computational models

Static molecular properties and dynamical phenomena which occur in the time scale from femtoseconds (10^{-15} s) to nanoseconds (10^{-9} s) are studied using classical and quantum *microscopic models and theories* like quantum Molecular Mechanics (QM), classical Molecular Mechanics (MM), classical Molecular Dynamics (MD), Quantum molecular Dynamics (QD) and Quantum-Classical Molecular Dynamics (QCMD), for review of the methods see e.g. [2-7]. Biological macromolecules can be called molecular machines. Explaining the dynamical structure, logic and function of these machines is a challenging problem.

The QM methods are used for solving the time-independent Schroedinger equation for calculations of molecular potential energy functions. The Born-Oppenheimer approximation which assumes parametrical dependence of the Hamiltonian on the positions of the nuclei is typically applied, see e.g.[2]. Precise determination of the potential energy functions is difficult, and usually requires extensive calculations. The methods applied for systems of biological relevance are usually not able to determine the energy differences with required accuracy below 1kT (k is the Boltzmann constant and T is the absolute temperature, kT = 2.5 kJ/mole at room temperature).

Let us assume the potential energy function is known. Usually the potential energy function of the whole macromolecular system is built up of its fragments. Knowledge of stable isomeric or conformational states requires <u>minimization of the potential energy function with respect to hundreds or thousands geometrical degrees of freedom</u>. Although the progress in this area is large, the problem has not been solved in a mathematically satisfactory way. Obviously local minimization techniques fail. Global minimization techniques, in particular those which deform the potential energy surface making it more flat, [8,9], provide some hope for finding satisfactory solutions.

The global minimization techniques are directly related to a <u>protein folding problem</u>. In practical terms, the task is to predict physiologically stable 3D protein structures from the knowledge of their amino-acid sequences. Apart of the exciting scientific task, the problem has its economic, financial impact. Knowledge of the 3D structures of key enzymes is always the beginning of rational drug design projects, usually of a large pharmacological interest. One should note that the dynamical folding process occurs on the free energy potential surface, which accounts for environmental (solvent) and temperature effects. The free energy simulations, and in particular <u>determination of the free energy density from microscopic, atomic-type interactions</u>, [10-11], is the barrier which limits a wide class of theoretical studies, in particular diffusive, structure formation processes (see e.g. [17]) occurring in a long time scale.

Studies of functions of biomolecules require <u>determination of the ionization states</u> of monomers which form the structures. Local effective atomic charges influence strongly the structure and dynamics of the macromolecules. The optimal "configuration" of the ionization states minimizes the total free energy of the system at given pH. Experimentally it is determined by titration procedures. "Titration" of the biomolecular structures in computers is one of the key and non trivial procedures. It is partially solved, [12].

Biological function is related directly to atomic motions. The mentioned above classical MD is based on the Newtonian equations (or other equivalent formulations of the classical mechanics), with the atoms treated as the classical particles. However, since the integration of the equations of motion is done with a very short time-step, usually in the range of femtoseconds (10^{-15} s), simulations of the atomic motions are able to "scan" only nanosecond (10^{-9} s) time-intervals. In order to cover areas in which biologically important events can occur one has to develop and apply non-standard methods. There are for example attempts to reduce the number of the dynamical degrees of freedom by introducing internal curvilinear or quaternion degrees of freedom, see e.g. [13]. Any reduction in the dimensionality of the dynamical problem causes creation of a new effective potential energy function, which in this case have to be generated from the elementary, atom-atom type interactions. Generation of the effective potential energy functions in a formal algorithmic way is not solved until now.

Let us assume the precise structure of a biomolecular system is known. We may think here of an enzyme or for example a photosynthetic molecular system. In fact, such biomolecular structures are quantum-classical molecular machines. Some parts of the structures undergo classical, Newtonian-type motions. Other, smaller fragments or selected particles like protons or electrons undergo time-dependent quantum motions governed by the time-dependent Schroedinger equation. Coupling of the motions must result from a formal, mathematically correct procedure which starts from the fully quantum-dynamical model and next comes with selected degrees of freedom to a classical limit.

One observes an increasing number of more and more sophisticated QCMD models, see e.g.[7,14,15]. However, <u>a mathematically satisfactory quantum-classical molecular dynamics model widely applicable to biomolecular systems doesn't exist. Such</u> a model must account also for the most important structural and dynamical properties of the thermal bath (solvent with counterions) surrounding the biomolecule. Usually it generates a non negligible electrostatic reaction field. This field can be evaluated based on the Poisson-Boltzmann equation [6,16]. Fig. 2 presents a functional model of an enzyme molecule in

Microscopic, quantum model

Enzyme active site - the quantum domain. Its dynamics is described by the time-dependent Schroedinger equation (QD). The potential energy function depends on instantaneous positions of the classical atoms as well as on the electrostatic potential generated by the environment (see below).

$$i\hbar \frac{\partial \Psi}{\partial t} = \left(T + V\left(\{\mathbf{x}_\alpha\}, \varphi\right)\right)\Psi$$

Microscopic, classical model

Classical atoms. The domain described by the Newtonian equations of motion (MD).

$$m_\alpha \ddot{\mathbf{x}}_\alpha = \mathbf{F}_\alpha(\{\mathbf{x}_\beta\}, \Psi)$$

Mesoscopic, classical model

Solvent treated as dielectric and ionic medium influences the classical and quantum domains with long-range electrostatic forces. Short-range, stochastic forces can also be included. The electrostatic potential φ is described by the nonlinear Poisson-Boltzmann equation (PB). ε is the scalar dielectric field. q_α are the effective atomic charges in the molecular structure. κ is a constant proportional to the concentration of the mobile ions in solvent.

$$\nabla(\varepsilon\nabla\varphi) + \sum_\alpha q_\alpha \delta(\mathbf{x}\text{-}\mathbf{x}_\alpha) + \kappa\,\sinh\left(\frac{e\varphi(\mathbf{x})}{kT}\right) = 0$$

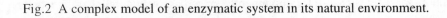

Fig.2 A complex model of an enzymatic system in its natural environment.

solution. The model accounts for time-dependent quantum-dynamical effects as well as allows for controlling the enzyme activity by changing environmental parameters. Thinking of relations between the quantum and classical molecular world one should note that some organisms, like viruses or bacteria, contain only one or a few copies of some enzymes, and the organisms "feel well". This means that description of the elementary biological processes related to these enzymes does not require ensembles or thermodynamical limits. The quantum-classical dynamics of one selected copy of the biomolecule should be sufficient to describe its macroscopic catalytic activity observed experimentally. Note that the classical molecular structure surrounding the quantum site plays a role of a classical measuring apparatus. This is a nano-scale apparatus which is governed by its own internal classical dynamics. If so, the question is if the motions of these classical atoms, coupled to the quantum particle(s), are deterministic or not ?! Physicists did not answer the question mentioned above. This potentially can have further consequences. Since neurological and psychological processes are obviously related to enzyme reactions one may ask how far macroscopic decision-type psychological processes depend on probabilistic quantum events?

Complex biomolecular systems are able to recognize each other. The molecular recognition processes are obviously related to the gain of the free energy due to the intermolecular interaction. Taking into account, however, that the interacting objects are not rigid, and in addition conformational changes are in some cases correlated with changes of the quantum proton dissociation/association processes, we should realize how far we are from a detail understanding of immunological processes, hormone-receptor interactions, activity of gene repressors/activators or mechanisms of structure formation in cells.

4. Conclusions

A brief overview of microscopic and mesoscopic models used in the theoretical studies of complex biomolecular systems was presented. Limitations of the models were indicated. It was pointed out that mechanisms of structure formation and mechanisms of biological function are often the result of coupled, quantum and classical elementary events. Models of real biological phenomena should have a hierarchical structure. A complex model of an enzymatic reaction, recently successfully implemented on a massively parallel architecture, was presented.

Acknowledgments: The research support was provided by the Polish State Committee for Scientific Research (project 8 T11F 006 09) and by ICM, Warsaw University.

References

[1]   E. Schroedinger, *What is Life*, Macmillan, New York, 1946.
[2]   A.Szabo and N.S.Ostlund, *Modern Quantum Chemistry. Introduction to Advanced Electronic Structure Theory*, McGraw-Hill, New York, 1989.
[3]   J.A. McCammon and S.C. Harvey, *Dynamics of Proteins and Nucleic Acids*, Cambridge University Press, Cambridge, 1987.
[4]   B.Lesyng and J.A.McCammon, *Molecular Modeling Methods. Basic Techniques and Challenging Problems*, in Pharmac. Ther., **60**, 149-167, 1993.
[5]   Ch.L.Brooks III, M.Karplus and B.M.Pettitt, *Proteins: Theoretical Perespectives of Structure, Dynamics and Thermodynamics*, Adv. in Chem. Phys., **LXXI**, 1-259, 1988.
[6]   J.M.Briggs and J.A.McCammon, Computers in Physics, **6**, 238-243, 1992.
[7]   P.Bala, P.Grochowski, B.Lesyng and J.A.McCammon, J.Phys.Chem., **100**, 2535-2545(1996)
[8]   L.Piela, J.Kostrowicki and H.A.Scheraga, J. Phys. Chem., **93**, 3339-3346,1989.
[9]   K.A.Olszewski, L.Piela and H.A.Scheraga, J. Phys.Chem. **96**, 4672-4676, 1992.
[10] J.A.McCammon, J. A., *Free Energy From Simulations.* Current Opinion in Structural Biology. **1**,196-200, 1991.
[11] T.P.Stratsma, T.P. and J.A.McCammon, *Computational Alchemy*, Annu. Rev. Phys. Chem., **43**, 407-435, 1992.
[12] J.Antosiewicz, J.A.McCammon and M.K.Gilson, J.Mol.Biol., **238**, 415-436, 1994.
[13] W.Rudnicki, B.Lesyng and S.C.Harvey, Biopolymers, **34**, 383-392, 1994.
[14] P.Bala, P.Grochowski, B.Lesyng and J.A.McCammon, in *Quantum Mechanical Simulation Methods for Studying Biological Systems*, M.Field., Ed., Les Houches Physics Series, 1995.
[15] R.Alimi, R.B.Gerber, A.D.Hammerich and R.J.Kosloff, J.Chem.Phys., **93**, 6484, 1990.
[16] M.E.Davis and J.A.McCammon, *Electrostatics in Biomolecular Structure and Dynamics*, Chem. Rev., **90**, 509-521, 1990.
[17] N. Kenmochi, M. Niezgódka, *Nonlinear System for Non-isothermal Diffusive Phase Separation,* Journal of Math. Analysis and Appl., **188**, 651-679, 1994.

Bogdan Lesyng

Interdisciplinary Centre for Mathematical and Computational Modelling (ICM), Warsaw University, ul. Pawińskiego 5A, 02-106 Warsaw, Poland

and

Department of Biophysics, Warsaw University
ul. Żwirki i Wigury 93, 02-089 Warsaw, Poland

email: lesyng@icm.edu.pl

J MASKAWA AND T TAKEUCHI

Phase separation in elastic bodies – pattern formation in gels

Abstract: We discuss pattern formation in gels during their large deformation such as volume phase transition. Patterns are consisting of swollen and collapsed phases,and are very different from the patterns in usual elastic bodies or in gels far from their volume phase transition points. A Ginzburg-Landau type theory is given for the pattern formation in shrinking gels. The phase diagram of patterns in some region of parameters which characterize the incompressibility and anisotropy of gels are derived from the linearized theory. Some numerical calculations for the evolution of patterns are also shown.

1   Introduction

Polymer gels undergo a volume phase transition with temperature, solvent composition or osmotic pressure being a external parameter[1]. The transitions are often accompanied by macroscopic regular patterns consisting of swollen and collapsed phases, which was first reported by Tanaka[2]. They are very different from the patterns (buckling etc) in usual elastic bodies which have no such phases or in gels far from their volume phase transition points.

The kinetic process of large volume change was studied in detail for swelling and shrinking cases by Matsuo and Tanaka[3][4]. Spherical swelling gels[3] transiently form a pattern like wrinkles of brain on the surface, whose cross-sectional view of the folded parts has cusp singularities. The pattern coalesces as the gel swells, and eventually disappear when the gel turn into the final equilibrium state. The same patterns are also observed in gel plates whose lower surfaces are clamped on substrates and whose upper surfaces are free to expand[5]. Some numerical models are proposed to show the evolution of patterns and the generation of cusp [6][7], and also theoretical frameworks which incorporate nonlinear elastic theories into the Ginzburg-Landau theory of phase transition were given for the pattern formation of gel plates[8][9].

Patterns in shrinking gels[3][4] are quite different from ones in swelling gels. Here we will just sketch out the phenomena according to the paper by Matsuo and Tanaka. Shrinkage starts from the boundary by the collective diffusion of polymer network[10]. In the early stage of volume phase transitions, gel forms a shrunken dense surface layer which prevents further shrinkage due to the impermeability to solvent. After a time bubbles of swollen state appear on the surface, and the gel resumes shrinking up to the final collapsed state. In the case of gels with cylindrical surface, more various patterns were observed, which was very stable and seemed to be permanent.

In this paper we start with a Gibbs free energy which is a natural extension of Flory's theory[11] to inhomogeneous gels as in previous works[9][7] to study the pattern formation of shrinking gels with cylindrical surface[12]. Some extra terms will also be introduced in order to take the effects of the shrunken surface layer into account. A dynamical elastic equation is derived from the functional derivative of the Gibbs free energy, which includes a dissipation term (first derivative in time) owing to the friction between polymers and solvent. The inertial term (second derivative in time) can be consistently neglected because the friction coefficient is huge ($\sim 10^{11} dyn \cdot sec/cm^4$)[13]. Phase diagram in a region of experimental parameters will be given from the stability analysis on the linearized equation around homogeneous states. The time development of an unstable mode will also be shown by the numerical calculation of our dynamical nonlinear equation.

2 Model

The order parameter of our system is the position vector $\mathbf{X} = X_r \mathbf{e_r} + X_\theta \mathbf{e_\theta} + X_z \mathbf{e_z}$ of volume elements which is originally palaced at $\mathbf{x} = r\mathbf{e_r} + z\mathbf{e_z}$ in a reference state. Throughout this paper we will assumme axial symmetry $X_r = X_r(r, z)$, $X_\theta = 0$ and $X_z = X_z(r, z)$ for simplicity. We propose the following functional G of $\mathbf{X}$ as the Gibbs free energy of the system,

$$(1) \qquad \frac{G[\mathbf{X}]}{k_B T \nu_0} = \int_{V_0} d\mathbf{x} \left[f(\frac{\phi}{\phi_0}) + \frac{1}{2} tr M^t M \right] + P_V V + h \int_{S_c} ds \left[\frac{1}{2} \tilde{u}_{\alpha\beta} \tilde{\sigma}_{\alpha\beta} \right],$$

where the first integral is taken over the volume of the reference state (the cylinder with the radius r_0 and the length L_0) in which the polymer chains take random-walk configuration , k_B is Boltzmann constant, T is the temperature and ν_0 represents the number of constituent chains per unit volume in the reference state. The integrand of the first integral is expressed in terms of the deformation matrix defined by $d\mathbf{X} = M d\mathbf{x}$, which is a natural extension of the Flory theory to inhomogeneous gels. ϕ is the polymer volume fraction of deformed states is related to the determinant of M by the relation $\phi_0/\phi = |M|$, where ϕ_0 is the polymer volume fraction of the reference state. In the numerical calculation, we will take a 4th order polynomial in $|M|$ as the function f. The second term describes the job by the external osmotic pressure imposed by the surface layer, P_V is the dimensionless quatity which represents the osmotic pressure divided by $k_B T \nu_0$ and V is the volume of the deformed gel. The third term provides the stretch energy of the surface layer, the integral is taken over the surface of a collapsed state (the surface of the cylinder with the radius r_c and the length L_c) and h is the thickness of the surface layer. The strain tensor $\tilde{u}_{\alpha\beta}$ and the stress tensor $\tilde{\sigma}_{\alpha\beta}$ are of the form

$$\tilde{u}_{\theta\theta} = \frac{1}{2}((\frac{X_r}{r_c})^2 - 1), \qquad \tilde{u}_{zz} = \frac{1}{2}((\frac{\partial X_r}{\partial z})^2 + (\frac{\partial X_z}{\partial z})^2 - 1), \tilde{u}_{z\theta} = 0,$$
$$\tilde{\sigma}_{\theta\theta} = \frac{E}{1-\sigma^2}(\tilde{u}_{\theta\theta} + \sigma \tilde{u}_{zz}), \tilde{\sigma}_{zz} = \frac{E}{1-\sigma^2}(\tilde{u}_{zz} + \sigma \tilde{u}_{\theta\theta}), \qquad \tilde{\sigma}_{z\theta} = 0,$$

401

where E and σ are the Young modulus divided by $k_B T \nu_0$ and the Poisson ratio of the surface layer respectively.

The dynamical elastic equation of the system is derived from the functional derivative of the Gibbs free energy (1) with respect to $\mathbf{X}$ as the following:

$$(2) \qquad f_r \frac{\partial \mathbf{X}}{\partial t} = \nabla \cdot \mathbf{\Pi} = -\frac{\phi}{\phi_0} \frac{\delta \mathbf{G}}{\delta \mathbf{X}}$$

where f_r is the friction coefficient between network polymers and solvent. $\mathbf{\Pi}$ is the stress tensor of the gel. Two boundary conditions on the surface and one on the each end of gel are derived in the same manner from the total derivative of the volume integration and the surface term as natural boundary conditions. The additional conditions

$$(3) \qquad X_r = 0, \frac{\partial X_z}{\partial r} = 0 (r = 0),$$

on the central axis of the cylinder are imposed by the assumption of the axial symmetry of deformations. When the length of the gel is experimentally controlled, the extra boundary conditions

$$(4) \qquad X_z = 0(z = 0), X_z = L(z = L_0),$$

are also imposed instead of natural boundary conditions on the ends of the cylinder.

3   Linearized Stability Analysis

For the purpose of studying the instability of homogeneous states $X_r = \alpha r, X_z = \beta z$, we derive the linearized form of the equation (2) for the fluctuation $u = X_r - \alpha r, v = X_z - \beta z$. Here α and β are the elongation ratios in the radial and axial directions of gel respectively. It is convenient to introduce a new coordinates $(r_h, z_h) = (\alpha r, \beta z)$.

In these coordinates the linearized equations are

$$(5) \qquad f_r \frac{\partial u}{\partial t} = (K + \frac{4}{3}\mu)(\frac{\partial^2 u}{\partial r_h^2} + \frac{1}{r_h}\frac{\partial u}{\partial r_h} - \frac{u}{r_h^2}) + (\frac{\beta}{\alpha})^2 \mu \frac{\partial^2 u}{\partial z_h^2} + (K + \frac{1}{3}\mu)\frac{\partial^2 v}{\partial r_h \partial z_h},$$

$$(6) \qquad f_r \frac{\partial v}{\partial t} = \mu(\frac{\partial^2 v}{\partial r_h^2} + \frac{1}{r_h}\frac{\partial v}{\partial r_h}) + (K + (\frac{1}{3} + (\frac{\beta}{\alpha})^2)\mu)\frac{\partial^2 u}{\partial z_h^2} + (K + \frac{1}{3}\mu)\frac{1}{r_h}\frac{\partial^2 (r_h u)}{\partial r_h \partial z_h}.$$

K and μ are respectively the bulk and shear moduli of the homogeneous state whose network density is $\phi_h = \phi_0/\alpha^2 \beta$:

$$K/k_B T \nu_0 = \frac{1}{(\alpha^2\beta)^3}f''\left(\frac{\phi_h}{\phi_0}\right) + \frac{1}{(\alpha^2\beta)^2}f'\left(\frac{\phi_h}{\phi_0}\right) - \frac{1}{3\beta},$$

$$\mu/k_B T \nu_0 = \frac{1}{\beta}.$$

We obtain the following equation from the 0th order of the natural boundary conditions on the surface,

$$(7) \qquad P_V = f''(\frac{\phi_h}{\phi_0}) - \frac{1}{\beta} - \frac{\tilde{E}}{2\beta r_0 (1 - \sigma^2)}(\alpha^2 - \alpha_c^2 + \sigma(\beta^2 - \beta_c^2)),$$

402

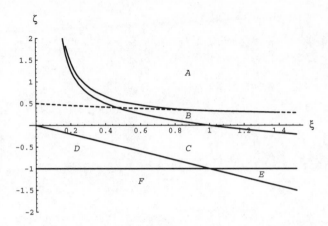

Fig. 1. Phase diagram of patterns in cylindrical gels in the case with E=0. Homogeneous states are stable in A. Other regions are unstable. In B and C patterns with transverse waves on the surface are expected. In B there is an wave number selection in the linear scheme, while in C there is no such a selection. Spinodal decomposition in the radial direction and in the axial direction ('bamboo') are expected in the region D and E respectively. In F the direction of spinodal decomposition will be arbitrary.

where $\tilde{E} = hE/\alpha_c^2$. Here we assume the dense layer of an isotropic collapsed state, $\alpha_c = \beta_c$ ($\alpha_c = r_c/r_0, \beta_c = L_c/L_0$). When the external pressure P_V is given, the radial elongation ratio α is obtained as a solution of (7):we must choose an appropriate value among the solutions, because the function α of P_V derived from the equation (7) is a multi-valued function in general. On the other hand, when the volume of gel is fixed, P_V works as a Lagrange multiplier. Using the equation (7), the 1st order of the natural boundary conditions are derived on the surface ($r_h = \alpha r_0$),

$$
\left(K + \frac{4}{3}\mu\right) u_r + \left(K - \frac{2}{3}\mu\right)\left(\frac{u}{r_h} + v_z\right)
$$

(8)
$$
+ \frac{\tilde{E}}{2\alpha\beta(1-\sigma^2)}\left[2\alpha^4\frac{u}{r_h^2} - \beta^2(\sigma(\alpha^2 - \alpha_c^2) + \beta^2 - \beta_c^2)u_{zz}\right.
$$
$$
\left. - \alpha^2(\alpha^2 - \alpha_c^2 - \sigma(\beta^2 + \beta_c^2))\frac{v_z}{r_h}\right] = 0\,,
$$

$$
\mu(u_z + v_r)
$$

(9)
$$
+ \frac{\tilde{E}}{2\alpha\beta(1-\sigma^2)}\left[-\beta^2(\sigma(\alpha^2 - \alpha_c^2) + 3\beta^2 - \beta_c^2)v_{zz}\right.
$$
$$
\left. + \alpha^2(\alpha^2 - \alpha_c^2 - \sigma(\beta^2 + \beta_c^2))\frac{u_z}{r_h}\right] = 0\,,
$$

and on the both ends of gel,

(10)
$$
\frac{\partial u}{\partial z} = 0\,.
$$

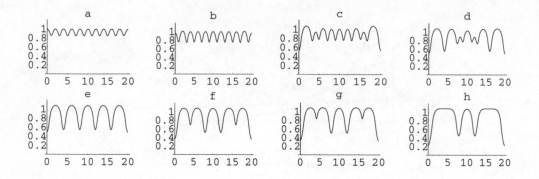

Fig. 2. An evolution of the surface profile $X_r(r = r_0)$. The horizontal axis is z. The time course of pictures is (a) → (h). The parameters in this numerical calculation are set as $L_0 = 20r_0$, $\alpha_c = \beta_c = 0.8$, $\alpha = 0.925$, $\beta = 1.0$ ($\xi = 1.055$, $\zeta = -0.015$), $\tilde{E} = 0.01$.

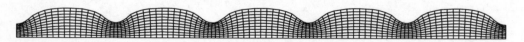

Fig. 3. The cross-sectional view of a pattern along the z axis (half plane). This corresponds to the snapshot (e) in Fig.2.

The linearized version of the equation (3) and (4)

(11) $$u = 0, v_r = 0 \quad (r_h = 0)$$

(12) $$\quad and \quad v = 0 \quad (z_h = 0, \beta L_0)$$

respectively.

We study the linearized stability analysis in the region of two parameters $\xi = (\beta/\alpha)^2$ and $\zeta = K/\mu + 1/3$. ξ is the degree of the anisotropy and ζ characterizes the incompressibility of gels [9]. We can derive the dynamic equation for the density fluctuation $\delta\phi = \phi_h \nabla u$ from the equation (5)(6),

(13) $$\frac{\partial}{\partial t}\delta\phi = D[(\zeta + 1)(\frac{\partial^2}{\partial r^2} + \frac{1}{r}\frac{\partial}{\partial r}) + (\zeta + \xi)\frac{\partial^2}{\partial z^2}]\delta\phi,$$

where $D = \mu/f_r$. As stated by Onuki in the reference [9], we can see from the above equation that in the region $\zeta < -1$ or $\zeta < -\xi$ spinodal decomposition will be expected to take place in the bulk of gels.

We expand the functions u and v in Fourier series with respect to z. Taking the boundary conditions (10)(12) into consideration, the expansion takes the following forms:

$$u(r, z, t) = \sum_n u_n(r, t) \cos k_n z \,, \tag{14}$$

$$v(r, z, t) = \sum_n v_n(r, t) \sin k_n z \,, \quad k_n = \frac{2\pi}{\beta L_0} n \,. \tag{15}$$

For the each Fourier component with wave number k, the eigenfunction with eigenvalue λ for the differential operator of the right-hand side of the equations (5) and (6) (divided by μ) is of the form

$$u = A\kappa_2 I_1(\kappa_2 kr) + B I_1(\kappa_1 kr) \,, \tag{16}$$

$$v = -A I_0(\kappa_2 kr) - B\kappa_1 I_0(\kappa_1 kr) \tag{17}$$

in the case with $\zeta + 1 > 0$ and $\zeta + \xi > 0$, where I_0 and I_1 are 0th and 1st modified Bessel functions respectively, and

$$\kappa_1 = (\xi + \frac{\lambda}{k^2})^{\frac{1}{2}} \,, \tag{18}$$

$$\kappa_2 = (\frac{\zeta + \kappa_1^2}{\zeta + 1})^{\frac{1}{2}} \,. \tag{19}$$

Integration constants A, B and λ are determined by the boundary conditions (8)(9). Here we consider a limit with Young modulus of the dense surface layer E vanishing (for more general consideration, see [12]). In this case the boundary conditions yield an equation for λ,

$$(\kappa_1^2 + 1)^2 \psi(\kappa_2 \tilde{k}) - 4\kappa_2^2 \psi(\kappa_1 \tilde{k}) - (\kappa_1^2 - 1)\kappa_2^2 = 0 \,, \tag{20}$$

where $\tilde{k} = \alpha k r_0$

$$\psi(x) = \frac{x I_0(x)}{2 I_1(x)} \,. \tag{21}$$

In Fig.1 A is the stable region where the equation (20) has no positive λ solution for any wave number k, while in the region B, C it has a positive solution on a set of positive number k. The boundary between A and B is a numerical curve which approaches to the curve

$$\zeta = -\frac{(\xi - 1)(\xi^2 + 6\xi + 1)}{\xi^3 + 5\xi^2 + 11\xi - 1} \,, \tag{22}$$

when ξ tends to the positive solution of the equation $x^3 + 5x^2 + 11x - 1 = 0$ which is approximately 0.087 and to the curve (dashed curve in Fig.1)

$$\zeta = \frac{1}{\xi + 2} \,, \tag{23}$$

when ξ tends to infinity. The eigen function (16)(17) in the region B and C indicates patterns with transverse wave on the surface ('bubble' pattern formation), since modified Bessel functions I_0 and I_1 monotonically increase with argument. In B the equation (20) has a positive solution only for long wavelength $(k < k_c(\xi, \zeta))$, that is, homogeneous states are stable against perturbations with short wavelength $(k > k_c(\xi, \zeta))$. On the other hands homogeneous states are unstable against arbitrary perturbations in C, that is, spinodal decomposition on the surface will occur. The curve (22) is the boundary between B and C, which was numerically checked. The appearance of the region B is an distinctive feature of three dimensional gels.

4   Numerical Calculation

We show the evolution of a pattern by the numerical calculation of the dynamical nonlinear equation (2). Here we take a 4th order polynomial in $|M|$ as the function f instead of Flory's expression,

$$(24) \qquad f = 0.25|M|^4 + 0.85|M|^3 + 1.0725|M|^2 - 1.5925|M| \,.$$

and the friction coefficient f_r is considered as approximately constants in the numerical calculation, which are the function of the network volume density ϕ $(f_r \propto \phi^{3/2}[13])$. We employed finite difference scheme for spatial dimensions and implicit scheme for time development. An example of the evolution of pattern in the case with the parameters ξ, ζ in the region B is shown in Fig.2. The initial condition is a homogeneous state with very small sinusoidal perturbation $(\sim 10^{-6} r_0)$. Fig.3 is the cross-sectional view of a pattern along the z axis (half plane). This corresponds to the snapshot (e) in Fig.2.

5   Summary

A model for pattern formation in shrinking gels has been constructed, which takes a external osmotic pressure and the stretch force owing to the dense layer on the surface of gel into account. The osmotic pressure prevent the volume of gel from further decreasing, and the stretch force suppresses the growth of fluctuations with short wave length. The phase diagram of patterns has been derived from a linearized analysis on our model. We have found the region in which homogeneous states are stable against perturbations with short wavelength. A numerical calculation of the evolution of pattern using finite difference scheme have given a fine 'bubble' pattern.

References

[1] T. Tanaka, Sci. Am. **244**, 124 (1981)

[2] T. Tanaka, Physica **140A**, 261 (1986)

[3] E. S. Matsuo and T. Tanaka, J. Chem. Phys **89**, 1695(1988)

[4] E. S. Matsuo and T. Tanaka, Nature **358**, 482 (1992)

[5] T. Tanaka, S.-T. Sun, Y. Hirokawa, S. Katayama, J. Kucera, Y. Hirose and T. Amiya, Nature **325**,796 (1987)

[6] T. Hwa and M. Kardar, Phys. Rev. Lett. **61**, 106 (1988)

[7] N. Suematsu, K. Sekimoto and K.Kawasaki, Phys. Rev. **A41**, 5751 (1990)

[8] K. Sekimot and K. Kawasaki, Physica **154A**, 384 (1989)

[9] A. Onuki, Phys. Rev. **39**, 5932 (1989)

[10] T. Tanaka and D.J.Fillmore, J. Chem. Phys **1**, 1214 (1979)

[11] P. J. Flory, *Principles of Polymer Chemistry* (Cornell Univ. Press, New York, 1953)

[12] J. Maskawa, T. Takeuchi, K. Maki, K. Tsujii and T. Tanaka, to appear

[13] M. Tokita and T. Tanaka, J. Chem. Phys. **95**, 4613 (1991)

Jun-Ichi Maskawa
Department of Human Life Study
Hijiyama Women's Junior College
Hiroshima 732,Japan

Toshiki Takeuchi
Department of Mathematics,Faculty of Engineering
The University of Tokushima
Tokushima 770,Japan

M RÓŻYCZKA, T PLEWA AND A KUDLICKI
Structure formation in cosmology

1. Introduction

At its present stage of development, structure formation in cosmology is very far from being a coherent mathematical theory comparable, for example, to General Relativity. It should be rather thought of as a busy node in an extended network of scientific activity, whose input links provide information concerning

1. global model of the Universe,
2. distribution of matter in the early Universe,
3. model of the material content of the Universe,
4. physical mechanism(s) responsible for structure formation, and
5. observational data directly related to the structure of the Universe at its present evolutionary epoch.

Based of the first four items of the above list, both the equations describing the structure evolution and the methods of their solution are developed at the node. The output from the node, i.e. the solutions of the equations, is confronted with observational data, thus providing a feedback that allows the input information to be updated. Independently of the main feedback loop, information concerning items (1–3), (2–4) and (3), respectively, is additionally updated according to

1. astrophysical observations *not directly related* to the structure of the Universe,
2. changes in current physical theories, and
3. results of physical experiments.

The present paper contains elementary reviews of input links and the node itself, as well as a short discussion of the output from the node and a brief report on structure–formation related research conducted by Warsaw cosmology group.

Before we proceed to the reviews, some introductory remarks are still needed. When cosmologists talk about the structure of the Universe, in most cases they mean large structures whose dimensions are measured in megaparsecs (Mpc; 1 Mpc $= 3 \times 10^{24}$ cm). On that scale the basic building blocks of the Universe, the galaxies, do not differ much from point masses. One should remember, however, that large galaxies with masses of the order of $10^{11} - 10^{12}$ $M_\odot$ ($M_\odot$ = mass of the Sun = 2×10^{33} g) are complicated aggregates of hundreds of millions of stars and various forms of the interstellar matter, whose dimensions may approach 0.1 Mpc.

Just a brief glance at a map of the sky with positions of galaxies indicated on it is sufficient to convince oneself that the distribution of galaxies *is not uniform*: they clearly tend to form groups, in astrophysics and cosmology referred to as *clusters*. The optical impression is confirmed by a rigorous statistical analysis from which the galaxy angular correlation function $w_{gg}(\theta)$ is derived (a nonzero value of the correlation function means that galaxy pairs with an angular separation θ are more often found on the sky than they would if the galaxies were distributed randomly). For small θ $w_{gg}(\theta)$ is a well-defined power-law function, while for $\theta > 5°$ it rapidly drops to zero [1]. That is, galaxies tend to group on angular scales smaller than $\sim 5°$, while on scales larger than that their distribution becomes nearly uniform. In other words, the Universe is isotropic in the sense that sufficiently large areas of the sky are statistically undistinguishable. This conclusion is strengthened by observed motions of galaxies which in all directions seem to escape from us with velocities proportional to their distances from the observer, according to the law first formulated by Hubble [2]. The isotropy of the Universe becomes even more obvious when the *cosmic microwave background radiation* (CMBR) is taken into account – a feeble, almost exactly uniform glow of the whole sky in the microwave domain of the electromagnetic spectrum.

The main task of cosmology is to explain these basic observational facts and to provide a coherent scenario for the formation of galaxies and higher order structures. The basic theoretical framework for this task is the equations of General Relativity, whose solutions of cosmological significance are commonly referred to as *cosmological models* or *universes* (to avoid confusion between a universe and the real Universe the latter is usually written with the capital U).

2. The model of the Universe

The contemporary standard cosmology is based on observed local isotropy of the Universe and on two arbitrary assumptions. According to the first assumption (the *Copernican Principle*), we do not occupy any special position in the Universe; i.e. what we observe locally is representative of the Universe as a whole. According to the second assumption the laws of physics formulated on the Earth have been valid and will remain valid always and everywhere in the Universe. Local isotropy and Copernican Principle taken together imply that the Universe is homogeneous and isotropic everywhere (note that the Copernican Principle alone allows for inhomogeneous or nonisotropic universes, e.q. with fractal structures on all scales or nonisotropic radial motions of galaxies). Since the observed isotropy of the Universe is only approximate (Sect. 1), the implied homogeneity of the Universe is also approximate in the sense that it can be observed on sufficiently large scales only (Sect. 3).

Let us neglect deviations from isotropy and homogeneity for a while. Then, from the point of view of General Relativity, the Universe becomes a four-dimensional differentiable manifold whose constant-time three-dimensional hypersurfaces have a uniform spatial curvature. The most general metric of a hypersurface conforming to that symmetry constraint is the Friedman-Robertson-Walker (FRW) metric

$$ds_3^2 = a^2(t) \left[\frac{dr^2}{1 - kr^2} + r^2(d\theta^2 + \sin^2\theta d\phi^2) \right], \qquad (1)$$

where $a(t)$, the *scale factor*, is a function of the time coordinate t only, and k, the *curvature parameter*, is a real number [3]. The function $a(t)$ is determined by the material content of the Universe through Einstein equations

$$R_{\mu\nu} - \frac{1}{2}g_{\mu\nu}R = \frac{8\pi G}{c^4}T_{\mu\nu} + \Lambda g_{\mu\nu}, \tag{2}$$

where the left-hand side containing the metric tensor $g_{\mu\nu}$ and its derivatives describes the evolving geometry of spacetime, the right-hand side containing the energy-momentum tensor $T_{\mu\nu}$ and the cosmological term $\Lambda g_{\mu\nu}$ describes the sources of gravitation, G is the gravitational constant and c is the speed of light. Λ, the *cosmological constant* originally introduced by Einstein, is a real nonnegative number, with a physical dimension of cm^{-2}.

If the content of the Universe is a fluid with energy density e and pressure p then with the help of (1) we can reduce the Einstein equations (2) to Friedman equations

$$\left(\frac{\dot{a}}{a}\right)^2 = \frac{8\pi G}{3c^2}e - \frac{kc^2}{a^2} + \frac{\Lambda c^2}{3} \tag{3}$$

$$\frac{\ddot{a}}{a} = -\frac{4\pi G}{3c^2}e + 3p + \frac{\Lambda c^2}{3} \tag{4}$$

From (4) it is evident that Λ represents a *universal repulsion*. Physically, the cosmological term is the effective contribution to the energy-momentum tensor from the vacuum state.

Friedman equations can be solved provided that the relation between e and p (the *equation of state*) is specified. For a realistic equation of state with $p > 0$ the solutions to equations (3) and (4) (*FRW universes*) have the following properties:

- Only under very special circumstances can a FRW universe be stationary. Without arbitrary, particular specification of Λ it has to either expand or contract in the sense that $a(t)$ is either increasing or decreasing.

- a FRW universe is both dense and hot when young.

An expanding, originally hot solution to Friedman equations, in astrophysics frequently referred to as the *Hot Big Bang (HBB) model* is the standard cosmological model of our Universe. The HBB model is widely accepted as the basic framework for contemporary cosmology because it successfully explains

- Hubble law of recession of galaxies (which can be derived from the FRW metric),

- properties of CMBR (microwave background is photons generated in early Universe which stopped interacting with matter when temperature and density became low enough, and which have been propagating freely since then), and

- abundances of light elements like ^{2}H, ^{3}He, ^{4}He and ^{7}Li (resulting from primordial nucleosynthesis in the early Universe).

In an expanding FRW universe photons emitted by distant sources are redshifted when they arrive to the observer. The redshift

$$z = \frac{\lambda_{obs}}{\lambda_{em}} - 1, \tag{5}$$

where λ_{em} and λ_{obs} are the emitted (laboratory) and observed wavelengths, is a single-valued function of a_{em}:

$$z = \frac{a_{obs}}{a_{em}} - 1, \tag{6}$$

where a_{em} and a_{obs} are values of a at the moments of emission and observation, respectively [3]. Thus, by measuring z we can tell how much the Universe has expanded since the photons we observe today left their source. Because a in turn is a single-valued function of t, we can also tell how long ago the emission took place, or, because the speed of light is a universal constant, how far away the source is located. According to the HBB theory, the oldest photons we can observe are the CMBR photons. Their redshift is ~ 1000, i.e. they were emitted when the Universe was only $1/1000$ of its present size.

When the Universe was still smaller (i.e. at epochs $z > 1000$), the pressure in it could not be neglected due to high radiation energy density and frequent interactions between photons and matter. At the present epoch the radiation energy density is negligible, photons and matter are entirely decoupled, and a good approximation to the equation of state is $p = 0$ and $e = \rho c^2$, where ρ is the rest-mass density [4]. With the help of such an approximation we can reduce Friedman equations to a single equation

$$\left(\frac{1}{H_0} \frac{dx}{dt} \right)^2 = \Omega_0 \frac{1}{x} + \lambda_0 (x^2 - 1) + 1 - \Omega_0, \tag{7}$$

in which x is the scale factor a normalized to its present value a_0. The three parameters H_0, Ω_0 and λ_0 are, respectively, the present value of the logarithmic derivative $a^{-1}da/dt$, the present value of the ratio of the mean density of the Universe to the critical density

$$\rho_{cr} = \frac{3H_0^2}{8\pi G}, \tag{8}$$

and the cosmological constant expressed in units $3H_0^2/c^2$ (is should be noted that in general both H and Ω are functions of t).

In principle, all three parameters are measurable, and, once their values are found, the equation (7) can be solved. However, the present observational limits on H_0, Ω_0 and λ_0 are weak: all we know is that $40 \le H_0 \le 100$ km/s/Mpc, $0.007 \le \Omega_0 \le 1$ and $\lambda_0 \le 0.7$ [5]. The basic cause of this situation is the still too small reach of cosmological observations. To put things bluntly, neither the past nor the future of the Universe as a whole can presently be specified with a satisfying accuracy (Fig. 1).

411

3. Basic observations relevant to structure formation

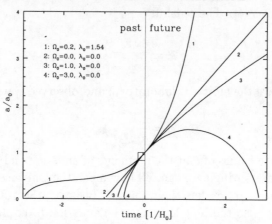

Fig. 1. Evolution of the dimensionless scale factor a/a_0 for several values of Ω_0 and λ_0. Small square: the present observational limit for the 3-D structure of the Universe. Big square: the expected limit of 3-D data from the Sloan Digital Sky Survey [32].

If we neglect the Doppler effect due to our motion with respect to the distant parts of the Universe [6] we may say that the flux of CMBR photons is truly remarkably isotropic. A convenient measure of that flux, the *radiation temperature* T_r, deviates from its all-sky mean value of 2.726 ± 0.010 K by no more than one part in 10^5 for angular scales $> 7°$ (corresponding to ~ 100 Mpc in the present evolutionary epoch of the Universe), with somewhat higher upper limits for smaller scales [7]. The fluctuations of T_r tell us that at $z = 1000$ the Universe was highly homogeneous, i.e. nearly structureless. Very unfortunately for the theory of structure formation there is a large gap in observational data between CMBR photons emitted at $z = 1000$ and the next oldest photons we are able to receive, which were emitted at $z \sim 5$. At that epoch there were already well defined, separate objects in the Universe (quasars, [8]).

For progressively smaller z normal galaxies begin to appear, and more and more complicated structures composed of galaxies can be identified (Fig. 2). For $z < 0.05$ at the large-scale end of the structural hierarchy *voids* (regions nearly devoid of luminous matter) and *walls* (elongated aggregations of galaxies) are observed, both extending for up to 100 Mpc. According to the present observational data *the scale set by the dimensions of voids and walls is the smallest scale at which the mature Universe can be regarded as homogeneous.* At smaller scales the matter is distributed nonuniformly, forming superclusters and clusters of galaxies. Superclusters and clusters have typical dimensions of up to several tens of Mpc and several Mpc, respectively. Within them the number density of galaxies, n_g (i.e. the number of galaxies per Mpc³) is markedly larger than the mean n_g on the 100 Mpc scale [9]. When the uniform Hubble expansion is subtracted from the observed motions of galaxies it turns out that both galaxies and higher-order structures move with respect to each other at residual velocities reaching several hundreds km/s [10]. Apart from galaxies, intergalactic matter can be observed either directly (hot gas in clusters of galaxies, [11]) or indirectly (clouds that absorb

412

photons emitted by quasars, [12]). Starting with initial conditions specified by both theoretical considerations and observations discussed in Sect. 4, a successful theory of structure formation must reproduce all this wealth of objects and motions at the right evolutionary epoch.

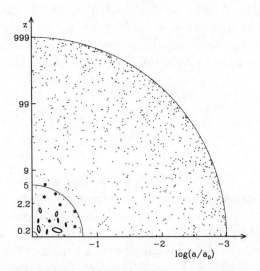

Fig. 2. Schematic view of the Universe. The observer is situated at the origin of the coordinate system. Dots: CMBR photons, stars: quasars, ellipses: galaxies (drawn not to scale), small circle at $z = 0.2$: the expected limit of 3-D data from the Sloan Digital Sky Survey [32]).

As between $z = 1000$ and the present epoch pressure effects have not been important (at least on scales much larger than the typical size of a galaxy), it is generally assumed that in that time the structure of the Universe evolved solely due to gravitational amplification of small deviations from homogeneity and isotropy. That is, the currently accepted physical mechanism of structure formation is based on *gravitational instability (GI)* naturally promoting the increase or decrease of density (relative to its global mean value) in initially overdense or underdense regions, respectively [13]. Hereafter, the deviations from homogeneity and isotropy at the evolutionary epoch corresponding to $z = 1000$ will be referred to as *initial perturbations*.

4. Standard equations of structure formation

According to the Correspondence Principle, General relativity converges to Newtonian mechanics when c approaches infinity. In that limit, the geodesic equation corresponds to the equation of motion in a Newtonian potential, and the Einstein equations (2) - to the Poisson equation for the potential [14]. In particular, for a static Minkowski spacetime perturbed with weak gravitational fields it is possible to expand the general-relativistic corrections to Newtonian mechanics into a series containing powers of Φ/c^2 and v/c, where Φ and v are Newtonian potential and velocity, respectively [14].

A similar procedure can be performed for a FRW universe in which perturbations in the above explained sense are introduced [15]. In this case, however, the zeroth order approximation to General Relativity is not Newtonian, but a pseudo-Newtonian mechanics, in which extra forces due to the global evolution of spacetime explicitly appear, and the source term in the Poisson equation is multiplied by momentary values of $H(t)$ and $\Omega(t)$ obtained from the unperturbed model. The lowest-order corrections to that mechanics are of order of Φ/c^2 and v^2/c^2, while the validity of the whole procedure is restricted to scales $L \ll c/H$ ($\approx 3-6$ thousands of Mpc at the present evolutionary epoch of the Universe). In the simplest, pressureless case the zeroth order GI equations governing the evolution of the perturbations take the form

$$\dot{\delta} + \nabla \vec{v} + \nabla(\vec{v}\,\delta) \;=\; 0 \tag{9}$$

$$\dot{\vec{v}} + 2H\,\vec{v} + (\vec{v}\,\nabla)\,\vec{v} \;=\; -\nabla\Phi_p \tag{10}$$

$$\nabla^2\Phi_p \;=\; \frac{3}{2}H^2\Omega\delta, \tag{11}$$

where $\delta = (\rho - \bar{\rho})/\bar{\rho}$, $\vec{x} = \vec{r}/a(t)$, $\vec{v} = (\vec{u} - \dot{a}\,\vec{x})/a$ and $\vec{u}$ are local relative density deviation from the global mean value, *comoving coordinates*, *peculiar velocity* and Newtonian velocity, respectively (note that from the equation (11) it is evident that the *peculiar potential* Φ_p differs from the ordinary Newtonian one).

The GI equations should be integrated over a period of time corresponding to the difference between $z = 1000$ and $z = 0$. However, before one attempts at integrating them the following questions should be answered:

- How (in)accurate is the pseudo-Newtonian, zeroth order approach?
- How large is the most appropriate domain to integrate the equations (9)-(11)?
- Is the backward integration (from relatively well known conditions at $z = 0$ to $z = 1000$) possible?
- If the backward integration fails – what are the initial and boundary conditions for the forward integration from $z = 1000$ to $z = 0$?

Since the recently measured fluctuations of T_r are directly related to potential fluctuations [7], we may be sure that at $z = 1000$ on scales defined in Sect. 3 the correction term Φ/c^2 was smaller than 10^{-5}. Today, the most compact objects among those considered in the present review, the galaxies, have $\Phi/c^2 < 10^{-6}$, while their maximal peculiar velocities of $\sim 10^3$ km/s result in $v^2/c^2 < 10^{-5}$. As for the most appropriate domain, it is clear that it must be large enough to accomodate the largest structures observed, and small enough to fulfill the $L \ll c/H_0$ requirement. A domain with the present size of $L \sim 100 - 200$ Mpc seems to be a reasonable compromise.

To answer the third question, let us obtain solutions to GI equations in the linear regime $\delta \ll 1$, in which they reduce to

$$\ddot{\delta} + 2H\dot{\delta} = \frac{3}{2}H^2\Omega\delta. \tag{12}$$

It may easily be found [13] that the general solution to (12) is composed not only of *growing*

$$\delta(\vec{x}, t) = D(t)\delta_i(\vec{x}) \tag{13}$$

414

(where $D(t) \longrightarrow \infty$ with $t \longrightarrow \infty$, and $\delta_i(\vec{x})$ is the initial spatial distribution of density perturbations), but also of *decaying* modes

$$\delta(\vec{x}, t) = S(t)\delta_i(\vec{x}) \tag{14}$$

(where $S(t) \longrightarrow 0$ with $t \longrightarrow \infty$). Due to the presence of the latter, any numerical noise would be rapidly amplified during the backward integration, rendering the recovery of δ_i for $z = 1000$ practically impossible.

Being left with the forward integration as the only viable opportunity, we still have to specify the boundary conditions at the borders of the integration domain. In most calculations that have been done so far the domain is a cube with periodic boundary conditions (which require the density and velocity distributions on opposite walls of the cube to be identical). It is obvious that such an approach results in wrong description of gravitational interactions on scales approaching the size of the domain, as they are severely influenced by "mirror images" of the cube. Nevertheless, the periodic boundary condition (in which the Universe endlessly repeats itself on scale L) seems to be much better than its only viable alternative, the vacuum boundary condition (in which the rest of the Universe is simply ignored).

To answer the fourth question we will have to discuss both the assumptions concerning the origin of perturbations at $z \gg 1000$, and the physical processes responsible for their evolution until the beginning of the pressureless epoch at $z = 1000$.

5. Primordial perturbations and the material content of the Universe

Before we begin the integration of equations (9)- (11) we must specify both the *initial density field* δ_i and the *initial velocity field* $\vec{v}_i$ at the epoch corresponding to $z = 1000$ (these fields should not be confused with analogous *primordial* fields δ_p and $\vec{v}_p$ at $z \gg 1000$). In practice, for calculations based on periodic boundary conditions it is more convenient to specify the Fourier transform

$$\hat{\delta}_i(\vec{k}) = \frac{1}{V} \int d^3x \delta_i(\vec{x}) \exp(i \, \vec{k} \vec{x}), \tag{15}$$

where V is the volume of the integration domain (it may easily be shown that $\vec{v}_i$ can be obtained from $\hat{\delta}_i$ provided that the amplitude of density perturbations is small [4]. It is generally assumed [4] that $\hat{\delta}_i$ can be represented as a product

$$\hat{\delta}_i(\vec{k}) = T_i(k)\hat{\delta}_p(\vec{k}), \tag{16}$$

where $\hat{\delta}_p$ is the Fourier transform of the primordial density perturbations $\delta_p(\vec{x})$ at $z \gg 1000$, and T_i (a function of the length k of the wave vector only) is the *transfer function* that describes the linear evolution of Fourier modes between $z \gg 1000$ and $z = 1000$.

It is further assumed that $\hat{\delta}_p$ originates from stochastic processes [16], which means that the Universe should be regarded as one of the many possible universes drawn

415

from an ensemble with a distribution specified by the underlying theory of primordial perturbations. Therefore, the question: *Is the model universe we obtain by integrating the GI equations up to $z = 0$ the same as the Universe we inhabit?* is a wrong question. The right question to ask is: *Are the model universe and our Universe members of the same ensemble?* As a consequence, our ability to discriminate between various hypotheses and assumptions underlying the scenario of structure formation is significantly reduced.

The simplest recipe for $\hat{\delta}_p$ is again based on assumptions which cannot be rigorously justified. It is proposed that in the early Universe

- there was no distinguished length scale (or, in Fourier space, no distinguished wavelength), and
- Fourier phases of different modes were entirely uncorrelated.

The only function satisfying the first requirement is a power-law function, traditionally written as

$$|\hat{\delta}_p(\vec{k})|^2 = Ak^n \tag{17}$$

(where A is a normalizing constant), and commonly referred to as the *power spectrum* $P(k)$. The second assumption and the Central Limit Theorem imply that the the primordial perturbations were Gaussian, i.e. that the ensemble of $\hat{\delta}_p(\vec{x})$ values had a Gaussian distribution [4].

According to the most widely accepted hypothesis, the primordial perturbations were generated at the end of the inflationary phase. Indeed, various inflationary scenarios easily produce Gaussian perturbations with a linear power spectrum ($n = 1$ or $P(k) = Ak$; [17],[18]. Independently of inflation, some physical arguments also point toward $n \sim 1$. It may be shown [13] that the rms potential fluctuation in an ensemble of identical spheres

$$R_{sph}(t) = a(t)x_{sph} = \left(\frac{M}{\frac{4}{3}\pi \, \bar{\rho}\,(t)}\right)^{1/3} \tag{18}$$

(where M is the mass of the sphere) is given by

$$\delta\phi_M = D(t)M^{\frac{1-n}{6}}, \tag{19}$$

with $D(t)$ defined by equation (13). From equation (19) it follows that for $n < 1$ large masses collapse first, producing too much structure on large scales. On the other hand, for $n > 1$ the power spectrum diverges on small scales, causing strong perturbations of the metric up to the formation of black holes [19]. Thus, the $n = 1$ case (known in cosmology as the *Harrison-Zeldovich spectrum* or *constant curvature scaling*) seems to be a rather reasonable choice. The post-inflationary perturbations are *adiabatic*, with a well-defined dependence between fluctuations of density and temperature. However, *isothermal* or *isocurvature* perturbations with baryon density fluctuations proceeding at essentially unperturbed temperature are also conceivable [20].

416

Since in the early Universe the condition $L \ll c/H$ is not fulfilled even for small wavelengths (let us recall that H is a decreasing function of t), it is clear that general-relativistic effects must be fully taken into account in order to determine the filter function $T_i(k)$. The mathematical procedure is simplified owing to the minute amplitudes of $\delta_p(\vec{x})$, but at the same time it is complicated by a large number of physical processes involved [21]. In particular, there are very good reasons to believe that the major part of the material content of the Universe is not the ordinary matter we deal with in the everyday life [22]. The latter is composed of well-known elementary particles (baryons and leptons) which take part in various interactions involving emission or absorption of photons, and as such it is relatively easily observable. However, provided that $\Omega_0 = 1$, its mean density cannot be greater that $\sim 20\%$ (and may be as low as $\sim 1\%$) of the total mean density of the Universe [23]. The remaining 90-99% is the so called *dark matter*, most probably composed of extremely weakly interacting particles, whose presence can only be detected due to the gravitational link they maintain with the baryonic component.

It is suggested that two basic forms of dark matter may be present in the Universe: the *cold dark matter (CDM)* and the *hot dark matter (HDM)*. The most likely CDM and HDM candidates are massive supersymmetric particles with low peculiar velocities, and low-mass neutrinos with high peculiar velocities, respectively [23]. As indicated in Sect. 2, at the present evolutionary epoch of the Universe the radiation energy density e_γ is much lower than the rest-mass energy density ρc^2. However, because e_γ has been declining with t faster than ρ, in the early Universe there was an epoch z_{eq} in which both densities were equal (most FRW models conforming to present observational constraints predict $z_{eq} \gg 1000$ [21]). For all $z > z_{eq}$ the principal source of gravitational potential was photons, and all gravitational interactions in the Universe were dominated by them. As a result, density perturbations in CDM and HDM were evolving in exactly the same way. In particular, in both CDM and HDM dissipative and diffusive processes caused the amplitude of small-scale perturbations to grow at an effectively lower rate than the amplitude of large-scale perturbations [21], causing $P(k)$ to drop below the original power-law at large k (i.e. at short wavelengths).

Because the dark matter interacts so weakly, it turned effectively pressureless already at $z > z_{eq}$, i.e. much earlier than the ordinary baryonic matter. At $z = z_{eq}$ it became the principal source of the potential, and the differences between CDM and HDM begun to matter. While peculiar velocities of CDM particles were small compared to c, those of HDM particles remained close to c. As a result, for $z < z_{eq}$ HDM perturbations were experiencing much stronger diffusion, due to which before $z = 1000$ short wavelengths were being removed more efficiently from the HDM power spectrum than from the CDM power spectrum [21]. To conclude our qualitative discussion we may say that

- at $z > z_{eq}$ short wavelengths of the primordial spectrum were damped at the same rate in both HDM and CDM components, while
- at $z_{eq} > z > 1000$ the effective damping was stronger in the HDM component.

The final conclusion is that the filter function $T_i(k)$ must not significantly differ from unity for small k (i.e. for long wavelengths, for which the primordial spectrum is hardly modified before $z = 1000$), and it must approach zero for large k (faster in HDM models; slower in CDM models). Therefore, $P_i(k)$ at $z = 1000$ cannot be a pure power-law function, but it has to have a maximum. Several initial power spectra based on various hypotheses concerning the material content of the Universe are shown in Fig. 3. All models except BDM are based on adiabatic perturbations, while the BDM model results from the isocurvature scenario. In an $\Omega_0 = 1$ universe the maxima of CDM and HDM spectra correspond to masses of $\sim 10^{15}\ M_\odot$.

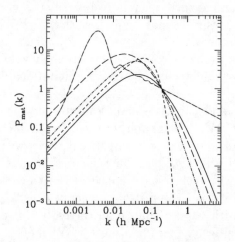

Fig. 3. The initial power spectra $P_i(k)$ at $z = 1000$ for several cosmological models with various combinations of baryonic matter, CDM and HDM, and with various physical interactions taken into account (adapted from [34]). The models are: CDM *(solid)*; HDM *(short dashed)*; MDM, i.e. a mixture of 70% CDM and 30% HDM *(dot-short-dashed)* (all three with $\Omega_0 = 1$); ΛCDM, i.e. CDM with $\Omega_0 = 0.2$ and $\lambda_0 = 0.8$ *(long-dashed)*; and BDM, i.e. pure baryonic with $\Omega_0 = 0.1$ and $\lambda_0 = 0$ *(dot-long-dashed)*.

The initial spectrum $P_i(k)$ has to be properly normalized before it is employed to generate the initial density and velocity fields (the normalization of the spectra in Fig. 3 is arbitrary). Unfortunately, the primordial amplitudes at $z \gg 1000$ are not known. The amplitudes at $z = 1000$ are constrained at long wavelengths (i.e. at small k) by measurements of CMBR fluctuations, and at short wavelengths by statistical analyses of the spatial distribution of galaxies. However, the error margin is still large, and at the present moment only an order-of-magnitude estimate of the normalizing factor can be made.

6. Cosmological simulations

Hitherto, most cosmological simulations have followed the evolution of the dark matter only. Instead of a *continuous medium* approach a *particle* approach has been almost exclusively applied, with the material content of the Universe represented by a set of point mass elements, and with equations (9)-(11) substituted by the set of equations describing motions of N pseudo-Newtonian particles

$$\frac{d\,\vec{x}_i}{dt} = \vec{v}_i$$

$$\frac{d\,\vec{v}_i}{dt} = -2H\,\vec{v}_i - \vec{F}_i \qquad (20)$$

$$i = 1, ..., N$$

where $\vec{x}_i$ and $\vec{v}_i$ are comoving coordinates and comoving velocity of the i-th particle, respectively, and $\vec{F}_i$ is the total gravitational force acting on i−th particle. The set (20) is integrated with the help of *particle computer codes* which calculate $\vec{F}_i$ either by solving equation (11) with a source term obtained by transforming the discrete distribution of mass elements into a continuous density distribution, or by a simple summation of appropriately scaled Newtonian forces, or by performing multipole expansion of the discrete distribution, or, finally, by various combinations of those methods [24],[25]. Both CDM and HDM are described this way, differing only in magnitude of peculiar velocities given to CDM and HDM particles at the beginning of simulations. In the largest current simulations the number of particles reaches tens of millions [26].

Some versions of particle codes (those involving SPH techniques; [27]) are able to simulate pressure effects, i.e. on relatively small scales (on which the latter become important for small z) parallel to the dark matter they are able to follow the baryonic component of the Universe. Recently, computer codes combining particle approach and fluid-dynamical approach based on generalized equations (9)-(11) begin to appear, which treat the baryonic matter as a continuous medium with nonvanishing pressure, account for its radiative cooling and/or heating, and, under suitable physical conditions, transform it into point masses representing single stars [28]. The progress in numerical techniques related to cosmology is indisputable, and both the accuracy and the capacity of cosmological codes grow on a time scale of about 2 years. At the present moment our Universe seems to be best matched by simulations based on initial spectra obtained from MDM, ΛCDM and BDM models ([29]; see the Fig. 3 for the explanation of abbreviations). Unfortunately, because of high nonlinearity of the problem the sensitivity of results to initial conditions is rather weak (especially on the smallest scales with the largest density fluctuations), which means that other models cannot be ruled out. In other words, the results of cosmological simulations in the *highly nonlinear* regime (for which we have the best observational data) are still far from being conclusive, Even worse than that: as one of the leading cosmologists wrote, "it may easily be the case that none of the theories presently being given serious consideration is correct" [30].

The problem of matching a cosmological model to the real Universe becomes even more complicated if we note that in universes with dark matter the spatial distributions of luminous baryons and nonluminous dark-matter particles may be different [31]. While some light may be soon shed on that problem owing to more accurate fluid-dynamical simulations, the most obvious way out of all those dilemmas is to extend the reach of cosmological observations. A radical increase of the amount of data concerning the 3-dimensional structure and dynamics of the Universe will soon be brought by the Sloan Digital Sky Survey [32] - a project aimed at mapping the spatial distribution and motions of galaxies up to $z \approx 0.2$ (i.e. up to scales of several hundred Mpc). As we already know, on scales that large the Universe is nearly uniform, i.e. the rms density fluctuations are significantly smaller than 1. From the theoretical point of view it means that the corresponding long-wave Fourier modes have been evolving almost independently, so that their evolution may be precisely described by an analytical theory based on the expansion of GI equations into a power series of δ up to a desired accuracy [4]. At the same time, in that *weakly nonlinear* regime the results of numerical simulations should be much more conclusive in the sense that their sensitivity to initial conditions should be much higher. Moreover, a detailed comparison of analytical and numerical predictions should allow for a successful debugging of both methods, thus lending a high credibility to the results.

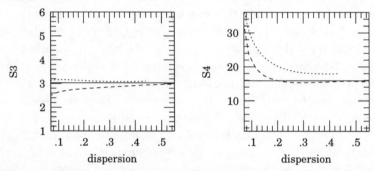

Fig. 4. Skewness and kurtosis of the cosmic density field plotted as a function of its dispersion for an $n = 1$ initial power spectrum [35] compared to the analytical predictions [36],[37]. Dotted: fluid-dynamical code; dashed: particle code; solid: analytical.

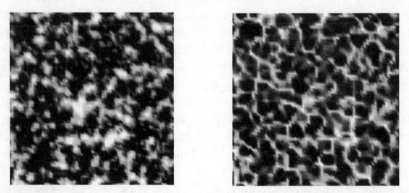

Fig. 5. Cosmic density field for $n = 1$ at $z = 2$. Simulation with a particle code (left) and with a fluid–dynamical code (right) [35].

This is precisely the way the Warsaw cosmology group intends to go. In addition to an extensive analytical research of the weakly nonlinear regime that has been performed

420

in Warsaw since early 80's [33], both particle and fluid-dynamical simulations have been started recently, focused primarily on the weakly nonlinear, large-scale regime. Preliminary results of those simulations are displayed in Figs. 4 and 5, in which the analytically predicted moments of δ distribution are compared to those obtained from fluid-dynamical, zero-pressure simulations, and an example of large-scale structures obtained from the same initial conditions with the help of a particle-code and a fluid-dynamical code is shown. Further numerical research will include baryonic matter with nonzero pressure with the aim to follow the evolution of the intergalactic medium and to estimate the differences in dark and baryonic matter distributions.

Acknowledgments. This work was supported by the Polish Committee for Scientifc Research through the grant 2P 304 017 07. The simulations whose preliminary results were reported in Sect. 6 are partly performed on Cray computers at the Interdisciplinary Centre for Mathematical and Computational Modeling in Warsaw, Poland.

REFERENCES

[1] Frieman, J.A. 1994, The Standard Cosmology (Fermilab-Conf-94/090-A;
    SISSA preprint server http://babbage.sissa.it/abs/astro-ph/9404040

[2] Peebles, P.J.E. 1993, Principles of Physical Cosmology (Princeton: Princeton Univ. Press)

[3] Ross, M. 1994, Introduction to Cosmology (Chichester: John Wiley & Sons)

[4] Strauss, M.A., and Willick, J.A. 1995, Physics Reports D 261, 271

[5] Primack, J.R., Status of Cosmological Parameters, in proceedings of the workshop on Particle and
    Nuclear Physiscs in the Next Millenium held in Snowmass, Colorado (1994), eds.
    E.W. Kolb & R. Peccei (World Scientific)

[6] Smoot, G.F. et al. 1992, Astrophys. J. Letters 396, L1

[7] Scott, D., Silk, J., and White, M. 1995, Science 268, 829

[8] Hartwick, F.D.A, and Schade, D. 1990, Ann. Rev. Astr. Astrophys. 28, 437

[9] Bahcall, N. 1988, Ann. Rev. Astr. Astrophys. 26, 631

[10] Dekel, A. and Bertschinger, E. 1991, Large-Scale Structure and Peculiar Motions in the Universe
    (Astron. Soc. Pacific Conference Series, Vol. 15)

[11] Bahcall, N. 1995, Large-Scale Structure of the Universe (World Scientific)

[12] Sargent, W.L.W., Young, P.J., Boksenberg, A., and Tytler, D. 1990, Astrophys. J. Suppl. 42, 41

[13] White, S.D.M. 1994, MPA preprint No. 831 (Formation and Evolution of Galaxies,
    1993 Les Houches Summer School Lecture)

[14] Will, C.M. 1993, Theory and Experiment in Gravitational Physics, (Cambridge: Cambridge University
    Press)

[15] Bertschinger, E. 1995, Cosmological Dynamics, Elsevier Sci. Publ.
    (1993 Les Houches Summer School Lecture)

[16] Bardeen, J.M., Bond, J.R. and Efstathiou, G. 1987, Astrophys. J. 321, 28

[17] Kolb, E.W. and Turner, M.S. 1990, The Early Universe (Redwood City: Addison–Wesley publ.)

[18] White, M., Scott, D., Silk, J. and Davis, M. 1995, Mon. Not. Royal Astron. Soc., submitted

[19] Peebles, P.J.E., and Yu, J.T. 1970, Astrophys. J. 162, 815

[20] Peebles, P.J.E. 1987, Nature 327, 210

[21] Padmabhanan, T. 1993, Structure Formation in the Universe (Cambridge: Cambridge Univ. Press)

[22] Einasto, J. 1990, Dark Matter in the Universe, NATO ASI Series (Dordrecht: Kluwer)

[23] Schramm, D.N. 1990, Dark Matter in the Universe, NATO ASI Series (Dordrecht: Kluwer)

[24] Bryan, G.L., Cen, R., Norman, M.L. Ostriker, J.P., and Stone, J.M. 1994, Astrophys. J. 241, 521

[25] Navarro, J.F., Frenk, C.S. and White, S.D.M. 1995, Mon. Not. Royal Astron. Soc., in press

[26] Norman, M.L. 1995, Grand Challenge Cosmology Consortium WWW server
http://zeus.ncsa.uiuc.edu:8080/GC3_Home_Page.html

[27] Hernquist, L. and Katz, N. 1989, Astrophys. J. Suppl. 70, 419

[28] Navarro, J.F. and White, S.D.M. 1993, Mon. Not. Royal Astron. Soc. 265, 271

[29] Strauss, M.A., Cen, R.Y., Ostriker, J.P., Lauer, T.R. and Postman, M. 1995, Astrophys. J. in press

[30] Ostriker, J.P. 1993, Ann. Rev. Astron. Astroph. 31, 689

[31] Frenk, C.S., Evrard, A.E., White, S.D.M. and Summers, F.J. 1995, MPA preprint No 870
(Astrophys. J., submitted)

[32] Bahcall, N. 1995, Publ. Astron. Soc. Pacific, in press

[33] Juszkiewicz, R. 1981, Mon. Not. Roy. Astr. Soc. 197, 93

[34] White, M., Scott, D., Silk, J. 1994, Ann. Rev. Astron. Astroph. 32, 319

[35] Kudlicki, A. 1995, in preparation

[36] Juszkiewicz, R., Bouchet, F., Colombi, S. 1993, Ap. J. 412, L9

[37] Łokas, E.L., Juszkiewicz, R., Weinberg, D.H., Bouchet, F.R. 1994, CAMK Preprint 281
http://www.camk.edu.pl/localinfo/index.html

Michał Różyczka[1,2]
Tomasz Plewa[3]
Andrzej Kudlicki[2]

[1] *Warsaw University Observatory, Al. Ujazdowskie 4, 00-478 Warszawa, Poland*
[2] *Nicolaus Copernicus Astronomical Center, Bartycka 18, 00-716 Warszawa, Poland*
[3] *Max-Planck-Institut für Astrophysik, Karl-Schwarzschild-Str. 1, 85740 Garching, Germany*

E WIMMER

Challenges for computational materials design

Abstract

With the development of advanced atomistic simulation methods and the unprecedented speed and availability of computer hardware, it seems that the design of novel materials by a purely computational approach has come within reach. However, a closer analysis reveals major unsolved problems, in particular the connection between the atomistic and the macroscopic scales bridging about 10 orders of magnitude in the length scale and about 20 orders of magnitude in the time scale. Furthermore, there is an urgent need for higher accuracy while maintaining computational efficiency to account for the subtle energy differences which govern so many materials properties. The present contribution reviews some of the key theoretical approaches in computational materials science with an emphasis on quantum mechanical methods. Illustrative examples include the calculation of the thermochemical stability of fluorinated compounds, the swelling of graphite upon Li intercalation, the prediction of oxygen diffusion in polymers for contact lenses, the adsorption of Ag atoms on a $MgO(001)$ surface, the control of the crystal morphology of chromia, the interpretation of the optical characteristics of ruby, and the prediction of the magnetic properties in artificially layered structures.

1. Introduction

The richness and variety of materials find their fundamental explanation in the fascinating interplay between the motions of the nuclei and the arrangement of the electrons. Quantum mechanics as formulated in the 1920's captures this complexity in an elegant mathematical framework. However, for many decades the solution of the fundamental equation of quantum mechanics, namely Schrödinger's equation, seemed far too complicated for any but the simplest systems such as the hydrogen atom. It is most remarkable that present theoretical and computational approaches together with the unprecedented capabilities of computer hardware have made it possible to solve the quantum mechanical equations with sufficient accuracy to allow quantitative predictions of materials properties such as molecular and crystallographic structures, binding energies, surface energies, elastic constants, and magnetic moments for fairly complex solids.

Given this remarkable capability, one might think that computational materials design has become a reality. While this is true to some extent - the design of magnetic multilayer structure being a good example - the properties of real materials depend in many cases on such subtle effects that it is extremely difficult or simply impossible with today's tools to achieve the necessary accuracy. For example, the energy difference between two crystal structures may be only 0.05 eV per unit cell while the total energy could be of the order of 100,000 eV. Besides the issue of accuracy, macroscopic materials properties such as ductility or fracture toughness depend on complicated mesoscopic phenomena involving, for

423

example, defect structures and grain boundaries which are difficult to characterize and to describe. Moreover, materials properties such as fatigue involve time-dependent phenomena spanning many orders of magnitude. Thus, bridging the gap between the atomistic and macroscopic scales represents perhaps the most fundamental challenge of computational materials science.

The length scale of atomistic phenomena is the nano-meter range (1 nm = 10^{-9} m = 10 Å). For example, most interatomic bond distances fall between 0.1 - 0.4 nm. The smallest time intervals relevant to atomistic processes are in the femto-second range. For example, a full cycle in the thermal vibration of a C-H bond in an organic compound takes about 10 femto-seconds whereas one might worry about the environmental stability of a material over hundreds or even thousands of years. Thus, about 10 orders of magnitude in the length scale and about 20 orders of magnitude in the time scale separate the atomistic domain from the macroscopic world.

The focus of this contribution is on the atomistic scale. To this end, the following section is devoted to an outline of the theoretical concepts and algorithmic approaches. By discussing specific examples from a range of different applications, the current capabilities and limitations are illustrated. In the concluding section some trends in computational materials science will be presented thus allowing projections of future possibilities.

2. Theoretical Concepts

All of chemistry and most of the physical properties of matter can be explained by the laws of quantum mechanics, statistical mechanics, and classical continuum physics. The fundamental quantity which drives any chemical transformation and which determines the thermochemical behavior of any molecule, liquid, or solid is the energy of the system as a function of the position of the nuclei. Excluding nuclear processes, the energy of any arrangement of atoms is due to the mass and motion of the nuclei, the kinetic energy of the electrons, and all electrostatic interactions between all the electrons and nuclei. Atomic nuclei are thousands of times heavier than an electron and for many chemical questions can thus be treated as classical particles following Newtonian mechanics. Because of their low mass, electrons follow the nuclei quasi instantaneously. At each moment in time, each nucleus has a kinetic energy

$$E_{kin} = \frac{1}{2} mv^2 \tag{1}$$

and a potential energy, E, which is due to the interactions between all atoms and possibly with external fields and forces.

$$E = E[R_1, R_2, ... R_n] + E_{ext} \tag{2}$$

Here, $R_1, R_2, ...$ denote the Cartesian coordinates of all atoms of the system under consideration and E_{ext} is the potential energy due to external influences such as electric fields. The derivative of the potential energy with respect to the displacements of an atom α along its Cartesian coordinates gives the force acting on this atom.

424

$$F_\alpha = -\left(\frac{\partial}{\partial x_\alpha}, \frac{\partial}{\partial y_\alpha}, \frac{\partial}{\partial z_\alpha}\right) E_{pot} \qquad (3)$$

with $E_{pot} = E$ from eq. (2) and $R_\alpha = (x_\alpha, y_\alpha, z_\alpha)$ being the Cartesian coordinates of atom α. If one knows the force on each atom, one can calculate the acceleration and hence solve the Newtonian equations of motion for any system of atoms. This is the foundation of molecular dynamics, which allows - at least in principle - to simulate, for example, the structural changes of the molecules in a polymer, the binding of molecules to a surface, but also the evolution of a chemical reaction such as a catalytic process.

The fundamental problem of computational molecular and materials science is the accurate and efficient calculation of the interatomic interactions. Three types of approaches are used today, empirical potentials (force fields), semi-empirical quantum mechanical methods and first-principles (or so-called ab initio) quantum mechanical methods.

2.1 Empirical potentials

One of the simplest forms to describe the total energy of an ensemble of atoms is the decomposition into pair-wise interactions as described, for example, by the so-called Lennard-Jones potential [1]

$$E = \sum_{i>j} V_{ij}, \qquad V_{ij} = \frac{A_{ij}}{r_{ij}^{12}} - \frac{C_{ij}}{r_{ij}^6} \qquad (4)$$

The parameters A and C, which describe the repulsive and attractive interactions between atoms i and j, have an empirical character. They can be chosen to fit experimental data such as the lattice spacing, the cohesive energy, and the bulk modulus of a system. This simple approximation to interatomic interactions is meaningful if all angular terms are averaged as, for example, in a gas or liquid of closed-shell atoms or nearly spherical molecules. The simulation of highly energetic collisions between atoms represents another area where such a simple description of interatomic interactions may be sufficient. Also closely packed arrangements in condensed phases can be described by such an approach provided that there is no significant covalent bonding in the system.

The description of ionic crystals and amorphous materials can be accomplished if the two-body terms of a Lennard-Jones or Buckingham potential are extended by a term that describes the polarization of the ions, for example in the form of a shell model [2]. Based on such models or similar approaches, zeolites [3] and some metal oxides [4] can be simulated if the potential parameters are properly adjusted. Typically, the parameters are fitted to experimental data. This limits the applications to the class of compounds which have been used in the fitting procedure. A conceptually more satisfying approach is the use of accurate ab initio quantum mechanical calculation as a basis for the fit, preferably in conjunction with experimental data.

In fact, this approach has been chosen for the development of new generations of valence force fields to describe the conformational properties of biopolymers such as proteins and the interactions of drug molecules and their receptors in an aqueous solution [5]. Given the chemical similarity between biopolymers such as proteins and synthetic polymers such as nylon, this type of valence force fields has been successfully extended to

425

describe synthetic polymers. In these materials, each atom has a well defined chemical bonding environment which allows the assignment of certain potential parameters from a library. As long as one is interested only in the simulation of conformational changes and weak intermolecular interactions, these potential parameters are reasonably transferable and properties such as low-energy molecular conformations, molecular packing in condensed phases, or dynamic quantities such as diffusion coefficients of small molecules can be obtained.

A typical valence force field has the following functional form:

$$E = \sum_{\text{bonds}} k_b (r_b - r_b^0)^2 + \sum_{\text{bond angles}} k_a (\theta_a - \theta_a^0)^2 + \sum_{\text{dihedrals}} k_d \left[1 - \cos\left(\tau_d - \tau_d^0 \right) \right]$$

$$+ \sum_{i>j} \left(\frac{A_{ij}}{r_{ij}^9} - \frac{C_{ij}}{r_{ij}^6} \right) + \sum_{i>j} \frac{q_i q_j}{r_{ij}}$$

(5)

The first term on the right hand side of eq. (5) represent the harmonic vibrations of a bond with $(r_b - r_b^0)$ being the difference between the bond distance at a specific time and the equilibrium distance. The parameter k_b describes the harmonic force constant corresponding to the bond stretching. The bond bending involving three atoms is described by the second term, and dihedral distortions defined by four atoms are represented by the third term. The fourth term is a Lennard-Jones-type interaction which describes the repulsion and attraction between non-bonded atoms and the last term captures the electrostatic interactions.

By construction, these potentials ignore the electronic structure. Hence, all effects which depend directly on a response of the electrons cannot be described by this approach. This includes any chemical reactions involving the making and breaking of bonds as found in many cases in materials research and processing, for example in the thermal decomposition of a polymer. But even the creation of a defect in a solid, the melting of silicon or the atomic rearrangements in a crack-tip involve the breaking and creation of bonds. Thus, a reliable description of such phenomena requires the inclusion of electronic degrees of freedom, i.e. one needs methods which incorporate quantum mechanical aspects.

The inclusion of some quantum mechanical effects has been proposed in the form of the effective medium theory [6] and the related embedded atom method [7]. In these approaches one uses the essence of density functional theory (which will be discussed below) to capture some electronic aspects of the bonding, and then adds empirical pair potentials which are fitted to reproduce experimental geometries, bulk moduli, and related data. These methods are intended for metallic systems. It overcomes some of the difficulties of pair potentials for metals while remaining computationally simple. Thus, the method lends itself to the study of dynamic phenomena of systems containing thousands of metal atoms. There is hope that this method and further extension will allow the study of complex processes such as failure mechanisms in the context of materials design. A review of this type of approach is given in ref. [8].

As pointed out in a recent overview [9], most of the atomistic simulations of materials are based on a rather simple model of interatomic interactions. In some cases, the results agree remarkably well with experiment. However, there is no systematic way to estimate the error in advance. By systematic comparisons with experiment, the validity of interatomic potentials needs to be carefully assessed before reliable quantitative predictions or even qualitative pictures are derived.

426

2.2 Semi-empirical quantum mechanical methods

One of the most widely used computational tools in the study of small organic molecules is a semi-empirical approach known as modified neglect of diatomic overlap (MNDO) [10]. This approach and its further developments are implemented in programs such as MOPAC [11]. Its usefulness comes from the right balance between theoretical rigor and algorithmic pragmatism, speed and accuracy, system-specific parameterization and generality. This method captures the essential aspects of a quantum mechanical approach including electronic levels, charge transfer, and spin polarization; it allows the making and breaking of chemical bonds; it provides geometric structures, heats of formation, infrared spectra, normal modes, as well as electronic charges, electrostatic potentials, dipole and multipole moments, polarizabilities, hyperpolarizabilities and, with the proper parameterization, also optical spectra. The heavy computational effort of ab initio calculations is avoided by a number of algorithmic simplifications such as minimal basis sets and the elimination of difficult integrals, which are either made small by proper mathematical transformations and then ignored or by using them as parameters to fit experimental data such as ground-state geometries, heats of formation, and ionization potentials. It is of course the latter aspect that limits the generality and transferability of the methods. In fact, most of the efforts in MNDO and related methods have been devoted to the description of organic molecules. Hence, this approach is useful for organic synthetic polymers and other organic materials.

In solid state physics, an equivalent semi-empirical approach became known as tight-binding theory. Similar to MNDO, the integrals can be parameterized, as was initially shown by Slater and Koster in 1954 [12]. This method has been successfully developed into powerful tools for the study of inorganic materials including, for example, the atomic and electronic structure of surfaces [13]. Using the concepts of the tight-binding approach, significant efforts are being made to improve the computational speed of this type of simulations in order to study dynamic processes such as the effect of irradiation on the stability of materials [14].

The tight-binding recursion method, which was developed in the early 1970's [15,16] presents a promising approach for the fast evaluation of total energies and forces in covalently bonded systems such as transition metals, their alloys, and their compounds. A novel scheme by Aoki [17] leads to a rapidly convergent bond order expansion, thus overcoming some of the earlier difficulties of this approach. This is a promising method for the investigation of structural and perhaps dynamic properties of novel materials, in particular alloys [18].

While pragmatic in their approach and very useful in their application, semi-empirical methods are limited in their capability to predict the behavior of novel compounds because of the inherent system-specific parameters. For these reasons, the most satisfactory and also most general atomistic approach in materials science is based on first-principles quantum mechanics ("ab initio" methods) which by definition do not contain any system-specific parameters.

2.3 Ab initio quantum mechanical approaches

Since the formulation of quantum mechanics in the 1920's, two major first-principles or "ab initio" quantum mechanical approaches have emerged, namely Hartree-Fock (HF) theory [19,20] and density functional theory (DFT) [21,22]. A third approach, quantum Monte Carlo (QMC) [23], is promising but, so far, has remained limited to rather small systems. Because of its applicability to a wide range of systems including metallic, semiconducting, and insulating materials and its good balance between accuracy and computational efficiency,

427

DFT has become the dominant approach for electronic structure calculations of solids and surfaces. For organic molecules Hartree-Fock based approaches have been very successful in describing the electronic structures, binding energies, vibrational frequencies and other molecular properties [24].

The central task of ab initio quantum mechanical methods is the accurate and efficient solution of Schrödinger's equation,

$$H\Psi = E\Psi \qquad (6)$$

to yield the total energy of any molecular or solid state system for any arrangement of the atoms. It is a major triumph of theoretical physical chemistry that today Schrödinger's equation can be solved for systems containing hundreds of atoms with sufficient accuracy to be useful to chemists and materials scientists. The solutions involve approximations, which are fairly well understood and controlled. However, it is very difficult to give rigorous estimates of the errors caused by these approximations.

For most cases encountered in materials science, one can use the so-called Born-Oppenheimer approximation which exploits the fact that the mass of the nuclei is much larger than that of the electrons as mentioned earlier. Therefore, the electrons are assumed to adjust instantly to any changes in the positions of the atoms. This decouples the motions of the electrons from those of the nuclei and one has to solve Schrödinger's equation only for the electrons assuming fixed positions of the atomic nuclei. The N electrons of a system are represented by a many-electron wave function

$$\Psi = \Psi(1,2,...N) \qquad (7)$$

where the arguments 1,2,...N denote the Cartesian coordinates and the spin coordinate of each electron. A direct solution for a realistic atomistic model containing hundreds of electrons requires approximations. As stated above, there are presently two major approaches which are compared below.

In Hartree-Fock (HF) theory [19,20], one uses the original Hamiltonian operator of Schrödinger's equation and seeks an approximation for the many-electron wave function. The simplest ansatz for such a wave function is a product of one-electron wave functions, ψ_i. However, such a wave function would violate Pauli's principle which requires that the total wave function of a system of Fermions such as electrons is antisymmetric, i.e. the wave function has to change its sign if the coordinates of two electrons are exchanged. This requirement can be fulfilled with the product wave functions by creating a linear combination of products with alternating signs corresponding to all possible exchanges of coordinates. For convenience, one can write such a wave function in the form of a so-called Slater determinant indicated by eq. (8a).

Now one has to find equations which allow the actual determination of the one-electron wave functions. This is done by using the variational principle which states that the expectation value of the total energy (10a) using any approximate many-electron wave function, such as a Slater determinant, is an upper bound for the exact total energy. Therefore, by varying each one-electron wave function such that it minimizes the total energy (10a), one obtains conditions for each wave function in the form of one-electron wave functions, which are known as Hartree-Fock equations (12a).

428

| Hartree-Fock | Density Functional |
| (1928, 1930) | (1964, 1965) |

$$\Psi(1,2,...N) \approx \psi_1(1) \cdot \psi_2(2) \cdot ... \cdot \psi_N(N) + ...$$

$$\Psi^*\Psi = \rho(r) = \sum_i \psi_i^* \psi_i \qquad (8ab)$$

$$E = E[\Psi]$$

$$E = E[\rho] \qquad (9ab)$$

$$E[\Psi] = \frac{\int \Psi^* H \Psi d\tau}{\int \Psi^* \Psi d\tau}$$

$$E[\rho] = T_o[\rho] + U[\rho] + E_{xc}[\rho] \qquad (10ab)$$

$$\frac{\delta E}{\delta \psi_i} = 0$$

$$\frac{\delta E}{\delta \psi_i} = 0 \qquad (11ab)$$

$$\Downarrow$$

$$\Downarrow$$

$$\left[-\tfrac{1}{2}\nabla^2 + V_C + \mu_x^i \right] \psi_i = \varepsilon_i \psi_i$$

$$\left[-\tfrac{1}{2}\nabla^2 + V_C + \mu_{xc} \right] \psi_i = \varepsilon_i \psi_i \qquad (12ab)$$

| Hartree-Fock equations | Kohn-Sham equations |

In density functional theory, one avoids dealing directly with the many-electron wave function and one focuses on the total electron density instead. It can be shown [21,22] that the total energy can be expressed as a functional of the total electron density by eqs. (9b) and (10b). This exact electron density of the interacting electron system is represented by a decomposition into one-particle densities which in turn are the square of one-particle wave functions given in eq. (8b). Through a variational procedure similar to that used in Hartree-Fock theory one arrives at a set of partial differential equations, known as Kohn-Sham equations (12b) which allow practical calculations.

At first glance, the Hartree-Fock equations and the Kohn-Sham equations are almost identical. Both have the form of single-particle eigenvalue problems. The first term in the square brackets is the kinetic energy operator of a single electron, the second term represents the electrostatic potential arising from all charged particles in the system. The third term in the Hartree-Fock equations is an exchange operator which keeps electrons of the same spin apart. The first and second terms of the Kohn-Sham equations are the same as in the Hartree-Fock equations. The third term includes both exchange effects as well as all other many-body interactions between electrons. While formally similar, the results from Hartree-Fock theory and density functional theory can be qualitatively different.

A key to successful density functional calculations is a practical approximation for the exchange-correlation energy. The exact form of this effective potential is not know, but the local density approximation (LDA) is an excellent first step [21,22]. In essence, the LDA rests on two basic assumptions: (i) the exchange and correlation effects come predominantly from the immediate vicinity of a reference point, r, and (ii) these exchange and correlation effects do not depend strongly on the variations of the electron density in the vicinity of r. If conditions (i) and (ii) are reasonably well fulfilled, then the contribution from a volume element dr would be the same as if this volume element were surrounded by a constant electron density of the same value as within dr. One can then use the known results of the exchange-correlation energy per electron, ε_0, of a system with constant electron density and write the exchange-correlation energy of the inhomogeneous systems approximately as

$$E_{xc} \approx \int \rho(r) \, \varepsilon_o[\rho(r)] \, dr \qquad (13)$$

Through the relationship

$$\mu_{xc} = \frac{\partial E_{xc}}{\partial \rho} \qquad (14)$$

an approximate exchange-correlation potential operator is defined for the Kohn-Sham equations.

The magnitude of the various terms in eq. (12b) is illustrated by the electronic structure of single atoms. The total energy (i.e. the energy gained by adding electrons to a bare nucleus up to charge neutrality) of an isolated C atom is approximately -1000 eV, that of a Si atom -8000 eV and that of a W atom -44000 eV. The kinetic energy and the Coulomb energy terms are of similar magnitude but of opposite sign. The exchange-correlation term is about 10% of the Coulomb term and attractive for electrons. The correlation energy is smaller than the exchange energy, but plays an important role in determining the details in the length and strength of interatomic bonds. In fact, compared with the total energy, the binding energy of an atom in a solid or on a surface is quite small and lies in the range of about 1 to 8 eV. Energies involved in changes of the position of atoms on a surface can be even smaller. For example, the rearrangement of the atoms at the surface of a semiconductor such as silicon or tungsten are driven by energy changes as small as about 0.03 eV per atom. It is a tremendous challenge for any theory to cope with such a range of energies. Density functional theory in its present implementations comes surprisingly close to this goal.

It would go beyond the scope of this contribution to discuss the various theoretical aspects of density functional theory in greater detail. The interested reader is referred to textbooks such as those by Parr and Wang [25]. An overview of computational implementations is given, for example, by Wimmer [26]. In the following sections, specific examples are discussed which illustrate the current status and the challenges for computational materials design.

3. Illustrative Applications

3.1 Thermochemical data

One of the most important features of ab initio quantum mechanical calculations is their high reliability and predictive power. This is due to the fact that no system-specific empirical or heuristic parameters are introduced in the calculations. Today, accurate ab initio quantum mechanical calculations are routinely possible for systems containing up to about 100 atoms per molecule or crystalline unit cell. A large number of technologically important compounds have thus become accessible by this level of theory. The primary result of a quantum mechanical calculation are the total energy as a function of the position of the atoms and detailed information on the corresponding electronic structure including dipole moments, polarizabilities, the character of the frontier orbitals and possible electronic excitations. Furthermore, the detailed knowledge of the total energy "hypersurface" around the equilibrium allows the calculation of the vibrational properties, which form an important component of thermodynamic quantities such as entropy at finite temperatures.

This capability of ab initio methods has been exploited by Dixon and coworkers at Dupont to calculate thermochemical data for a range of compounds. For example, Dixon and Smart [27] have determined the relative energies of the three isomers of diaminobenzene and calculated heats of formation for CFC replacements [27]. Most significantly, the calculated values rival the experimental values in accuracy. In fact, in a series of similar compounds, it is likely that the calculated results are more consistent and reliable than the experimental data. This, of course, requires that the quantum mechanical calculations are done carefully at a high level of accuracy. The accuracy of this approach can be seen, for example, by comparing the calculated and experimental relative formation enthalpies for the three isomers of difluorobenzenes (cf. Fig.1).

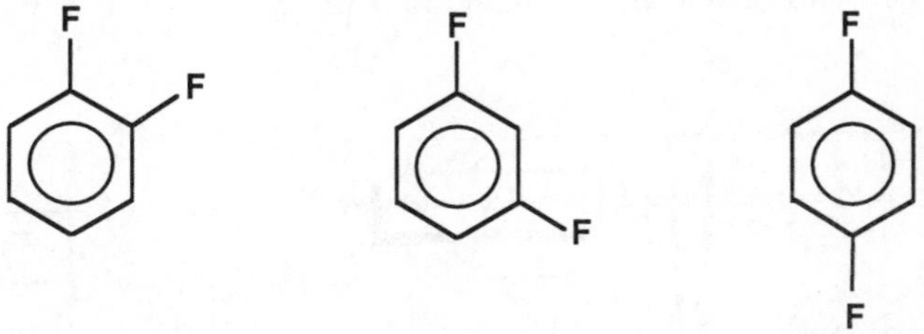

ΔH^o(kcal/mol)			
calculated	4.23	0.0	0.53
experiment	3.7 ± 0.5	0.0	0.6 ± 0.5

Figure 1. Relative energies (kcal/mol) for difluorobenzenes. The calculations were carried out using Hartree-Fock theory with second-order perturbation theory (MP2). After Dixon and Smart [27].

In this case of thermochemical data of molecules, the link between the atomistic and macroscopic scales is overcome by the facts that (i) no time-dependent phenomena have to be considered and (ii) the relative heats of formation obtained for a single isolated molecule allow an accurate extrapolation to the behavior of macroscopic quantities of these molecules in the gas phase. In the design of solid state materials, the link between the atomistic and macroscopic scales are not so obvious.

3.2 Li intercalation in graphitic layers
The Li-graphite system plays an important role in the design of rechargeable batteries. One of the problems in this type of batteries is the volume change associated with the operational cycle. Upon charging a graphite electrode with Li, the crystallites in the electrode expand. Upon discharge, the material shrinks. The repeated volume changes lead to fracture of the crystallites and eventually to a destruction of the electrode, thereby limiting the lifetime of the battery. A similar problem persists in the nickel-hydride batteries where the host material is derived from $LaNi_5$ and the battery cycles involve charging and discharging of H.

It would be desirable to develop materials which retain the necessary electrochemical properties, but show less volume changes during the operating cycles. In a preliminary study [28] using LDA theory and numerical atomic basis functions as implemented in the DMol method [29], the volume changes of a Li-graphite model were studied by total energy electronic structure calculations. The graphite host was modeled by two small C_{30} sheets

431

(see Fig. 2) in AA stacking mode. The dangling bonds on the borders of these graphitic sheets were saturated with H atoms.

The total energy as a function of the in-plane distance (see upper left panel of. Fig. 2) gives a sharp minimum at 1.41 Å which is very close to the experimental C-C distance of bulk graphite of 1.42 Å. The in-plane force constant is calculated to be 43 eV/Å^2 per C-C bond. For this model system, the calculated equilibrium interplanar distance is 3.53 Å, which is larger than the experimental value of 3.35 Å found in bulk graphite. In part, this discrepancy could be due to the fact that bulk graphite exists in an AB stacking whereas the model system has AA stacking. The force constant in the direction perpendicular to the graphitic sheets is 0.283 eV/Å^2 per C-C pair in this model. Intercalation of one Li atom increases the interplanar equilibrium distance by about 5% to 3.71 Å while making the bond between the layers significantly stiffer. In fact, upon intercalation with Li, the interplanar force constant increases from 0.283 to 0.753 eV/Å^2 per C-C pair.

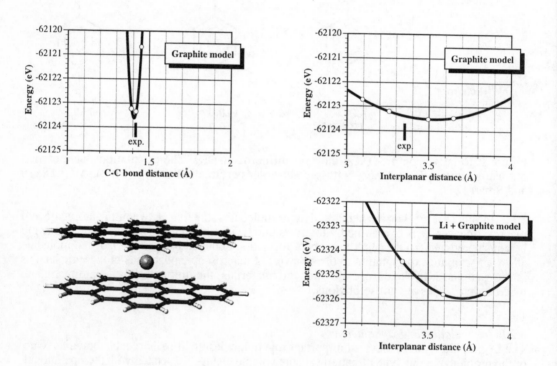

Figure 2. Models for graphite and Li intercalated in graphite. The three panels show in the upper left panel the dependence of the total energy as a function of the in-plane C-C distance, in the upper right panel as a function of the interplanar distance of the pure graphite model and in the lower right panel the interplanar distance with one Li atom intercalated between the two graphitic sheets. All calculations were performed on the local density functional level using a numerical atomic orbital basis as implemented in the DMol method [29].

This model calculation gives valuable insight into the possible changes of elastic constants of a graphite lattice upon intercalation of Li atoms. The use of an ab initio approach ensures the credibility to the results. However, the actual materials design questions are much more complex. In a real material such as a graphite electrode, the preparation and the processing has a critical influence on the microstructure, which influences the Li uptake and release through its complicated surface and interface structure. Moreover, impurities and their diffusion properties can be expected alter the performance. All of these aspects are not included in these model calculations. However, the computational approach allows the investigation of various effects separately, which would be difficult or impractical experimentally.

3.3 Oxygen diffusion in polymers

This example demonstrates the use of empirical valence force field methods in the design of polymers for contact lenses. Since the discovery of the hydrophilic monomer hydroxyethyl methacrylate (HEMA), hydrogel contact lenses have become an important ophthalmic product. Hard contact lenses, based on methyl methacrylate (MMA) are being replaced by the softer and more comfortable hydrogel materials. Continuing improvements in lens materials beyond HEMA are now being sought to further enhance comfort and extend wear. However, progress has been slow. One of the reasons is the difficulty to find the right balance between the optical properties, the swelling in tear fluid, mechanical strength, resistance to the adsorption of lipids or proteins and oxygen permeability. In order to increase the biocompatibility of a contact lens (so that one can keep, for example, the contact lenses in the eye over night), the oxygen permeability has to be high. This can be achieved by increasing the swelling ratio of the polymer, yet this comes at the expense of other required properties such as mechanical stability and durability. Alternatively, one might search for a polymer that does not necessarily undergo high swelling, but which itself is highly permeable.

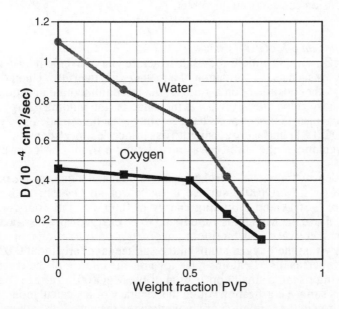

Figure 3. Diffusion coefficients of water and molecular oxygen in aqueous solutions of polyvinylpyrrolidinone (PVP) at 293 K as a function of the weight fraction of polymer, obtained from molecular dynamics calculations using a force-field approach (after ref. [30]).

In a recent study [30] it has been shown how molecular modeling techniques can help solving this materials design problem. One aspect which is amenable to simulations is the permeability of oxygen. The permeability of a gas through a barrier (such as a polymer membrane) is defined as the product of the solubility and the diffusion coefficient. Molecular dynamics methods using force-fields can be used to calculate diffusion coefficients. However, molecular dynamics runs are limited to relatively short time spans of a few nanoseconds as mentioned earlier. Therefore, only rapidly diffusing species can be reasonable well computed by this approach. For example, Müller-Plathe et al. [31] simulated the diffusion of He, H_2 and O_2 molecules in polyisobutylene at 300 K and calculated values for the diffusion constant of 30, 9.2, and $0.047-0.169 \times 10^{-6}$ cm^2/s compared with experimental values of 5.93, 1.52, and 0.081×10^{-6} cm^2/s, respectively. The calculated results give the correct trends, but the absolute numbers differ from the experimental values. In a materials design problem, this situation is acceptable since often one has a reference compound which can be used for calibration and the important aspect is the correct prediction of relative trends.

In fact, the calculations reveal [30] that in aqueous polyvinylpyrrolidinone (PVP) solutions, the diffusion coefficient of O_2 molecules does not increase significantly beyond a water content of about 50% (cf. Fig. 3). Thus, there is no point in trying to design polymers with water contents beyond this level, since the diffusion of O_2 would not increase while probably the mechanical and optical properties would degrade compared with denser polymers. Furthermore, the drying of the swelled polymer, for example in a dessert climate, would also be a serious issue if the water content of the polymer is high. In fact, to control this design aspect, one also needs to know the diffusion of water molecules in the polymer, which can be calculated as well as shown in Fig. 3.

The well chosen force-field parameters and molecular dynamics methods underlying this study lead to semi-quantitative results. The trends are properly reproduced, but the absolute values for diffusion constants are incorrect.

3.4 Adhesion of Ag atoms on a MgO(001) surface

Magnesium oxide, MgO, with its simple rock salt structure is a convenient system for the study of surface processes such as the formation of a metal/oxide interface. Upon deposition of a metal atom or any other adsorbate on a surface, the three immediate questions are: (i) where is the preferred adsorption site, (ii) what is the value of the binding energy, and (iii) what is the growth mechanism. These aspects have recently been investigated by Spiess [32] for Ag atoms on a MgO(001) surface by using a first-principles local density functional approach. In these calculations, the surface was modeled by a finite cluster of about 100 atoms.

Total energy local density functional calculations revealed that the most stable position of an adsorbed Ag atom on a MgO(001) surface is on top of an O atom with an Ag-O equilibrium distance of 2.36 Å and a binding energy of 0.66 eV. The four-fold hollow positions and the sites above the Mg atoms were found to be energetically less favorable by 0.13 and 0.30 eV, respectively. Furthermore, geometry optimizations of the clean MgO(001) surface show that the oxygen atoms in the surface layer are about 0.03Å above the ideal crystal plane whereas the Mg atoms relax towards the interior of the crystal by the same amount, thus giving rise to a small corrugation of the MgO(001) surface. Adsorption of Ag does not alter this surface corrugation within the accuracy of the calculation.

An important question in the formation of an overlayer is the actual growth mode. Bulk Ag crystallizes in a face-centered cubic lattice with an Ag-Ag distance which is about 3% smaller than the O-O distance on the MgO(001) surface. Therefore, an Ag overlayer on this surface could grow either epitaxially by retaining the on-top position of each Ag atom above an O atom or the overlayer could grow non-epitaxially if the Ag-Ag interactions are stronger than the Ag-O bonding. This intriguing question was also addressed by Spiess [32] by considering an island of five Ag atoms on the MgO cluster (a) in an epitaxial geometry and (b) in a non-epitaxial structure with the Ag-Ag distance fixed at the value of bulk Ag. The calculations show that the non-epitaxial geometry is slightly more stable. Furthermore, the Ag-O distance is found to increase from 2.36 Å (single Ag atom) to 2.47 Å (five Ag atoms) indicating a weakening of the adsorbate-substrate interaction with increasing Ag coverage.

3.5 Control of crystal morphology

In many chemical processes, the control of the particle morphology is of critical importance. In areas such as in the cement technology, the morphology of the crystallites has direct implications for the rheological and mechanical properties. Therefore, the understanding and control of crystal morphologies has become an active and fruitful area for atomistic simulations.

An example in this field is related to corrosion. Steel is rendered stainless by protective oxide layers, such as chromia. The efficiency of such barrier oxide films depends on their adhesion and coverage. Cr_2O_3 crystallizes in a corundum structure leading to rather boxy crystal morphologies. The morphology is controlled by the energies of the various surfaces and their growth rates. Atomistic simulations using ionic potentials [33] have provided surface segregation energies and relative surface stabilities for Al-substituted chromia. Aluminum surface substitution stabilizes the basal (0001) surfaces, promoting a beneficial plate-like morphology.

3.6 Color and optical properties

Ruby crystals are appreciated as gem stones because of their beautiful red color, their great hardness and their chemical resistance. Technologically, ruby is exploited in lasers. The optical properties of ruby are directly related to its electronic structure and thus lend themselves to a computational study. Ruby crystals consist of an α-Al_2O_3 lattice in which some Al atoms are replaced by Cr atoms. In α-Al_2O_3 the oxygen atoms form a close-packed hexagonal structure with the Al cations occupying octahedral interstitial sites (cf. Fig. 4). Cr and Al ions have similar atomic radii, hence a substitution of Al by Cr does not cause significant distortions of the Al_2O_3 lattice. The Cr atoms in ruby are surrounded by six oxygen atoms in a slightly distorted octahedral coordination.

Fig. 4 shows a unit cell of α-Al_2O_3 in which one Al atom has been replaced by Cr. The electronic structure of this system was calculated [34] using local density functional theory and the augmented spherical wave (ASW) method [35] with the atomic sphere approximation (ASA). Pure α-Al_2O_3 is an insulator which is reflected in the large separation of the valence band and the conduction band (see Fig. 4).

The replacement of an Al atom by a Cr atom introduces two narrow energy bands separated by about 2 eV. The lower band is partially occupied and the higher band is empty. The partial density of states of Cr, shown in the panel to the right, identifies these flat bands as Cr-$3d$ states. The chromium atom in ruby has a formal charge of 3+, which amounts to an electronic configuration of s^0d^2. As can be seen from the Cr-d partial density of states (see Fig. 4), the lowest part of the Cr-d bands is occupied. The calculations give an energy difference between the partially occupied and the unoccupied Cr-d states of about 2 eV. This energy difference can be interpreted as a crystal field splitting. Since optical transitions between states of the same angular quantum number are symmetry forbidden, the life-time of

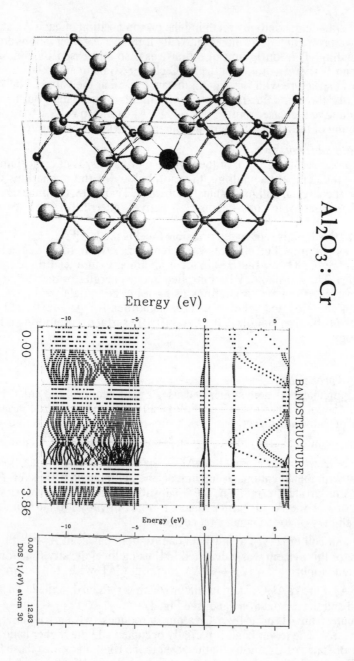

Figure 4. Crystal structure, energy bands, and Cr-*d* projected partial density of states of a model of ruby (Cr-doped Al_2O_3). The larger light spheres represent oxygen atoms and the smaller spheres are Al atoms. One Al atom in the *a*-Al_2O_3 unit cell is replaced by a Cr atom (shown as dark sphere). The electronic structure calculations were performed with the augmented spherical wave method [35] using the atomic sphere approximation.

the excited state is long and it is possible to pump electrons on many Cr atoms into these states. This effect allows the exploitation of ruby as material for lasers.

The energy of the optical transition is about 2 eV, which corresponds to the red part of the visible spectrum. The energy differences between local density functional (LDA) one-particle energies should not be quantitatively interpreted as excitation energies. However, in the case of a crystal field splitting the LDA eigenvalues allow an almost quantitative interpretation of optical excitation energies.

3.7 Magnetic ordering in Cu/Co multilayers

Recording technology is one of the major driving forces for the search for novel magnetic materials. Transition metal surfaces, atomic overlayers, sandwich structures, and artificially layered materials present particularly fascinating possibilities since the magnetic properties can be tailored through the control of the epitaxial process. First principles local spin density functional calculations have been very successful in the prediction of the magnetic properties of systems with reduced dimensionality [36], especially using the very accurate full-potential linearized augmented plane wave method [37].

The interest in magneto-optical recording and the discovery of the giant magnetoresistance effect have recently intensified the interest in these systems and their investigation by electronic structure calculations [38]. An example of the present computational capabilities is the magnetic ordering in Co/Cu multilayers [39]. Depending on the thickness of the non-magnetic Cu layers, the Co layers couple either ferromagnetically or anti-ferromagnetically. In the hexagonal Co structure, nearest-neighbor Co-Co distance is 2.51 Å compared with 2.54 Å for fcc copper. In the calculations, an fcc superlattice of Co and Cu is created by using the Cu lattice constant. For multilayer systems with 2 Co layers and 2, 3, 4, and 5 Cu layers, the total energy difference for ferromagnetic (FM) and antiferromagnetic (AF) coupling has been calculated with the ASW-ASA approach using the ESOCS program [40]. The calculations reveal an oscillation with the systems containing 2, 3 and 6 layers being preferentially ferromagnetic whereas the systems with 4 and 5 layers of Cu spacer are predicted to be antiferromagnetic in their ground states.

4. Summary and Outlook

The accurate calculation of forces between atoms together with the ability to determine the corresponding total energy has opened the possibility to predict a great variety of properties of critical importance to the design of materials. In this contribution, this capability is illustrated for the prediction of the relative stability of difluorobenzene molecules, the widening and stiffening of bonding between graphitic layers upon intercalation of graphite, the diffusion of oxygen in polymers as a function of water content, the determination of the adsorption geometry and energy of Ag atoms on a MgO(001) surface, and the relationship between surface energy, crystal growth, and morphology in Al-doped chromium oxide.

Once the position of the atoms are known, the calculation of the electronic structure provides the key for predicting electrical, optical, and magnetic properties. This is demonstrated in this article for the optical properties of ruby, where electronic structure calculations give a clear explanation of the color and the laser properties in terms of the electronic states of Cr atoms which are embedded in an aluminum oxide lattice. The last example illustrates the ability to predict the relative stability of ferromagnetic versus antiferromagnetic ordering in Co/Cu multilayer structures as a function of the thickness of the non-magnetic Cu spacer. These predictions are helpful in the design of materials for magnetic recording.

From these examples it is clear that atomistic computations based on ab initio quantum mechanics and on empirical potential energy functions can make a major contribution to materials design. In fact, such computational studies can help to isolate and identify critical features in the atomistic architecture which are related to macroscopic materials properties such as thermodynamic stability, elastic constants, diffusion constants, the adhesion of metals on an oxide such as MgO, and crystal morphology. By manipulating these critical features in atomistic simulations, it is then possible to help in the design of novel compounds and materials. Furthermore, calculations of the electronic structure provide detailed insight into the fundamental processes which govern macroscopic properties such as electrical conductivity, absorption and emission of electromagnetic radiation (e.g. color), and magnetic ordering.

However, at present there is no general, robust, and practical procedure that would lead in a systematic way from the atomistic length and time scales to reliable and accurate prediction of macroscopic materials properties. The major challenge appears to be the gap in the time scales between the atomistic domain and the macroscopic world. This is perhaps most evident in the case of structural materials such as metal alloys, where the entire history of the processing like casting, rolling, and tempering have a major influence on the mechanical and corrosive properties of a material through the evolution of the microstructure. No practical simulation method has been developed yet that would allow to follow the evolution of the materials properties through the manufacturing process by including simultaneously atomistic processes, the microstructure evolution, and the macroscopic thermomechanical properties. Clearly, excellent research by a number of groups is being pursued to address the various issues such as microstructure formation during casting.

The situation may actually be more favorable in electronic, optical, and magnetic materials. Often the processing of these materials such as semiconducting devices is controlled almost to the atomistic level, so that one has a detailed knowledge of the atomistic structure. From that knowledge it is possible to calculate energy gaps, band offsets across heterojunctions, and dielectric properties which are directly related to the macroscopic performance characteristics of the material. Similar arguments hold for optical materials as well as for magnetic materials. However, here the quantitative predictions of properties such as electrical conductivity is difficult and one has to return to semi-quantitative statements. On the other hand, properties such as optical absorption edges can be predicted quite accurately from advanced first principles calculations.

In the case of optical properties, the link between the atomistic and macroscopic scales is possible because (i) the atomic nuclei can be treated in a quasi-static picture (i.e. there is not the problem of bridging the gap in time scales) and (ii) the crystalline order of system eliminates the need for considering microstructures (at least in a first pass). In the case of molecular materials (especially liquids) and perfectly randomized organic polymers, the link between the atomistic and macroscopic properties may also be accessible because of the lack of long range order. Given this situation, it is intriguing to speculate on future developments.

Progress in computational materials design will be driven by the following factors: (i) developments and improvements of theoretical and computational approaches, (ii) higher speed, larger memories, and faster network connections of computer hardware, (iii) progress in the development of graphical user interfaces and computational environments which allow easy access to a wide range of sophisticated modeling and simulation methods, (iv) linking of experimental databases into the molecular modeling environments allowing

438

direct juxtaposition of experimental and calculated data, and (v) an increasing number of well trained scientists who understand the capabilities of atomistic simulations.

We see a steady progress in all of these areas and thus computational materials design has a promising future. However, compared with the progress in computer hardware and software, advances in theoretical and computational approaches are comparatively slow. Therefore, even at the current rate of technological innovation in the computer industry, the gaps in length and time scales which separate the atomistic domain from the macroscopic world will not be closed simply by applying more compute power. Fundamental progress in physical theory and applied mathematics are called for. In this context, universities, research institutions, and especially interdisciplinary centers will play a critical role to ensure the further progress of the field and to train a new generation of materials scientists and engineers.

Acknowledgments. It is a great pleasure to acknowledge the help and fruitful discussions with many colleagues and friends, especially Jan Andzelm, Bernard Delley, David Dixon, Bruce Eichinger, Arthur Freeman, Clive Freeman, Catalina Guerra, John Harris, Dominic King-Smith, Jürgen Kübler, John Newsam, Kastriot Seiti, Mike Stapleton, and Jürgen Sticht.

References

1. J. E. Lennard-Jones, Proc. Roy. Soc. London A **106**, 463 (1924).
2. B. G. Dick and A. W. Overhauser, Phys. Rev. **112** (1958) 90.
3. C. R. A. Catlow, editor, Modelling of Structure and Reactivity in Zeolites, Academic Press, London (1992).
4. P. J. Cragg and M. G. B. Drew, J. Comp.-Aided Mat. Design **1**, 149 (1993) .
5. see, for example, J. Maple, U. Dinur, and A. T. Hagler, Proc. Nat. Acad. Sci. USA **85**, 5350 (1988).
6. L. P. Nielsen, F. Besenbacher, I. Stensgaard,E. Lægsgaard, C. Engdahl, P. Stoltze, K. W. Jacobsen, and J. K. Nørskov, Phys. Rev. Lett. **71**, 754 (1993) and references therein.
7. M. I. Baskes, Proceedings of the International Conference on Computer-assisted Materials Design and Process Simulation, Tokyo, The Iron and Steel Institute of Japan, p. 219 (1993) and references therein.
8. T. J. Raeker and A. E. DePristo, Inter. Rev. in Phys. Chem. **10**, 1 (1991).
9. Materials Research Society Bulletin **21**, February 1996.
10. M. J. S. Dewar and W. Thiel, J. Am. Chem. Soc. **99**, 4899 (1977).
11. J. J. P. Stewart, J. Computer-Aided Mol. Design **4**, 1 (1990).
12. J. C. Slater and G. F. Koster, Phys. Rev. **94**, 1498 (1954).
13. M. Lannoo and P. Friedel, Atomic and Electronic Structure of Surfaces. Theoretical Foundations, Springer Series in Surface Science **16**, ed. M. Cardona, Springer Verlag (1991).
14. E. Salomons, P. Bellon, F. Soisson, and G. Martin, Phys. Rev. B **45**, 4582 (1992).
15. R. Haydock, V. Heine, and M. J. Kelly, Phys. C **5**, 2845 (1972); ibid. **8**, 2591 (1975).
16. P. Turchi and F. Ducastelle, The Recursion Method and Its Applications, D. G. Pettifor and D. L. Weaire, editors, Springer-Verlag, Berlin, p.104 (1985).
17. M. Aoki, Phys. Rev. Lett. **71**, 3842 (1993).
18. M. Aoki, P. Gumbsch, and D. G. Pettifor, Computer Aided Innovation of New Materials, M. Doyama, J. Kihara, M. Tanaka, and R. Yamamoto, editors, North-Holland, Amsterdam, p. 1457 (1993).

19. D. R. Hartree, Proc. Camb. Phil. Soc. **24**, 89 (1928).
20. V. Fock, Z. Phys. **61**, 126 (1930) and ibid. **62**, 795 (1930).
21. P. Hohenberg and W. Kohn, Phys. Rev. **136**, B864 (1964).
22. W. Kohn and L. J. Sham, Phys. Rev. Phys. Rev. **140**, A1133 (1965).
23. L. Mitás in *Computer Simulation Studies in Condensed-Matter Physics V*, D. P. Landau, K. K. Mon, K. K. and H. B. Schuttler, Springer, Berlin (1993) and references therein.
24. W. J. Hehre, L. Radom, P. von Schleyer, and J. A. Pople, *Ab initio molecular orbital theory*, John Wiley & Sons, New York (1986).
25. R.G. Parr and W. Yang, *Density-Functional Theory of Atoms and Molecules*, Oxford University Press, New York, (1989).
26. E. Wimmer, J. Comp.-Aided Mat. Design **1**, 215 (1993).
27. D. A. Dixon and B. E. Smart, Chem. Eng. Comm. **98**, 173 (1990).
28. C. Guerra and E. Wimmer, unpublished.
29. B. Delley, J. Chem. Phys. **94**, 7245 (1991).
30. D. R. Rigby, unpublished; B. E. Eichinger, D. R. Rigby, and M. H. Muir, submitted to J. Computational Polymer Science.
31. F. Müller-Plathe, S. C. Rogers, and W. F. van Gunsteren, J. Chem. Phys. **98**, 9895 (1993).
32. L. Spiess, Surf. Rev. Lett. (submitted).
33. S. C. Parker, P. J. Lawrence, C. M. Freeman, S. M. Levine, and J. M Newsam, Catal. Letters **15**, 123 (1992).
34. J. Sticht, unpublished.
35. A. R. Williams, J. Kübler, and J. R. Gelatt, Phys. Rev. **B 19**, 6094 (1979).
36. D.-S. Wang, A. J. Freeman, and H. Krakauer, Phys. Rev. B **26**, 1340 (1982) and references therein.
37. E. Wimmer, H. Krakauer, M. Weinert, and A. J. Freeman, Phys. Rev. B **24**, 864 (1981).
38. H. J. F. Jansen, Physics Today **48**, 50(1995).
39. P. Oppeneer and J. Kübler, editors, Physics of Transition Metals vol 1 and 2, World Scientific (1993), reprinted from the Int. J. of Modern Physics B **7** (1993).
40. ESOCS User Guide, version 2.0, Biosym Technologies, San Diego, CA, Feb. 1995.

Biosym / Molecular Simulations
Parc Club Orsay Université, 20 rue Jean Rostand, 91893 Orsay, France

D WRZOSEK

On an infinite system of reaction–diffusion equations in the theory of polymerization and sol-gel transition

The following initial boundary value problem is a generalization of the *coagulation equations* derived by Smoluchowski (1917).

$$\frac{\partial u_1}{\partial t} = d_1 \Delta u_1 - a_1 u_1 \sum_{j=1}^{\infty} a_j u_j \,,$$

$$\frac{\partial u_k}{\partial t} = d_k \Delta u_k + \frac{1}{2} \sum_{i=1}^{k-1} a_i a_{k-i} u_i u_{k-i} - a_k u_k \sum_{j=1}^{\infty} a_j u_j \,, \qquad (1.a)$$

$$k = 2, 3, \ldots .$$

on $\Omega \times (0, T)$, where Ω is a bounded open set in R^n, $n \geq 1$, with smooth boundary $\partial\Omega$; d_k, a_k are positive constants. We impose homogeneous Neumann boundary conditions

$$\frac{\partial u_k}{\partial \nu} = 0 \text{ on } \partial\Omega \times (0, T) \quad k = 1, 2, \ldots, , \qquad (1.b)$$

and initial data

$$u_1(x, 0) = U_0(x), \; u_k(x, 0) = 0 \text{ for } k \geq 2, \; x \in \Omega, \qquad (1.c)$$

where U_0 is a given function on Ω satisfying

$$U_0 \in L^{\infty}(\Omega), \quad U_0 \geq 0 \,.$$

The problem (1) has the following interpretation. Let us consider a system of the infinite number of species $S_1, S_2, \ldots$, where S_k is a population of molecules that consist of k identical particles bonded together (e.g. polimers). Since the number of monomers in a polimer is not limited *a priori* the infinite number of populations is considered. The variables u_k in (1) represent concentrations of S_k. The substances S_k, S_l react according to the following scheme:

$$S_k + S_l \xrightarrow{r_{kl}} S_{k+l}$$

where $r_{kl} = a_k a_l$ is a reaction constant. In this model it is assumed that:
−   fragmentation of polymers is absent , so the process is unidirectional and

irreversible

- bonding between polymers takes place only at their surface,
- molecules of each species may diffuse according to the Fick law.

The majority of papers on the topic treat, so far, only kinetic aspects of the process neglecting the space distribution of variables (see [3], [4] and references given there). We refer to [2] for physical justification of the coagulation-diffusion model. In [1] we generalize some results from [5] where space-homogeneous as well as radially symetric stationary solutions to (1.a) has been studied.

The dynamics of the system heavily depends on the assumptions imposed on the asymptotic behaviour of the coefficients a_k as $k \to +\infty$.

We study (1) under two different sets of hypotheses:

(H1) $\lim_{k \to \infty} \frac{a_k}{k} = 0$

and the second

(H2)   i) there exists $M \geq 1$ such that $d_k = d = $ constant, for $k \geq M$.

ii) $\limsup_{k \to \infty} \frac{a_k}{k} < \infty$.

In [3] the coefficients a_k are taken as

$$a_k = (Ak + B)^\omega \quad \text{with} \quad A \geq 0, B > 0 \tag{2}$$

and $0 \leq \omega \leq 1$. These coefficients satisfy the assumption (H2.ii). The case $\omega = 1$ precludes the intramolecular bonds as it is in the classical Flory–Stockmayer theory of polymerization [3]: for a monomer having f ($f \geq 2$) identical reactive groups responsible for bimolecular bonds one obtains the number of free reactive groups in k−mer equal to $c_k = (f - 2)k + 2$ and then a_k is proportional to c_k .

It is worth pointing out that this assumption implies structural phase transition known as *sol-gel transition* (see [3] and [4] for details).

The case $0 < \omega < 1$ corresponds to the intramolecular bonding (see [3]). In this case a part of free groups in a polymer molecule is responsible for intramolecular bonds [3]. The case $A = 0$ corresponds to a situation when the reaction rates are independent of the number of monomers in the interacting polymers.

Notice that we should expect the diffusion coefficients to decrease as the size of polymers increases. Hypothesis (H1) enables us to assume that $d_k \to 0$ as $k \to \infty$, whereas, hypothesis (H2.i) corresponds to the situation when diffusion coefficients are the same for large clusters.

Solutions of (1.a)-(1.c) are constructed as the limit of solutions of finite systems of reaction-diffusion. The finite system being the N−th approximation of (1.a)-(1.c) consists of the first $2N$ equations of (1) where a_k is replaced by 0 for $k > N$. In [1] we prove the following result.

442

Theorem 1. *Under assumption (H1), the problem (1.a)-(1.c) has a nonnegative global solution* $\{u_k\}_{k=1}^{k=\infty}$: *for* $k = 1, 2, \ldots$

$$u_k \in C([0, \infty[; L^2(\Omega))$$
$$u_k \geq 0, \quad u_k \in L^\infty(\Omega \times (0, \infty)),$$
$$u_k \in W_{loc}^{1,2}(]0, \infty[; L^2(\Omega)) \cap L_{loc}^2(]0, \infty[; D(L)),$$
$$\sum_{k=1}^{\infty} a_k u_k \in L^2(\Omega \times (0, \infty))$$

and equations (1.a) are satisfied a.e. on $\Omega \times (0, \infty)$.

We do not know if there, is uniqueness of solutions to (1) in the general case. The solutions $\{u_k\}_{k=1}^{k=\infty}$ also satisfy:

$$\lim_{t \to \infty} \|u_k(t)\|_\infty = 0 \quad \text{for} \quad k = 1, 2, \ldots \tag{3}$$

$$\sum_{k=1}^{\infty} k \|u_k(t)\|_1 \leq \|U_0\|_1 \quad \text{for} \quad t > 0, \tag{4}$$

$$\|\sum_{k=1}^{\infty} a_k u_k\|_{L^2(\Omega \times (0, \infty))} \leq \|U_0\|_1 .$$

The property (3) reflects the fact that larger and larger particles are formed from smaller one. Therefore, their concentration decreases in time. Notice that expression

$$\sum_{k=1}^{N} \int_\Omega k u_k(x, t) dx$$

defines the total mass of elementary units contained in the populations $u_1, \ldots, u_N$ at a time t. The collection of all k−mers ($k = 1, \ldots, \infty$) can be identified as a *sol phase*. The property (4) means that the total mass of soll at a moment t does not excess the initial mass.

Hypothesis (H2) enables us to prove existence of solutions preserving the total mass of *sol* on some interval of time $[0, T)$

Theorem 2. *Under assumptions (H2), there exists* $T \in]0, \infty]$ *such that the problem (1.a)-(1.c) has a nonnegative solution* $\{u_k\}_{k=1}^{k=\infty}$ *on* Ω_T *such that for* $k = 1, 2, \ldots$

$$u_k \in C([0, T[; L^2(\Omega)),$$
$$u_k \geq 0, \quad u_k \in L^\infty(\Omega_T) \cap W_{loc}^{1,2}(]0, T[; L^2(\Omega)),$$
$$u_k \in L_{loc}^2(]0, T[; D(L)),$$
$$\sum_{k=1}^{\infty} a_k u_k \in L^\infty(\Omega_T).$$

The equations of (1.a) are satisfied a.e. on $\Omega \times (0, T)$ and

$$\sum_{k=1}^{\infty} \int_{\Omega} k u_k(x, t) dx = \int_{\Omega} U_0(x) dx \quad \text{for } t \in [0, T[. \tag{5}$$

Moreover, if

$$\limsup_{k \to \infty} \frac{a_k}{k^{\frac{1}{2}}} < \infty, \tag{6}$$

this is also true for $T = \infty$.

It is worth pointing out that in the case (2) the property (6) corresponds to $0 \leq \omega \leq \frac{1}{2}$; notice also that, using existence result from Theorem 1, for $\frac{1}{2} < \omega < 1$ existence of global in time solutions which preserve locally in time the total mass of *sol* is proved. It is not yet clear whether this property holds also globally in time. On the other hand some physical interpretations (see e.g.[3], [4]) suggest that it does not.

The violation of mass conservation of *sol* is interpreted as an appearance of an infinite molecule that does not give any contribution to the expression on the left hand side of (5). Those molecules are identified as a new phase - *gel*. It is proved that if $0 \leq \omega \leq \frac{1}{2}$, i.e. the number of intramolecular bonds is sufficiently large, the mass of sol is preserved and *gel* does not appear. In this case large molecules resemble clusters rather then a network-like structure typical for *gel*.

It is worth pointing out that mass conservation breaks down abruptly at a finite time for the system of o.d.e. in the case $\omega = 1$ in (2), a phenomenon related to the mentioned *sol-gel* irreversible structural phase transition (see [3], [4]) . In the Flory-Stockmayer theory of polymerization, this effect is interpreted as the interchange of mass between *sol* and *gel*.

In our case, similarly as in [3] and [4] for o.d.e. system, gelation time t_g can be defined as

$$t_g = \sup\{t : \|\sum_{k=1}^{\infty} k^2 u_k(\cdot, t)\|_{\infty} < \infty\}$$

The convergence of the last series is connected with the preservation of the total mass of *sol*. Since it is proved that the above set is nonempty, there is a conjecture, that a set of blow-up points of $\sum_{k=1}^{\infty} k^2 u_k$ determines an interphase between *sol* and *gel*. Moreover, a lower bound on t_g can be derived in terms of a_k, $(1 \leq k \leq M)$, M, $\|U_0\|_{\infty}$ and C, where C is a constant such that $a_k \leq Ck$ for $k \geq M$.

In the case $\omega = 1$ in (2) we also prove uniqueness of solutions satisfying the mass conservation (5): actually, this follows from the following theorem which states that the infinite system can be reduced, in a sense, to finite ones.

Theorem 3. *Under assumptions (H2.i) and $\omega = 1$ in (2), let $\{u_k\}_{k=1}^{k=\infty}$ be a solution of (1.a)-(1.c) defined on Ω_T and satisfying the mass conservation (5). Let $\varrho :=$*

$\sum_{k=K}^{\infty} a_k u_k$, for any $K \geq M$, then

$$\varrho \in C([0,T]; L^2(\Omega)), \quad \varrho(\cdot,0) = \begin{cases} a_1 U_0, & \text{if } K = 1 \\ 0, & \text{if } K > 1 \end{cases},$$

$$\varrho \in L^\infty(\Omega_T) \cap W_{loc}^{1,2}(]0,T[; L^2(\Omega)),$$

$$\varrho \in L_{loc}^2(]0,T[; D(L)).$$

and $\{u_1, u_2, \ldots, u_{K-1}, \varrho\}$ satisfy the following system of reaction-diffusion equations

$$\frac{\partial u_k}{\partial t} = d_k \Delta u_k + \frac{1}{2} \sum_{i=1}^{k-1} a_i a_{k-i} u_i u_{k-i} - a_k u_k \left(\sum_{i=1}^{K-1} a_i u_i + \varrho \right)$$

$$\text{for } k = 1, \ldots K-1, \tag{7}$$

$$\frac{\partial \varrho}{\partial t} = d\Delta \varrho + \tfrac{1}{2} \sum_{i,j<K \leq i+j} a_i u_i a_j u_j a_{i+j} + A\left(\sum_{i=1}^{K-1} i a_i u_i\right)\varrho - \tfrac{B}{2}\varrho^2.$$

Since there is uniqueness of solutions for finite system (7) of K equations for any $K \geq M$ it is also true for the problem (1.a)-(1.c). This result also yields C^∞ regularity of solutions.

The model does not concern the particular mechanisms of bonding which are typical for biomolecules (enzymatic reactions).

It seems to be interesting to take also into account the fragmentation of polymers and investigate an interplay of coagualation and fragmentation in relation to the transition *sol-gel*. The above mentioned aspects lead to a much more complicated description which is a subject of the next paper being in a preparation.

References

1. Ph. Bénilan, D. Wrzosek, *On an infinite system of reaction-diffusion equations*, to appear in Adv. Math. Sc. Appl..
2. P. G. J. van Dongen, *Spatial Fluctuations in Reaction-Limited Aggregation*, J. Stat. Phys. **54** (1989), 221-267.
3. E.M. Hendriks, M.H. Ernst, and R.M.Ziff, *Coagulation Equations with Gelation*, J. Stat. Phys. **31** (1983), 519-563.
4. F. Leyvraz and H.R. Tschudi, *Singularities in the kinetics of coagulation processes*, J. Phys. A: Math.Gen. **14** (1981), 3389-3405.
5. M. Slemrod, *Coagulation-Diffusion Systems: Derivation and Existence of Solutions for the Diffuse Interface Structure Equations*, Physica D **46** (1990), 351-366.

Dariusz Wrzosek
Institute of Applied Mathematics and Mechanics,
Warsaw University,
ul. Banacha 2,
02-097 Warsaw, Poland

445